BIOINORGANIC CHEMISTRY OF COPPER

BIOINORGANIC CHEMISTRY OF COPPER

Edited by
Kenneth D. Karlin
Zoltán Tyeklár

CHAPMAN & HALL
New York • London

First published in 1993 by
Chapman & Hall
29 West 35 Street
New York, NY 10001-2299

Published in Great Britain by
Chapman & Hall
2-6 Boundary Row
London SE1 8HN

© 1993 Chapman & Hall, Inc.

Printed in the United States of America

Library of Congress Cataloging-in-Publication Data

Bioinorganic chemistry of copper / edited by Kenneth D. Karlin and Zoltán Tyeklár.
    p. cm.
    Includes bibliographical references and index.
    ISBN 0-412-03631-2(HB)
    1. Copper—Physiological effect. I. Karlin, Kenneth D., 1948-
II. Tyeklár, Zoltán.
QP535.C9B56 1993
574.19'214—dc20   93-12323
                                                        CIP

British Library Cataloguing in Publication Data also available.

CONTRIBUTORS

Elinor T. Adman
University of Washington

El H. Alilou
Université d'Aix-Marseille III

William E. Allen
The University of North Carolina, Chapel Hill

Edith Amadéi
Université d'Aix-Marseille III, France

William E. Antholine
Medical College of Wisconsin

James K. Bashkin
Washington University

Joseph A. Berry
Carnegie Institution of Washington

Ninian J. Blackburn
Oregon Graduate Institute of Science and Technology

Joseph E. Bradshaw
Rice University

Doreen E. Brown
Amherst College

Nigel L. Brown
University of Birmingham

James W. Bryson
Northwestern University

Jim Camakaris
University of Melbourne

Joan E. Carpenter
Case Western Reserve University

Susan M. Carrier
University of Minnesota

Luigi Casella
Università di Pavia

Sunney I. Chan
California Institute of Technology

Danae Christodoulou
National Cancer Institute, Frederick

Alexander W. Clague
Amherst College

Haim Cohen
Ben-Gurion University of the Negev, NRCN

Valeria C. Culotta
The Johns Hopkins University

Gidon Czapski
The Hebrew University of Jerusalem

C. T. Dameron
University of Utah Medical Center

I. G. Dance
University of New South Wales

David M. Dooley
Amherst College

Tambra M. Dunams
National Cancer Institute, Frederick

Franck Eydoux
Université d'Aix-Marseille III

C. Fan
Northwestern University

Jaqui A. Farrar
University of East Anglia

J. A. Fee
Los Alamos National Laboratory

M. C. Feiters
University of Nijmegen

Ben L. Feringa
University of Groningen

Scott Flanagan
Rice University

Stephen Fox
Rutgers, The State University of New Jersey

Francoise C. Fredericks
Purdue University

John C. Freeman
The Pennsylvania State University

Yasuhiro Funahashi
Nagoya University

Martha L. Garrity
The University of North Carolina, Chapel Hill

Clifford George
Naval Research Laboratory

G. N. George
Stanford University

Sara Goldstein
The Hebrew University of Jerusalem

Jorge A. González
Rice University

Edith B. Gralla
University of California, Los Angeles

Michele Gullotti
Università di Milano

R. J. Gurbiel
Northwestern University

Kenneth J. Haller
University of Notre Dame

Brooke L. Hemming
Stanford University

B. M. Hoffman
Northwestern University

Kenneth D. Karlin
The Johns Hopkins University

D. H. W. Kastrau
Universität Konstanz

Larry K. Keefer
National Cancer Institute, Frederick

Jyllian N. Kemsley
Amherst College

P. J. A. Kenis
University of Nijmegen

Nobumasa Kitajima
Tokyo Institute of Technology

R. J. M. Klein Gebbink
University of Nijmegen

Judith P. Klinman
University of California, Berkeley

Spencer Knapp
Rutgers, The State University of New Jersey

Anne Kotchevar
The Pennsylvania State University

H. Koteich
Medical College of Wisconsin

P. M. H. Kroneck
Universität Konstanz

Raviraj Kulathila
Unigene Laboratories, Inc.

P. Kyritsis
The University, Newcastle upon Tyne

Paula Lapinskas
The Johns Hopkins University

T. O. LeeBarry
University of Melbourne, Australia

Hsiupu D. Lee
Purdue University

Mary E. Lidstrom
California Institute of Technology

Xiu F. Liu
The Johns Hopkins University

Yi Lu
University of California, Los Angeles

Karen A. Magnus
Case Western Reserve University

Chris M. Maragos
National Cancer Institute, Frederick

Dale W. Margerum
Purdue University

Arthur E. Martell
Texas A&M University

C. F. Martens
University of Nijmegen,

Hideki Masuda
Nagoya Institute of Technology

Cynthia D. McCahon
Amherst College

Michael R. McDonald
Purdue University

Michele A. McGuirl
Amherst College

W. S. McIntire
Veterans Administration Medical Center, San Francisco

Rached Menif
Texas A&M University

David J. Merkler
Unigene Laboratories, Inc.

Albrecht Messerschmidt
Max-Planck-Institut für Biochemie

Dan Meyerstein
Ben-Gurion University of the Negev

vi

Deborah Morley
Temple University

Durgesh Nadkarni
Case Western Reserve University

F. Neese
Universität Konstanz

Hiep-Hoa T. Nguyen
California Institute of Technology

Patrick M. Ngwenya
Texas A&M University

R. J. M. Nolte
University of Nijmegen

M. Nordling
Chalmers University of Technology and University of Göteborg

Thomas V. O'Halloran
Northwestern University

Akira Odani
Nagoya University

Jack Peisach
Albert Einstein College of Medicine

I. J. Pickering
Stanford University

Marcel Pierrot
Université d'Aix-Marseille III

F. Christopher Pigge
The University of North Carolina, Chapel Hill

Joseph A. Potenza
Rutgers, The State University of New Jersey

K. Veera Reddy
Case Western Reserve University

Marius Réglier
Université d'Aix-Marseille III

Joseph L. Richards
The University of North Carolina, Chapel Hill

David A. Rockcliffe
Texas A&M University

James A. Roe
Loyola Marymount University

David E. Root
Stanford University

Duncan A. Rouch
University of Birmingham

Christy E. Ruggiero
University of Minnesota

Joann Sanders-Loehr
Oregon Graduate Institute of Science and Technology

Lawrence M. Sayre
Case Western Reserve University

W. Robert Scheidt
University of Notre Dame

A. P. H. J. Schenning
University of Nijmegen,

William M. Scheper
Purdue University

Harvey J. Schugar
Rutgers, The State University of New Jersey

Andrew K. Shiemke
California Institute of Technology

Edward I. Solomon
Stanford University

Thomas N. Sorrell
The University of North Carolina, Chapel Hill

Gábor Speier
University of Veszprém

David M. Stanbury
Auburn University

K. Surerus
Los Alamos National Laboratory

A. G. Sykes
The University, Newcastle upon Tyne

Wei Tang
Case Western Reserve University

Andrew J. Thomson
University of East Anglia

Gaochao Tian
University of California, Berkeley

William B. Tolman
University of Minnesota

Hoa Ton-That
Case Western Reserve University

Stewart Turley
University of Washington

Petra N. Turowski
Amherst College

Zoltán Tyeklár
The Johns Hopkins University

Joan S. Valentine
University of California, Los Angeles

Joseph J. Villafranca
The Pennsylvania State University

Bernard Waegell
Université d'Aix-Marseille III

Lihua Wang
Purdue University

J. L. Ward
The Johns Hopkins University

M. Werst
Northwestern University

James W. Whittaker
Carnegie Mellon University

Lon J. Wilson
Rice University

D. R. Winge
University of Utah Medical Center

David A. Wink
National Cancer Institute, Frederick

Osamu Yamauchi
Nagoya University

S. Young
*Chalmers University of Technology and University
of Göteborg*

Stanley D. Young
Unigene Laboratories, Inc.

Andreas D. Zuberbühler
University of Basel

W. G. Zumft
Universität Karlsruhe

CONTENTS

Natural and Synthetic Regulation of Gene Expression

Hemocyanin and Copper Monooxygenases

Copper-Mediated Redox/Oxidative Pathways

Dioxygen-Binding and Oxygenation Reactions

Nitrogen Oxide (NO_x) Chemistry and Biochemistry

Copper Oxidases

PREFACE

The interplay between inorganic & biochemistry, molecular biology and medicinal chemistry is the crux of the discipline of *bioinorganic* chemistry. In this Volume, we present thirty-nine articles detailing many of the major developments in the chemistry of copper ion as it pertains to biological systems, since copper plays a vital role in organisms ranging from bacteria to mammals. Its major importance is as an essential metalloenzyme active site "cofactor", many of which are involved in bioenergetic processes. Reactions catalyzed include electron transfer, dioxygen (O_2) transport, substrate oxygenation/oxidation coupled to O_2 reduction to H_2O_2 or water, and reduction of nitrogen oxides. In excess, copper ion is a toxic heavy metal; there has been recent progress in beginning to understand the control of its free ion concentration (i.e., homeostasis) through transport mechanisms and metalloregulatory processes. As a redox active metal, copper ion is also known to be capable of mediating biochemical reactions which generate toxic oxygen species. Copper complexes are capable of oxidatively cleaving or hydrolyzing nucleic acids, and these chemistries either have already found or may potentially provide for future applications in biochemistry and molecular biology.

It is our particular interest and approach to the field to present simultaneously in one place, papers dealing with both the chemistry and biology of copper. There is an increased realization that researchers from either subject critically need to learn about the others' sub-discipline, in order to better understand their own. For example, in order to accurately and usefully interpret biochemical or biophysical information determined for enzyme or other biochemical systems, it is essential to further develop our understanding of the basic coordination chemistry of copper. Yet, traditionally, the structural, spectroscopic, reactivity and mechanistic aspects of copper chemistry as exists for low molecular weight copper coordination compounds has been separated out from the more biological scientific discussions, meetings and literature. Researchers from established disciplines now find it necessary to join forces, in order to better develop our understanding of copper in biology, determine structure-activity relationships, and develop applications to molecular biology and medicine. A better knowledge of certain biochemical processes may also contribute to the development of new copper chemistry and thus new practical catalysts, e.g., for the mild, selective oxidation of organics, or reduction of NO_x pollutants. As can be seen in a number of the present contributions, very chemical and biological studies may both occur in the same work.

Recent exciting advances in copper bioinorganic chemistry have been numerous, and most of these are described here. For example, chemical and spectroscopic investigations along with X-ray structural studies have now revealed the presence of completely new (i.e., unexpected) structures, including a new binding mode for dioxygen to copper, active site copper clusters, or novel cofactors. This Volume is divided into eight parts. While each section contains a basic theme, there is however a considerable overlapping of ideas and approaches and even the same exact chemical system or protein may be discussed in articles from different sections. This in itself reflects the considerable excitement and interdisciplinary nature of investigations by researchers interested in copper.

The first article is by E. I. Solomon, a leading figure in copper biophysical chemistry, who overviews electronic structural aspects of copper proteins containing a dinuclear (e.g., hemocyanin and tyrosinase) or trinuclear (e.g., laccase and ascorbate oxidase) active site copper center. Solomon's work also highlights the importance and interplay between detailed electronic structural elucidation of well-defined small molecule complexes and the application of these results to aid the understanding of protein structure and spectroscopy. Then, Peisach demonstrates how pulsed electron spin resonance (EPR) spectroscopy can be usefully applied to elucidate subtle or detailed aspects of the local environment around copper in proteins, such as ligand identification and hydrogen bonding effects. Schugar follows with a detailed discussion of the structures and spectroscopy of novel Cu(II) macrocycles possessing disulfide ligand moieties.

The focus of the next section is the spectral and electron transfer properties of "blue" protein or model compound Cu centers. Sanders-Loehr discusses electronic structure of mutants of the protein azurin, using resonance Raman spectroscopy, while Valentine actually generates "blue" copper proteins by site-directed mutagenesis, thus the protein redesign of Cu-Zn superoxide dismutase. Sykes studies electron-transfer properties and mechanisms using plastocyanin mutants, while electron transfer rates of model complexes studied by Wilson and Stanbury are used to provide insights which may apply to "blue" proteins. Aspects of "blue" copper also surface in later articles by Adman (Ch. 31), Kroneck and Antholine (Ch. 33), Messerschmidt (Ch. 38) and Fee (Ch. 39).

Chapters 8-11 discuss aspects of the regulation, an area that is not necessarily new, but one that has become exceptionally exciting, with the realization of important direct effects of metals on gene expression. This topic has become much more molecular, thus of greater interest to chemists. To start, O'Halloran discusses the genetics and regulation of copper resistance in *E. coli*, while Winge and Dance discuss Cu(I)-thiolate clusters, emphasizing their basic chemistry, and their occurrence in biological macromolecules which function in copper ion buffering, signal transduction and copper ion storage. Culotta studies copper ion homeostasis in yeast, and Bashkin is interested in a synthetic system that involves RNA hydrolysis by copper(II) complexes.

Chapters 12-16 outline exciting new advances in proteins involved in O_2-binding or monooxygenation of biological substrates. Firstly, Magnus, reports the long-awaited unambiguous X-ray structural determination of oxy-hemocyanin, found to have a novel side-on peroxo coordination, bridging the two copper atoms. This was actually predicted from a model compound study of Kitajima (Ch. 20), attesting to the value of the synthetic biomimetic approach. Klinman then describes an elegant study which provides a new tool for detailed mechanistic investigations for systems which involve O_2 binding to metals or O-O cleavage processes; here the application is to dopamine β-monooxygenase. Blackburn continues with structural (by EXAFS spectroscopy) and ligand-probe (e.g., azide or carbon monoxide bound protein derivatives) UV-Vis or infra-red spectroscopic studies on the same enzyme, plus the pterin-dependent phenylalanine hydroxylase. Critical new insights have been obtained and these also relate to the peptidyl-glycine α-amidating enzyme, PAM, also discussed in Ch. 16. In Ch. 15, Chan breaks new ground with a report that the membrane bound form of methane monooxygenase in methanogenic bacteria is a Cu enzyme with a trinuclear Cu cluster active site. This is an exciting finding, since it means that a trinuclear copper cluster is capable of mediating either (i) O_2-reduction to water (i.e., in laccase and ascorbate oxidase), or (ii) O_2-

activation for cleavage of the very strong carbon-hydrogen bond in methane. As is found for porphyrin-iron (i.e., heme) enzymes, nature appears to modulate certain repeating structural motifs, in order to effect varying specific functions. In Ch. 16, Merkler elaborates on the pharmacologically important enzyme PAM, now confirmed as a true monooxygenase.

Beginning with Ch. 17, there is a change in emphasis to much more chemical, with this and the subsequent two chapters being separated out as having a particular focus on oxidative mechanisms. Margerum discusses oxidative decomposition reactions of copper(III) complexes, the latter oxidation state being stabilized by the deprotonated amide groups of peptide ligands. Free radicals cause many of the deleterious effects in biological systems, and Goldstein and co-workers describe how reactions of organic radicals R· with copper(I) or copper(II) ions can lead to transients with Cu-carbon bonds; their subsequent decomposition may account for actual biological pathways observed. The article by Sayre addresses detailed mechanistic pathways which may take place in the oxidation of amines, such as are known to occur in a variety of published chemical systems, as well as in enzymes like PAM.

The next eleven chapters deal with copper-dioxygen chemistry in synthetically derived systems. As mentioned above, the chapter by Kitajima summarizes what might have to be the most important chemical breakthrough in the field in many years, pertaining to biological or chemical dioxygen utilization by copper ion. Model chemistry developed by this research group actually confirmed an unknown binding mode for dioxygen to copper, found subsequently to occur in the natural system. Zuberbühler's contribution surveys recent kinetic/thermodynamic studies on O_2-binding to Cu(I) centers, of critical fundamental importance to an overall understanding of biological O_2-utilization. We then survey our own contributions to the field, largely the identification and characterization of chemical systems with reversibly bind dioxygen and/or activate it for hydroxylation of an unactivated arene substrate. Casella has also discovered a variety of dinuclear copper complex systems which undergo ligand hydroxylation, and the trends observed provide additional mechanistic insights. Feringa considerably expands upon the use of dinuclear Cu complexes involved in O_2-activation, elaborating and characterizing new systems (with Cu and Ni) which can be used for self-assembly of helical polymers, oxidative demethylation and demethyoxylation or dehalogenation, catalytic dehydrogenation, and in the generation of chiral molecules with potential practical uses. Then, Martell describes very interesting variations in the chemistry when employing dinucleating macrocyclic ligands, while Sorrell's focus is the new synthesis and resulting chemistry of complexes with imidazole donor groups, an important pursuit because this is the ligand found in the proteins. In Ch. 27, Réglier and Waegell describe a number of novel copper-ligand complex systems, which further develop the chemistry related to O_2-activation by tyrosinase and dopamine β-monooxygenase. Yamauchi describes chemistry of copper ion associated with the redox-active pteridine nucleus; this is a relatively unexplored new area, but directly relevant to the O_2-activation chemistry of phenylalanine hydroxylase. Martens and Feiters then report on new organic molecules generated as host-guest systems, and designed as closer functional mimics for enzymes effecting hydroxylation reactions. Finally, Speier describes _di_oxygenation chemistry, which takes full advantage of oxygen atoms in O_2, and mimics quercetin dioxygenase; the chemistry involved appears to involve 'substrate activation', for attack by dioxygen.

The next section involves nitrogen oxide chemistry, a relatively speaking new and exciting sub-discipline of bioinorganic copper chemistry. Anaerobic denitrifying bacteria employ several copper-containing enzymes in the reduction of NO_x species. In Ch. 31, Adman describes the X-ray crystal structure of a nitrite reductase, which contains both a Type 1 'blue' (green) copper center for electron transfer, plus an anion binding, presumably functional Type 2 Cu active site. Tolman then presents very new copper coordination chemistry involving nitrite (NO_2^-) and nitric oxide (NO), including reversible binding in the latter case. Nitrous oxide reductase ($N_2O \rightarrow N_2$) is as yet a relatively poorly understood but structurally and spectroscopically fascinating enzyme, which has striking similarities to cytochrome c oxidase ($O_2 \rightarrow H_2O$); Kroneck and Antholine describe EPR studies, from which an intriguing mixed-valence dinuclear copper site model emerges, suggested to be common to both of these proteins. In the last chapter in this section, Christodoulou describes some very different and novel chemistry, with copper chelates as potential nitric oxide (NO) releasing agents. NO has just recently been recognized as an important naturally occurring bioregulatory agent, critical to a variety of processes and of pharmacological utility. This type of chemistry seems surely destined for considerable elaboration, since NO is well known to readily react with copper proteins.

The last section is on copper oxidases, enzymes which reduce dioxygen either to hydrogen peroxide or water, while oxidizing some biological substrate. Villafranca starts out by discussing mechanistic and spectroscopic studies of galactose oxidase and phenoxazinone synthase; the latter catalyzes the last step in the biosynthesis of the drug actinomycin D, and until now its copper make-up has not been characterized. Whittaker focuses exclusively on galactose oxidase, summarizing his detailed spectroscopic and chemical studies which lead to the prediction of the presence of an unusual radical stabilizing cofactor; the results are compared to the actual active-site structure, now known through a X-ray crystallographic study. Dooley reviews structure and reactivity aspects of copper amine oxidases, determined largely through detailed spectroscopic studies; these enzymes also contain a novel redox-active quinone cofactor, directly involved in its function. Messerschmidt then details elegant X-ray crystallographic studies on a number of ascorbate oxidase derivatives, all generated with chemistry directly on single crystals. These provide great insight to both structure and possible mechanism of action. Finally, Fee presents an extremely useful and concise overview of the structure, spectroscopy and function of heme and copper containing cytochrome c oxidase. Studies from bacterial enzymes have contributed greatly to recent advances and new results bearing on the ligation at the so-called Cu_A and Cu_B sites are presented.

The idea of this Volume actually developed out of the Editors' interest in holding an international get-together on the bioinorganic chemistry of copper; such a Conference was in fact held at Johns Hopkins University in August of 1992. Thus articles from invited speakers and others were solicited; with the obvious occurrence of so many new and exciting developments in the field, the publication of a Volume such as this appeared to be very favorable. We would like to acknowledge the sponsors of the Hopkins Copper Conference, and it is their generous support which contributed to making both the meeting and this book possible. Their interest further attests to the significant wide-spread basic scientific interest in the discipline, and for the potential applications such as in pharmacy and other industrial chemical applications. We are particularly grateful for the backing of the International Copper Association, Ltd., since they have been supportive of our efforts for the last twelve years. In addition, private and government agency support for the recent meeting came from: Bioanalytical Systems, Inc., Bristol-Myers Squibb Company, Chapman and Hall, Inc., ChemGlass, Inc., Desert Analytics, E. I. du Pont de Nemours & Co., General Electric Company, Hoechst-Roussel Pharmaceuticals, Inc., ICI Americas Inc., Icon Services, Inc., Lab Glass, Inc./Wilmad Glass, Mallinckrodt Medical Inc., Merck Research Laboratories, Millipore Corporation, Molecular Structure Corporation, Monsanto Company, National Institutes of Health, National Science Foundation, Office of Naval Research, On-line Instrument Systems, Inc., Petroleum Research Fund of the American Chemical Society, Shimadzu Scientific Instruments, Inc., Sterling Drug Inc., and Union Carbide Corporation.

We would also like to acknowledge the students and postdoctorals in the Karlin research group at Johns Hopkins University, for their help in the smooth running of the Cu '92 Conference. Particular thanks goes to Dr. Rebecca R. Conry, for aiding the assembly of this Volume. Finally, we wish to dedicate this Volume to our wives Nancy and Klara, and our families, for their support and great patience.

January 1993

Kenneth D. Karlin and Zoltán Tyeklár
Department of Chemistry
The Johns Hopkins University
Baltimore, Maryland (USA)

Copper Proteins and Complex Spectroscopy

ELECTRONIC STRUCTURES OF ACTIVE SITES IN COPPER PROTEINS: COUPLED BINUCLEAR AND TRINUCLEAR CLUSTER SITES

Edward I. Solomon, Brooke L. Hemming and David E. Root

Department of Chemistry, Stanford University, Stanford, CA 94305

INTRODUCTION

It has now been ten years since a general review of our Group's research in the field of copper proteins has been prepared.[1] Over this period, our understanding of the electronic structure of the active sites in these proteins has strongly evolved, and is providing significant insight into the reactivity of these sites in biology. The first goal of my sabbatical was to generate an overview of our present understanding of this field, which has now appeared in *Chemical Reviews* [2] and will be summarized in two parts.[3] Part I of this summary is being published as our contribution to the Proceedings of the International Conference on the Chemistry of the Copper and Zinc Triads, held in Edinburgh, Scotland July 13-16, 1992. Part II is submitted for the Proceedings of the Symposium on Copper Coordination Chemistry: Bioinorganic Perspectives, August 3-7, 1992.

Many of the most important classes of active sites in copper proteins exhibit unique spectral features compared to simple high-symmetry transition metal complexes. These derive from the unusual geometric and electronic structures which can be imposed on the metal ion in a protein site. It has been the general goal of our research to understand these electronic structures and evaluate their contributions to the reactivities of these active sites in catalysis.

Our contributions to four topics will be summarized. First, if one is to understand the origin of unique spectral features, one must first understand the electronic structure of normal high symmetry transition metal complexes. For us, square planar cupric chloride has served as an electronic structural model complex. It is now one of the most well-understood molecules in inorganic chemistry[4], and its spectral features and associated electronic structure will be briefly described in Part I[3a]. Then the unique spectral features of the blue copper active site will be addressed and used to provide insight into ground and excited state contributions to long-range electron transfer in these proteins. In Part II[3b] of this summary we focus on the coupled binuclear copper proteins, hemocyanin and tyrosinase, which have similar active sites that generate the same oxy intermediate involving peroxide bound to two copper(II)'s. The hemocyanins reversibly bind dioxygen while the tyrosinases have highly accessible active sites which bind phenolic substrates and oxygenate these to *ortho*-diphenols. Their oxy sites exhibit unique excited state spectral features which reflect new peroxide copper-bonding interactions which make a significant electronic contribution to the binding and activation of dioxygen by these sites. Finally, the

From K.D. Karlin and Z. Tyeklár, Eds., *Bioinorganic Chemistry of Copper*
(Chapman & Hall, New York, 1993).

3

multicopper oxidases laccase, ascorbate oxidase and ceruloplasmin catalyse the four electron reduction of O_2 to water. Spectral studies on these enzymes will also be summarized in Part II which demonstrate that they contain a fundamentally different coupled binuclear copper site (called Type 3) which, in fact, is part of a trinuclear copper cluster which plays the key role in the multielectron reduction of dioxygen by this important class of enzymes.

NORMAL COPPER COMPLEXES AND BLUE COPPER PROTEINS: SEE PART I (ref. 3a)

COUPLED BINUCLEAR COPPER PROTEINS

We now turn to the binuclear copper proteins hemocyanin and tyrosinase which reversibly bind dioxygen, and in the case of tyrosinase activate it for hydroxylation of phenol to ortho-diphenol and further oxidize this to ortho-quinone (Figure 1).[5] Both proteins have essentially the same oxy active sites[6a] which involve two copper (II) from X-ray absorption

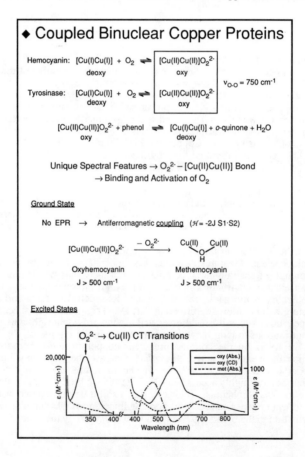

Figure 1. Coupled binuclear copper proteins, ground and excited state spectral features.

4

edges[6b,c] (vide infra) and bound peroxide from the unusually low O-O stretching frequency of 750 cm^{-1} observed in the resonance Raman spectrum.[7] As will be summarized in this section, the unique vibrational, ground and excited state electronic spectral features of this oxy site are now well understood and generate a detailed description of the peroxide-copper bond which provides fundamental insight into the reversible binding and activation of dioxygen by this site.

The ground state of oxyhemocyanin exhibits no EPR signal which is understood and involves a strong antiferromagnetic coupling of the two copper(II)'s (J > 500 cm^{-1}) hence its classification as a <u>coupled</u> binuclear copper site.[8] Displacement of the peroxide produces a met derivative which also has two Cu(II)'s that are strongly antiferromagnetically coupled (J > 500 cm^{-1}).[9] Thus there must be an endogenous bridge

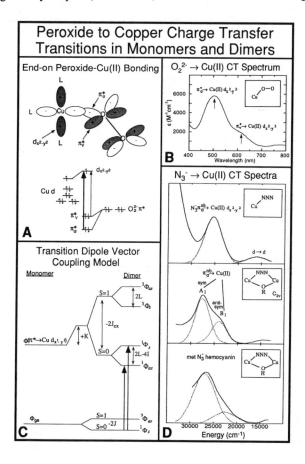

Figure 2. Charge transfer transitions in copper monomers and dimers: A) Orbital interactions involved in end-on peroxide-copper bonding. and predicted charge transfer transitions (thickness of arrow indicates relative intensity). B) Charge transfer absorption spectrum of peroxide bound to a single Cu(II).[13] C) Ground state and charge transfer excited state splittings due to dimer interactions in a peroxide bridged copper dimer. K is the coulomb dimer interaction, J_{ex} is the excited state magnetic exchange, and I and L are the coulomb and exchange contributions to the excitation transfer between halves of the dimer. D) Azide-to-copper charge transfer spectra of model complexes and met azide hemocyanin.[14]

present in the met derivative, which from the crystal structure of the deoxyhemocyanin[10] is hydroxide. With respect to the excited state spectroscopy, oxyhemocyanin exhibits a reasonably intense band in the absorption spectrum at ~ 600 nm (ε ~ 1000 M^{-1} cm^{-1}) and an extremely intense band at ~ 350 nm (ε ~ 20,000 M^{-1} cm^{-1}). Displacement of peroxide on going to the met derivative (solid to dashed spectrum in Figure 1 bottom) eliminates these features as well as a band at 480 nm which is present in the CD but not the absorption spectrum.[11] These features can then be assigned as peroxide to copper charge transfer transitions and will be seen to provide a detailed probe of the peroxide-cupric bond. We are particularly interested in 1) the fact that there are three charge transfer bands, 2) the selection rules associated with the presence of a band in the CD but not absorption spectrum and 3) the high energy and intensity of the 350 nm band.

We first consider peroxide bound end-on to a single copper (II) (Figure 2A). The valence orbitals of peroxide involved in bonding are the π^* set which split into two non-degenerate levels on bonding to the metal ion. The π^*_σ orbital is oriented along the Cu-O bond and has strong overlap with the $d_{x^2-y^2}$ orbital producing a higher energy intense charge transfer transition. The peroxide π^*_v orbital is vertical to the Cu-O bond and weakly π interacting with the copper thus producing a lower energy relatively weak transition. Thus, end-on peroxide bonding is dominated by the σ donor interaction of the O_2^{2-} π^*_σ orbital with $d_{x^2-y^2}$. This predicted low energy weak/high energy intense charge transfer spectrum is just what is observed experimentally for the complex prepared by Prof. Karlin[12] which has O_2^{2-} end-on bound to a single Cu(II) based on isotope effects on its resonance Raman spectrum (Figure 2B).[13] One notes, however, that there are only two bands in the model spectrum, and the π^*_σ is at considerably lower energy (500 nm) and weaker in intensity (ε~ 5000 M^{-1} cm^{-1}) than the 350 nm O_2^{2-} charge transfer band in oxyhemocyanin.

The fact that there are three peroxide to copper charge transfer transitions led us to consider the spectral effects of bridging peroxide between two Cu(II). A transition dipole vector coupling model was developed which predicts that each charge transfer band in a monomer will, in fact, split into four states in a dimer (Figure 2C).[11,14,15] First, there is a singlet/triplet antiferromagnetic splitting in the excited state just as there is in the ground state but considerably larger. In addition, both the singlet and triplet states are further split into two states which correspond to symmetric and antisymmetric combinations of the O_2^{2-} $\rightarrow$ Cu(II) charge transfer transition to each copper in the bridged dimer. As the antiferromagnetically coupled ground state is a singlet, only the two transitions to the singlet excited states should have absorption intensity. This predicted splitting into two bands is indeed observed[14] in a series of azide model complexes prepared by Profs. Sorrell[16], Reed[17] and Karlin[18] (Figure 2D). Azide bound to a single Cu(II) produces one charge transfer transition which is the equivalent of the π^*_σ charge transfer of the peroxide. As predicted, bridging the azide in a cis μ-1,3 geometry between two copper(II)'s results in a splitting of this charge transfer transition into two bands, the symmetric (or A_1 in the C_{2v} dimer symmetry) and antisymmetric (or B_1) components of the π_σ charge transfer transition. Note in Figure 2D bottom that binding N$_3^-$ to the met hemocyanin derivative produces the same A_1/B_1 charge transfer intensity pattern indicating that azide also bridges in a cis μ-1,3 geometry in met hemocyanin.[14]

This transition dipole vector coupling model can then be applied to predict the energy splittings and symmetries hence selection rules for the charge transfer spectrum of peroxide bridged between two copper(II)'s for different possible structures of oxyhemocyanin. Initially the three end-on peroxide bridging structures in Figure 3 were considered(μ-1,1, trans μ-1,2 and cis μ-1,2), and only the cis μ-1,2 structure predicted spectral features which could be consistent with oxyhemocyanin. This is also an attractive structure as it involves replacing the cis μ-1,3 N$_3^-$ in met hemocyanin with a cis μ-1,2 peroxide. However, in 1989, Prof. Kitajima obtained a new side-on bridging structure for peroxide (μ-$\eta^2\eta^2$) in transition metal chemistry.[19] This structure also predicts and exhibits spectral features very similar to those of oxyhemocyanin.[20] For both structures bridging the peroxide results in a splitting of the O_2^{2-} π^*_v level into a low energy component which is electric dipole allowed and should appear in the absorption spectrum, but with limited intensity as it is a π^*_v charge transfer transition. This can be associated with the 600 nm

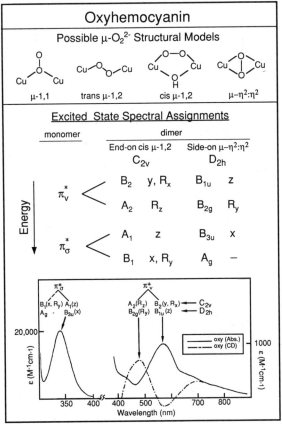

Figure 3. Possible structural models for peroxide bridged coppper dimers and corresponding excited state spectral assignments for the cis μ-1,2 and μ-η^2:η^2 possible structures for oxyhemocyanin.

absorption band. The second component of π^*_v for both structures is predicted to be only magnetic dipole allowed meaning it should contribute to the CD but not absorption spectrum, and the 480 nm CD feature can be associated with this transition. For both structures the π^*_σ splits into two bands with the lower energy component having dominant absorption intensity and this can be associated with the 350 nm absorption band in oxyhemocyanin. Thus the transition dipole vector coupling model requires that the peroxide bridge the two Cu(II)'s. This produces the three observed charge transfer transitions with one being present in the CD but not the absorption spectrum. However, one must still account for the high intensity and energy of the 350 nm O_2^{2-} π^*_σ charge transfer transition and the low vibrational frequency of the O-O stretch. Thus we proceed to quantitatively evaluate the electronic structures associated with these end-on and side-on peroxide bridging structures both theoretically and experimentally.

Broken symmetry-spin unrestricted-SCF-Xα-SW calculations were performed to describe the electronic structures associated with both geometric structures.[21] These calculations are appropriate for antiferromagnetically coupled dimers.[22] In Figure 4 we focus on the interaction of the HOMO and LUMO, which are symmetric and antisymmetric combinations of $d_{x^2-y^2}$ orbitals on each copper, with the valence orbitals of the peroxide.

For the end-on bridged structure the bonding is consistent with the qualitative description presented earlier. The peroxide π^*_σ is stabilized through a bonding interaction with the LUMO on the coppers. Thus in the end-on bridged geometry the peroxide acts as a σ donor ligand with one bonding interaction with each of the two coppers. A very different bonding description is obtained for the side-on bridged peroxide. In this structure, π^*_σ is again stabilized hence involved in a σ donor interaction with the copper LUMO. In the side-on structure the bonding/antibonding interaction of the π^*_σ is larger than in the end-on structure as the peroxide now occupies two coordination positions on each of the two coppers. Thus peroxide behaves as a stronger σ donor in the side-on structure. Further, the side-on peroxide is predicted to have an additional bonding interaction with the $d_{x^2-y^2}$ orbitals on the coppers which has not been previously considered for peroxide. This involves stabilization of the HOMO through its interaction with the high energy unoccupied σ^* orbital on the peroxide. This additional bonding interaction shifts some copper electron density into the peroxide, thus it also acts as a π acceptor ligand using this highly antibonding σ^* orbital.

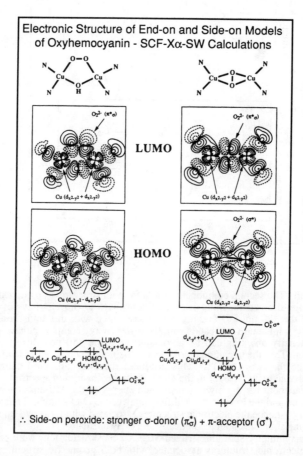

Figure 4. Electronic structures of the end-on cis μ-1,2 (C_{2v}) and side-on μ-η^2:η^2 (D_{2h}) models of the oxyhemocyanin active site: wavefunction contours of the HOMO and LUMO and energy level diagrams showing dominant orbital contributions.

It is next of critical importance to experimentally evaluate this unusual electronic structure description for a side-on bridged peroxide and its relation to oxyhemocyanin. This was accomplished through a series of studies of the charge transfer and vibrational spectral features of end-on[23] and side-on[20] bound peroxide-copper model complexes prepared by Profs. Karlin[24] and Kitajima[19]. The σ donor ability of the peroxide can be related to the intensity (and energy[15b]) of the $\pi^*_\sigma \rightarrow$Cu charge transfer transition (Figure 5). The idea here is that as the wavefunction of the occupied O_2^{2-} π^*_σ orbital gains copper character, α, its σ donor interaction with the copper increases. This wavefunction results in an expression for the ligand-to-metal charge transfer intensity which (along with some geometric factors) is proportional to α^2. Thus the peroxide π^*_σ charge transfer intensity increases as its σ donor interaction with the copper increases. If we normalize to the π^*_σ charge transfer intensity of the end-on peroxide monomer complex shown earlier, the O_2^{2-} charge transfer intensity of the Karlin trans μ-1,2 end-on bridged complex increases by a factor of two, consistent with peroxide binding to each of two coppers. While no cis model complex exists, our Xα calculations indicate the peroxide binding in this geometry should have a similar σ donor interaction with the coppers to peroxide bridged in the trans complex. Alternatively, the side-on bridged complex of Kitajima exhibits an extremely intense π^*_σ charge transfer transition, very similar to that of oxyhemocyanin, which quantitates to ~ 4 times the σ donor interaction of peroxide bound to a single copper(II). This is consistent with the Xα calculations and the fact that peroxide has four bonding interactions with the two copper(II). The extremely high intensity of the 350 nm band in oxyhemocyanin also quantitates to having ~ 4 σ donor interactions with the binuclear copper site strongly supporting the side-on peroxide bridged structure of the Kitajima complex for oxyhemocyanin.

One can probe the π acceptor ability of the peroxide through a study of its intraligand stretching force constant hence O-O bond strength, which is obtained from a normal coordinate analysis of vibrational spectra (Figure 5). The idea here is that one would expect

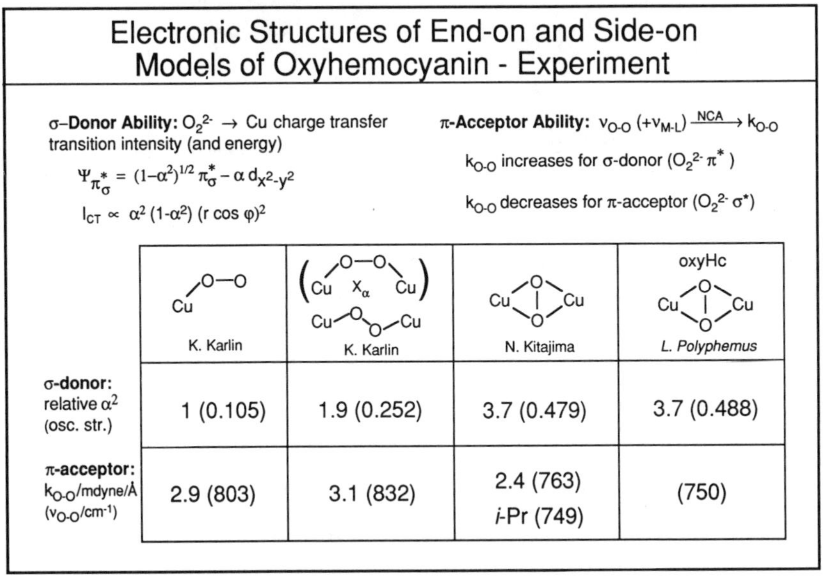

Figure 5. Electronic structures of end-on and side-on peroxide bridged models of oxyhemocyanin: comparison of experimentally determined peroxide σ-donor and π-acceptor abilities.

this force constant to increase as the σ donor interaction of the peroxide with the copper increases, as this removes the electron density from a π antibonding orbital on the peroxide increasing its intra-ligand bond strength. This is experimentally observed in going from the end-on monomer to the trans end-on dimer, where the O-O vibrational frequency increases from 803 to 832 cm^{-1} consistent with the latter having σ donor interactions with two coppers.[23] However, on going to the side-on bridged peroxide, the O-O stretching frequency goes way down as in oxyhemocyanin, yet in this geometry the peroxide is the strongest σ donor from the high charge transfer intensity associated with four bonding interactions to the two coppers. While the mechanical coupling in this geometry is complicated, the normal coordinate analysis on the side-on bridged model complex gives a significantly lower O-O force constant hence weaker O-O bond.[20] This is direct experimental evidence for the Xα calculated prediction that side-on bridged peroxide also participates in a π acceptor interaction with the coppers which shifts some electron density into the strongly antibonding σ* orbital on the peroxide.

Having determined the unique electronic structure of the side-on bridged peroxy-binuclear cupric complex and likely oxyhemocyanin and oxytyrosinase, it is now possible

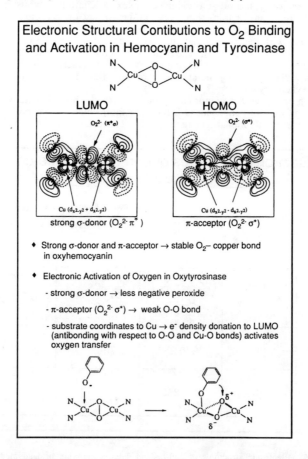

Figure 6. Electronic structural contributions to oxygen binding and activation in hemocyanin and tyrosinase.

to evaluate electronic structure contributions to the functions of these protein active sites[21] (Figure 6). The combination of strong σ donor and π acceptor interactions with the coppers leads to a very strong dioxygen-copper bond in the side-on structure which would contribute to reversible O_2 binding and in particular stabilize the bound peroxide with respect to decay to the inactive met site in hemocyanin. These bonding interactions further provide an electronic mechanism of dioxygen activation in oxytyrosinase. The strong σ donor interaction with the coppers results in a less negative peroxide while the π acceptor contribution to the bonding shifts a small amount of electron density into the peroxide σ^* orbital which leads to a weak O-O bond activating it for cleavage. As pictured in Figure 6, substrate coordination[25] to the copper would then contribute electron density into the LUMO, which is antibonding with respect to both the O-O and Cu-O bonds, and thus further initiate oxygen transfer in catalysis.

MULTICOPPER OXIDASES

The final section of this Review focuses on the multicopper oxidases (Figure 7) which utilize at least four copper ions, grouped into three types based on spectral properties, to

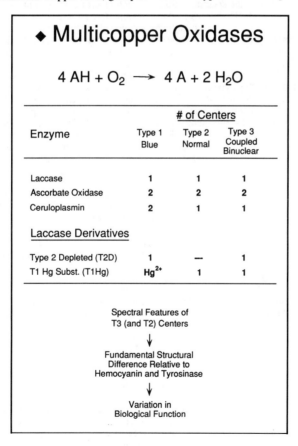

◆ **Multicopper Oxidases**

$$4\,AH + O_2 \longrightarrow 4\,A + 2\,H_2O$$

Enzyme	# of Centers		
	Type 1 Blue	Type 2 Normal	Type 3 Coupled Binuclear
Laccase	1	1	1
Ascorbate Oxidase	2	2	2
Ceruloplasmin	2	1	1
Laccase Derivatives			
Type 2 Depleted (T2D)	1	---	1
T1 Hg Subst. (T1Hg)	Hg^{2+}	1	1

Spectral Features of
T3 (and T2) Centers

↓

Fundamental Structural
Difference Relative to
Hemocyanin and Tyrosinase

↓

Variation in
Biological Function

Figure 7. Multicopper Oxidases: Enzymes and derivatives.

couple four one-electron oxidations of substrate to the four electron reduction of dioxygen to water. Two of the coppers form an antiferromagnetically coupled pair which is referred to as a Type 3 center. The Type 1 center is a blue copper site as discussed in Part I[3a] of this review, and the Type 2 copper is "normal" in the sense of having a tetragonal Cu(II) EPR spectrum ($g_\parallel > g_\perp > 2.00$, $A_\parallel > 140 \times 10^{-4} cm^{-1}$) and weak ligand field absorption features in the visible spectrum.[26] Laccase is the simplest of the multicopper oxidases containing one of each type of center for a total of four copper ions in the native enzyme.[1,2,27] This is still a complex problem and two derivatives have served to simplify this system; the Type 2 depleted (T2D) form where the Type 2 is reversibly removed leaving the Type 1 and Type 3 centers[28], and a Type 1 mercury substituted derivative (T1Hg) where the Type 1 is replaced by a spectroscopic and redox innocent mercuric ion.[29] The goal of our research on the multicopper oxidases has been to determine the spectral features of the the Type 3 (and Type 2 centers), to use these to define geometric and electronic structural differences relative to hemocyanin and tyrosinase, and to understand how these structural differences contribute to their variation in biological function where the hemocyanins and tyrosinases reversibly bind and activate dioxygen while the multicopper oxidases catalyse its four electron reduction to water.

We start by defining the spectral features of each type of copper in native and Type 2 depleted laccase (Figure 8). The native enzyme exhibits two EPR signals (Figure 8A), one with a large and a second with a small parallel hyperfine splitting associated with the Type 2 and Type 1 coppers, respectively. The Type 3 is EPR nondetectable as in hemocyanin and tyrosinase, hence can be initially be considered to be a coupled binuclear copper site. The absorption spectrum (Figure 8B) exhibits an intense thiolate S $\rightarrow$ Cu charge transfer transition at 600 nm associated with the Type 1 center.[3a] The only spectral feature which had been associated with the Type 3 center is an absorption band at 330 nm ($\varepsilon \sim 3000$ M^{-1} cm^{-1}) which reduces in intensity with the addition of two electrons at the same potential. This spectral region should contain histidine and hydroxide to Type 3 copper charge transfer transitions.[30] However, this assignment was complicated by the spectral features observed for the Type 2 depleted derivative. The EPR spectrum is straightforward in that the Type 2 contribution is eliminated leaving a single Cu(II) signal with a small $A_\parallel$ value associated with the Type 1 center.[31] The absorption spectrum also shows that the 600 nm band associated with the Type 1 remains. However, the 330 nm band is not present in T2D laccase (Figure 8B) which led to much confusion concerning its origin. Early on we discovered a key reaction which clarified this system. Addition of peroxide to T2D laccase leads to the reappearance of this 330 nm band.[32] This seemed to indicate that the Type 3 site in T2D laccase was in fact reduced even in the presence of dioxygen and oxidized by peroxide. We were able to confirm this through X-ray absorption spectral studies (Figure 8C) at the Cu-K edge near 9000 eV.[33] The Type 2 depleted derivative exhibits a peak at 8984 eV which is characteristic of Cu(I) in a three coordinate site and its magnitude could be quantitated using a normalized edge method we developed to determine that the Type 2 depleted derivative had a fully reduced Type 3 site. Peroxide eliminates this 8984 eV feature indicating that the site is fully oxidized (i.e., a met Type 3 center).

We were then able to study the Type 3 site in the absence of the Type 2 copper and compare this to the coupled binuclear site in hemocyanin (Figure 8D). First, as demonstrated from the X-ray edges in Figure 8C, the fully reduced Type 3 site is strikingly different from that of hemocyanin in that it does not react with dioxygen.[32,33] Peroxide does oxidize the site and we can further compare this met Type 3 center in laccase to met hemocyanin. For this derivative the sites are similar in that both are strongly antiferromagnetically coupled[26] indicating the presence of an endogenous bridge which must also be hydroxide for the Type 3 site in the multicopper oxidases based on the results of crystallography.[34]

One-electron reduction of both met derivatives produces the mixed valent half-met sites which exhibit dramatic differences (Figure 9). In particular, half-met hemocyanin has very unusual coordination chemistry with respect to exogenous ligand binding. For example, azide binds to this half-met active site with an equilibrium binding constant which is more than two orders of magnitude greater than that of azide binding to aqueous copper(II) and binds to produce quite unusual mixed valent spectral features.[35] We have

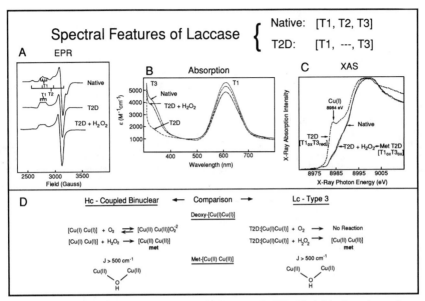

Figure 8. Spectral features of native and T2D laccase: A) EPR; B) visible absorption, and C) x-ray absorption spectra of native, T2D, and T2D laccase following reaction with hydrogen peroxide; D) Comparison of the reactivity and magnetism of deoxy and met hemocyanin and the laccase type 3 copper sites.

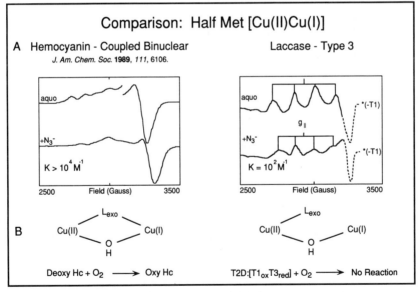

Figure 9. Comparison of half-met hemocyanin and the half-met Type 3 laccase copper sites: A) EPR spectra and binding constants of exogenous azide binding; B) Spectroscopically effective structural models for exogenous ligand binding to the half-met derivatives and their relation to differences in dioxygen reactivity.

studied this unusual half-met hemocyanin chemistry and spectroscopy in some detail[35] and determined that these derive from the fact that exogenous ligands bridge between the Cu(II) and Cu(I) of this mixed valent site (Figure 9B, left). Alternatively, the half-met Type 3 site in T2D laccase exhibits normal Cu(II) EPR spectra for all exogenous ligand bound forms and has a normal equilibrium binding constant consistent with the exogenous ligand binding terminally to the Cu(II) of the half-met Type 3 site.[30] This difference in bridging vs. terminal exogenous ligand binding modes directly correlates with differences in O_2 reactivity of these binuclear copper sites as described earlier in that only the hemocyanin site reversibly binds dioxygen (Figure 9B).

The combination of the Type 3 with the Type 2 center does, of course, react with dioxygen in the native enzyme and this led us to consider exogenous ligand interactions with both the Type 3 and Type 2 coppers in native laccase. A most appropriate spectral method to study the interaction of exogenous ligands with each center is low temperature MCD spectroscopy as this allows us to correlate excited state features with ground state

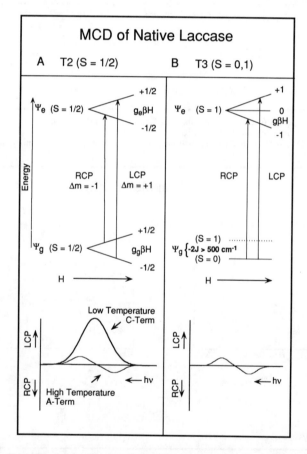

Figure 10. MCD of native laccase: A) Transitions and band profiles associated with type 2 copper; B) Transitions and band profiles associated with type 3 copper. Note the difference in temperature dependence of the MCD signal as described in the text.

properties.[36] In particular, the paramagnetic Type 2 copper will exhibit very different low temperature MCD features relative to an antiferromagnetically coupled Type 3 center (Figure 10). For the Type 2 center, both the ground and excited states have $S = 1/2$ and split in a magnetic field. The selection rules for MCD spectroscopy predict that there should be two transitions to a given excited state of equal magnitude but opposite sign. As the Zeeman splitting will be on the order of 10 cm^{-1} and absorption bands are on the order of a few thousand cm^{-1} broad these will mostly cancel and produce a broad weak derivative shaped MCD signal known as an A-term. This is observed if both components of the ground state are equally populated. However, as one lowers the temperature, the Boltzmann population of the higher energy component is reduced, cancellation no longer occurs, and one observes intense low temperature MCD signals known as C-terms. These can be two to three orders of magnitude more intense than the high temperature MCD signals. Alternatively, for the Type 3 center, the antiferromagnetic coupling leads to a $S = 0$ ground state which cannot split in a magnetic field. Thus, this site cannot exhibit C-term intensity and the low temperature MCD spectrum of native laccase will be dominated by the intense C-terms associated with the paramagnetic Type 2 copper center.[36a,b]

Low temperature MCD spectroscopy was used to probe azide binding to native laccase.[36a,b] Titration of the native enzyme with azide produces two N$_3^-$→Cu(II) charge transfer transitions, one at 500 nm and a second, more intense band at 400 nm (Figure 11A). The intensity of the 400 nm band as a function of azide concentration is plotted as a dashed line in the titration figure (Figure 11C). One can use low temperature MCD to correlate these excited states features to specific copper centers. The 500 nm absorption has a negative low temperature MCD signal associated with it which increases in magnitude with increasing azide concentration (Figure 11B). The intensity of this MCD feature is plotted as

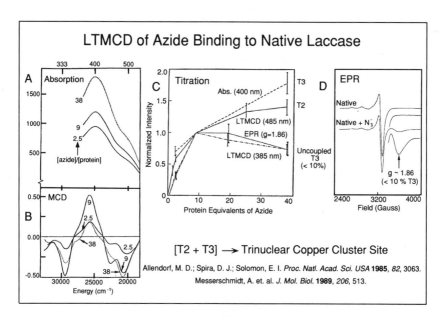

Figure 11. Low temperature MCD of azide binding to native laccase: A) Difference electronic absorption spectra at 298 K (See Ref. 36a); B) Difference MCD spectra at 4.9 K (5 T); C) Changes in EPR, LTMCD, and absorption intensities plotted as a function of increasing azide concentration; D) EPR of laccase titrated with azide. Arrow indicates the new signal present at g = 1.86 and 8 K.

a solid line in Figure 11C. As the 500 nm absorption band has a corresponding low temperature MCD signal, it must be associated with azide binding to the paramagnetic Type 2 center. There is also an MCD signal in the region of the 400 nm absorption band; however, it does not exhibit the same behavior as the absorption intensity (Figure 11B). The 385 nm MCD signal first increases, then decreases in intensity with increasing azide concentration. Its magnitude is plotted as a broken line in Figure 11C. While the low temperature MCD signal does not correlate with the 400 nm absorption band, it does correlate with an unusual g=1.86 signal in the EPR spectrum (Figure 11D) which we have shown to be associated with < 10% of the Type 3 sites which get protonatively uncoupled on binding azide. Thus the intense 400 nm absorption band has no low temperature MCD signal associated with it and it must correspond to azide binding to the coupled Type 3 center.

These low temperature MCD and absorption titration studies have determined that azide binds to both the Type 2 and Type 3 centers with similar binding constants; a series of chemical perturbations and stoichiometry studies have, in fact, shown that these are associated with the same azide demonstrating that this ligand bridges between the Type 2 and Type 3 centers.[36a] Thus MCD spectroscopy first defined the presence of a trinuclear copper cluster active site in biology. Recently, Messerchmidt, et al. have found from crystallography that a similar trinuclear copper cluster site is also present in ascorbate oxidase.[34]

Having demonstrated that the Type 3 center must in fact be viewed as part of a trinuclear copper cluster, including the Type 2 center, it was important to determine which coppers are required for the reactivity of the multicopper oxidases with dioxygen. We had already demonstrated using X-ray absorption edges in Figure 8C that a reduced Type 3 in the presence of an oxidized Type 1 center does not react with O_2.[32,33] We next looked at the reactivity of the fully reduced T2D [$T1_{red}$ $T3_{red}$] derivative with O_2 as this had been

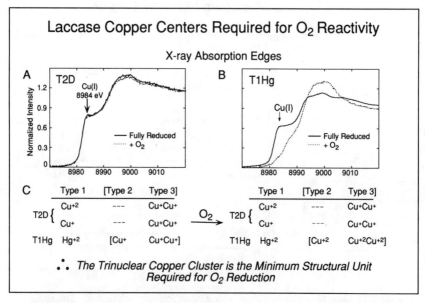

Figure 12. Laccase copper centers required for dioxygen reactivity: A) XAS of deoxy T2D laccase and deoxy T2D laccase following exposure to dioxygen; B) XAS of reduced T1Hg laccase and reduced T1Hg laccase following exposure to dioxygen; C) Summary of the reactivity of deoxy T2D, met T2D and reduced T1Hg laccase with oxygen.

generally viewed as the combination of copper centers in laccase required for dioxygen reactivity in the mechanistic proposals in the literature.[37] From Figure 12A it is clear that the 8984 eV reduced copper edge peak does not change on exposure of fully reduced T2D to O_2 indicating that the Type 2 center is required for dioxygen reactivity. Thus the T1Hg derivative was investigated which contains a valid T2/T3 trinuclear copper cluster. From Figure 12B the fully reduced trinuclear copper cluster site rapidly reacts with O_2 eliminating the 8984 eV peak and thus this trinuclear copper cluster is the minimum structural unit required for O_2 reduction.[38]

Since the mercuric ion is redox inactive, this derivative has one less equivalent available for O_2 reduction than native laccase and this property enabled us to stabilize an oxygen intermediate in T1Hg laccase. A combination of low temperature MCD and XAS has demonstrated that two coppers of the trinuclear cluster are oxidized in this intermediate.[39] Thus two electrons have been transferred to dioxygen and this species corresponds to a peroxide level intermediate which can be compared to the peroxo-binuclear

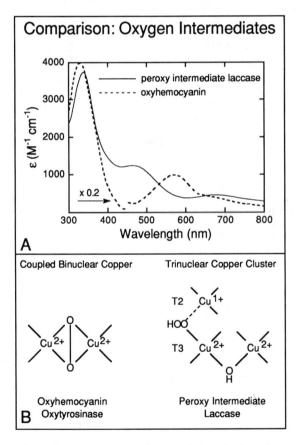

Figure 13. Comparison - oxygen intermediates: A) Electronic absorption spectra of the peroxy-intermediate in laccase versus oxyhemocyanin; B) Proposed structural differences between peroxide binding in oxyhemocyanin and oxytyrosinase relative to the peroxide intermediate at the trinuclear copper cluster in laccase.

cupric sites in oxyhemocyanin and oxytyrosinase. As is clear from Figure 13A, the peroxide intermediate in laccase has a strikingly different charge transfer spectrum from that of oxyhemocyanin and oxytyrosinase requiring a different geometric and electronic structure for this peroxy-trinuclear copper cluster site.[39a] Detailed spectral studies on this intermediate are presently underway. However, our most reasonable suggestion at this time is that it corresponds to a hydroperoxide bound end-on to one oxidized Type 3 copper and likely bridging to a reduced Type 2 copper center (Figure 13B). This difference in peroxide binding relative to hemocyanin and tyrosinase appears to play a key role in its irreversible, further reduction to water at the trinuclear copper cluster site.

Summary

At this point the unique spectral features associated with the major classes of active sites in copper proteins are reasonably well understood and define active site electronic structures which provide significant insight into their reactivities in biology. For the blue copper sites we have determined that the unique spectral features derive from a ground state wavefunction which has a high anisotropic covalency involving the thiolate ligand. This provides a very efficient superexchange pathway for long range electron transfer. For the coupled binuclear copper active sites we have seen that the unique spectral features of the oxy site correspond to a new bridging peroxide electronic structure which has very strong σ donor and π acceptor properties. These appear to make significant contributions to the reversible binding and activation of dioxygen by these active sites. In the multicopper oxidases, our spectral studies have determined that the Type 3 center is fundamentally different from the coupled binuclear copper site in hemocyanin and tyrosinase, that it is in fact part of a trinuclear copper cluster and that this trinuclear copper cluster is the structural unit required for O_2 reactivity. We have now defined a peroxide level intermediate at this trinuclear copper cluster site which is strikingly different from the peroxide bound in oxyhemocyanin and oxytyrosinase. Our spectral studies presently underway should provide important insight into the geometric and electronic structure differences which are indicated by these spectral differences and their contribution to differences in biological function.

ACKNOWLEDGEMENTS

This research has been supported by the NIH (DK-31450). EIS wishes to express his sincere appreciation to all his students and collaborators who are listed as co-authors in the literature cited for their commitment and contributions to this science.

REFERENCES

1. E. I. Solomon, K. W. Penfield, D. E. Wilcox, *Structure and Bonding*, **53**, 1 (1983).
2. E. I. Solomon, M. J. Baldwin, M. D. Lowery, *Chem. Rev.*, **92**, 521 (1992).
3. (a) E. I. Solomon, M. D. Lowery, "Electronic Structure of Active Sites in Copper Proteins: I. Blue Copper Sites", *Proceedings of the International Conference on the Chemistry of the Copper and Zinc Triads*, Royal Society of Chemistry (1992), in press. (b) This chapter.
4. E. I. Solomon, *Comments on Inorg. Chem.*, **3**, 227 (1984).
5. R. L. Jolly, Jr., L. H. Evans, N. Makino, H. S. Mason, *J. Biol. Chem.*, **249**, 335 (1974).
6. (a) R. S. Himmelwright, N. C. Eickman, C. D. LuBien, K. Lerch, E. I. Solomon, *J. Am. Chem. Soc.*, **102**, 7339 (1980). (b) G. L. Woolery, L. Powers, M. Winkler, E. I. Solomon, T. G. Spiro, *J. Am. Chem. Soc.*, **106**, 86 (1984). (c) G. L. Woolery, L. Powers, M. Winkler, E. I. Solomon, K. Lerch, T. G. Spiro, *Biochemica et Biophysica Acta*, **788**, 155 (1984).
7. (a) T. B. Freedman, J. S. Loehr, T. M. Loehr, *J. Am. Chem. Soc.*, **98**, 2809 (1976). (b) J. A. Larrabee, T. G. Spiro, *J. Am. Chem. Soc.*, **102**,

4217 (1980). (c) N. C. Eickman, E. I. Solomon, J. A. Larrabee, T. G. Spiro, K. Lerch, *J. Am. Chem. Soc.*, **100**, 6529 (1978).

8. (a) E. I. Solomon, D. M. Dooley, R. H. Wang, H. B. Gray, M. Cerdonio, F. Mogno, G. L. Romani, *J. Am. Chem. Soc.*, **98**, 1029 (1976). (b) D. M. Dooley, R. A. Scott, J. Ellinghaus, E. I. Solomon, H. B. Gray., *Proc. Natl. Acad. Sci. U.S.A.*, **75**, 3019 (1978).

9. D. E. Wilcox, T. D. Westmoreland, P. O. Sandusky, E. I. Solomon, unpublished results.

10. A. Volbeda, W. G. J. Hol, *J. Mol. Biol.*, **209**, 249 (1989).

11. N. C. Eickman, R. S. Himmelwright, E. I. Solomon, *Proc. Natl. Acad. Sci. U.S.A.*, **76**, 2094 (1979).

12. K. D. Karlin, R. W. Cruse, Y. Gultneh, A. Farooq, J. C. Hayes, J. Zubieta, *J. Am. Chem. Soc.*, **109**, 2668 (1987).

13. J. E. Pate, R. W. Cruse, K. D. Karlin, E. I. Solomon, *J. Am. Chem. Soc.*, **109**, 2624 (1987).

14. J. E. Pate, P. K. Ross, T. J. Thamann, C. A. Reed, K. D. Karlin, T. N. Sorrel, E. I. Solomon, *J. Am. Chem. Soc.*, **111**, 5198 (1989).

15. (a) P. K. Ross, M. D. Allendorf, E. I. Solomon, *J. Am. Chem. Soc.*, **111**, 4009 (1989). (b) F. Tuczek, E. I. Solomon, to be published.

16. T. N. Sorrel, C. J. O'Conner, O. P. Anderson, J. H. Reibenspies, *J. Am. Chem. Soc.*, **107**, 4199 (1985).

17. (a) V. McKee, J. V. Dagdigian, R. Bau, C. A. Reed, *J. Am. Chem. Soc.*, **103**, 7000 (1981). (b) V. McKee, M. Zvagulis, J. V. Dagdigian, M. G. Patch, C. A. Reed, *J. Am. Chem. Soc.*, **106**, 4765 (1984).

18. K. D. Karlin, B. I. Cohen, J. C. Hayes, A. Farooq, J. Zubieta, *J. Inorg. Chem.*, **26** 147 (1987).

19. (a) N. Kitajima, K. Fujisawa, Y. Moro-oka, K. Toriumi, *J. Am. Chem. Soc.*, **111**, 8975 (1989). (b) N. Kitajima, K. Fujisawa, C. Fujimoto, Y. Moro-oka, S. Hashimoto, T. Kitagawa, K. Toriumi, K. Tatsumi, A. Nakamura, *J. Am. Chem. Soc.*, **114**, 1277 (1992).

20. M. J. Baldwin, D. E. Root, J. E. Pate, K. Fujisawa, N. Kitajima, E. I. Solomon, submitted to *J. Am. Chem. Soc.*

21. (a) P. K. Ross, E. I. Solomon, *J. Am. Chem. Soc.*, **112**, 5871 (1990). (b) P. K. Ross, E. I. Solomon, *J. Am. Chem. Soc.*, **113**, 3246 (1991).

22. (a) L. Noodleman, J. G. Norman, Jr., *J. Chem. Phys.*, **70**, 4903, (1979). (b) L. Noodleman, *J. Chem. Phys.*, **74**, 5737, (1981).

23. M. J. Baldwin, P. K. Ross, J. E. Pate, Z. Tyeklár, K. D. Karlin, E. I. Solomon, *J. Am. Chem. Soc.*, **113**, 8671 (1991).

24. R. R. Jacobson, Z. Tyeklár, A. Farooq, K. D. Karlin, S. Liu, J. Zubieta, *J. Am. Chem. Soc.*, **110**, 3690 (1988).

25. (a) M. E. Winkler, K. Lerch, E. I. Solomon, *J. Am. Chem. Soc.*, **103**, 7001 (1981). (b) D. E. Wilcox, A. G. Porras, Y. T. Hwang, K. Lerch, M. E. Winkler, E. I. Solomon, *J. Am. Chem. Soc.*, **107**, 4015 (1985).

26. J. L. Cole, P. A. Clark, E. I. Solomon, *J. Am. Chem. Soc.*, **112** 9534 (1990).

27. (a) B. G. Malmström, in *New Trends in Bio-inorganic Chemistry*, R. J. P. Williams and J. R. R. F. Da Silva, Eds., (Academic Press, New York, 1978), Chapter 3. (b) J. A. Fee, *Struct. Bonding (Berlin)*, **23**, 1 (1975). (c) B. Reinhammer, in *Copper Proteins and Copper Enyzmes*, R. Lonti, Ed.,(CRC Press, Boca-Raton, FL, 1984), Vol III, Chapter 1. (d) A. Finazzi-Agró, *Life Chem. Rep.*, **5**, 199-209 (1987). e) L. Rydén, in *Copper Proteins and Copper Enzymes*, (CRC Press, Boca-Raton, FL, 1984), Vol III, Chapter 2.

28. M. T. Graziani, L. Morpurgo, G. Rotilio, B. Mondovi, *FEBS Lett.*, **70**, 87 (1976).

29. M. M. Morie-Bebel, M. C. Morris, J. L. Menzie, D. R. McMillian, *J. Am. Chem. Soc.*, **106**, 3677 (1984).

30. D. J. Spira-Solomon, E. I. Solomon, *J. Am. Chem. Soc.*, **109**, 6421 (1987).
31. B. Reinhammer, *Biochim. Biophys. Acta*, **275**, 245 (1972).
32. C. D. LuBien, M. E. Winkler, T. J. Thamann, R. A. Scott, M.S. Co, K. O. Hodgson, E. I. Solomon, *J. Am. Chem. Soc.*, **103**, 7014 (1981).
33. L. -S. Kau, D. J. Spira-Solomon, J. E. Penner-Hahn, K. O. Hodgson, E. I. Solomon, *J. Am. Chem. Soc.*, **109**, 6433 (1987).
34. (a) A. Messerschmidt, A. Rossi, R. Ladenstein, R. Huber, M. Bolognesi, G. Gatti, A. Marchesini, R. Petruzelli, A. Finazzi-Agró, *J. Mol. Biol.*, **206**, 513 (1989). (b) A. Messerschmidt, R. Huber, *Eur. J. Biochem.*, **187**, 341 (1990).
35. T. D. Westmoreland, D. E. Wilcox, M. J. Baldwin, W. B. Mims, E. I. Solomon, *J. Am. Chem. Soc.*, **111**, 6106 (1989).
36. (a) M. D. Allendorf, D. J. Spira, E. I. Solomon, *Proc. Natl. Acad. Sci. USA*, **82**, 3063 (1985). (b) D. J. Spira-Solomon, M. D. Allendorf, E. I. Solomon, *J. Am. Chem. Soc,* **108**, 5318 (1986). (c) P. J. Stephens, *Adv. Chem, Phys.*, **35**, 197 (1976). (d) S. B. Piepho and P. N. Schatz, in *Group Theory in Spectroscopy*, (John Wiley and Sons, Inc, New York, 1983).
37. (a) B. Reinhammer, Y. Oda, *J. Inorg. Biochem.*, **11**, 115 (1979). (b) O. Farver, M. Goldberg, I. Pecht, *Eur. J. Biochem.*, **104**, 71 (1980).
38. J. L. Cole, G. O. Tan, E. K. Yang, K. O. Hodgson, E. I Solomon, *J. Am. Chem. Soc.*, **112**, 2243 (1990).
39. (a) J. L. Cole, D. P. Ballou, E. I. Solomon, *J. Am. Chem. Soc.*, **111**, 8544 (1991). (b) J. L. Cole, B. L. Hemming, M. L. Kirk, D. E. Root, J. A. Eisenberg, E. I Solomon, unpublished results.

PULSED EPR STUDIES OF COPPER PROTEINS[1]

Jack Peisach

Department of Molecular Pharmacology, Albert Einstein College of Medicine, Bronx, New York, NY 10461, USA

CONTINUOUS WAVE EPR AND ENDOR

For more than 30 years, continuous wave EPR spectroscopy has been used to characterize active site structures of paramagnetic metalloproteins. For some metal centers, attempts to identify metal ligands from correlative information derived from EPR investigations of model compounds[1-5] constitutes some of the earliest examples of how bioinorganic chemistry can be used to mimic physical properties of metal-containing biomolecules.

An early method used to identify metal ligand structures in Cu(II) proteins[4] is one where the relationship between the EPR parameters $g_{||}$ and $A_{||}$ for Cu(II) model compounds is related to equatorial metal ligand composition and overall charge. (Axial ligand effects on EPR parameters are much smaller and are ignored in this treatment.) With the aid of this type of "truth diagram", a number of conclusions were drawn about copper protein structures that in many cases were subsequently substantiated by X-ray crystallographic analysis. As originally noted, a major shortcoming of the approach, though, is that it could not be used in instances, such as for type 1 copper proteins, where the site symmetry of the metal center deviates significantly from near square planar.

When Cu(II) is coordinated to an atom with a nuclear spin, one can sometimes observe superhyperfine splittings in the EPR spectrum that arise from the magnetic interaction of the electron spin of the metal ion with the nuclear magnetic moment of the coordinated nucleus. The superhyperfine pattern, then, provides a signature for the kind and number of interacting nuclei. For endogenous ligands in copper proteins, the only nucleus whose contributing magnetic interaction has thus far been spectroscopically resolved in frozen solution samples is ^{14}N.

[1]This work was supported by United States Public Health Service grants RR-02853 and GM-40168.

From K.D. Karlin and Z. Tyeklár, Eds., *Bioinorganic Chemistry of Copper*
(Chapman & Hall, New York, 1993).

(Interactions with protons are considerably weaker but their effects can be recognized from line width differences in samples exchanged against D_2O.) The nuclear spin of ^{14}N is unity and one expects to observe a splitting of individual features of the EPR spectrum corresponding to the number of ^{14}N nuclei coordinated to Cu(II). For a single, equatorially coordinated ^{14}N, one is entitled to observe such splitting as a triplet having intensities in the ratio 1:1:1. For multiple, equivalently coupled ^{14}N, the patterns become more complicated. For two such nuclei, a five line pattern is expected with lines having the ratio of intensities of 1:2:3:2:1; for three ^{14}N, a seven line pattern with the ratio of intensities of 1:3:6:7:6:3:1; while for four ^{14}N the intensities are 1:4:10:16:19:16:10:1.

Although the presence of superhyperfine lines in a Cu(II) protein EPR spectrum can be taken as proof of ^{14}N coordination, quantifying the number of interacting nuclei can be problematic. In some cases, spectral resolution is poor because of strain broadening, but here, marked improvements are sometimes achieved by making measurements at microwave frequencies away from the commonly employed X-band, 9 GHz[6,7].

A second problem arises from the potential overlap of superhyperfine patterns in the EPR spectrum, especially in the region of $g_\perp$, leading to erroneous structural assignment. If the Cu(II) site has a symmetry away from strict D_{4h}, slight differences in g_x and g_y will sometimes be sufficient to increase the apparent number of superhyperfine lines at $g_\perp$ in the spectrum due to overlap of features. For this reason, ^{14}N assignments are best obtained from an examination of a $g_{||}$ feature of the spectrum, where such ambiguities are not found. Another difficulty in assigning structure from an analysis of superhyperfine patterns comes about when more than a single ^{14}N nucleus is coupled to Cu(II). The expected 5-, 7- or 9-line patterns may suffer both in intensity and number of features when ^{14}N couplings are not equivalent. A unique interpretation of the spectrum may then not be possible.

The ability to make structural assignment ultimately rests in the ability to resolve spectral components. Even with a multifrequency EPR approach, one may not always recognize superhyperfine contributions in the spectrum. The reason for this is attributed, in part, to short intrinsic coherence times of many of the systems under investigation leading to inhomogeneous broadening of the EPR line. Even at low temperatures, where lattice relaxation times no longer effect line widths, this broadening may preclude spectral resolution, especially in frozen solution samples. Here, though, one may take other approaches, including ENDOR spectroscopy in its continuous wave or pulsed forms[8,9], or electron spin echo envelope modulation (ESEEM) spectroscopy.

With ENDOR, one can clearly differentiate ^{14}N from other nuclei having a nuclear moment, in particular ^1H and ^{13}C. (The latter is a minority species in natural abundance but sometimes high enough in concentration to recognize by ENDOR[10].) By making measurements at different microwave frequencies, one can differentiate overlapping spectral features arising from ^{14}N and ^1H. For pulsed ENDOR, with different spectrometer conditions (radio frequency power and pulse widths) one can emphasize one or another spectral component[11-13]. In all these studies, the regime of magnetic coupling useful for investigation is generally large and the measurement cannot be easily used to determine the specific number of nuclei coupled to Cu(II).

ESEEM SPECTROSCOPY

The problem of inhomogeneous line broadening as a consequence of loss of phase coherence is endemic to continuous wave EPR and is overcome with pulsed EPR. ESEEM spectroscopy with pulsed rather than continuous wave EPR is a complement to ENDOR that is particularly well suited for examining weak electron nuclear interactions. The method has been successfully employed in investigations of nearly every paramagnetic center of biological interest and has also been used to examine magnetic interactions with both natural and exogenously added nuclei including ^1H, ^2H, ^{13}C, ^{14}N, ^{15}N, ^{23}Na, ^{31}P, ^{95}Mo (as MoO_4) and ^{133}Cs [14-16].

The details of a home built pulsed EPR spectrometer used in many of these studies have been described elsewhere[17]. (It should be noted as well that a commercial spectrometer is now available.) In one procedure, two high powered (~1 kw), short duration (10 ns), microwave pulses, separated by time τ, are applied to a paramagnetic sample in an appropriate cavity (Fig. 1A).

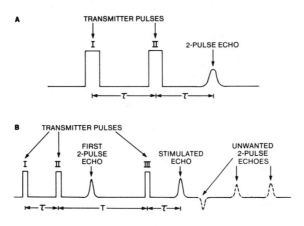

Fig. 1. (A) 2-pulse and (B) 3-pulse electron spin echo sequences. In (A), the echo envelope is obtained by incrementing τ and observing the echo intensity as a function of τ. In (B), τ is set and T is incremented. The echo envelope is then obtained from the observation of the echo intensity as a function of $T+\tau$. The ESEEM spectrum is obtained by Fourier transformation of the echo envelope.

At time τ after the second pulse, a spin echo is generated. As τ is incremented, the spin echo does not decrease monotonically but instead modulations in the echo envelope are obtained. These arise from the magnetic interaction of the electron spin of the paramagnetic with nuclear spins in the vicinity. The pattern of modulations is related to the precession frequencies of these nuclei.

The modulation envelope obtained by this method consists of periodicities that are related to a superhyperfine frequencies associated with the various electron spin states in a spin manifold. In the simplest case, with an S = 1/2 electron interacting with an I = 1/2 nuclear spin, one observes two periodicities corresponding to sum and differences (ENDOR) frequencies of the spin manifold. The Fourier transformation of the modulation envelope obtained this way is the ESEEM spectrum. For the case cited, the spectrum contains two lines arising from nuclei coupled through a dipole interaction. These lines correspond to the nuclear Zeeman interaction and twice this value.

A major shortfall of the 2-pulse method is that the phase memory is rather short, thereby limiting the time when modulations can be obtained. In cases where the periodicity is very slow or where the modulations are quickly damped, the lines in the Fourier transform tend to be broad. With more than a single type of nucleus coupled to the electron spin, spectral resolution may therefore be compromised. Additionally, a small contact interaction may broaden lines. The spectra obtained may thus be difficult to analyze.

An alternative procedure is one utilizing three pulses (Fig. 1B). Here, two fixed microwave pulses separated by time τ are applied to a paramagnetic sample. A third pulse is applied at a time T after the second pulse, and the spin echo is observed at time τ after the third pulse. The value of T is incremented and the echo intensity is once again modulated with periodicities related to the magnetic interaction of the electron spin with nuclei in the vicinity. By setting the value of τ corresponding to a whole number multiple of the periodicity of protons in the sample, one can suppress their spectral contribution.

There are some advantages of the three-pulse over the two pulse procedure. Most important is that the phase memory limitations associated with a two-pulse experiment is largely obviated. Thus, periodicities in the echo envelope extend to longer times, leading to narrower lines in the Fourier transform. Another advantage is that the lines obtained correspond to the superhyperfine frequencies, and not their sums and differences.

HISTIDINE IMIDAZOLE AS A Cu(II) LIGAND

For Cu(II) proteins, ESEEM spectroscopy has proven useful in three types of experiments, the first to identify and quantify the number of interacting imidazoles at a Cu(II) site, the second to determine the effect of H-bonding or even metal coordination at the remote or amino nitrogen of histidine imidazole coordinated to Cu(II), and finally, to identify an axially or equatorially coordinated water ligand to Cu(II). Only the first two will be addressed here.

ESEEM was used to verify the already known structures of azurin, plastocyanin and superoxide dismutase and to predict the Cu(II)-ligand structures of laccase, dopamine β-hydroxylase, ceruloplasmin and stellacyanin[17-19]. The original assignments of imidazole ligation at the type 1 and type 2 copper sites of ascorbate oxidase[20] and at the copper site of galactose oxidase[21] have been subsequently verified by x-ray crystallographic analysis[22,23]. The recognition of the zero field quadrupolar frequencies, near 0.7, 0.7 and 1.4 MHz, attributable to the remote, protonated or amino nitrogen of imidazole in the ESEEM spectrum is the single spectral characteristic that has allowed us to make the structural assignment of an histidine side chain as a Cu(II) ligand in both natural and artificial copper proteins[24] (Fig. 2A,B). Thus, the spectrum of stellacyanin bears strong resemblances to that of a Cu(II) diethylenetriamine imidazole model [25,26].

(Nitrogen that is directly coordinated equatorially makes no spectral contribution as the electron-nuclear coupling is large and is outside the range for ESEEM observation.)

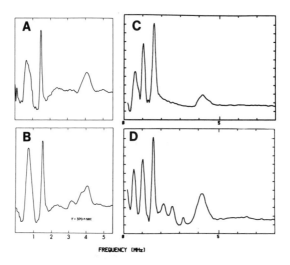

Fig. 2. ESEEM spectra of (A) the model compound Cu(II) diethylenetriamine imidazole (B) stellacyanin (C) the model Cu(II) diethylenetriamine 2-methylimidazole and (D) *Chromobacterium violaceum* phenylalanine hydroxylase. The sharp, low frequency lines in the spectra arise from the zero field quadrupolar frequencies of the remote, protonated ^{14}N of Cu(II)-coordinated imidazole.

Based on an analysis of ESEEM spectra, including those carried out by spectral simulation, we have been able to determine the nuclear quadrupole parameters for the remote ^{14}N of coordinated imidazole (either N1 or N3) in a number of Cu(II)-proteins. We have been able to show in Cu(II)-imidazole model compounds[27] that the nuclear quadrupole parameters can be altered by changes in the local environment of the remote nitrogen, such as through hydrogen bonding, we can reconcile the quadrupolar changes that occur with metal bridging, and have suggested a mechanism for metal coordination to imidazole at N3 as determined by structural features at N1. The mechanism is equally valid when Cu(II) is coordinated to N1, such as in Type 1 copper sites.

NUCLEAR QUADRUPOLE INTERACTION

As noted above, ESEEM is particularly well suited for the study of weak electron nuclear coupling. For this reason, the interaction between Cu(II) and an equatorially coordinated imino nitrogen of imidazole, is too large to be observed by this method, but can be studied by ENDOR[28]. It is the remote ^{14}N of coordinated imidazole that makes its ESEEM spectral contribution as follows:

Let us consider that the spin Hamiltonian for ^{14}N contains three terms, the nuclear Zeeman interaction, the electron-nuclear superhyperfine interaction and the nuclear quadrupole interaction (nqi) (Fig. 3). For measurements made at X-band,

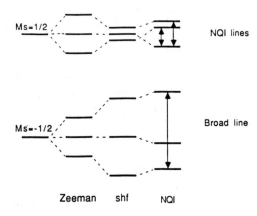

Fig. 3. Electron spin energy diagram for the remote ^{14}N of Cu(II)-coordinated imidazole. The three low frequency lines in the ESEEM spectrum of Fig. 2C,D correspond to the nuclear quadrupole transitions in the upper manifold. The broad line near 4 MHz corresponds to a $\Delta M = 2$ transition of the lower manifold.

the nuclear Zeeman term is approximately equal to one-half the electron-nuclear superhyperfine term. For one of the ^{14}N spin manifolds, the nuclear Zeeman term is nearly canceled by the electron-nuclear superhyperfine term so that the energy level splitting is dominated by the nuclear quadrupole terms giving rise to transitions at ν_o, ν_+, ν_-, the nuclear quadrupole frequencies (Fig. 2C,D). For the other manifold, the Zeeman and superhyperfine terms are additive and these, in turn, add to the nuclear quadrupole term. Although in Cu(II)-doped single crystal measurements one observes individual $\Delta M = 1$ transitions from within this manifold[29], because of their strong orientation dependence, they are not resolved in a powder pattern spectrum, as might be obtained for frozen solution samples. However, one does observe a $\Delta M = 2$ transition giving rise to a broad line in the spectrum at approximately four times the nuclear Zeeman frequency. It is from an analysis of the spectrum based on spectral simulation from which one obtains the nuclear quadrupole interaction, e^2qQ, and an asymmetry parameter, η, which varies from zero to unity, so that

$$\nu_\pm = 3/4 \; e^2qQ(1 \pm \eta/3)$$
and
$$\nu_o = 1/2 \; e^2qQ\eta.$$

Other information that can be obtained by spectral simulation[30], in particular for data collected at discrete points across the EPR line, includes the isotropic electron nuclear coupling parameter, the effective distance between the unpaired electron and its interacting nucleus and the relative orientation of the principal axis of the electric field gradient tensor of the ^{14}N in question to that of the g tensor for Cu(II). Further, from an analysis of nuclear quadrupole results, one may determine the electron orbital occupancy of a particular bond associated with ^{14}N[27].

For Cu(II)-imidazole systems, it is fortunate that the magnitude of the electron nuclear coupling is comparable to the Zeeman interaction at X-band, as this allows for a simple explanation of the ESEEM spectrum. Nevertheless, for systems where the magnitudes of electron nuclear coupling are considerably smaller, such as for Cu(II)(benzac)$_2$(pyr), one can nevertheless determine the quadrupole coupling constants as well as the electron nuclear coupling parameters by spectral simulation, using a knowledge of the g tensor orientation with respect to the structural axis and from a reasonable assignment of the direction of the electric field gradient tensor of the Cu(II)-ligated axial pyridine nitrogen[30].

For the remote nitrogen of a Cu(II)-coordinated substituted imidazole, one can radically alter the nuclear quadrupole parameters by making substitutions for protons on the ring carbons or on the remote ^{14}N[27]. By and large, the greatest effect on e^2qQ is obtained with substitution of an alkyl or aryl group for the remote ^{14}N proton. Ring nitrogen substitution reduces η and concomitantly increases e^2qQ. These differences in quadrupole coupling parameters are a resultant of alteration of the electric field gradient at the nitrogen leading to a reduction of orbital occupancy of the remote N-H (or N-C) bond. For imidazole, nuclear quadrupole parameters and thus electron orbital occupancy reflect the local environment of the nitrogen in question.

ELECTRON ORBITAL OCCUPANCY AND HYDROGEN BONDING

The relation of orbital occupancy to the structure of imidazole can be easily recognized in systems where nuclear quadrupole parameters can be related to alteration of hydrogen bonding. In the simplest case, one may cite the large differences in nuclear quadrupole parameters for imidazole in the gas phase and in the solid state [31,32] (Table 1). The parameters for pyridine show much less

Table 1 - Nuclear quadrupole parameters for nitrogen bases in the gas phase and in the solid state.

	GAS PHASE		SOLID STATE	
	e^2qQ	η	e^2qQ	η
imidazole	2.54	0.18	1.42	0.99
pyridine	4.87	0.42	4.92	0.40

change[33]. Based on the Townes-Dailey theory[34], it can be shown that the sp^2 orbital occupancy of the N-H bond in imidazole, as assessed from the nuclear quadrupole parameters, is markedly increased in the solid state where imidazole forms a hydrogen bonded structure. As pyridine is incapable of intermolecular hydrogen bonding, the nuclear quadrupole parameters do not show as large a change.

In another example, an infrared and NQR study was carried out on crystalline $Zn(II)(imid)_2X_2$ (X = Cl, Br, I)[35-37]. Here, the remote N-H of Zn(II)-coordinated imid is in close proximity to a halide on an adjoining molecule. The infrared stretching frequency of the N-H bond increases with the size of the halide. This has been interpreted as representing the effect of hydrogen bonding capabilities of the various halides on the strength of the closeby N-H bond. Chloride forms the best H-bond, thereby the weakest the N-H bond. From the NQR measurement it was shown that the electron occupancy of the N-H sp^2 orbital is correlated with a decrease of the N-H stretching frequency. It may therefore be argued that for imidazole, the magnitude of the N-H electron orbital occupancy is directly related to the strength of the hydrogen bond that is formed. As a corollary, it is suggested that the N-H bond length increases with electron orbital occupancy. This idea is verified from neutron diffraction and NQR studies for a series of crystalline imidazole-containing complexes, demonstrating the linear dependence between N-H valence orbital occupancy and the inverse cube of the N-H bond length[38].

HYDROGEN BONDING IN PROTEINS

For some Cu(II) proteins, including galactose oxidase[21], phenylalanine hydroxylase[39] (Fig. 2D) and amine oxidase[17], the low frequency lines in the ESEEM spectrum (near 0.5, 1.0 and 1.5 MHz) are so far removed from those for the Cu(II)- diethylenetriamine imid model (Fig. 2A) as to suggest that there is a change in the sp^2 orbital occupancy of the remote NH of Cu(II)-coordinated imidazole. In model studies, it was shown that such changes can come about from chemical substitution on imidazole (2-methylimidazole is a model that mimics features in the protein spectrum) (Fig. 2C)[39], but structural alterations of this type are not found in proteins. Another suggestion is that the local environment of the remote nitrogen may be altered by local effects, such as those related to hydrogen bonding. This suggestion was verified from model studies[27] where it was shown that in the presence of high concentrations of formate, a good hydrogen bonding anion, the nuclear quadrupole frequencies of a Cu(II)-coordinated substituted imidazole are altered in a way indicative of increased sp^2 electron orbital occupancy at N-H. In contrast, in the presence of D_2O, the changes in the ESEEM spectrum suggest reduced electron orbital occupancy, as would be expected as deuterium bonding is weaker than hydrogen bonding. It would appear, then, that nuclear quadrupole measurements of Cu(II) proteins can provide information about hydrogen bonding at the remote N-H of a Cu(II) coordinated histidine imidazole side chain. Further, from an analysis of the ESEEM spectrum, one can relate nuclear quadrupole parameters to electron orbital occupancy and further, relate this to the N-H bond length.

The effects of N-H bond polarization brought about by hydrogen bonding is expected to affect the association between Cu(II) and N3 of coordinated imidazole. Let us consider that when Cu(II) binds to imidazole, N1 donates its lone-pair electron to the metal ion and in this way acquires a partial positive charge

(Fig. 4). This charge can delocalize onto the π system of the heterocycle. In the presence of a nucleophile, the charge can delocalize further onto N1 through the formation of a hydrogen bond whose presence polarizes the N-H bond. With increased bond polarization at N1, there is an increased affinity for the metal ion

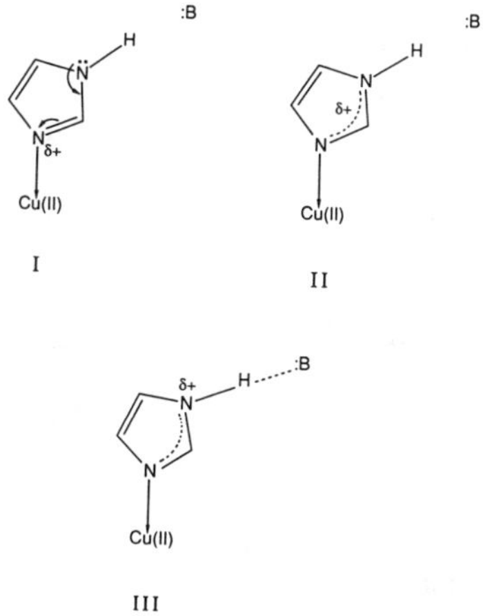

Fig. 4. The effect of Cu(II) coordination to imidazole on N-H bond polarization. See text for a description.

at N3. Where hydrogen bonding capabilities are reduced, the affinity of imidazole for Cu(II) is likewise reduced.

Other perturbants at the remote or amino nitrogen of Cu(II)-coordinated imidazole affect the electric field gradient and thus, the nuclear quadrupole parameters. A case in point is Cu,Zn superoxide dismutase[40] where the replacement of the proton by Zn(II) on the remote nitrogen of Cu(II)-coordinated imidazole can be recognized as the alteration of the 0.7, 0.7, 1.4 MHz characteristic spectrum so that the lines now appear at 0.4, 1.1 and 1.5 MHz. Under conditions where imidazole is no longer bonded to Zn(II), either at low pH or where Zn(II) is removed, the altered spectral features are not observed.

QUANTITATION OF LIGANDS

One of the strengths of the ESEEM method, as compared to ENDOR, is the ability to quantify the number of coupled nuclei associated with a paramagnetic center. For example, where more than a single imidazole ^{14}N is coupled to Cu(II), the result is the appearance of combination lines in the ESEEM spectrum[39]. For example, when the electron spin of Cu(II) is coupled to a single 2-methyl imidazole (as in a Cu(II) diethylenetriamine complex), a four line spectrum is seen containing three zero field quadrupolar line and a broader component at higher frequency, corresponding to a $\Delta M = 2$ transition (Fig. 2C). As the number of interacting, equivalently coupled nuclei is monotonically increased from one to four, new lines appear in the spectrum whose frequencies are the combination of the nuclear quadrupole frequencies of the mono 2-methyl imid complex (Fig. 5). In addition, one notes that the relative intensities of the $\Delta M = 2$ transition with respect to the nuclear quadrupole transitions is altered as well.

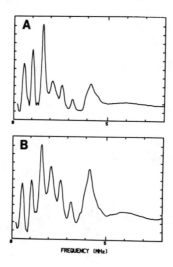

Fig. 5. ESEEM spectra of Cu(II) coordinated to (A) two and (B) four 2-methylimidazole molecules. Note the combination frequencies in A and B not found in the spectrum from a single 2-methylimidazole coordinated to Cu(II) (Fig. 2C).

Based on a comparison of ESEEM spectral data, one can determine the number of equivalently coupled nuclei to Cu(II). This method is particularly useful in differentiating populations containing a single as compared to multiple ^{14}N coordination. Differentiating between 2 and 3 or 3 and 4 nuclei becomes more difficult as this is based on comparisons with appropriate models. The ability to use ESEEM spectral simulations for structures not completely defined in solution as, say, in the crystalline state where ligand rotations are hindered, introduces ambiguities in the analysis. Nevertheless, even with these limitations, reasonable structural assignments may be made.

Where inequivalent populations of ^{14}N couple to Cu(II), one can recognize their differences. For example, when Cu(II) binds to isopenicilline synthase, the continuous wave EPR spectrum, in particular in the region of $g_\perp$, exhibits superhyperfine splittings attributable to ^{14}N ligation to metal[41]. In the presence of substrate, the EPR spectrum is changed, but again ^{14}N coordination is sustained.

The ESEEM spectrum of the Cu(II) protein contains lines characteristic of the remote ^{14}N of coordinated imidazole. In addition, one observes combination lines indicative of multiple imidazole coordination (Fig. 6). The spectrum can be simulated assuming coupling to two equivalent ^{14}N. When substrate is bound to

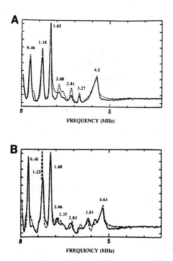

Fig. 6. ESEEM spectra of Cu(II) isopenicilline synthase (A) in the absence and (B) in the presence of substrate δ-(L-α aminoadipoyl)-L-cysteine-D-valine. The spectra contain combination lines indicative of multiple imidazole coordination. When substrate is added, new lines appear in the spectrum that are assigned to $\Delta M = 2$ transitions of ^{14}N from populations of imidazole with different electron nuclear coupling. Solid lines, Fourier transform of ESEEM envelopes. Dotted lines, simulated spectra.

the enzyme, the change in the EPR spectrum is indicative of an altered structure at the metal binding site. ESEEM indicates that the double imidazole structure is retained. However, the electron-nuclear coupling has been altered so that the binding to individual imidazoles is no longer equivalent. These can easily be recognized from alterations in the $\Delta M = 2$ transitions and can be verified by spectral simulation.

CONCLUSIONS

In summary, we have shown how ESEEM spectroscopy can be used to identify imidazole as a metal ligand in copper proteins and to identify the number of imidazoles equatorially coordinated. Further, we note how the method is particularly useful in describing the structure of the local environment of the remote or amino ^{14}N of coordinated imidazole, relating nuclear quadrupole parameters to the extent of hydrogen bond polarization or to the N-H bond length.

REFERENCES

1. W.E. Blumberg and J, Peisach, in *Bioinorganic Chemistry*, R.F. Gould, Ed. (American Chemical Society Publications, 1971), pp. 3342-3395.
2. W.E. Blumberg and J. Peisach, in *Probes of Structure and Function of Macromolecules and Membranes, Vol. II, Probes of Enzymes and Hemoproteins*, B. Chance, T. Yonetani, A.S. Mildvan, Eds. (Academic Press, New York, 1971), pp. 231-239.
3. W.E. Blumberg and J. Peisach, *Arch. Biochem. Biophys.*, **162**, 502 (1974).
4. J. Peisach and W.E. Blumberg, *Arch. Biochem. Biophys.*, **165**, 691 (1974).
5. J. Peisach and W.B. Mims, *Biochemistry*, **16**, 2795 (1977).
6. J.S. Hyde and W. Froncisz, *Annu. Rev. Biophys. Bioeng.*, **11**, 391 (1982).
7. W. Froncisz and P. Aisen, *Biochim. Biophys. Acta*, **700**, 55 (1982).
8. B.M. Hoffman, R.J. Gurbiel, M.M. Werst, M. Sivaraja, in *Advanced EPR Applications in Biology and Biochemistry*, A.J. Hoff, Ed. (Elsevier, Amsterdam, 1989), pp. 541-591.
9. G.H. Rist, J.S. Hyde, T. Vänngård, *Proc. Natl. Acad. Sci. USA*, **67**, 789 (1970).
10. A.L.P. Houseman, B.-H. Oh, M.C. Kennedy, C. Fan, M.M. Werst, H. Beinert, J.L. Markley, B.M. Hoffman, *Biochemistry*, **31**, 2073 (1992).
11. E.R. Davies, *Phys. Lett.*, **47A**, 1 (1974).
12. W.B. Mims, *Proc. Roy. Soc.*, **A283**, 452 (1965).
13. C. Fan, P.I. Doan, R.F. Davoust, B.M. Hoffman, *J. Mag. Res.* **44**, 114 (1992).
14. W.B. Mims and J. Peisach, in *Biological Applications of Magnetic Resonance*, R.G. Shulman, Ed., (Academic Press, New York, 1979), pp. 221-269.
15. W.B. Mims and J. Peisach, in *Biological Magnetic Resonance, Vol. 3*, L.J. Berliner and J. Reuben, Eds., (Plenum Press, New York, 1981), pp. 213-263.
16. W.B. Mims and J. Peisach, in *Advanced EPR: Applications in Biology and Biochemistry*, A.J. Hoff, Ed., (Elsevier, Amsterdam, 1989), pp. 1-57.
17. J. McCracken, J. Peisach, D.M. Dooley, *J. Am. Chem. Soc.*, **109**, 4064 (1987).
18. B. Mondovi, M.T. Graziani, W.B. Mims, R.Oltzik, J. Peisach, *Biochemistry*, **16**, 4198 (1977).
19. J. McCracken, P.R. Desai, N.J. Papadopoulos, J.J. Villafranca, J. Peisach, *J. Am. Chem. Soc.*, **110**, 1069 (1988).
20. L. Avigliano, J.L. Davis, M.T. Graziani, A. Marchesini, W.B. Mims, B. Mondovi, J. Peisach, *FEBS Lett.*, **136**, 80 (1981).
21. D.J. Kosman, J. Peisach, W.B. Mims, *Biochemistry*, **19**, 1304 (1980).
22. A. Messerschmidt, A. Rossi, R. Ladenstein, R. Huber, M. Bolognesi, G. Gartti, A. Marchesini, R. Petruzzelli, A. Finazzi-Agro, *J. Mol. Biol.*, **206**, 513 (1989).
23. N. Ito, S.E.V. Phillips, C. Stevens, Z.B. Ogel, M.J. McPherson, J.N. Keen, K.D.S. Yadav, P.F. Knowles, *Nature*, **350**, 87 (1991).
24. W.B. Mims, J. Peisach, *J. Chem. Phys.*, **69**, 4921 (1978).
25. W.B. Mims, J. Peisach, *Biochemistry*, **15**, 3863 (1976).
26. W.B. Mims, J. Peisach, *J. Biol. Chem.*, **254**, 4321 (1979).
27. F. Jiang, J. McCracken, J. Peisach, *J. Am. Chem. Soc.*, **112**, 9035 (1990).
28. J.E. Roberts, J.F. Cline, V. Lum, H. Freeman, H.B. Gray, J. Peisach, B. Reinhammar, B.M. Hoffman, *J. Am. Chem. Soc.*, **106**, 5324 (1984).
29. M.J. Colaneri, J.A. Potenza, H.J. Schugar, J. Peisach, *J. Am. Chem. Soc.*, **112**, 9451 (1990).

30. J. Cornelius, J. McCracken, R. Clarkson, R.J. Belford, J. Peisach, *J. Phys. Chem.*, **94,** 6977 (1990).

31. G.L. Blackman, R.D. Brown, F.R. Burden, I.R. Elsum, *J. Mol. Spectrosc.*, **60,** 63 (1976).

32. M.J. Hunt, A.L. Mackay, D.T. Edmonds, *Chem. Phys. Lett.*, **34,** 473 (1975).

33. Y.N Hsieh, G.V. Rubenacker, C.P. Cheng, T.L. Brown, *J. Am. Chem. Soc.*, **99,** 1384 (1977).

34. C.H. Townes, B.P. Dailey, *J. Chem. Phys.*, **17,** 782 (1949).

35. C.I.H. Ashby, C.P. Cheng, T.L. Brown, *J. Am. Chem. Soc.*, **100,** 6057 (1978).

36. C.I.H. Ashby, C.P. Cheng, E.N. Duesler, T.L. Brown, *J. Am. Chem. Soc.*, 100, 6063 (1978).

37. C.I.H. Ashby, W.F. Paton, T.L. Brown, *J. Am. Chem. Soc.*, **102,** 2990 (1980).

38. M.J. Colaneri and J. Peisach, *J. Am. Chem. Soc.*, **114,** 5335 (1992).

39. J. McCracken, S. Pember, S.J. Benkovic, J.J. Villafranca, R.J. Miller, J. Peisach, *J. Am. Chem. Soc.*, **110,** 1069 (1988).

40. J.A. Fee, J. Peisach, W.B. Mims, *J. Biol. Chem.*, **256,** 1910 (1981).

41. F. Jiang, J. Peisach, L.-J. Ming, L.W. Que, V.J. Chen, *Biochemistry,* **30,** 11437 (1991).

COPPER(II) COMPLEXES OF BINUCLEATING MACROCYCLIC BIS(DISULFIDE) TETRAMINE LIGANDS

Stephen Fox, Joseph A. Potenza, Spencer Knapp, and Harvey J. Schugar

Department of Chemistry, Rutgers, The State University of New Jersey, New Brunswick, NJ 08903, USA

INTRODUCTION

The few well-characterized examples of Cu(II)-disulfide bonding are those involving synthetic low molecular weight complexes. Suggestions in the chemical literature for the existence of disulfide-containing copper protein active sites have been limited to stellacyanin, a blue copper protein that is devoid of methionine residues and thus cannot have the customary axial thioether ligation. This possibility of disulfide ligation has been considered in detail and shown to be unlikely based upon a structural study of cucumber basic blue protein, which shares considerable sequence homologies with stellacyanin.[1] Spectroscopic studies have led to suggestions that an amide group substitutes for the usual methionine ligation,[1] and a recent pulsed endor study[2] of the high pH form of stellacyanin indicates that the active site unit is Cu(II)[N(His)$_2$]S(Cys)N(deprotonated amide). Structurally characterized examples of Cu(II)-disulfide bonding are limited to three binuclear complexes[3-5] and one polymeric complex[6] that all feature Cu(II) ions weakly interacting with apical disulfide groups. The reported Cu(II)-S(disulfide) bond distances are 3.057(10) and 3.138(9)Å,[3] 3.16(1) and 3.28(10)Å,[4] 2.721(3)Å,[5] and 2.678(2)Å.[6] On the basis of an electronic absorption at 330 nm assigned as Cu(II)→disulfide CT, equatorial Cu(II)-disulfide bonding was proposed for the solution complex formed from CuCl$_2$ and bis[2-(2-pyridylmethylamino)ethyl]disulfide.[7] This ligand is pentadentate towards Ni(II). The equatorial Ni-(S)disulfide bond distance is 2.472(5)Å, and the sixth coordination site of Ni(II) is occupied by a chloride ion.[8] Similarly-bound Cl⁻ in the Cu(II) analogue is expected to result in a near uv CT absorption[9], and this possibility clouds the spectroscopic assignment described above.

This chapter deals with the synthesis, structure, and spectroscopic properties of macrocyclic Cu(II) complexes that provide the first fully characterized examples of the short (approximately 2.5Å) equatorial Cu(II)-disulfide bonding.

From K.D. Karlin and Z. Tyeklár, Eds., *Bioinorganic Chemistry of Copper* (Chapman & Hall, New York, 1993).

34

PRECURSOR CuN_2S_2 COMPLEXES: TYPE II COPPER ALIPHATIC THIOLATES

The syntheses of the macrocyclic bis(disulfide)tetramine ligands feature the oxidation of diaminodithiolate complexes by I_2 (Figure 1).

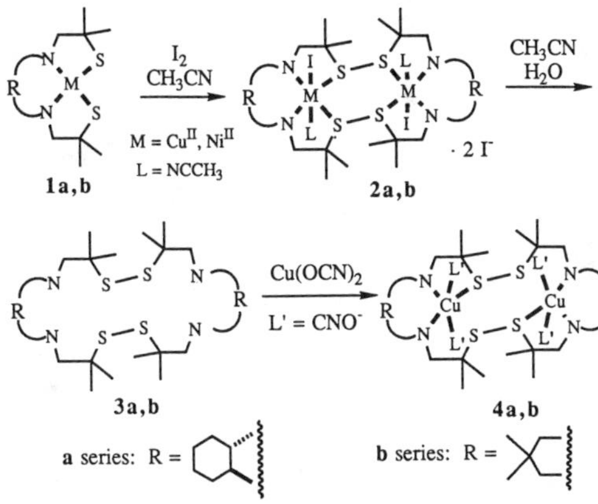

Figure 1. Overview of the synthesis of the mononuclear metal aminothiolates and the binuclear macrocyclic bis(disulfide) tetramine ligands and complexes that are the subjects of this chapter.

Our initial experiments utilized the Cu(II) complex of ligand **1a** (M=Cu). As examples of redox stable crystalline Cu(II) aliphatic thiolates are rare, the reader may be interested in the rationale of our ligand design. Also, the structural and spectroscopic properties of four- and five- coordinate type II Cu(II)thiolate models allow some of the corresponding features reported for the native type I blue protein sites to be placed in better perspective. We have developed two strategies for obtaining Cu(II) aliphatic thiolates that are stable enough at room temperature for conventional crystallization and spectroscopic studies. These strategies are based on the premise that the redox decomposition of Cu(II)RS⁻ complexes to yield RSSR and two Cu(I) ions is a particularly facile two electron process for bis(thiolato)-bridged Cu(II) dimers. In one case, the kinetically non-labile Cu(II) complex of the macrocyclic tetramine "tet a" was used so that addition of a single thiolate such as ⁻$SCH_2CH_2CO_2^-$ would afford the coordinatively saturated Cu(II)N₄S species **5** (Figure 2a).[10] The likelihood of the second thiolate bond that is required for dimer formation is most likely reduced by the completeness of the N₄S donor set and the non-lability of the N₄ donor set. Solutions of the corresponding complex with ⁻$SCH_2CO_2^-$ are less stable, and deposit the novel persulfide chromophore Cu(II)N₄ (⁻$SSCH_2CO_2^-$)**6** (Figure 2b).[11] The mechanism of this redox decomposition process is not understood, but conceivably is associated with the type of constrained coordination geometry noted above.

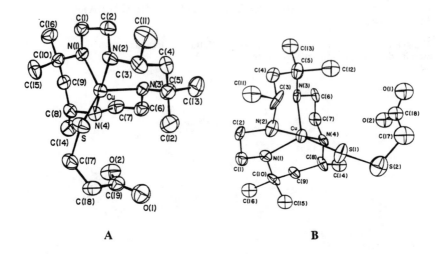

A   B

Figure 2. (a) ORTEP view of **5**.[10] Two methanol solvate molecules have been omitted for clarity. This approximately trigonal-bipyramidal complex has equatorial Cu-N(1) and Cu-N(3) distances of 2.153(4) and 2.169(4)Å, respectively, and axial Cu-N(2) and Cu-N(4) distances of 2.031(4) and 2.001(4)Å; the equatorial Cu-S distance is 2.314(2)Å. (b) ORTEP view of **6**[11], whose equatorial Cu-N(1) and Cu-N(3) distances are 2.128(13) and 2.125(12)Å, respectively, and whose axial Cu-N(2) and Cu-N(4) distances are 2.009(13) and 2.000(11)Å, respectively. The equatorial Cu-S(1) distance is 2.328Å and the S(1)-S(2) distance of the persulfide is 2.017(7)Å.

In a second case, the well-known redox instability of the Cu(II)-cysteine system was suppressed by linking two cysteine units together.[12] The dimethyl ester of N,N'-ethylenebis(L-cysteine) is a potent linear tetradentate N_2S_2 ligand for Cu(II); it rapidly and quantitatively displaces macrocyclic tetramines from their Cu(II) complexes (which themselves have formation constants[13] exceeding 10^{28}). The resulting chiral cis-Cu(II)N_2S_2 complex **7** (Figure 3) may be intrinsically so stable that there is no driving force for further sterically-allowed ligation.

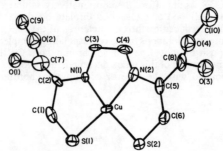

Figure 3. ORTEP view of **7**. Noteworthy structural parameters include the Cu-S(1) and Cu-S(2) distances [2.230(5) and 2.262(4)Å], the Cu-S(1)-C(1) and Cu-S(2)-C(6) angles [97.0(5) and 96.8(6)°], and the S(2)-Cu-S(1)/N(2)-Cu-N(1) dihedral angle (21.0°).[12]

The unusual stability of this cis $Cu(II)N_2S_2$ complex led us to study complexes of other linear tetradentate aminothiolate ligands whose preparation is outlined in Figure 4.

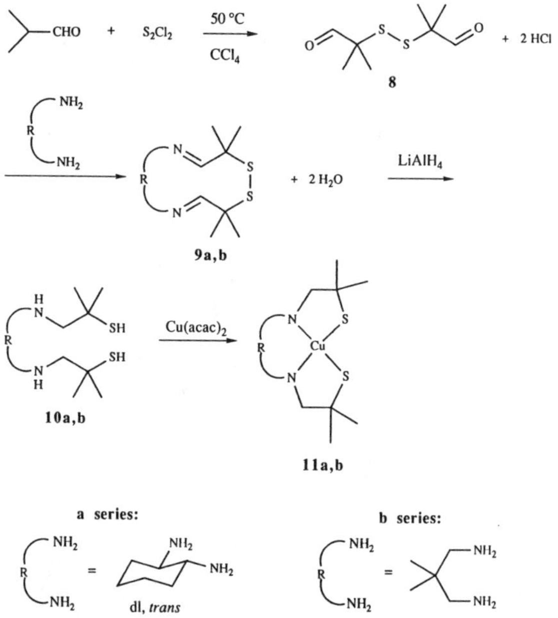

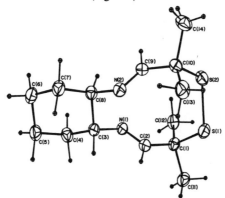

Figure 4. Elaboration of <u>dl</u>-trans-1,2-diaminocyclohexane and 2,2-dimethyl 1,3-diaminopropane to diaminodithiol ligands and their complexes.

The dialdehydedisulfide presursor **8** is easily obtainable in good yield by using a published procedure.[14] Condensation of **8** with trans-1,2-diaminocyclohexane afforded a crystalline macrocyclic Schiff's base **9a** (Figure 5).

Figure 5. ORTEP view of **9a**. Noteworthy structural parameters include the S(1)-S(2) distance (2.0229(6)Å), the C(9)-N(2) and C(2)-N(1) distances [1.254(2) and 1.254(2)Å], and the C(1)-S(1)-S(2)-C(10) torsion angle of 90.8(1)° which falls within the range reported (78.6-101°) for presumably unstrained acyclic disulfides.[15]

The free diaminodithiol ligand **10a** was obtained by reduction of **9a** with LiAlH$_4$, and was isolated as a pale yellow oil that crystallized slowly in the freezer. This ligand may be stored under Ar in a freezer, but is best used promptly. Ligand exchange of **10a** with Cu(acac)$_2$ conducted in acetonitrile results in high yields of the racemic cis-Cu(II)N$_2$S$_2$ complex **11a** that deposited as well-formed crystals from the hot (70-80°C) reaction mixture.[16] The cis-Cu(II)N$_2$S$_2$ unit present in this complex (see Figure 6) structurally is similar to that produced by the linked cysteine analogue **7** (Figure 3).

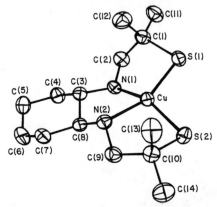

Figure 6. ORTEP view of **11a**. Noteworthy structural parameters include the Cu-S(1) and Cu-S(2) distances of 2.2317(7) and 2.2296(6)Å, respectively, the Cu-N(1) and Cu-N(2) distances of 2.040(2) and 2.047(2)Å, respectively, and the S(1)-Cu-S(2)/N(1)-Cu-N(2) dihedral angle of 32.77(6)Å.[16]

The pseudotetrahedral twist as measured by the N-Cu-N/S-Cu-S dihedral angle is larger for complex **11a** (32.8°) than for complex **7** (21.0°). Despite the presence of two aliphatic thiolate ligands, both Cu(II) chromophores exhibit nearly identical conventional axial epr spectra; measured and simulated spectra for glassed solutions of **11a** are presented in Figure 7.[17] .

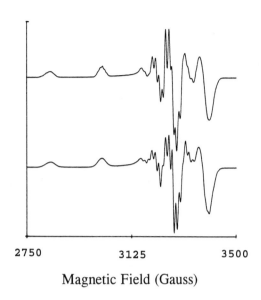

Figure 7. Measured (upper) and simulated (lower) epr spectra of **11a** in glassed methanol at 77°K.[17] Parameters used in the simulation are: $g_{//}$ =2.113, $g_{\perp}$ =2.018, $A_{//}$ (Cu)=176x10^{-4}cm^{-1}, $A_{\perp}$(Cu)=40.1x10^{-4}cm^{-1}, and $A_{//}$ (N)=7.0x10^{-4}cm^{-1}.

The half-occupied Cu(II) orbital for both complexes is $x^2 - y^2$, most likely oriented with its lobes along the Cu-ligand bonds. The observed Cu-S-C angles 94.30(8)° and 96.30(7)° for **11a** and 97.0(5)° and 96.8(6)° for **7** suggest that the HOMO and SHOMO of each sulfur are σ- and π-bonded respectively to the Cu(II) d-vacancy (Figure 8). The "third" lone pair of sulfur (often depicted as an sp^3-hybrid) actually is a stable core-like orbital that is strongly localized on sulfur and has neither the energy nor spatial extension required for prominent metal bonding interactions.[18] The electronic-structural implications of this situation are that a strong $\sigma(S) \rightarrow Cu$ CT band is flanked at lower energy by a weaker $\pi(S) \rightarrow Cu$ CT band associated with the poorer overlap and consequently less stabilization of the π-symmetry sulfur lone pair. The electronic spectra of the chiral cis-Cu(II)N$_2$S$_2$ chromophore **7** consist of a LF band at 18500 cm^{-1} accompanied by overlapping S$\rightarrow$Cu(II) CT bands at 24600(sh) and 29200 cm^{-1} and a N$\rightarrow$Cu(II) CT absorption at approximately 40000 cm^{-1}. Spectral deconvolution reveals that the higher energy S$\rightarrow$Cu(II) band is approximately 3.9 times as intense as the low energy shoulder, a result in harmony with the intensity/energy pattern noted above. Furthermore, the assignment is supported by the CD spectra exhibited by **7**. Both CT bands are well resolved, and the measured $\Delta\varepsilon$ values mean that the Kuhn factor ($\Delta\varepsilon/\varepsilon$) of the lower energy CT is approximately 2.6 times larger than that of the more intense S$\rightarrow$Cu(II) CT band. As indicated in Figure 8, $\sigma(S) \rightarrow Cu(II)$ CT involves a linear charge displacement that imparts electric dipole allowedness and thus dominance of conventional absorption spectra. On the other hand, $\pi(S) \rightarrow Cu(II)$ CT involves substantial rotary charge displacement and thus magnetic dipole allowedness.

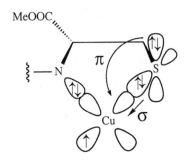

Figure 8. A view of the chromophore **7** showing the interactions of the σ- and π-symmetry S(thiolate) orbitals with the Cu(II) d-vacancy. Ligand to metal charge-transfer excitation originating from the σ-thiolate orbital involves a linear charge redistribution (electric dipole allowed), whereas that from the π-symmetry orbital involves a linear plus rotary charge-redistribution that confers prominence to the CD spectra and thus the Kuhn factor.

We note several features of the type II Cu(II)-aliphatic thiolate model complexes. First thiolate ligation does not result in the large covalency reported for the type I blue copper sites whose electronic ground states[18] exhibit considerable S(thiolate) character. Second, the presence of two aliphatic thiolate ligands does not transform the epr spectra of the cis-Cu(II)N$_2$S$_2$ chromophores **11a** and **7** into the type I spectra having unusually small $^{Cu}A_{//}$ values. Third, as the coordination number of these type II models shrinks from five (CuN$_4$S, **5**), to four (cis-CuN$_2$S$_2$, **7**, **11a**), the Cu-S bond distance shrinks from 2.314 to about 2.24Å (see Figures 2, 3, and 6). The effective coordination number for type I Cu(II) having a "trigonal N$_2$S plus one" ligand set recently has been estimated as approximately 3.3.[19] The reported Cu(II)-S(thiolate) distances for type I sites average to about 2.15Å, with an uncertainty of +/-0.2Å.[20] These bond distances are considered abnormally short relative to those reported for type II Cu(II)-thiolate models, and are said to represent unusually strong bonding. Hovever, when corrected for the usual decrease of metal-ligand bond length with diminishing coordination number, the reported distances of about 2.15Å seem normal. A similar trend is exhibited by Cu(I) thiolates: Cu(I)-S bond distances average about 2.27Å for trigonal complexes and shrink to about 2.16Å for linear complexes.[21] Tetrahedral Cu(I)-S complexes having various types of sulfur-donor ligands exhibit average Cu-S distances of about 2.31Å.[22] Finally, a refined interpretation of the type I S(thiolate)→Cu(II) CT spectra indicates they are unique in having an intensity pattern opposite from that shown by type II models. The type I sites show the strong CT band at low energy and the flanking band at higher energy. Combined single-crystal epr, single-crystal electronic spectral, and calculational studies have been used to substantiate a Cu(II)→S(thiolate) interaction that is summarized in Figure 9.[18]

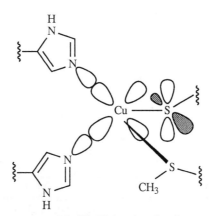

Figure 9. A representation of Cu(II)-thiolate bonding for type I blue copper sites where the major interaction is π-bonding between a S(3p) orbital and the Cu(II) $x^2 - y^2$ d-vacancy [18]; the major Cu-S(thiolate) bonding interaction for type II (normal copper) analogues is the σ-symmetry type shown in Figure 8.

The Cu(II) d-vacancy is positioned approximately in the plane of the trigonal N_2S donor set, and is rehybridized (reoriented) such that its dominant bonding interaction is strong π-overlap with a sulfur lone pair. Since the Cu(II)-S bond distance seems to be normal, the bonding view presented in Figure 9 becomes an even more attractive vehicle for understanding the unusual covalency and spectroscopic features of type I protein chromophores.

MACROCYCLIC BIS(DISULFIDE) TETRAMINE LIGANDS AND COMPLEXES

Oxidative coupling of the racemic cis-Cu(II)N_2S_2 complex **1a** (Figure 1) with iodine in acetonitrile affords the novel binuclear complex **2a** [M=Cu(II)], shown in Figure 10.[23]

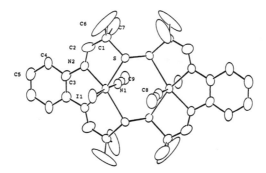

Figure 10. ORTEP view of the isomorphous complexes **2a** whose six-coordinate metal ion sites are occupied by two Ni(II) or by two Cu(II) ions; the two lattice I^- ions have been omitted for clarity. Preliminary bond distances at R_f=0.08 are comparable for both complexes. Selected values are: M-S (2.52Å), M-N(amine) (2.05Å), M-N(nitrile) (2.04Å), M-I (2.77Å).

Noteworthy structural features include the twenty-membered ring macrocyclic bis(disulfide) ligand, the unprecedented shortness (2.53Å) of the Cu(II)-S(disulfide) bonds, and the bridging roles of both disulfide groups. Further characterization of this complex by spectroscopic and magnetic measurements has proven to be difficult. The reaction solution deposited a mixture of well-formed shiny black crystals of which **2a** appears to be only a minor component. This became apparent after the Ni(II) analogue (**1a**, M=Ni(II)) was oxidatively coupled by the same procedure to afford in good yield the corresponding binuclear Ni(II) complex (**2a**, M=Ni(II)) as well-formed green crystals. Careful microscopic examination revealed that this emerald green product is obtained as a single crystalline phase, and a second complete crystallographic study established that these binuclear Ni(II) and Cu(II) complexes are intimately isostructural.[23] The single ORTEP picture in Figure 10 serves well to represent both structures; selected bond distances are listed in the caption and are rather similar for both complexes. The Ni(II) version of **2a** exhibited rich and intense x-ray powder diffraction spectra whereas those obtained for the isostructural Cu(II) analogue were considerably weaker. Apparently, the Cu(II) analogue represents only a minor component of the black crystalline reaction product. These complications may arise from the redox instability of Cu(II) iodides. To characterize the spectroscopic features of Cu(II)-disulfide chromophores, we attempted to prepare analogues of **2a** that are more stable and that can be obtained as single phases. Even if the Cu(II) version of **2a** could be obtained as a single phase, the strongly reducing nature of the iodide ligand coupled with its large spin-orbit coupling might obscure S(disulfide)→Cu(II) CT absorptions. While these absorptions could be identified in analogues of **2a** described below, we were unable to duplicate the bridging disulfide feature and quantitate its mediation of magnetic coupling between the Cu(II) ions. The magnetic properties of the Ni(II) version of **2a** are summarized in Figure 11.[23] The magnetic moment per Ni(II) steadily decreases from $2.86\mu_B$ at 290°K to $0.284\mu_B$ at 5 °K, and the susceptibility passes through a maximum at about 40°K. The data were fit to the model based upon the usual $H = -2JS_1 \cdot S_2$ Hamiltonian including terms for interdimer interactions and zero field splitting.[24] Considerable exploration of parameter space indicates that the best fits are obtained for g=2.06(2), J=-13.0(2)cm^{-1}, TIP=130-300x10^{-6} cgs per Ni(II), while allowing for the presence of 0.6 to 1.1% monomeric Ni(II) impurity. Owing to the strong correlation between the interdimer interaction (0.2 to -0.9cm^{-1}) and zero field splitting (4.4 to +7.3 cm^{-1}), only the indicated ranges rather than absolute values can be determined using this approach. The extent of magnetic coupling between the Ni(II) ions mediated by two bridging disulfide groups is substantial, and comparable in magnitude to that generally reported for two halide or hydroxo bridges.[25]

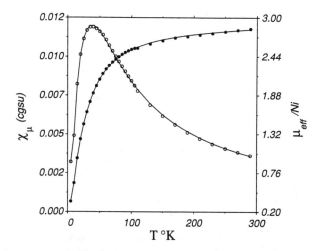

Figure 11. Temperature dependence of the magnetic susceptibility per mole of Ni(II) (χ, open circles) and magnetic moment (μ, solid circles). The circles are the experimental values, and the curves are calculated using the parameters and equations described in the text.

The free macrocyclic ligand **3a** may be readily displaced from the Ni(II) version of **2a** (see Figure 1) and obtained as colorless crystals in good yield (Figure 12).[23]

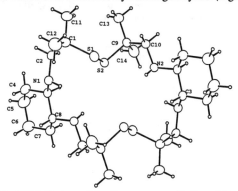

Figure 12. ORTEP view of the free macrocyclic ligand **3a**.

Reaction of **3a** with two equivalents of $Cu(ClO_4)_2 \cdot 6H_2O$ and four equivalents of KOCN led in good yield to the binuclear Cu(II) complex **4a** as blue crystals. Cyanate, a linear pseudohalide counterion, was chosen because the $OCN^- \rightarrow Cu(II)$ CT absorptions should be well removed toward higher energy and because the structure of the binuclear complex might be analogous to that of **2a**, with cyanate replacing the ligating and lattice iodides. However, the crystal structure of **4a** (Figure 13) revealed that the Cu(II) ions are pentacoordinate rather than hexacoordinate, that all four cyanates are ligating, and that the disulfide groups are ligating (Cu(II)-S(1), 2.505Å) but not bridging (Cu(II)-S(2),

43

3.253Å).[23] The electronic spectra of **4a** and the homologous complex **4b** are presented below.

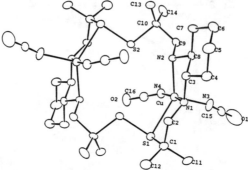

Figure 13. ORTEP view of **4a**. Noteworthy structural parameters include: Cu-S(1), 2.505(2)Å; Cu-S(2), 3.253Å; Cu-N(1), 2.032(5)Å; Cu-N(2), 2.118(5)Å; Cu-N(3), 2.064(7)Å; Cu-N(4), 1.928(6)Å.

To probe the effects of the ring size on the coordination structure and electronic spectra, the binuclear Cu(II) complex of the twenty two-membered macrocycle **3b** was synthesized and characterized. The Schiff's base **9b** (Figure 4) was reduced to the aminothiol **10b** which then was converted to the $Ni(II)N_2S_2$ complex **1b** (Figure 1). Oxidative coupling with I_2 afforded the macrocyclic bis(disulfide) tetramine complex **2b**, which was demetallated to the free ligand **3b**. Remetallation with two equivalents of $Cu(ClO_4)_2 \cdot 6H_2O$ followed by addition of four equivalents of KOCN yielded **4b** whose structure is presented in Figure 14. The hexacoordinate Cu(II) ions of this binuclear complex have highly distorted geometry. The Cu-S(1) and Cu-S(2) bonds (2.932 and 3.103Å, respectively) are substantially longer than the perhaps limiting distance of 2.505Å found for **4a**. These cis apical Cu-S bonds impart a cis rather than the common trans (tetragonal) distortion of the Cu(II) coordination geometry.

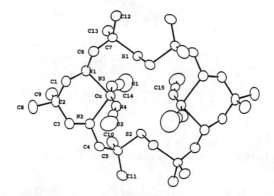

Figure 14. ORTEP view of **4b**. Noteworthy structural parameters for refinement to a R_f value of 0.04 are: Cu-S(1), 2.932Å; Cu-S(2), 3.103Å; Cu-N(1), 2.156Å; Cu-N(2), 2.163Å; Cu-N(3), 1.872Å; Cu-N(4), 1.862Å.

The solution electronic spectra of **4a** and **4b** in CH_2Cl_2 are presented in Figures 15 and 16, respectively. The similarities between the band positions in these solution spectra with those observed for mineral oil mull spectra (not shown) of **4a** and **4b** suggests that their coordination structures, as revealed by x-ray crystallographic studies, are little changed in CH_2Cl_2 solution. Frozen (80°K) solutions of **4a** and **4b** in CH_2Cl_2 exhibit broad featureless epr spectra (not shown) typical of binuclear Cu(II) complexes having intramolecular spin-spin interactions. A weak antiferromagnetic Cu(II)···Cu(II) coupling for **4a** was quantified by our analysis of the variable-temperature magnetic susceptibility exhibited by the polycrystalline complex. The gradual decrease of the magnetic moment per Cu(II) from $1.67\mu_B$ at 290°K to $1.60\mu_B$ at 5°K can be fit to the usual dimer model for which g=1.91, J=0.52cm^{-1}, TIP=44x10^{-6}cgsu, and the presence of 0.3% by weight paramagnetic (mononuclear) impurity.

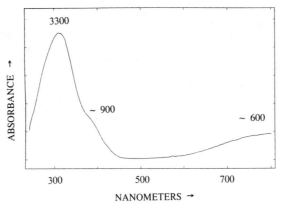

Figure 15. Electronic spectra of **4a** measured in CH_2Cl_2 at 298°K; λ (max) values are given in the text and ε's per mole of Cu(II) are shown in the Figure.

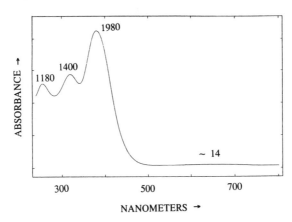

Figure 16. Electronic spectra of **4b** measured in CH_2Cl_2 at 298°K; λ (max) values are given in the text and ε's per mole of Cu(II) are shown in the Figure.

Owing to the low symmetry of the Cu(II) chromophores in **4a** and **4b**, a detailed interpretation of their electronic absorptions can not be made from the available spectroscopic data. Regarding the general features of Figure 15, the highest energy LF absorption corresponds to the broad band centered at approximately 800 nm. The only source of the near uv absorptions at 310 nm and approximately 380 nm is S(disulfide)→Cu(II) CT. Disulfide and cyanate are spectroscopically transparent in this spectral region, while amine → and cyanate → Cu(II) CT are well-removed towards higher energies. The intensity pattern of these near uv CT bands is that of a relatively intense σ-symmetry absorption flanked at lower energy by a weaker π-symmetry absorption. The electronic spectra of **4b** differ from those of **4a** in several interesting ways. First, the highest energy LF absorption of **4b** at approximately 670 nm (barely apparent on Figure 16) is weak (ε=14). Second, the pattern of the overlapping near uv absorptions is reversed, with the lowest energy absorption at 380 nm, assigned as above to S→Cu(II) CT, carrying more intensity than the higher energy CT absorption at 320 and 250 nm. A more detailed analysis of these CT absorptions will require a full description of the Cu(II) electronic ground states, spectral deconvolution of the uv region, and perhaps preparation of chiral macrocyclic analogues so CD measurements can be made. Disulfide→Cu(II) CT absorptions have not been reported for the four structurally characterized Cu(II)-disulfide complexes described in the literature[3-6]. An absorption at 330 nm exhibited by solutions of $CuCl_2$ and bis[2-(2-pyridylmethylamino)ethyl] disulfide has been assigned as S→Cu(II) CT.[7] As noted in the Introduction, this assignment is complicated by the possibility of Cl⁻→Cu(II) CT in the same spectral region.

ACKNOWLEDGEMENT

We thank the National Science Foundation for support of research (Grant CHE 8919120) and the National Institutes of Health for an instrumentation grant (Grant 1510 RRO 1486).

REFERENCES AND NOTES

1. J. M. Guss, E. A. Merritt, R. P. Phizackerley, B. Hedman, M. Murata, K. O. Hodgson, H. C. Freeman, *Science* **241**, 806 (1988).

2. H. Thomann, M. Bernardo, M. J. Baldwin, M. D. Lowery, E. I. Solomon, *J. Am. Chem. Soc.* **113**, 5911 (1991).

3. J. A. Thich, D. Mastropaolo, J. Potenza, H. J. Schugar, *J. Am. Chem. Soc.* **96**, 726 (1974).

4. K. Miyoshi, Y. Sugiura, K. Ishizu, Y. Iitaka, H. Nakamura, *J. Am. Chem. Soc.* **102**, 6130 (1980).

5. P. N.-H Dung, B. Viossat, A. Busnot, J. M. Gonzalez-Perez,. J. Niclos-Gutierrez, *Acta Crystallogr.* **C41**, 1739 (1985).

6. M. L. Brader, E. W. Ainscough, E. N. Baker, A. M. Brodie, D. A. Lewandoski, *J. Chem. Soc., Dalton Trans.*, 2089 (1990).

7. O. Yamauchi, H. Seki, T. Shoda, *Bull. Chem. Soc. Jpn.* **56**, 3258 (1983).

8. P. E. Riley, K. Seff, *Inorg. Chem.* **11**, 2993 (1972).

9. V. M. Miskowski, J. A. Thich, H. J. Schugar, *J. Am. Chem. Soc.* **98**, 8344 (1976).

10. E. John, P. K. Bharadwaj, J. A. Potenza, H. J. Schugar, *Inorg. Chem.* **25**, 3065 (1986).

11. E. John, P. K. Bharadwaj, K. Krogh-Jespersen, J. A. Potenza, H. J. Schugar, *J. Am. Chem. Soc.* **108**, 5015 (1986).

12. P. K. Bharadwaj, J. A. Potenza, H. J. Schugar, *J. Am. Chem. Soc.* **108**, 1351 (1986).

13. D. K. Cabbiness, D. W. Margerum, *J. Am. Chem. Soc.* **92**, 2151 (1970).

14. J. J. D'Amico, W. E. Dahl, *J. Org. Chem.* **40**, 1224 (1975).

15. P. K. Bharadwaj, J. A. Potenza, H. J. Schugar, *Acta Crystallogr.* **C44**, 763 (1988).

16. P. K. Bharadwaj, R. Fikar, M. N. Potenza, X. Zhang, J. A. Potenza, H. J. Schugar, unpublished results.

17. The epr simulation was performed in our laboratory by Dr. Mark Brader using the program QPOWA obtained from Prof. R. L. Belford, Illinois ESR Research Center, 506 S. Matthews St., Urbana, IL 61801.

18. E. I. Solomon, M. J. Baldwin, M. D. Lowery, *Chem. Revs.* **92**, 521 (1992).

19. M. D. Lowery, E. I. Solomon, *Inorg. Chim. Acta,* in press.

20. E. T. Adman, in *Advances in Protein Chemistry*, C. B. Anfinsen, J. T. Edsall, F. M. Richards, and D. S. Eisenberg, Eds., (Academic Press, New York, 1991), Vol. 42, pp. 145-197).

21. I. G. Dance, *Aust. J. Chem.*, **31**, 2195 (1978).

22. B. J. Hathaway, in *Comprehensive Coordination Chemistry*, G. Wilkinson, R. D. Gillard, and J. A. McCleverty, Eds., (Pergamon Press, New York, 1987), Vol. 5, p. 551.

23. S. Fox, J. A. Potenza, S. A. Knapp, H. J. Schugar, unpublished results.

24. Professor C. J. O'Connor, Department of Chemistry, University of New Orleans, New Orleans, LA 70148.

25. A. P. Ginsberg, R. L. Martin, R. W. Brookes, R. C. Sherwood, *Inorg. Chem.* **11**, 2884 (1972).

"Blue" Copper Proteins and Electron Transfer

INVESTIGATION OF TYPE 1 COPPER SITE GEOMETRY BY SPECTROSCOPY AND MOLECULAR REDESIGN

Joann Sanders-Loehr

Department of Chemical and Biological Sciences, Oregon Graduate Institute of Science & Technology, Beaverton, OR 97006-1999, USA

INTRODUCTION

Resonance Raman (RR) spectra of type 1 Cu proteins (cupredoxins) show a multiplicity of vibrational fundamentals between 250 and 500 cm^{-1} that is ascribed to kinematic coupling of the Cu-S(Cys) stretch with deformations of the Cys and His ligand side chains. A similar set of vibrational frequencies is observed for 11 different cupredoxins. These findings suggest that all cupredoxins have a highly conserved Cu(His)$_2$Cys geometry including (i) a trigonal planar array for the three Cu ligands and (ii) a *coplanar arrangement of the Cu-S-C$_\beta$-C$_\alpha$-N atoms* in the Cu-cysteinate moiety.

A number of cupredoxin-type proteins have a strong absorption band near 460 nm in addition to their characteristic absorption near 600 nm. Excitation within either absorption band yields a similar RR spectrum. Thus, both electronic transitions are likely to have (Cys)S → Cu(II) charge transfer character. Increased absorptivity at 460 nm in proteins such as pseudoazurin and nitrite reductase appears to be signaling *a more tetrahedral Cu site geometry*. It is accompanied by (i) decreased absorptivity at 600 nm, (ii) increased rhombicity in the EPR spectrum, and (iii) a strengthening of the bond between Cu and an axial ligand, thereby pulling the Cu away from the Cu(His)$_2$(Cys) trigonal plane.

Molecular redesign of the copper site in azurin has been accomplished by replacing the copper ligand, His 117, with glycine. Addition of monodentate exogenous ligands such as imidazole and chloride regenerates a cupredoxin with absorption, EPR, and RR properties characteristic of a *type 1 Cu site*. Addition of bidentate exogenous ligands such as histidine and histamine converts H117G to a non-blue copper protein with absorption and EPR properties characteristic of a *type 2 Cu site*. The intense RR modes previously at 360-430 cm^{-1} have shifted to a lower energy range of 260-350 cm^{-1}. These results suggest that the conversion to a type 2 site by the addition of a fourth strong ligand is accompanied by a lengthening of the Cu-S(Cys) bond.

From K.D. Karlin and Z. Tyeklár, Eds., *Bioinorganic Chemistry of Copper* (Chapman & Hall, New York, 1993).

CONSERVED STRUCTURAL FEATURES

In type 1 copper sites, the copper ion is coordinated to one cysteine and two histidine ligands in a trigonal planar array.[1] These sites also contain one or two weaker axial ligands (methionine S, peptide C=O) leading to an overall geometry that is tetrahedral or trigonal bipyramidal, respectively. Since the primary function of the type 1 copper site is to mediate long-range electron transfer, proteins containing these sites are categorized as cupredoxins.[2] All of the cupredoxins possess a β-sheet structure with the type 1 Cu being located at one end of the β-barrel adjacent to a hydrophobic patch. There are now a number of cases in which the electron-exchange partner is known to bind to the hydrophobic patch, and the intervening histidine ligand clearly facilitates electron transfer between the Cu and the surface of the protein (Figure 1). In addition, the side-chain of the cysteine ligand has an extended conformation[7] that leads to a second electron-transfer site (Figure 1). It is likely that the cysteine ligand plays a role in facilitating electron transfer between the Cu and this additional electron-transfer site.

A consequence of the presence of only three strong ligands in cupredoxins is a shortening of the Cu-S(Cys) bond to 2.13 Å ($\pm$ 0.05 Å).[7] This is accompanied by the appearance of an intense (Cys)S → Cu(II) CT band near 600 nm and a characteristic narrowing of the EPR hyperfine splitting in the $A_{||}$ region.[8] As will be shown below, several types of modifications of type 1 sites are possible. A strengthening of the bond to one of the axial ligands creates a more tetrahedral site with a more rhombic EPR spectrum, but with a small $A_{||}$ splitting still characteristic of type 1 Cu. Addition of a fourth strong ligand from results in conversion to a type 2 Cu site. This is most likely due to a lengthening of the Cu-S(Cys) bond and a tetragonal distortion of the four-ligand set. Each of these types of structural change is accompanied by a change in the (Cys)S → Cu(II) electronic transitions that can be verified by resonance Raman spectroscopy.

RESONANCE RAMAN SPECTRA OF CUPREDOXINS

Exposure of cupredoxins to laser excitation within their 600-nm absorption band leads to remarkably rich resonance Raman (RR) spectra. Whereas a single Cu-S stretching vibration would have been expected to be enhanced by the S → Cu CT transition, more than 10 vibrational fundamentals are observed between 200 and 500 nm (Figure 2). Although the spectral peaks vary considerably in intensity (due to variations in the extent of vibronic coupling), they occur at a surprisingly similar set of frequencies. Thus, azurin and amicyanin each have their four most intense peaks between 375 and 430 cm^{-1} and three of these peaks are within 2 cm^{-1} of one another (but with a marked difference in intensity for the 411 cm^{-1} peak). The constancy of vibrational frequencies implies that similar copper site structures are involved. However, the structural features responsible for the multiplicity of vibrational modes has been the subject of much conjecture.

Normal coordinate analysis (NCA) calculations originally showed that the multiplicity of RR modes in cupredoxin spectra could be explained by a kinematic coupling of the Cu-S stretch with other ligand vibrations.[11] Copper and deuterium isotope substitution experiments suggested cysteine-ligand deformations and Cu-imidazole deformations as likely candidates for this coupling.[12,13] For example, replacement of ^{63}Cu by ^{65}Cu in azurin causes at least six vibrational modes to shift to lower energy by 0.5 to 1.0 cm^{-1} (Figure 3A). Since a pure Cu-S stretch would be expected to show a shift of -3 cm^{-1} in a single mode, these multiple shifts are indicative of coupled modes. We have obtained definitive evidence for Cu-imidazole participation through the use

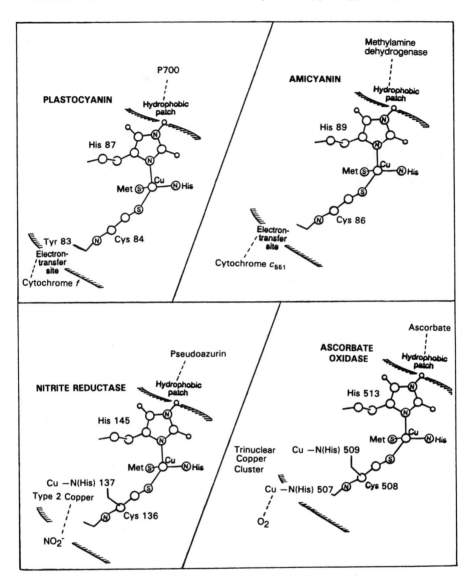

Figure 1. Common features of type 1 copper sites in cupredoxins (plastocyanin and amicyanin) and cupredoxin-like domains (nitrite reductase and ascorbate oxidase). Dotted lines indicate electron donor or acceptor that is believed to bind at the hydrophobic patch or the electron transfer site. Data from X-ray crystallographic and chemical studies for plastocyanin,[3] amicyanin,[4] nitrite reductase,[5] and ascorbate oxidase.[6]

Figure 2. Resonance Raman spectra of cupredoxins obtained with 647-nm excitation. Azurin from *Alcaligenes denitrificans*[9] was at 90 K. Amicyanin from *Paracoccus denitrificans*[10] was at 15 K.

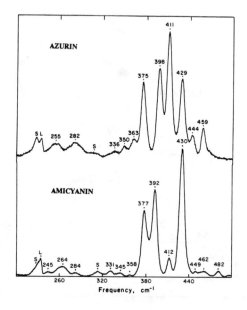

of an azurin mutant in which the histidine-117 ligand was replaced with glycine.[14,15] Addition of exogenous imidazole regenerates a type 1 Cu site whose RR spectrum is indistinguishable from that of the wild-type protein.[15] Substitution of ^{15}N-labeled imidazole results in shifts of -0.5 to -1.0 cm^{-1} in at least seven different vibrational modes (Figure 3B). However, production of a type 1 site in the H117G mutant by the addition of chloride instead imidazole still yields a RR spectrum similar to Figure 2 (upper).[15] This indicates that the RR spectrum of azurin is derived mainly from the cysteinate moiety.

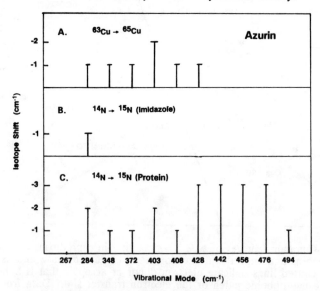

Figure 3. Shifts in RR frequencies upon isotopic substitution of *Pseudomonas aeruginosa* azurin. A. H117G azurin reconstituted with [63]Cu or [65]Cu and exogenous imidazole. B. H117G azurin reconstituted with normal Cu and [14]N- or ^{15}N-labeled exogenous imidazole. C. Wild-type azurin from cells grown in [14]N- or ^{15}N-labeled ammonium chloride.

Table I. Conserved Cysteine Dihedral Angles

Cupredoxin	$\tau(Cu\text{-}S\text{-}C_\beta\text{-}C_\alpha)$	$\tau(S\text{-}C_\beta\text{-}C_\alpha\text{-}N)$
plastocyanin	-168	169
pseudoazurin	-178	166
cucumber basic	-171	166
azurin	-169	173

A significant advance in our understanding of the RR spectra of cupredoxins came from the realization[7] that the cysteine ligand has a conserved conformation in all of the known cupredoxin structures. As can be seen in Table I, the dihedral angle between the $Cu\text{-}S\text{-}C_\beta$ and $S\text{-}C_\beta\text{-}C_\alpha$ planes is close to 180°; a similar $\approx$ 180° angle is observed for the $S\text{-}C_\beta\text{-}C_\alpha$ and $C_\beta\text{-}C_\alpha\text{-}N$ planes. Thus, a total of 5 atoms are coplanar (see Cu, S, C_β, C_α, and N in Figure 4).

Extensive studies on $Fe_2S_2(Cys)_4$ ferredoxins have indicated that cysteine ligand coplanarity greatly increases the tendency for kinematic coupling because deformations such as the S-C-C bend are now in line with the metal-sulfur stretch.[16] Evidence favoring the application of this argument to the cupredoxins has been obtained by growing bacteria on ^{15}N-ammonia. This results in complete labeling of all of the nitrogens in the protein (imidazoles as well as amides of the polypeptide backbone).[15] In this case, the RR spectra show much larger shifts of -2 to -3 cm^{-1} in at least 6 different modes (Figure 3C), which were generally less affected by either copper or imidazole isotope substitution. Such nitrogen-isotope dependence is best explained by ligand deformation modes that include the amide nitrogen of cysteine.

A diagrammatic representation of the RR spectra of 11 different cupredoxins is shown in Figure 5. All spectra exhibit 1-3 peaks between 240 and 280 cm^{-1} and another 5-9 peaks between 350 and 450 cm^{-1}. Although the relative peak intensities are even more variable than for the two proteins in Figure 2, the peaks still occur at a fairly constant set of frequencies. Since vibrational frequencies are a property of the electronic ground state,[18] this indicates that all cupredoxins have a common ground state structure. The conserved features needed to generate such a similar set of Raman modes are (1) a short Cu-S(Cys) bond of 2.15 Å, (2) a trigonal planar orientation of the (His)$_2$Cys ligand set, and (3) a coplanar cysteine side chain. The fact that the coplanar cysteine conformation appears to be so highly conserved suggests that this conformation may provide a favored pathway for electron transfer, as indicated in Figure 1.

Figure 4. Coplanarity of cysteine ligand in cupredoxins. The planes containing Cu-S-C_β, S-C_β-C_α, and C_β-C_α-N are all congruent with one another.

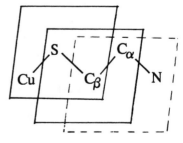

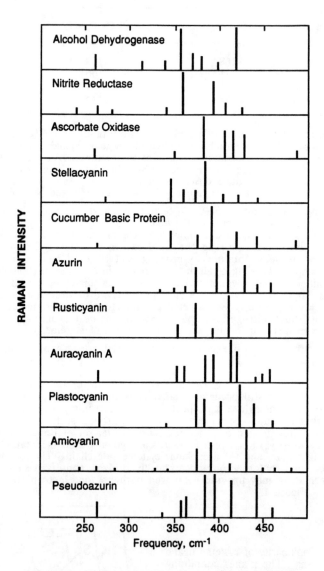

Figure 5. Diagrammatic representation of the resonance Raman spectra of cupredoxins. Bar graph denotes frequencies of vibrational fundamentals and relative peak intensities. Cu-substituted alcohol dehydrogenase[17] is a synthetic type 1 site with a $(Cys)_2His$ ligand set, including a short Cu-S bond and a coplanar cysteine moiety. Figure reprinted from reference 7.

MULTIPLE CYS → COPPER CHARGE TRANSFER BANDS

The presence of a strong absorption band near 600 nm (ϵ > 2000 M^{-1} cm^{-1}) has been the hallmark of a cupredoxin, leading to its description as a blue copper protein. However, this spectral feature is generally accompanied by a second feature near 460 nm that in certain cases is even more intense than the 600-nm absorption. The electronic spectrum of pseudoazurin, for example, has a prominent band at 450 nm in addition to its major band at 595 nm (Figure 6B). Based on our Raman experiments,[19] both of these electronic transitions have considerable (Cys)S → Cu CT character.

As can be seen in Figure 6A, excitation within either the 450- or the 595-nm absorption band of pseudoazurin produces essentially the same RR spectrum, with multiple vibrational modes characteristic of a cupredoxin. The constancy of vibrational frequencies implies that both of the electronic transitions arise from the copper-cysteinate moiety. The small variability in RR intensities (e.g., peak at 363 cm^{-1}) indicates that a somewhat different electronic transition is responsible for the 450-nm absorption. Nevertheless, the similarity of the RR spectrum strongly suggests that it is also a (Cys)S → Cu(II) CT band. The similar origin of the two electronic transitions is more dramatically demonstrated by RR excitation profiles (Figure 6B). These were obtained by quantitating the intensity of the 363- and 445-cm^{-1} modes of pseudoazurin (relative to the 230-cm^{-1} mode of ice as an internal standard). Both of these vibrational modes (as well as all of the other peaks in the RR spectrum) are clearly in resonance with the 450- and 595-nm absorption bands.

Nitrite reductase provides an interesting variation. The X-ray structure[5] shows a standard type 1 copper site (Figure 1), and the RR spectrum (Figure 7A) is characteristic of a cupredoxin. However, the electronic spectrum of the type 1 site in nitrite reductase is unusual: the 460 nm absorption is actually more intense than the 585 nm absorption (Figure 7B). In keeping with this difference in absorptivity, the RR excitation profiles exhibit greater enhance-

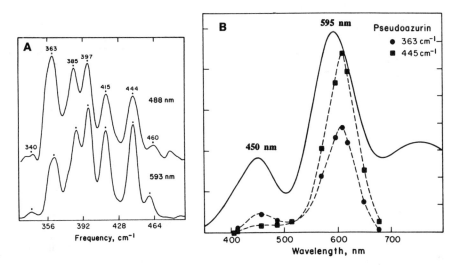

Figure 6. A. RR spectra of *Alcaligenes faecalis* pseudoazurin obtained at two excitation wavelengths: 488 and 593 nm. B. Absorption spectrum (−) and RR excitation profiles (---) for pseudoazurin vibrational modes at 363 and 445 cm^{-1}.

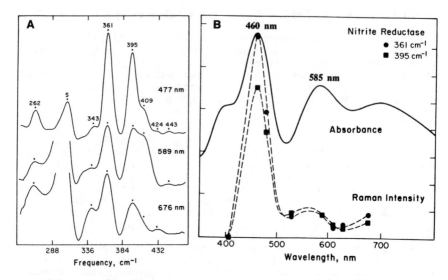

Figure 7. A. RR spectra of *Achromobacter cycloclastes* nitrite reductase obtained at three excitation wavelengths: 477, 589, and 676 nm. S denotes peak from frozen solvent. B. Absorption spectrum (—) and RR excitation profiles (---) for nitrite reductase vibrational modes at 361 and 395 cm⁻¹.

ment within the 460-nm absorption band (Figure 7B). The fact that the same vibrational modes are observed for excitation at 477, 589, or 676 nm shows that all three electronic transitions are due to a single Cu-cysteinate moiety and that all three have some (Cys)S → Cu(II) CT character.

A reexamination of the RR excitation profile for azurin from *P. aeruginosa* also reveals an additional (Cys)S → Cu CT band associated with its very weak absorption at 460 nm.[19] Similar findings apply to the H80C mutant of yeast superoxide dismutase in which a His ligand in the zinc site has been converted to Cys and the resulting Cu-substituted product has the properties of a type 1 Cu site.[20] The RR spectrum of the mutant exhibits multiple peaks between 300 and 450 nm, indicative of a short Cu-S(Cys) bond and a coplanar copper-cysteinate conformation.[19] Like nitrite reductase, the H80C mutant is green in color due to more intense absorption at 460 than 595 nm. The RR excitation profiles again track both absorption bands showing that both electronic transitions are primarily (Cys)S → Cu CT in character.

Based on other types of spectroscopic analyses (including the absorption of polarized light by oriented single-crystals), Solomon and coworkers have developed a set of assignments for plastocyanin[8] as a representative cupredoxin. They have proposed that the 600-nm band is due to charge transfer from a sulfur p orbital with π overlap to a copper $d_{x^2-y^2}$ orbital, but that the 460-nm band is predominantly His → Cu CT in character. Our Raman experiments are certainly in agreement with the assignment for the 600-nm band, but are more difficult to reconcile with a His → Cu assignment at 460 nm. Although the terminal copper orbital appears to have significant sulfur character based on Xα calculations,[8] the accompanying change in histidine electron density should have resulted in increased RR enhancement of histidine ring modes with 460-nm excitation. This has not been observed. Furthermore, there appears to be an inverse relationship between the absorption intensities at 460 and 600 nm.[19] Cupredoxins such as azurin and plastocyanin have a large absorption at 600 nm ($\epsilon \simeq$ 5,000 M^{-1}cm^{-1}) and a weak absorption at 460 nm ($\epsilon \simeq$ 500 M^{-1}cm^{-1}). Cupredoxins such as pseudoazurin and stellacyanin have weaker absorption at 600 nm ($\epsilon \simeq$ 3,000 to 4,000 M^{-1}cm^{-1}) and stronger absorption at 460 nm ($\epsilon \simeq$ 1,000 M^{-1}cm^{-1}). This inverse relationship implies that both electronic transitions involve S $\rightarrow$ Cu CT and that when the orbital overlap is optimal for one transition, it is less than optimal for the other transition.

It has long been known that type 1 Cu sites can be subdivided into two categories on the basis of the EPR spectrum: either predominantly axial or predominantly rhombic in character. It now appears that there is a correlation between the intensity of the absorption at 460 nm and EPR rhombicity. As can be seen in Table II, cupredoxins with axial EPR signals tend to have a low ratio ≤ 0.1 for the ϵ values at 460 and 600 nm, whereas cupredoxins with rhombic EPR signals tend to have a higher ratio ≥ 0.2 for the ϵ values at 460 and 600 nm. Previously it was suggested that rhombic EPR character was indicative of stronger binding of the Cu(His)$_2$(Cys) moiety to an axial ligand.[21] This type of structural difference could also explain the change in orbital overlap that gives rise to a more intense absorption at 460 nm. There is evidence (Table II) that sites with axial EPR character tend to have a longer Cu-S(Met) distance of ≥ 2.8 Å or their Cu within 0.2 Å of the NNS plane of the (His)$_2$(Cys) ligand set. In contrast, sites with rhombic EPR character tend to have a shorter Cu-S(Met) distance of ~2.6 Å or a greater displacement of the Cu by 0.3 to 0.4 Å from the NNS plane. Thus, the geometry of the Cu site varies from predominantly trigonal planar for the axial-EPR cupredoxins to more tetrahedral for the rhombic-EPR cupredoxins.

Support for this hypothesis comes from the M121Q mutant of *P. aeruginosa* azurin in which the axial methionine has been replaced by glutamine.[22] The X-ray crystal structure of this mutant shows a fairly strong bond to the carbonyl oxygen of the glutamine ligand and a movement of Cu away from the NNS plane (Table II), in agreement with a the more rhombic EPR and increased ϵ_{460} relative to native azurin. Recently, a model complex has been synthesized that for the first time effectively models the EPR properties of type 1 Cu sites.[23] In this complex the Cu(II) is coordinated to a tris-pyrazolylborate ligand and a

Table II. EPR, Electronic Spectra and Copper Site Geometry[19]

Cupredoxin	$\dfrac{\epsilon_{460}}{\epsilon_{600}}$	Cu-X (axial) in Å	Cu$\cdots$(NNS) in Å
Axial EPR			
azurin	0.11	3.13 (S)	0.12
amicyanin	0.11		
plastocyanin	0.06	2.82 (S)	0.36
Cu(Pz$_3$)(SC$_6$H$_5$)	0.11	2.12 (N)	0.34
Rhombic EPR			
azurin-(M121Q)	0.20	2.27 (O)	0.26
stellacyanin	0.27		
pseudoazurin	0.41	2.76 (S)	0.43
cucumber basic	0.43	2.62 (S)	0.39
SOD-Cu$_2$-(H80C)	1.03		
nitrite reductase	1.34	~2.6 (S)	~0.5

benzenethiolate yielding Cu-N distances of 1.93 and 2.04 Å and a Cu-S distance of 2.18 Å in the NNS ligand plane. Although the axial nitrogen distance of 2.12 Å (Table II) is fairly short, the EPR and ϵ_{460} properties would have suggested a trigonal planar geometry. Clearly more structural data is required for an adequate description of type 1 copper sites.

INTERCONVERSION OF TYPE 1 AND TYPE 2 COPPER

The H117G mutant of *P. aeruginosa* azurin[14] has provided an opportunity to explore the potential flexibility of type 1 Cu sites. The crystal structures of several apo-cupredoxins, including apo-azurin, have revealed that there is surprisingly little change in the disposition of the ligating amino acids upon removal of copper.[1] This led to the idea that the protein imposes a fairly rigid geometry on the copper site. It was of some surprise to discover that this favored trigonal planar geometry is maintained even after removal of one of the histidine ligands, as long as an appropriate exogenous ligand is provided to complete the trigonal ligand set.[24] The H117 ligand in azurin is in the same location as H89 in amicyanin, connecting the Cu to the hyrodrophobic patch on the surface of the protein (Figure 1). After converting residue 117 to glycine and adding exogenous imidazole, the resulting cupredoxin is essentially indistinguishable from wild-type. This applies to optical and EPR properties (Table III), as well as RR spectra (Figure 8A,B). Thus, the copper site geometry has been conserved including not only the short Cu-S(Cys) bond and coplanar cysteinate ligand, but also the orientation of the imidazole ring at the H117 site and the position of the Cu with respect to axial ligands and the NNS plane.

Addition of an anionic ligand such as chloride (instead of imidazole) to the H117G mutant still generates a type 1 Cu site (Table III). The RR spectrum of the chloro adduct is close to that of the aqua adduct shown in Figure 8C. The only significant changes are increased intensity at 301 and 397 cm⁻¹, with essentially no changes in vibrational frequency. Thus, the ground state structure of the Cu-cysteinate chromophore has not been perturbed. The most likely explanation is that these mutant sites still have three strong ligands in a trigonal planar array with N,S,Cl for the chloro adduct and N,S,O for the aquo (water-1) adduct. In contrast, addition of a weaker ligand such as bromide

Table III. Properties of Mutant Azurin (H117G) with Exogenous Ligands[24]

Species	Absorption Maxima (nm)		EPR Spectra	
	λ_1	λ_2	$A_{||}(10^{-4}cm^{-1})$	Type
Type 1 Cu				
His 117(wt)	465	628	58	axial
imidazole	460	625	85	axial
chloride	470	648	17	axial
bromide	485	683	52	rhombic
water-1		628		
histidine-1		634		
Type 2 Cu				
water-2	420		139	
histidine-2	400		156	

60

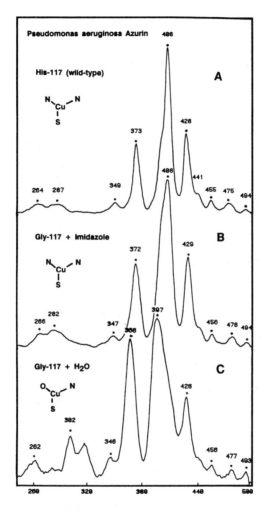

Figure 8. RR spectra of wild-type and mutant azurins obtained with 647-nm excitation. A. Wild-type azurin. B. H117G mutant plus exogenous imidazole. C. H117G mutant at pH 6.0 with water as the exogenous ligand.

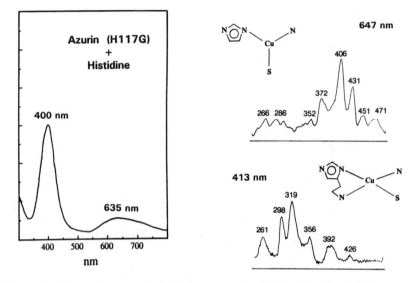

Figure 9. Spectral properties of mutant azurin (H117G) with histidine as an exogenous ligand. Left: Electronic spectrum. Right: RR spectra of histidine-1 form (647-nm excitation) and histidine-2 form (413-nm excitation).

leads to a more rhombic EPR spectrum (Table III). This suggests a more tetrahedral geometry with stronger binding to one of the axial ligands in addition to the N,S,Br ligand set.

The H117G azurin mutant is also capable of binding larger exogenous ligands such as the amino acid histidine. The resultant product is actually a mixture of species, histidine-1 and histidine-2, with absorption maxima at 635 and 400 nm, respectively (Figure 9, left). Excitation within the 635-nm absorption band leads to a RR spectrum (Figure 9, upper) closely resembling that of the wild-type protein (Figure 8A). This suggests that for the histidine-1 species only the imidazole portion is coordinated, yielding a trigonal planar N,N,S ligand set. Excitation within the 400-nm absorption band produces a *completely different RR spectrum* (Figure 9, lower) with the most intense vibrational modes appearing between 260 and 320 cm^{-1} instead of the 350- to 450-cm^{-1} range characteristic of cupredoxins. The EPR spectrum of the histidine-2 species, which is the major component in the sample, is indicative of a normal, type 2 Cu site (Table III). This species is most likely generated by the bidentate coordination of histidine, involving the N's of both the imidazole and the amino group. The resultant N,N,N,S ligand set presumably forms a more *tetragonal array* with a concomitant lengthening of the Cu-S(Cys) bond. This would explain the observed shift of the S → Cu(II) CT band to higher energy (400 nm), as well as the shift of the RR modes of the Cu-cysteinate chromophore to lower energy.

Similar mixtures of type 1 and type 2 Cu sites can be generated in the H117G azurin mutant by the addition of histamine or water as exogenous ligands.[15] In both cases, the type 2 species dominates the optical and EPR spectrum (Table III). These studies show that although type 1 Cu is favored in the presence of a monodentate ligand, type 2 Cu can readily form in the presence of a bidentate ligand. Thus, there appears to be more flexibility in the coordination potential of type 1 sites than had been previously suspected.

REFERENCES

1. E. T. Adman, *Adv. Protein Chem.* **45**, 145 (1991).
2. E. T. Adman, in *Metalloproteins*, P. M. Harrison, Ed., (Verlag Chemie, Weinheim, 1985), Part I, pp. 1-42.
3. V. A. Roberts, H. C. Freeman, A. J. Olsen, J. A. Tainer, and E. D. Getzoff, *J. Biol. Chem.* **266**, 13431 (1991).
4. L. Chen, R. Durley, B. J. Poliks, K. Hamada, Z. Chen, F. S. Mathews, V. L. Davidson, Y. Satow, E. Huizinga, F. M. D. Vellieux, and W. G. J. Hol, *Biochemistry* **31**, 4959 (1992).
5. J. W. Godden, S. Turley, D. C. Teller, E. T. Adman, M. Y. Liu, W. J. Payne, and J. LeGall, *Science* **253**, 438 (1991).
6. A. Messerschmidt, R. Ladenstein, R. Huber, M. Bolognesi, L. Avigliano, R. Petruzelli, A. Rossi, and A. Finazzi-Agro, *J. Mol. Biol.* **224**, 179 (1992).
7. J. Han, E. T. Adman, T. Beppu, R. Codd, H. C. Freeman, L. Huq, T. M. Loehr, and J. Sanders-Loehr, *Biochemistry* **30**, 10904 (1991).
8. E. I. Solomon, M. J. Baldwin, and M. D. Lowery, *Chem. Rev.* **92**, 521 (1992).
9. E. W. Ainscough, A. G. Bingham, A. M. Brodie, W. R. Ellis, H. B. Gray, T. M. Loehr, J. E. Plowman, G. E. Norris, and E. N. Baker, *Biochemistry* **26**, 71 (1987).
10. K. D. Sharma, T. M. Loehr, J. Sanders-Loehr, M. Husain, and V. L. Davidson, *J. Biol. Chem.* **263**, 3303 (1988).
11. T. J. Thamann, P. Frank, L. J. Willis, and T. M. Loehr, *Proc. Natl. Acad. Sci. U.S.A.* **79**, 6396 (1982).
12. L. Nestor, J. A. Larrabee, G. Woolery, B. Reinhammar, and T. G. Spiro, *Biochemistry* **23**, 1084 (1984).
13. D. F. Blair, G. W. Campbell., J. R. Schoonover, S. I. Chan, H. B. Gray, B. G. Malmström, I. Pecht, B. I. Swanson, W. H. Woodruff, W. K. Cho, A. M. English, H. A. Fry, V. Lum, and K. A. Norton, *J. Am. Chem. Soc.* **107**, 5755 (1985).
14. T. den Blaauwen, M. van de Kamp, and G. W. Canters, *J. Am. Chem. Soc.* **113**, 5050 (1991).
15. T. den Blaauwen, G. W. Canters, J. Han, T. M. Loehr, and J. Sanders-Loehr, manuscript in preparation.
16. S. Han, R. S. Czernuszewicz, and T. G. Spiro, *J. Am. Chem. Soc.* **111**, 3496 (1989).
17. W. Maret, A. K. Shiemke, W. D. Wheeler, T. M. Loehr, and J. Sanders-Loehr, *J. Am. Chem. Soc.* **108**, 6351 (1986).
18. P. R. Carey in *Biochemical Applications of Raman and Resonance Raman Spectroscopies* (Academic Press, New York, 1982) p. 47.
19. J. Han, T. M. Loehr, Y. Lu, J. S. Valentine, B. A. Averill, and J. Sanders-Loehr, manuscript submitted.
20. Y. Lu, E. B. Gralla, J. A. Roe, and J. S. Valentine, *J. Am. Chem. Soc.* **114**, 3560 (1992).
21. A. A. Gewirth, S. L. Cohen, H. J. Schugar, and E. I. Solomon, *Inorg. Chem.* **26**, 1133 (1987).
22. A. Romero, C. W. G. Hoitink, H. Nar, R. Huber, A. Messerschmidt, and G. W. Canters, manuscript submitted.
23. N. Kitajima, K. Fujisawa, M. Tanaka, and Y. Moro-oka, manuscript submitted.
24. T. den Blaauwen and G. W. Canters, manuscript submitted.

METALLOPROTEIN LIGAND REDESIGN: CHARACTERIZATION OF COPPER-CYSTEINATE PROTEINS DERIVED FROM YEAST COPPER-ZINC SUPEROXIDE DISMUTASE

Yi Lu,[a] James A. Roe,[b] Edith Butler Gralla,[a] Joan Selverstone Valentine[a]

[a]Department of Chemistry and Biochemistry, University of California, Los Angeles, Los Angeles, California 90024, USA,
[b]Department of Chemistry and Biochemistry, Loyola Marymount University, Los Angeles, California 90045, USA.

Introduction

Design vs. Redesign of a Protein

Designing proteins to have specific structures and functions is a particularly challenging goal because of the complexity of protein structures. Recently, several studies have been carried out with this goal, including some *de novo* designs of a few common protein motifs such as four α helix bundle[1,2], β sheet[3] and α/β barrel.[4] These studies have showed that a peptide chain, designed with minimal complexity based on the current knowledge of the relationship of primary, secondary and tertiary structures, and synthesized either *in vitro* or *in vivo*, can fold itself into the predicted three dimensional structure under certain conditions. A zinc-binding site has also been introduced into a designed protein.[5] These studies have demonstrated that protein design is indeed achievable.

Successful design of proteins with specific, predicted structures and activities requires an understanding of the factors which govern protein folding, which, unfortunately, are not yet well understood.[6] Protein *re*design, on the other hand, starts with a known and stable protein structure and changes one or several amino acid residues at a time to make a new protein. Using this approach, one can learn the specific structural and property changes associated with changes in the amino acid sequence.

Redesign of a Metallo- vs. a Nonmetallo-Protein

The number of reported successful protein redesigns has been steadily increasing, especially after site-directed mutagenesis became a routine technique in the early 1980's.[7] Several reports of redesign dealt with non-metalloproteins. For example, replacing residues around the substrate binding site and two of the surface loops of trypsin with the analogous residues from chymotrypsin created a protein with chymotrypsin-like activity.[8] Lactate dehydrogenase was also redesigned into a malate dehydrogenase, which is even more effective than the malate dehydrogenase isolated from the host organism.[9] After redesign of glutathione reductase, its coenzyme specificity was changed from NADP+ to NAD+.[10] However, redesigning a *metallo*protein has certain advantages over redesigning a *non-metallo*protein for two reasons: (1) Metal centers in metalloproteins often have rich spectroscopies, which can be utilized to detect subtle changes in the proteins that can otherwise be unnoticed if non-metals are involved; (2) Functionally, metals in metalloproteins play key roles, serving either catalytic or structural functions. The effect of any change on functions of metalloproteins due to changes in residues around the metal is more

From K.D. Karlin and Z. Tyeklár, Eds., *Bioinorganic Chemistry of Copper*
(Chapman & Hall, New York, 1993).

dramatic than that of other non-metal parts. This fact is demonstrated by a study by Sligar and Egeberg[11] in which one of the two axial histidines of cytochrome b5 (an electron transfer protein) was changed into a methionine, resulting in a protein with increased peroxidase and catalytic demethylase activity. The resulting changes were monitored by spectroscopic methods such as electronic absorption and resonance Raman, which are very sensitive to the nature of the ligand environments around the heme iron.

Background on Copper Proteins

Generally copper proteins are classified into three types according to their spectroscopic properties. Type 1 copper proteins, commonly called "blue copper" proteins, usually have an intense blue color due to a strong absorption around 600 nm in their electronic absorption spectra. They also have a very small $A_{//}$, which is the hyperfine splitting with the molecular z axis (d_{z^2} in the case of Cu(II) proteins) oriented parallel to the external magnetic field, in electronic spin resonance spectroscopy (ESR). Type 2 copper proteins have spectroscopic properties similar to those of simple copper complexes. Type 3 copper proteins are binuclear copper proteins that have no ESR signal in the oxidized state due to the antiferromagnetic coupling of the two neighboring Cu(II)'s.

One of the most important questions in the field of copper protein chemistry is centered on the unusual structural and spectroscopic properties of type 1 copper proteins, as compared with small copper complexes. Most copper complexes display a pale blue color due to Cu(II) d-d transitions around 600-800 nm in electronic absorption spectra (extinction coefficient $\varepsilon \sim 50$ M^{-1} cm^{-1}). They also have an ESR hyperfine coupling constant of 0.013-0.020 cm^{-1} due to the interaction of a Cu(II) unpaired electron with the Cu(II) nucleus itself (I=2/3). Type 2 copper proteins, which may have a distorted geometry around copper due to the rigid protein network, display a moderate increase in absorption around the same region with an extinction coefficient (ε) up to 300 M^{-1} cm^{-1}. By contrast, however, type 1 copper proteins have a deep blue color with an ε almost 100 times higher than that of low molecular-weight copper complexes (2000-5000 M^{-1} cm^{-1}) and an $A_{//}$ that is approximately half that of low molecular-weight copper complexes (0.003-0.006 cm^{-1}). These unusual properties are believed to be related to the function of type 1 copper proteins, most of which are involved in biological electron transfer.

Understanding the relationship of the unusual structures, spectroscopic properties and electron transfer reactivities of type 1 copper proteins is one of the primary goals for much of the research in this area.[12] The essential features of type 1 copper proteins appear to be two histidine imidazoles and a cysteine thiolate coordinated to Cu(II) in a trigonal planar (or distorted tetrahedral) geometry with one or two additional weak axial ligands such as methionine.[13] Most of the classic type 1 copper proteins are "blue copper" proteins due to the strong electronic absorption around 600 nm, which was assigned to a cysteine S p$\pi \rightarrow$Cu(II) charge transfer.[14] However some type 1 copper proteins such as stellacyanin,[15] pseudoazurin,[16] and cucumber basic blue protein[17] also have a moderately strong absorption around 460 nm in addition to the band at 600 nm. The most extreme example of increased 460 nm band intensity is the recently discovered nitrite reductase,[18] whose X-ray structure is now available.[19] Its 460 nm and 600 nm bands are of similar intensity (ε_{458nm}= 2200 $M^{-1}cm^{-1}$, ε_{585nm}= 1800 $M^{-1}cm^{-1}$), resulting in a green rather than a blue protein. Nonetheless, the type 1 Cu(II) in this protein has other characteristics of type 1 copper proteins. The nitrite reductase also has a type 2 Cu(II) that is 12.5 Å away from the type 1 Cu(II). The understanding of these "green copper" proteins and their relationship with "blue copper" proteins will provide further insight into the spectroscopic and functional properties of type 1 copper proteins. In terms of the function of type 1 copper proteins, the distorted tetrahedral geometry of type 1 copper proteins, which is a compromise between the square planar geometry preferred by Cu(II) with strong field ligands[20,21] and the tetrahedral geometry preferred by Cu(I),[20] was postulated to be responsible for the fast electron transfer rate because it takes the least amount of reorganization energy to access both oxidation states.[22] This postulation remains to be tested.

Model Studies of Type 1 Copper Proteins

To design a type 1 copper protein, one needs to know the minimal structural requirements for a copper ion to exhibit such unusual spectroscopic properties and such fast electron transfer rates. These requirements have yet to be determined. In the past, scientists have relied on model-making to mimic the spectral and redox characteristics of proteins. The primary obstacle to modeling type 1 copper proteins with low molecular weight complexes is that the thiolate anion is rapidly oxidized to disulfide in the presence of Cu^{2+}:[23]

$$2 Cu^{2+} + 2 RS^- \rightarrow 2 Cu^+ + RSSR$$

Another obstacle is the preference of Cu(II) to maintain a square planar geometry.[21] If the ligands are not held in a rigid arrangement, Cu(II) will dictate the geometry surrounding it. In 1979, Schugar et al.[24] succeeded in preparing the first stable CuN_4S(mercaptide) complex, $[Cu(tet\ b)(o\text{-}SC_6H_4CO_2)]\cdot H_2O$. However, inclusion of the thiolate sulfur was not sufficient to produce the blue copper properties. The copper tet b complex was spectroscopically a "normal copper", with no intense 600 nm absorption band and a normal ESR spectrum, and determination of its structure revealed a long 2.36 Å Cu – S bond. Recently, Kitajima et al.,[25] succeeded in preparing a series of tris(pyrazolylborate) copper(II) thiolate complexes which successfully mimic the intense blue color and decreased copper $A_{||}$ values of the blue copper site. Moreover, the Cu–S bond length is 2.13 Å, the same value that is observed in the blue copper protein x-ray structures. However, these complexes are very unstable and can be studied only at low temperature (-20 °C) and in non-aqueous solvents, which makes characterization and functional studies such as electron transfer extremely difficult or impossible.

Redesign of Copper Proteins

The use of an unrelated protein as the initial scaffolding for models of type 1 copper proteins can provide thiolate ligands with protection from oxidation and give the copper site a rigid structure. Most metalloproteins are known to form their final three dimensional structures even before the metals are in place, and the metal binding site undergoes little structural change due to metal incorporation.[26] Thus the rigidity of the metal binding site imposed by the protein usually outweighs the geometric preferences of the metal ion to a large extent, and the protein does not rearrange to suit the metal. Metals are usually buried relatively deep within a metalloprotein. Although there may be a solvent accessible channel in order to allow substrates to approach the metal, the channel has usually evolved to accept only certain substrates. Therefore the thiolate ligand can be protected. This principle was utilized in a study in which a thiolate ligand, pentafluorothiophenol, was incorporated into Cu(II)-substituted human insulin and a type 1 copper mimic was obtained.[27] Type 1 copper-like spectra were also observed when Cu(II) replaced Zn(II) in horse liver alcohol dehydrogenase.[28]

A number of fascinating studies have appeared describing the redesign of the type 1 protein azurin[29]. This protein contains two histidines and one cysteine in a distorted trigonal planar geometry, with distant axial interactions to a methionine and a backbone carbonyl oxygen. Karlsson et al.[29a] and Chang et al.[29b] have reported that replacing the methionine (Met121) with other non-sulfur amino acids had little effect on the spectral properties of the azurin, indicating methionine was not needed to create a type 1 copper site. One of the histidines (His 117) was also replaced with a glycine by Blaauwen et al.,[29c] resulting in a type 2 copper spectrum. The type 1 copper spectrum could be restored by addition of N-methylimidazole.

Redesign of CuZnSOD from *Saccharomyces cerevisiae*

We approached the redesign of copper proteins from a different direction, i.e. our goal was to redesign the metal binding site of a non-type 1 copper protein in order to convert it to a type 1 center. We chose as our starting point the type 2 or "normal" copper- and zinc-containing protein copper-zinc superoxide dismutase (CuZnSOD) from the simple yeast *Saccharomyces cerevisiae*. We considered this protein an excellent candidate because (1) CuZnSOD is one of the best characterized type 2 copper proteins, (2) we had already cloned its gene,[30] and (3) we can carry out *in vivo* experiments in yeast that are difficult in higher organisms.

CuZnSOD is a dimeric enzyme (MW 31,900 for yeast CuZnSOD) with two identical subunits, each of which contains one copper ion and one zinc ion. Its structure has been characterized by a variety of spectroscopic techniques[31] as well as by X-ray crystallography.[32] The overall structure is a greek-key β barrel, which is very similar to known type 1 copper proteins.[13] The metal binding site shown in Figure 1 is composed of a copper(II) ion ligated by four histidines and a water in a distorted 5-coordinate geometry and a zinc(II) ion ligated by three histidines and an aspartate in a distorted tetrahedral geometry. Histidine 63 forms a bridge between the copper and the zinc. The Cu(II) sits at the bottom of a channel, where many anions such as CN^-, N_3^-, F^- as well as the substrate of the enzyme, O_2^-, enter, replace the coordinated water and react with Cu(II). Zn(II) is solvent inaccessible and is believed to play a structural role. A variety of metals have been substituted into either the copper or zinc site to make metal derivatives such as Cu_2Cu_2, Ag_2Cu_2, Cu_2Co_2[33] in addition to the native form Cu_2Zn_2.[31] Since Cu(II) can be selectively bound to either the copper or the zinc site, each site can be redesigned individually in the attempt to mimic type 1 copper proteins.

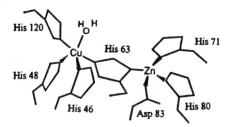

Figure 1. Metal binding site region of bovine CuZnSOD. The residues have been renumbered to represent the corresponding conserved residues in CuZnSOD from *Saccharomyces cerevisiae*.

Structural Characterizations of Cysteine Mutants

Copper Site Cysteine Mutants

Two copper site His-to-Cys mutants were constructed (H46C and H120C)[34,35,36] Instead of displaying an intense blue color, they all have an intense yellow color due to a strong absorption around 400 nm (ε_{379nm}=1940 $M^{-1}cm^{-1}$ for H46C-Cu_2Zn_2 and ε_{406nm}=1250 $M^{-1}cm^{-1}$ for H120C-Cu_2Zn_2, Figure 2A), which we assign to cysteine-to-copper charge transfer transitions. The energy of this transition is much higher than that of type 1 copper proteins, which is typically observed around 600 nm. While the high energy of the charge transfer band may seem surprising initially, its position is in fact consistent with modeling studies, since four-coordinate tetragonal and five-coordinate Cu(II)-thiolate and mercaptide model complexes typically display high energy charge transfer transitions.[21,24] According to the crystal structure of wild type YSOD,[32b] the geometry of the copper site of CuZnSOD is distorted square planar and our previous studies with yeast CuZnSOD mutants (H46C, H120C and H80C) have demonstrated that the geometry of the site is not changed.[34,35,36] This, together with the result that the mutants display typical type 2

67

copper ESR spectrum (see Figure 2B), shows that cysteine itself is not enough to make a structural type 1 copper mimic.

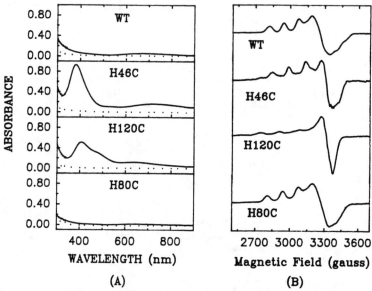

Figure 2. Electronic Absorption Spectra (A) and Electronic Spin Resonance Spectra (B) of the Cu_2Zn_2 Derivative of the Cysteinate Mutants and the Comparison with that of the Recombinant Wild Type. Dotted line: apoprotein; solid line: Cu_2Zn_2. Subunit concentration: WT, 0.40 mM; H46C, 0.48 mM; H120C, 0.46 mM; H80C, 0.42 mM. Instrument settings for ESR experiment: mirowave frequency, 9.5 GHz, microwave power, 20 mW, sample temperature, 90K.

Zinc Site Cysteine Mutant

His80Cys was constructed. Characterizations by electronic absorption (UV-Vis), magnetic circular dichroism (MCD), resonance Raman (RR) and electronic spin resonance (ESR) spectroscopies demonstrated that H80C is a new member of the type 1 copper protein family.[35,37]

UV-Vis of Copper Derivatives: Stepwise addition of four equivalents Cu(II) into apo-WT, a dimeric enzyme, resulted in Cu(II) filling the copper site first and then the zinc site to give WT-Cu_2Cu_2[38]. This is evidenced by the observation of absorption appearing first at 664 nm (ε=156 $M^{-1}cm^{-1}$) and then at 810 nm (ε=214 $M^{-1}cm^{-1}$) (see Figure 3A), since the distorted square planar copper site results in higher d-d transition energy (664 nm) than the distorted tetrahedral zinc site (810 nm). The Cu(II) access to the zinc site can be blocked by adding Zn(II) before adding Cu(II), in which case only the absorption of Cu(II) in the copper site can be observed at 670 nm (ε=147 $M^{-1}cm^{-1}$, see Figure 2A). Addition of Cu(II) to apo-H80C resulted in two strong absorptions (ε_{459nm}=1460 $M^{-1}cm^{-1}$, ε_{595nm}=1420 $M^{-1}cm^{-1}$) and one weak absorption ($\varepsilon_{810\ nm}$=370 $M^{-1}cm^{-1}$, see Figure 3A). None of the three absorptions could be observed if two equivalents Zn(II) were added before the addition of two equivalents Cu(II) (see Figure 2A), because the zinc site was blocked by the Zn(II).

The strong absorptions at 459 nm and 595 nm are assignable as cysteine→Cu(II) charge transfer bands due to their high intensities, and the assignment was confirmed by resonance Raman spectroscopy[40] (*vide infra*). Most type 1 copper proteins have only one strong absorption around 600 nm, which has

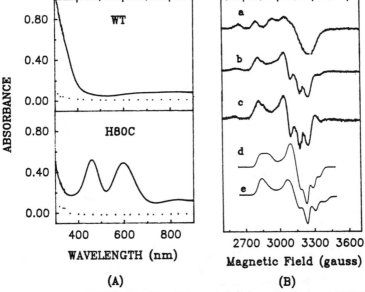

Figure 3. Electronic Absorption Spectra (A) and Electronic Spin Resonance Spectra at 77 K (B) of the Cu_2Cu_2 Derivative of the Cysteinate Mutant H80C and the Comparison with that of Wild Type. (A). Dotted line: apoprotein; solid line: Cu_2Cu_2. Subunit concentration: WT, 0.40 mM; H80C, 0.42 mM. (B). a. Cu_2E_2; b. Cu_2Cu_2; c. difference spectrum, b-a; d. stellacyanin, pH8; e. stellacyanin, pH11. Subunit concentration of H80C: 1 mM. Instrument settings for ESR experiment: mirowave frequency, 9.1GHz, microwave power, 0.5 mW. Spectra d and e are adapted from Figure 1 of reference 15b with permission from Elsevier Science Publishers.

been assigned as a cysteine S $p_\pi \rightarrow$ Cu(II) charge transfer.[14] However, some type 1 copper proteins such as stellacyanin,[15] pseudoazurin[16] and cucumber basic blue protein[17] display a moderately strong absorption around 460 nm in addition to the band around 600 nm. The most extreme example of increased 460 nm band intensity is the recently discovered copper-containing nitrite reductase from *Achromobacter cycloclastes*.[18] Its 460 nm and 600 nm bands are of similar intensity (ε_{458nm}= 2200 $M^{-1}cm^{-1}$, ε_{585nm}= 1800 $M^{-1}cm^{-1}$). Nonetheless, this copper center in this multicopper protein has other characteristics of type 1 copper proteins.[18,19] The H80C electronic spectrum is strikingly similar to that of nitrite reductase.

UV-Vis of Cobalt Derivatives: Cobalt derivatives of type 1 copper proteins have their own distinctive features and have been used to characterize the detailed structure of type 1 copper proteins. The UV-Vis spectrum of the cobalt derivative of H80C (see Figure 4B) looks very similar to those of type 1 copper proteins in general and stellacyanin in particular.[39] The similarity can be best depicted in Figure 4D as difference spectra of H80C-Co_2Co_2 and apoprotein, in which the apo-H80C spectrum has been subtracted from the spectra in Figure 4B to exclude the strong absorptions of proteins in that region. In contrast, similar depiction of the cobalt derivative of the wild type does not result in a spectrum that resembles the cobalt derivatives of type 1 copper proteins (Figure 4A, C). For H80C, the absorptions at 301 nm (ε=2900 $M^{-1}cm^{-1}$) and its shoulder at 360 nm (ε=788 $M^{-1}cm^{-1}$) are assignable to cysteine$\rightarrow$Co(II) charge transfer transitions. The three bands at 545 nm (ε=517 $M^{-1}cm^{-1}$), 579 nm (ε=625 $M^{-1}cm^{-1}$) and 608 nm (ε=863 $M^{-1}cm^{-1}$) are assigned to d-d transitions of Co(II) in a tetrahedral geometry. This result further supports the conclusion that H80C is a type 1 copper protein.

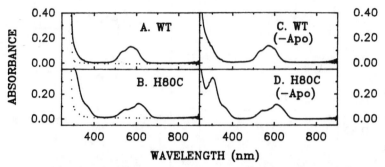

Figure 4. Electronic Absorption Spectra of the Co$_2$Co$_2$ Derivative of the Zinc Site Mutant H80C and the Comparison with Recombinant Wild Type. A: WT; B: H80C; C: subtraction of apo-WT (dotted line) from the spectrum in Panel A; D: subtraction of apo-H80C (dotted line) from the spectrum in Panel B. The subunit concentration of the protein samples: WT, 0.18 mM; H80C, 0.15 mM. The samples were in 50 mM potassium phosphate, pH 7.8.

Electron Spin Resonance: The ESR spectra of H80C-Cu$_2$E$_2$ ($g_\perp$=2.05, $g_{//}$=2.26, $A_{//}$=15.3 x 10^{-3} cm^{-1}, see Figure 3Ba) and H80C-Cu$_2$Zn$_2$ (g_x=2.02, g_y=2.07, g_z=2.26, A_z=13.9 x 10^{-3} cm^{-1}, see Figure 2B) are virtually identical to those of the corresponding WT derivatives, indicating that the mutation in the zinc site had little effect on the nature of the copper site. However, addition of Cu^{2+} to H80C-Cu$_2$E$_2$ resulted in a different ESR spectrum (Figure 3Bb) from that of WT-Cu$_2$Cu$_2$. No evidence of antiferromagnetic coupling in the H80C-Cu$_2$Cu$_2$ derivative could be obtained since no half field ESR signal was observed (data not shown). Subtraction of

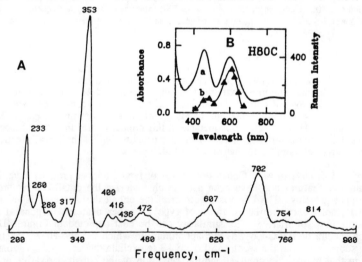

Figure 5. Resonance Raman Spectrum (A) and the Resonance Raman Enhancement Profile (B) of H80C-Cu$_2$Cu$_2$. Subunit concentration of the protein: 1.5 mM. A. Resonance Raman spectrum at 59 K with a 610-nm excitation (80 mW), a resolution of 7.5 cm^{-1}, a scan rate of 0.5 cm^{-1}/s and an average of 12 scans; B. Absorption spectrum (a) and resonance Raman enhancement profile (b). Data were obtained at nine excitation wavelengths on a single sample under spectral conditions as in A. Enhancement was measured as the area of the 353 cm^{-1} peak relative to the area of the ice peak at 233 cm^{-1}.

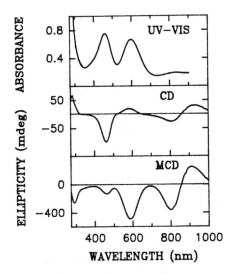

Figure 6. Electronic Absorption, Circular Dichroism and Magnetic Circular Dichroism Spectra of the Cu_2Cu_2 Derivatives of the Zinc Site Mutant Protein H80C at 40 kG. Subunit concentration: 1.0 mM. Sample temperature: 4 K.

the spectrum of H80C-Cu_2E_2 from that of H80C-Cu_2Cu_2 produced a spectrum (Figure 3Bc, $g_x=2.02$, $g_y=2.05$, $g_z=2.31$ and $A_z \leq 1.5 \times 10^{-3}$ cm^{-1}) that appears to be a more stellacyanin-like type 1 copper spectrum (Figure 3Bd, e), especially that of the high pH form of stellacyanin[15b] (Figure 3Be, pH11 $g_{//} = 2.31$ and $A_{//} \leq 1.7 \times 10^{-3}$ cm^{-1}), which was shown[45] to have an amide nitrogen as the fourth ligand besides the two histidines and one cysteine. (The fourth ligand in the zinc site of H80C mutant protein is an aspartate. Neither H80C nor stellacyanin have methionine in the copper binding site, in contrast to those classic type 1 copper proteins such as plastocyanin and azurin.) The apparent similarities of the ligand identities and the ESR spectra between the H80C mutant protein and the high pH form stellacyanin is consistent with the conclusion that the Cu(II) in the zinc site is a type 1 copper.

Resonance Raman: RR has been used extensively to characterize type 1 copper proteins because of its ability to identify charge transfer transitions. The RR spectrum of H80C-Cu_2Cu_2 (see Figure 5A) is typical of a type 1 copper protein with two vibrational fundamentals at 259 and 280 cm^{-1} and six higher energy fundamentals at 341, 352, 398, 415, 435 and 468 cm^{-1}.[40] Excitation within either the 458 nm or the 595 nm band (see Figure 5B) yields the same RR spectrum, indicating both absorptions have S(Cys)-to-Cu(II) charge transfer character. The extent of resonance enhancement (as judged by the intensity of the protein Raman peaks relative to the ice mode at 230 cm^{-1}) is also comparable to other type 1 copper proteins.

Magnetic Circular Dichroism: The MCD spectra of the Cu_2Cu_2 derivatives of H80C are shown in Figure 6, which displayed two negative bands at 459 nm and 595 nm, assignable to cysteine→Cu(II) charge transfer transitions based on RR. In the d-d transition region, there are two bands, one at 810 nm (negative) and one at 910 nm (positive). They are low-energy d-d transitions. It has been well established[41] that one of the key differences between a Cu(II) in a tetragonal site and a Cu(II) in a tetrahedral site is that the d-d transitions are at a much lower energy, usually around

71

1000 nm, which have become a hallmark of type 1 copper proteins.[14] Therefore the discovery of low energy d-d transitions in the spectrum of H80C-Cu$_2$Cu$_2$ further supports that a type 1 copper protein has been constructed. A careful comparison indicates that the MCD spectrum of H80C-Cu$_2$Cu$_2$ (Figure 6) is remarkably similar to the MCD spectrum of the type 1 copper protein stellacyanin (not shown), where the charge transfer transitions occurred at 455 nm (negative), 606 nm (negative) and the d-d transitions occurred at 769 nm (negative), 926 nm (positive) and 1136 nm (negative). The detailed comparison and absorption band assignments have been discussed elsewhere.[37]

Characterization of the Reactivity of His-to-Cys Mutants

The success of modeling a metalloprotein metal binding site must be judged not only by its ability to mimic the structural and spectroscopic properties of the target protein, but also by its ability to mimic its reactivity. On one hand, the target protein of our study is a type 1 or "blue" copper protein, whose function is believed to be electron transfer. On the other hand, our starting protein, wild type CuZnSOD, is known to react very slowly with reducing agents[42] such as ascorbate in spite of the fact that its reduction potential is higher than some of the type 1 copper proteins (see Figure 7A). We found that introduction of the cysteine either into the copper site (Figure 7B, 7C) or into the zinc site (Figure 7D) increases the redox reactivity of the protein. Because the redox potentials of the mutant proteins have not yet been measured, we cannot determine whether the change in redox activity is a thermodynamic effect or a kinetic effect. However, the data in Figure 7 are consistent with the thiolate providing an efficient superexchange pathway for electron transfer.

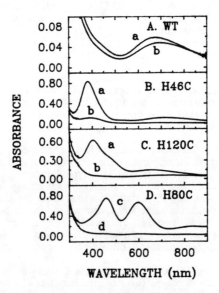

Figure 7. Electronic Absorption Spectra of the Cysteinate Mutants and the Wild Type upon Addition of Ascorbate. a. Cu$_2$Zn$_2$; b. about 2 min. after addition of 2 eq. ascorbate (per protein dimer) to a.; c. Cu$_2$Cu$_2$; d. about 2 min. after addition of 4 eq. ascorbate (per protein dimer) to c. Protein concentration in subunit: WT, 0.40 mM; H46C, 0.48 mM; H120C, 0.46 mM; H80C, 0.15 mM.

Correlation of 460 nm Electronic Absorption Band and Rhombicity of ESR Spectrum

As shown above, the introduction of a cysteine into a distorted tetrahedral geometry resulted in a type 1 copper protein that has electronic absorption and resonance Raman spectra that are similar to those of nitrite reductase, and MCD and ESR spectra, as well as the electronic aborption spectrum of the cobalt derivative, that are similar to those of stellacyanin.

The most unusual feature of the type 1 copper in H80C is probably the anomalously high intensity of the 460 nm band, comparing with those classic type 1 copper proteins such as plastocyanin and azurin. However, careful examination of all the type 1 copper proteins reveals that there is in fact a 460 nm band in almost all the type 1 copper proteins, but it is of variable intensity (see Table I). For example, $\varepsilon_{460\ nm}$ varies from as small as 300 $M^{-1}cm^{-1}$ in plastocyanin to as large as 2200 $M^{-1}cm^{-1}$ in nitrite reductase. On the other hand, $\varepsilon_{600\ nm}$ varies from as small as 1800 $M^{-1}cm^{-1}$ in nitrite reductase to as large as 5200 $M^{-1}cm^{-1}$ in plastocyanin. The consistency of the sum of $\varepsilon_{460\ nm}$ and $\varepsilon_{600\ nm}$ has been noted.[40] The X-ray structure determination and the spectroscopic characterization of nitrite reductase, the protein that has the highest 460 nm absorption band, indicate that it has all the typical type 1 copper properties.

Another interesting result from the characterization of H80C-Cu_2Cu_2 is the rhombicity of its ESR spectrum. This result may be surprising, since some classic type 1 copper proteins, such as plastocyanin and azurin, display axial ESR spectra that are similar to those of type 2 copper proteins, except that $A_{//}$ is much smaller. However, ESR spectra of type 1 coppers are expected to be more rhombic than those of type 2 copper because type 1 coppers are in distorted tetrahedral sites that are less symmetric than tetragonal type 2 copper sites. Indeed, since the characterization of plastocyanin and azurin, more and more type 1 copper-containing proteins that have been studied have shown rhombic ESR spectra, including stellacyanin,[15] pseudoazurin,[16] cucumber basic blue protein[17] and nitrite reductase from *Achromobacter cycloclastes*.[43] Analysis of the literature has led us to the conclusion that the ratio (R_L) of $\varepsilon_{460\ nm}/\varepsilon_{600\ nm}$ correlates very well with the rhombicity of the ESR spectrum. This correlation can be seen from Table I. From Table I we can see that: (1) The ratios (R_L) between the absorption around 460 nm and the absorption around 600 nm range from as low as 0.06 in plastocyanin to as high as 1.22 in the type 1 copper center of nitrite reductase of *Achromobacter cycloclastes*; (2) all those that have rhombic ESR spectra display appreciable absorption at 460 nm in addition to the 600 nm absorption ($R_L > 0.29$), though to different extents. This correlation occurs even in the same type of enzyme obtained from different organisms. For example, the nitrite reductase from *Alcaligenes* sp. NCIB 11015[44] has an axial ESR spectrum rather than a rhombic ESR spectrum as in nitrite reductase from *Achromobacter cycloclastes*; it also has a very weak 460 nm absorption (R_L=0.17 versus R_L=1.22). The correlation occurs even in the different mutants of the same protein. Wild type azurin has one of the weakest 460 nm absorptions (R_L=0.11) and an axial ESR spectrum. However, some of its mutants[29a] in which the methionine ligand (Met121) was changed into Asn, Asp, Gln, Cys, His and End (the polypeptide from methionine on was truncated by changing the methionine codon into a stop codon), have stellacyanin-like rhombic ESR spectra. An inspection of the electronic absorptions of the above mutants reveals that they also have stronger 460 nm absorptions. Those mutants that have WT-azurin-like axial ESR, such as Met121Thr, Met121Leu, Met121Ala, Met121Val, Met121Ile and Met121Trp, also have a 460 nm absorption as weak as WT-azurin. The above observations suggest that both the rhombicity of ESR spectra and the increase of the 460 nm electronic absorption band have a similar structural origin.

A possible explanation for the differences between the classic blue copper proteins with low R_L and axial ESR spectra and the "green" blue copper proteins with high R_L and rhombic ESR spectra is the strength of the Cu(II)-L bond of the fourth ligand. From Table I, pseudoazurin, cucumber blue protein and stellacyanin have large

Table I. Correlation between 460 nm Absorption Bands and Rhombic ESR Spectra of Type 1 Copper Centers in Copper Proteins

Protein	Source	Electronic Absorptions $(M^{-1}cm^{-1})$	R_L [†]	ESR	Cu-Met (Å)
Plastocyanin	P. nigra[a]	$\varepsilon_{460\ nm}=300; \varepsilon_{597\ nm}=5200$	0.06	axial	2.90
Umecyanin	A. lapathifolia[b]	$\varepsilon_{460\ nm}=300; \varepsilon_{595\ nm}=3400$	0.09	axial	?
Azurin	A. denitrificans[c]	$\varepsilon_{460\ nm}=580; \varepsilon_{619\ nm}=5100$	0.11	axial	3.13
Amicyanin	P. denitrificans[d]	$\varepsilon_{464\ nm}=520; \varepsilon_{595\ nm}=4610$	0.11	axial	?
Amicyanin	T. versutus[e]	$\varepsilon_{460\ nm}=524; \varepsilon_{596\ nm}=3900$	0.13	axial	?
Nitrite Reductase	A. Sp. NCIB 11015[f]	$\varepsilon_{470\ nm}=640; \varepsilon_{594\ nm}=3700$	0.17	axial	?
Stellacyanin	R. vernicifera[g]	$\varepsilon_{448\ nm}=1150; \varepsilon_{604\ nm}=4000$	0.29	rhombic*	(amide)
Auracyanin	C. aurantiacus[h]	$\varepsilon_{455\ nm}=900; \varepsilon_{596\ nm}=2900$	0.31	rhombic*	?
Mung Bean Blue Protein	P. aureus[i,j]	$A_{450\ nm}=0.03;^{\#} A_{598\ nm}=0.09^{\#}$	0.33	rhombic*	?
Mavicyanin	C. pepo medullosa[k]	$\varepsilon_{445\ nm}=1900; \varepsilon_{600\ nm}=5000$	0.38	rhombic*	?
Pseudoazurin	A. faecalis S-6[l]	$\varepsilon_{450\ nm}=1180; \varepsilon_{595\ nm}=2900$	0.41	rhombic	2.69
Rusticyanin	T. ferrooxidans[m,n]	$\varepsilon_{450\ nm}=1060; \varepsilon_{597\ nm}=1950$	0.54	rhombic*	?
Cucumber Basic Blue Protein	S. oleracea[o]	$\varepsilon_{443\ nm}=2030; \varepsilon_{597\ nm}=3400$	0.60	rhombic*	2.62
Nitrite Reductase	R. sphaeroides forma sp. denitrificans[p]	$\varepsilon_{464\ nm}=3660; \varepsilon_{584\ nm}=4860$	0.75	rhombic	?
CuZnSOD H80C	S. cerevisiae[q]	$\varepsilon_{459\ nm}=1460; \varepsilon_{595\ nm}=1420$	1.03	rhombic*	(Asp)
Nitrite Reductase	A. cycloclastes[r,s]	$\varepsilon_{458\ nm}=2200; \varepsilon_{585\ nm}=1800$	1.22	rhombic	?

[†] R_L is the ratio between the absorption around 460 nm and the absorption around 600 nm. * The ESR spectra of those proteins are very similar to that of stellacyanin. ESR spectra of other proteins are less rhombic. # The extinction coefficient was not available from reference i.

a. K. W. Penfield, R. R. Gay, R. S. Himmelwright, N. C. Eickman, V. A. Norris, H. C. Freeman, E. I. Solomon, J. Am. Chem. Soc. 103, 4382 (1981). b. K.-G Paul, T. Stigbrand, Biochim. Biophys. Acta 221, 255 (1970). c. E. W. Ainscough, A. G. Bingham, A. M. Brodie, W. R. Ellis, H. B. Gray, T. M. Loehr, J. E. Plowman, G. E. Norris, E. N. Baker, Biochemistry 26, 71 (1987). d. M. Husain, V. L. Davidson, J. Biol. Chem. 260, 14626 (1985). e. T. Houwelingen, G. W. Canters, G. Stobbelaar, J. A. Duine, J. Frank, A. Tsugita, Eur. J. Biochem. 153, 75 (1985). f. M. Masuko, H. Iwasaki, T. Sakurai, S. Suzuki, A. Nakahara, J. Biochem. 96, 447 (1984). g. (1) J. Peisach, W. G. Levine, W. E. Blumberg, J. Biol. Chem. 242, 2847 (1967). (2) B. G. Malmström, B. Reinhammar, T. Vänngård, Biochim. Biophys. Acta 205, 48 (1970). h. J. T. Trost, J. D. McManus, J. C. Freeman, B. L. Ramakrishna, R. E. Blankenship, Biochemistry 27, 7858 (1988). i. H. Shichi, D. P. Hackett, Arch. Biochem. Biophys. 100, 185 (1963). j. J. Peisach, unpublished result. k. A. Marchesini, M. Minelli, H. Merkle, P. M. H. Kroneck, Eur. J. Biochem. 101, 77 (1979). l. T. Kakutani, H. Watanabe, K. Arima, T. Beppu, J. Biochem. 89, 463 (1981). m. J. C. Cox, D. H. Boxer, Biochem. J. 174, 497 (1978). n. J. C. Cox, R. Aasa, B. G. Malmström, FEBS Lett. 93, 157 (1978). o. V. T. Aikazyan, R. M. Nalbandyan, FEBS Letters. 55, 272 (1975). p. W. Michalski, D.J.D. Nicholas, Biochim. Biophys. Acta 828, 130 (1985). q. Y. Lu, E. B. Gralla, J. A. Roe, J. S. Valentine, J. Am. Chem. Soc. 114, 3560 (1992). r. M.-Y. Liu, M.-C. Liu, W. J. Payne, J. LeGall, J. Bacteriology 166, 604 (1986). s. H. Iwasaki, S. Noji, S. Shidara, J. Biochem. 78, 355 (1975).

R_L values and more rhombic ESR spectra relative to plastocyanin and azurin. X-ray structures exist for the first two proteins which show significantly shorter $Cu\text{-}S_{met}$ bond lengths of 2.62 and 2.69Å relative to plastocyanin and azurin (2.90 and 3.1Å, respectively). There is currently no crystal structure available for stellacyanin; however, it contains no methionine in its sequence and pulsed ENDOR studies[45] have led to the conclusion that there is a Cu(II)-amide bound in its high pH form that would provide a stronger field than the thioether sulfur of the plastocyanin blue copper sites. (A three-dimensional model of stellacyanin,[46] which has been derived by computer graphics, energy minimization and molecular dynamics techniques and which can rationalize its spectroscopic, redox and electron-transfer properties, predicts that the fourth ligand is Gln97, which undergoes a switch from Cu-O(amide) at low pH to Cu-N(amide) at high pH.) The fourth ligand in Cu_2Cu_2 SOD-H80C is Asp83 which, as an oxygen ligand, is also expected to be a stronger ligand for Cu(II) than methionine. Furthermore, among the Met 121 mutants that have been generated, those that have a stronger 460 nm band and rhombic ESR spectrum have axial ligands such as Asp, Cys, His, Asn and Gln that are capable of stronger metal ligand bonding interactions than the methionine they replace. Those that have WT-like weak 460 nm bands and axial ESR spectra have non-coordinating axial residues replacing the methionine (Thr, Leu, Ala, Val, Ile, and Trp).

The conserved structural feature of classic blue copper proteins such as plastocyanin and azurin seem to be the Cu(II) in the trigonal plane formed by two histidines and one cysteine, with a weak fourth axial ligand (called pseudotetrahedral). One of the consequences of the stronger axial bonding would be the Cu(II) being pulled out of the trigonal plane, forming a near true tetrahedral geometry. Table II summarizes the three cases of Cu(II)-thiolate spectral features and their corresponding structural origins.

Table II Summary of Cu(II)-thiolate Centers

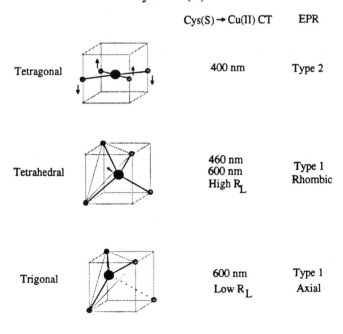

		Cys(S) → Cu(II) CT	EPR
Tetragonal		400 nm	Type 2
Tetrahedral		460 nm 600 nm High R_L	Type 1 Rhombic
Trigonal		600 nm Low R_L	Type 1 Axial

A model has been developed[47] to explain the rhombic splitting in stellacyanin, a protein having very similar spectral features to the perturbed blue copper site in H80C-Cu_2Cu_2SOD. This model invokes d_{z^2} mixing into the $d_{x^2-y^2}$ ground state of the copper site, which lowers g_x, raises g_y, and increases A_x. A small amount of d_{z^2} mixing (2-5%) will reproduce the ground state spectral features of stellacyanin. A ligand field calculation further showed that this mixing is achieved by starting from the plastocyanin geometry and increasing the ligand field strength of the axial ligand. These ligand field calculations show that this increased ligand field strength along the axial direction rotates the z axis of the $d_{x^2-y^2}$ orbital by ~15° and further rotates the g_x and g_y directions by a similar amount. This rotation leads to some overlap of the $d_{x^2-y^2}$ orbital with the Cys pseudo σ orbital and reduces its overlap with the Cys π level. Thus, the Cys pseudo σ to Cys π charge transfer intensity ratio (which relates to R_L) should correlate with the rhombic splitting as both would reflect the d_{z^2} mixing due to the increased ligand field strength of the axial ligand. These ligand field calculations now are strongly supported by the experiments mentioned above.

Acknowledgment

We thank Drs. J. Sanders-Loehr, T. M. Loehr, E. I. Solomon, J. Peisach for fruitful collaborations and helpful discussions. We would also like to thank the following people for their permissions to publish the following figures: Ms. Jane Han, Dr. T. M. Loehr and Dr. Joanne Sanders-Loehr of Oregon Graduate Institute (Figure 5), Mr. Louis B. LaCroix, Dr. Michael D. Lowery and Dr. Edward I. Solomon of Stanford University (Figure 6). Financial support from PHS grant GM28222 (JSV) and a New Faculty Grant from Loyola Marymount University (JAR) are gratefully acknowledged. Y. L. also acknowledges a Hortense Fishbaugh Memorial Scholarship, a Phi Beta Kappa Alumni Scholarship Award, and a Product Research Corporation Prize for Excellence in Research.

References and Footnotes

1 (a) L. Regan, W. F. DeGrado, *Science* **241**, 976 (1988); (b) W. F. DeGrado, Z. R. Wasserman, J. D. Lear, *Science* **243**, 622 (1989).

2 M. H. Hecht, J. S. Richardson, D. C. Richardson, R. C. Ogden, *Science* **249**, 884 (1990).

3 (a) J. S. Richardson, D. C. Richardson, in *Protein Engineering* D. L. Oxender, C. F. Fox, Eds. (Liss, New York, 1989), pp. 233-234; (b) J. S. Richardson, D. C. Richardson, *Trends in Biochem. Sci.* **14**, 304 (1989).

4 For a recent review see T. Handel, *Protein Engineering* **3**, 233 (1990).

5 (a) L. Regan, N. D. Clarke, *Biochemistry* **29**, 10878 (1990); (b) T. Handel, W. F. DeGrado, *J. Am. Chem. Soc.* **112**, 6710 (1990).

6 J. S. Richardson, D. C. Richardson, in *Prediction of Protein Structure and the Principles of Protein Conformation* G. D. Fasman, Ed. (Plenum Press, New York, 1989), pp. 1-98.

7 (a) M. Smith, *Phil. Trans. R. Soc. Lond. A* **317**, 295 (1986); (b) R. Leatherbarrow, A. R. Fersht, *Protein Engineering* **1**, 7 (1986); (c) J. R. Knowles, *Science* **236**, 1252 (1987).

8 L. Hedstrom, L. Szilagyi, W. J. Rutter, *Science* **255**, 1249 (1992).

9 A. R. Clarke, T. Atkinson, J. J. Holbrook, *Trends Biochem. Sci.* **14**, 145 (1989).

10 N. S. Scrutton, A. Berry, R. N. Perham, *Nature* **343**, 38 (1990).

11 S. G. Sligar, K. D. Egeberg, *J. Am. Chem. Soc.* **109**, 7896 (1987).

12 H. B. Gray, E. I. Solomon, in *Copper Proteins* T. G. Spiro, Ed. (Wiley, New York, 1981), vol. 53, pp. 1-57.

13 E. T. Adman, *Advances in Protein Chemistry* **42**, 145 (1991).

14 A. A. Gewirth, E. I. Solomon, *J. Am. Chem. Soc.* **110**, 3811 (1988).

15 (a) J. Peisach, W. G. Levine, W. E. Blumberg, *J. Biol. Chem.* **242**, 2847 (1967); (b) B. G. Malmström, B. Reinhammar, T. Vänngård, *Biochim. Biophys. Acta* **205**, 48 (1970).

16 T. Kakutani, H. Watanabe, K. Arima, T. Beppu, *J. Biochem.* **89**, 463 (1981).

17 Another name of this protein is plantacyanin, the spectral characterization of which was reported by: V. Ts. Aikazyan, R. M. Nalbandyan, *FEBS Letters* **55**, 272 (1975).

18 M.-Y. Liu, M.-C. Liu, W. J. Payne, J. LeGall, *J. Bacteriology* **166**, 604 (1986).

19 J. W. Godden, S. Turley, D. C. Teller, E. T. Adman, M.-Y. Liu, W. J. Payne and J. LeGall, *Science* **253**, 438 (1991).

20 F. A. Cotten, G. Wilkinson, in *Advanced Inorganic Chemistry* (Wiley, New York, 1980), p. 799.

21 A. B. P. Lever, *Inorganic Electronic Spectroscopy* (Elsevier, New York, 1986), p. 553.

22 A. G. Sykes, *Adv. Inorg. Chem.* **36**, 377 (1991).

23 P. Hemmerich, in *Biochemistry of Copper* J. Peisach, P. Aisen, Eds. (Academic Press, New York, 1966), pp. 15-32.

24 E. John, P. K. Bharadwaj, J. A. Potenza, H. J. Schugar, *Inorg. Chem.* **25**, 3065 (1986).

25 N. Kitajima, K. Fujisawa, Y. Moro-oka, *J. Am. Chem. Soc.* **112**, 3210 (1990).

26 W. H. Armstrong, in *Metal Clusters in Proteins* L. Que, Jr., Ed. (American Chemical Society, 1988), vol.
372, pp. 1-27.

27 M. L. Brader, M. F. Dunn, *J. Am. Chem. Soc.* **112**, 4585 (1990).

28 W. Maret, H. Dietrich, H.-H. Ruf, M. Zeppezauer, *J. Inorg. Biochem.* **12**, 241 (1980).

29 (a) B. G. Karlsson, M. Nordling, T. Pascher, L.-C. Tsai, L. Sjölin, L. G. Lundberg, *Protein Engineering* **4**, 343 (1991); (b) T. K. Chang, S. A. Iverson, C. G. Rodrigues, C. N. Kiser, A. Y. C. Lew, J. P. Germanas, J. H. Richards, *Pro. Natl. Acad. Sci. USA* **88**, 1325 (1991); (c) T. den Blaauwen, M. van de Kamp, G. W. Canters, *J. Am. Chem. Soc.* **113**, 5050 (1991).

30 O. Bermingham-McDonogh, E. B. Gralla, J. S. Valentine, *Proc. Natl. Acad. Sci. U.S.A.* **85**, 4789 (1988).

31 For a general review, please see: (a) J. S. Valentine, M. W. Pantoliano, in *"Copper Proteins"* T. G. Spiro, Ed., *Metals in Biology* Vol. 3 (Wiley, New York, 1981), pp. 291-357; (b) E. M. Fielden, G. Rotilio, in *Copper Proteins and Copper Enzymes* R. Lontie, Ed. (CRC Press, Florida, 1984), vol. 2, pp. 27-61.

32 (a) Bovine Cu, ZnSOD: J. A. Tainer, E. D. Getzoff, K. M. Beem, J. S. Richardson, D. C. Richardson, *J. Mol. Biol.* **160**, 181-217 (1982); (b) Yeast Cu, ZnSOD: K. Djinovic, G. Gatti, A. Coda, L. Antolini, G. Pelosi, A. Desideri, M. Falconi, F. Marmocchi, G. Rotilio, M. Bolognesi, *Acta Cryst.* **B47**, 918 (1991).

33 Abbreviations: M₂M'₂SOD, M- and M'-substituted superoxide dismutase with M in the copper site and M' in the zinc site (an E in the above derivatives represents an empty site); UV-Vis, electronic absorption spectroscopy in the ultraviolet and visible range; MCD, magnetic circular dichroism; RR, resonance Raman; ESR, electron spin resonance; ESEEM, electron spin echo envelpoe modulation; NMR, nuclear magnetic resonance.

34 Y. Lu, E. B. Gralla, J. A. Roe, J. S. Valentine, *J. Am. Chem. Soc.* **114**, 3560 (1992).

35 Y. Lu, Ph.D. dissertation, University of California: Los Angeles, 1992.

36 Y. Lu, E. B. Gralla, J. A. Roe, J. S. Valentine, C. Bender, J. Peisach, L. B. LaCroix, M. D. Lowery, E. I. Solomon, L. Banci, I. Bertini, L.-J. Ming, H. E. Parge, J. A. Tainer, manuscript in preparation.

37 Y. Lu, L. B. LaCroix, M. D. Lowery, E. I. Solomon, C. Bender, J. Peisach, J. A. Roe, E. B. Gralla, J. S. Valentine, submitted to *J. Am. Chem. Soc.*

38 M. W. Pantoliano, J. S. Valentine, L. A. Nafie, *J. Am. Chem. Soc.* **104**, 6310 (1982).

39 (a) D. R. McMillin, R. C. Rosenberg, H. B. Gray, *Proc. Natl. Acad. Sci. USA*, **71**, 4760 (1974); (b) E. I. Solomon, J. Rawlings, D. R. McMillin, P. J. Stephens, H. B. Gray, *J. Am. Chem. Soc.* **98**, 8046 (1976).

40 J. Han, Y. Lu, J. S. Valentine, B. A. Averill, T. M. Loehr, J. Sanders-Loehr, submitted to *J. Am. Chem. Soc.*

41 A. B. P. Lever, *Inorganic Electronic Spectroscopy* (Elsevier, New York, 1986), p. 555.

42 C. S. St. Clair, H. B. Gray, J. S. Valentine, *Inorg. Chem.* **31**, 925 (1992), and references therein.

43 H. Iwasaki, S. Noji, S. Shidara, *J. Biochem.* **78**, 355 (1975).

44 M. Masuko, H. Iwasaki, T. Sakurai, S. Suzuki, A. Nakahara, *J. Biochem.* **96**, 447 (1984).

45 H. Thomann, M. Bernardo, M. J. Baldwin, M. D. Lowery, E. I. Solomon, *J. Am. Chem. Soc.* **113**, 5911 (1991).

46 B. A. Fields, J. M. Guss, H. C. Freeman, *J. Mol. Biol.* **222**, 1053 (1991).

47 A. A. Gewirth, S. L. Cohen, H. J. Schugar, E. I. Solomon, *Inorg. Chem.* **26**, 1133 (1987).

ELECTRON TRANSFER REACTIVITY OF MUTANTS OF THE BLUE COPPER PROTEIN PLASTOCYANIN

A G Sykes,[a] P Kyritsis,[a] M Nordling,[b] and S Young,[b]

[a]Department of Chemistry, The University, Newcastle upon Tyne, NE1 7RU, England, UK.
[b]Department of Biochemistry and Biophysics, Chalmers University of Technology and University of Göteborg, S-4129, Göteborg, Sweden.

ABSTRACT:

This contribution focuses on the electron-transfer (ET) reactivity of plastocyanin in the PCu(I) state with the oxidants $[Fe(CN)_6]^{3-}$ and $[Co(phen)_3]^{3+}$. A programme of study involving five mutant forms obtained by site directed mutagenesis has been commenced in which effects of mutations at the adjacent and remote reaction sites are considered. For two of the mutants Tyr83His and Leu12Glu the variation of rate constants (25°C; I = 0.100M NaCl) with pH is explored. Protonation at the imidazole of His83 (pK_a 8.4) influences the reduction potential and hence reactivity. The results for Leu12Glu at high pH when the 1- glutamate form is present ($pK_a \sim 7.0$) indicate a sharp decrease in reactivity with $[Fe(CN)_6]^{3-}$, whereas the converse applies with $[Co(phen)_3]^{3+}$. The very striking electrostatic influence of the 1- charge is noted. Effects observed for Tyr83Phe are low key, but the reaction of this mutant and *Scenedesmus obliquus* plastocyanin (which has a Tyr82Phe83 sequence) with cytochrome f suggest that the Tyr62 in *S.obliquus* plastocyanin (and similar 57 and 58 depleted forms) may have a significant role to play in ET at the remote site. The mutation Leu12Asn benefits $[Fe(CN)_6]^{3-}$ (~ 5-fold), but surprisingly no change in reactivity is obtained for the Asp42Asn mutant with $[Co(phen)_3]^{3+}$.

INTRODUCTION:

Plastocyanins (M_r 10,500; normally 99 amino acids) are type 1 single Cu proteins which are involved in photosynthetic electron transport in higher plants and algae. Extensive structural information is available particularly from X-ray diffraction studies on the Cu(II) and Cu(I) oxidation states of poplar plastocyanin, and the Cu(II) protein from the green algae *Enteromorpha prolifera*.[1,2]

From K.D. Karlin and Z. Tyeklár, Eds., *Bioinorganic Chemistry of Copper*
(Chapman & Hall, New York, 1993).

Details of the various properties and reactivities have been reviewed.[3-4] A feature of reactivity patterns to date is the identification of adjacent (to the Cu) hydrophobic and remote acidic patch regions on the surface of the protein for ET. This type of duality of mechanism for ET has not so far been identified in any other protein. It remains however to develop further both an understanding of the precise specificities of the two sites, the nature of the binding, and of intramolecular ET to and from in particular the remote the site with different redox partners. The expression of the spinach PCu gene in *Escherischia coli* has led to the successful isolation of singly-modified variants by site-directed mutagenesis.[5] Using such variants it is possible to explore further possible influences on reactivity.

TECHNIQUES:

Small inorganic complexes are useful in probing the redox reactivity of plastocyanin. In particular the couples $[Fe(CN)_6]^{3-/4-}$ (410mV) and $[Co(phen)_3]^{3+/2+}$ (370mV), the reduction potentials of which are close to that

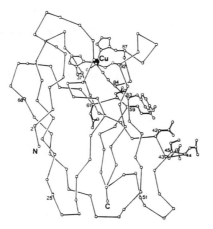

Figure 1 The structure of plastocyanin showing the Cu active site and location of adjacent (to the Cu) and remote binding sites. The carboxylate side chains indicated at 42-45 and 59-61 are as for spinach plastocyanin.

of PCu(II)/PCu(I) (375mV at pH 7.5), have been proved to be particularly useful.[6-7] The physiological reductant for PCu(II) is cytochrome f(II) which has a reduction potential of 350mV. By having the inorganic complex present in large excess (at least 10-fold) it is possible to monitor reactions of both PCu(I) and PCu(II) to ≥95% completion.

Investigations of effects of pH variations can consume large amounts of protein, and the pH-jump method is often used therefore. In this procedure the pH of the protein solution (with small ~2mM concentrations of buffer), has no controlling influence as compared to that of the inorganic complex (38mM buffer), and the latter determines the final pH. In this way one protein solution can be used for studies at a number of pH's. The approach requires that all related pH-dependent changes are relatively rapid.

EXISTING pH INFLUENCES:

Effects of pH have revealed valuable information regarding the reactivity of plastocyanin, more so than for most proteins. In Newcastle, plastocyanins from parsley[9] and spinach leaves,[10] as well as from the green algae *S.obliquus*,[11] and the blue-green algae *Anabaena variabilis*,[7,12] have been studied. The reactivity of French bean and poplar PCu's have also been investigated.[13,14] The overall charge on PCu(I) is generally in the range -9 ± 1 at pH ~ 7.5, indicating an excess of acidic Asp and Glu over basic Lys, Arg and uncoordinated His residues. The positive charge on *A.variabilis* plastocyanin (+1) is at present quite unique.[12] There is a 'switch-off' in reactivity e.g. with $[Fe(CN)_6]^{3-}$ as oxidant, observed for all PCu(I) forms, but not for PCu(II) with $[Fe(CN)_6]^{4-}$ as reductant, as the pH is decreased from 7.5 to <5.0. Rate constants of close to $10^5 M^{-1} s^{-1}$ decrease to values close to if not actually at zero (a point which it is difficult to establish with certainty). From X-ray crystallography on PCu(I) the effect has been assigned as protonation (and dissociation) of His87 from the Cu to give a redox inactive trigonal planar coordinated PCu(I) form.[16] Such studies have further established that rotation of the imidazole occurs, Figure 2, by identifying H-bonding of the ring

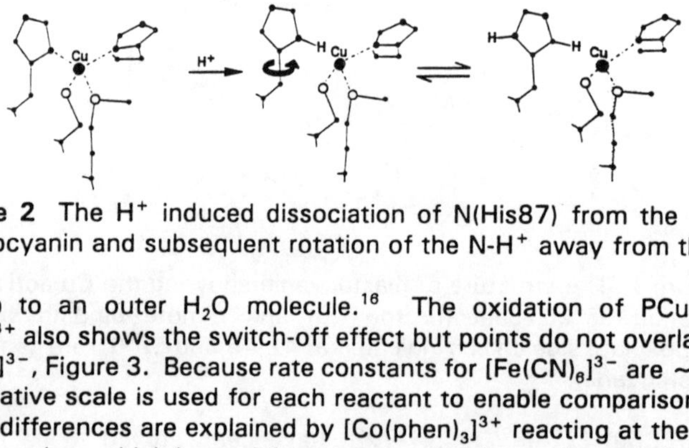

Figure 2 The H^+ induced dissociation of N(His87) from the Cu(I) of plastocyanin and subsequent rotation of the N-H^+ away from the Cu.

N-H^+ group to an outer H_2O molecule.[16] The oxidation of PCu(I) with $[Co(phen)_3]^{3+}$ also shows the switch-off effect but points do not overlay those for $[Fe(CN)_6]^{3-}$, Figure 3. Because rate constants for $[Fe(CN)_6]^{3-}$ are ~30-fold greater a relative scale is used for each reactant to enable comparisons to be made. The differences are explained by $[Co(phen)_3]^{3+}$ reacting at the remote acidic patch region, which has acidic properties which supplement the effects observed for the active site. Two pK_a fits can be carried out for the $[Co(phen)_3]^{3+}$ data. The different PCu(I)'s pK_a's can be summarised as in Figure 4. As a result of the 'switch-off' the reduction potential for the PCu(II)/PCu(I) couple increases to >430mV at pH's < 7.5.[7] Active site pK_a's fall into two ranges the higher one (e.g. for parsley and *S.obliquus*) are for plastocyanins with deletions at positions 57 and 58.[3] All such plastocyanins have 97 instead of 99 amino acids. An additional effect of pH is that observed for the reduction of PCu(II)'s with $[Co(phen)_3]^{2+}$ (an effect not detected with $[Fe(CN)_6]^{4-}$). These studies give the pK_a of the remote acidic patch for the oxidised protein, Figure 4. As compared to the remote site pK_a's for PCu(I) the latter appear to be shifted to slightly lower values.[15] Line-broadening effects of paramagnetic redox inactive $[Cr(CN)_6]^{3-}$ and $[Cr(phen)_3]^{3+}$ complexes on ^1H NMR spectra

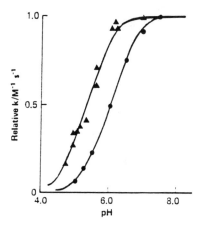

Figure 3: Rate constant $/M^{-1}s^{-1}$ trends illustrating the H^+-induced active site switch- off in reactivity of parsley PCu(I) with $[Fe(CN)_6]^{3-}$ (▲) and $[Co(phen)_3]^{3+}$ (●) as oxidant. A relative scale is used to enable this comparison. The points for $[Co(phen)_3]^{3+}$ do not overlay those for $[Fe(CN)_6]^{3-}$ because a remote site pK_a is also effective.

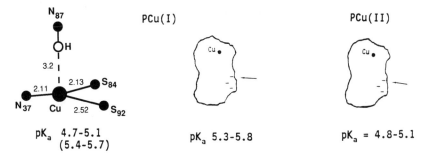

Figure 4: Summary of active site and remote binding site pK_a's for PCu(I) and PCu(II) plastocyanins, (see Table 2 in ref 3).

of PCu(I) further support assignments made with anionic and cationic complexes associating preferentially at the adjacent and remote sites respectively.[16]

From competitive inhibition studies the reaction of $[Co(phen)_3]^{3+}$ at the remote site is ~70% effective in the case of spinach and ~50% in the case of parsley PCu(I) of the total reaction. The remaining reaction is assumed to be at the adjacent site.[10,17] Previously the adjacent and remote sites have been referred to in geographic terms as north and east respectively. Although these terms have their usefullness they do have the disadvantage of being ambiguous, and slightly restricting since they are dependent on a fixed orientation of the plastocyanin structure as in Figure 1. We also note that *A.variabilis* plastocyanin is basic and has no pronounced acidic patch. The two remaining carboxylates in this region from Asp42 and Glu85 are bridged by the basic chain of Arg88. There is now no remaining evidence for the reaction of cations e.g. $[Co(phen)_3]^{3+/2+}$ at this remote site,[7] and $[H^+]$-dependences as in Figure 3 overlay.

OTHER EFFECTS OF pH:

Most plastocyanins have just the two coordinated His residues. *S.obliquus* and *A.variabilis* are unusual in that they have an additional (uncoordinated) His at position 59 at the surface. Examination of PCu sequences (25 are now known)[4] confirms that all PCu's except *A.variabilis* have a substantial number of acidic residues, at 42-45 (A) and 59-61 (B), either side of the conserved surface Tyr83 residue, Figure 5. While it is customary to

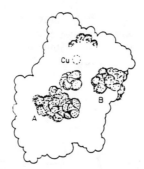

Figure 5: Perspective view of plastocyanin illustrating the relative positions of the acidic residues 42-45 (A) and 59-61 (B) either side of Tyr83. The positions of the (buried) Cu and exposed edge of His87 are also indicated.

consider A as more important than B this has not been confirmed, and for the present it is probably safer to regard both as constituting the acidic patch. His59 is a component of B, and its influence and the properties displayed are therefore of interest. Interestingly the rate constants for both the $[Fe(CN)_6]^{3-}$ and $[Co(phen)_3]^{3+}$ oxidations of *S.obliquus* PCu(I) respond to a second pK_a (7.8) alongside and separated from the active site effects already indicated.[11] The trend is the same in both cases (increasing k with increasing pH), which can only be accounted for as an effect of protonation/deprotonation of His59 on the reduction potential. Since the amplitude of the change in rate constants is greater in the case of $[Co(phen)_3]^{3+}$ it can be concluded that there is in addition an electrostatic influence of the acidic patch.   Unexpectedly *A. variabilis* plastocyanin does not behave in the same way and at pH > 7.0 rate constants are not always reproducible and can be upto ~66% greater.[7,12] The results suggest that different forms of the protein may be present.

PREPARATION OF MUTANTS:

Recombinant wild-type plastocyanin and the mutants were prepared using the previously described system for overexpression of plastocyanin in *E.coli* employing the expression-vector pUG101t,.[18] The mutant protein was constructed using polymerase chain reaction (PCR) amplification according to the method of Landt et al[19] with the modifications described.[18] Growth and fractionation of *E.coli* cells and purification of plastocyanin was made as in reference 5 with the following exceptions. The bacterial strain used was *E.coli* RV308 (ATCC 31608).[20] A Sepharose HP (26/10) (Pharmacia) FPLC column was used in the last ion-exchange chromatography and an ordinary Sephacryl S-100 column was used in the final gel-filtration step.   The plastocyanin containing fractions after each of these steps were pooled and concentrated by dialysis against dry polyethylene glycol. A communication of some of the work described herein has appeared.[21]

RESULTS

A summary of kinetic data obtained at pH 7.5 is given in Table 1.

Table 1

Rate constants at 25°C/$M^{-1}s^{-1}$ for the $[Fe(CN)_6]^{3-}$ (k_{Fe}) and $[Co(phen)_3]^{3+}$ (k_{Co}) oxidation of spinach PCu(I) native, wild type and mutant forms at pH 7.5, I = 0.100M (NaCl).

Protein	$10^{-5}k_{Fe}$	$10^{-3}k_{Co}$	k_{Fe}/k_{Co}
Native	0.85	2.54	33
Wild-type	0.71	2.24	32
Tyr83His	0.37	0.88	42
Leu12Glu	0.11	4.8	2.3
Tyr83Phe	0.71	2.64	27
Leu12Asn	3.7	2.50	148
Asp42Asn	0.85	2.24	38

These results will now be discussed under separate headings.

STUDIES ON THE Tyr83His MUTANT:

The variation of rate constants with pH for the reactions of $[Fe(CN)_6]^{3-}$

and $[Co(phen)_3]^{3+}$ with PCu(I), Figure 6, are very similar to those reported for *S.obliquus* PCu(I). From the current best fit of data at the higher pH's with $[Fe(CN)_6]^{3-}$ as oxidant a pK_a of ~7.9 is obtained, and with $[Co(phen)_3]^{3+}$ the corresponding value is ~8.5, in satisfactory agreement with an NMR determined value of 8.4, see Table 2. Again the relative amplitude from the increase in rate constant is greater for $[Co(phen)_3]^{3+}$ than $[Fe(CN)_6]^{3-}$ consistent with the electrostatic benefits which result from $[Co(phen)_3]^{3+}$ reacting at the remote site. The high pK_a value (a His residue might normally be expected to have a pK_a ~6.0) suggests a sharing of the proton with a nearby carboxylate at 42-45 or 59-61.

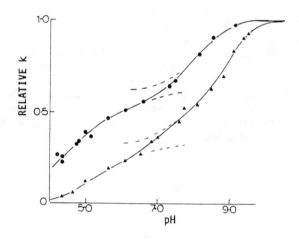

Figure 6: The effect of pH on relative rate constants $/M^{-1}s^{-1}$ (25°C) for the $[Fe(CN)_6]^{3-}$ (●) and $[Co(phen)_3]^{3+}$ (▲) oxidation of the Tyr83His mutant, $I = 0.100M$ (NaCl).

STUDIES ON THE Leu12Glu MUTANT:

Here the effect of 1- glutamate has a very pronounced inhibitory effect on the $[Fe(CN)_6]^{3-}$ reaction, Figure 7. It may well be that at very high pH the reactivity approaches zero. This represents a quite dramatic effect indicating a remarkable specificity of $[Fe(CN)_6]^{3-}$ for residue 12 at the adjacent site. A fit of data at the higher pH's gives a pK_a of 6.8, Table 2. With $[Co(phen)_3]^{3+}$ as oxidant the 1- glutamate results in more extensive contribution to reaction at the adjacent site, Figure 8, and from the k's at the higher pH's a pK_a in this case 7.0 is obtained. The two average at 6.9 and indicate an unusually high pK_a for a carboxylate residue, suggesting H-bonding with a nearby residue to retain the glutamic acid protein to higher pH's. Two possibilities are either Ser11 or His87. In both cases the behaviour observed at the lower pH's is as previously described. However an additional (minor) effect is that reactions are

Table 2

Summary of acid dissociation constant pK_a estimated from accompanying pH dependencies of rate constants at 25°C, I = 0.100M (NaCl). Values in parentheses determined by NMR.

	pK_1	pK_2	pK
$[Fe(CN)_6]^{3-}$ + Glu12	5.3 (4.9)	6.8	
$[Co(phen)_3]^{3+}$ + Glu12	5.7	7.0	
$[Fe(CN)_6]^{3-}$ + His83	4.9		7.9 (8.4)
$[Co(phen)_3]^{3+}$ + His83	5.8		8.5 (8.4)
	Cu site	Glu12	His83

biphasic. The behaviour observed suggests that in the case of $[Fe(CN)_6]^{3-}$ some of a less reactive deprotonated form and for $[Co(phen)_3]^{3+}$ a less reactive protonated form may be held back, and conversion to the reactive form is rate determining. This effect requires further study.

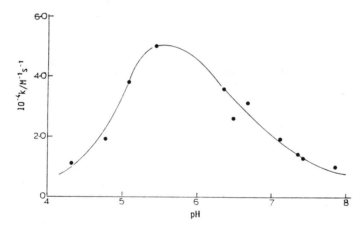

Figure 7: The effect of pH on rate constants (25°C) for the $[Fe(CN)_6]^{3-}$ oxidation of the Leu12Glu spinach plastocyanin mutants, I = 0.100M (NaCl).

STUDIES ON THE Tyr83Phe MUTANT:

As compared to native protein no effect on rate constants is observed

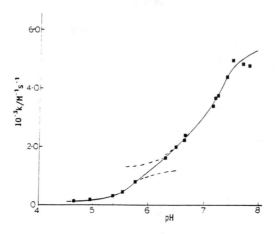

Figure 8: The effect of pH on rate constants (25°C) for the $[Co(phen)_3]^{3+}$ oxidation of the Leu12Glu spinach plastocyanin mutant, I = 0.100M (NaCl).

with $[Fe(CN)_6]^{3-}$ or $[Co(phen)_3]^{3+}$ as oxidants. However studies with the physiological partner cytochrome f produce interesting effects. It has already been concluded from competitive inhibition studies and effects of pH that the reaction of cytochrome f(II) with PCu(II) is at the remote site.[22] From their studies on the Tyr83Phe mutant He et al have reported an 8-fold effect on the overall rate constant for ET.[23] It has been demonstrated that this arises from an 8-fold decrease in K_a for association prior to ET, with the rate constant k_f for the intrinsic ET process remaining essentially unchanged at $60s^{-1}$, Table 3. This suggests that the phenolic OH group has a beneficial influence, most likely by H-bonding to the cytochrome f reactant. What is particularly interesting is that no similar effect is observed on the overall rate constants for the reaction of *S.obliquus* (as compared to parsley and spinach) PCu(II) with cytochrome f(II), Table 4.[7] The *S.obliquus* protein has a sequence Tyr82Phe83, Figure 9, in which the extensively conserved Phe82 and Tyr83 residues are interchanged. In addition there are deletions at 57 and 58, which result in a tightening of the kink of which residues 59-61 are very much a part, Figure 10. The six known plastocyanin sequences with deletions at 57 and 58 have an additional feature which is a surface Tyr at 62, Figure 11. These results suggest that in the reaction of *S.obliquus* PCu(II) with cyt f(II) the Tyr62 residue may compensate in some way for a less effective Phe83 in *S.obliquus* plastocyanin.

Table 3

Summary of mechanism and parameters for the reaction of oil-seed rape cytochrome f(II) with pea plastocyanin PCu(II) at 300K, pH 6.0, I = 0.10M (NaCl) reference 23.

$$PCu(II) + Cyt\ f(II) \quad \overset{K_a}{\underset{k_f}{\rightleftarrows}} \quad PCu(II),\ Cyt\ f(II)$$

$$PCu(II),\ Cyt\ f(II) \quad \rightarrow \quad PCu(I) + Cyt\ f(III)$$

	K_a/M^{-1}	k_f/s^{-1}
wild type	9890	62×10^3
Tyr83Phe	1270	58×10^3

Table 4

Summary of rate constants for the reaction of cytochrome f(II) (*Brassica*) with different PCu(II) plastocyanins, 10°C, pH 5.0, I = 0.20 (NaCl), reference 7.

PCu(II)	$10^{-5}k/M^{-1}s^{-1}$
Spinach	115
Parsley	110
S. obliquus	78
A. variabilis	2.8

	82 83 84
A. variabilis	Phe – Tyr – Cys
S. obliquus	Tyr – Phe – Cys
Poplar	Phe – Tyr – Cys
Spinach	Phe – Tyr – Cys
Fr. bean	Phe – Tyr – Cys
Parsley	Phe – Tyr – Cys

Figure 9: Sequence information for plastocyanins.

Figure 10: Effect of deletions at 57 and 58 and tightening of the peptide chain. Residue 62 is adjacent to or a component of the acidic patch B in Figure 5.

Figure 11: Location of Tyr62 relative to Tyr83 and remote and adjacent sites on plastocyanin.

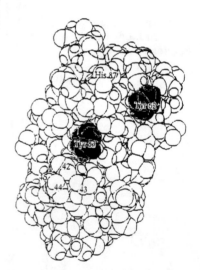

STUDIES ON OTHER MUTANTS:

The mutant Leu12Asn gives enhanced reactivity (x5) with $[Fe(CN)_6]^{3-}$, and there seems to be a greater compatibility of this oxidant with Asn as compared to Leu. However in the case of the Asp42Asn mutant there is surprisingly no apparent effect in the reaction with $[Co(phen)_3]^{3+}$. The latter is difficult to understand and must await the availability of other mutants.

CONCLUSION:

The use of mutants has helped to further establish the reactivity of plastocyanin at the adjacent and remote sites. Whereas residue 12 appears to have a critical role to play which is understandable in terms of its proximity to His87 and the Cu site, the studies with inorganic complexes are less helpful in indicating precise specificities at the remote site. From studies on the reaction with cytochrome f Tyr83 has an important role, and Tyr62 when it is present, may also have beneficial effects.

ACKNOWLEDGEMENTS:

We thank the State Scholarship Foundation of Greece for a Scholarship (to PK), and the Swedish Research Council.

REFERENCES

1.   (a) J.M. Guss and H.C. Freeman, *J.Mol.Biol.*, **169**, 521 (1983) (b) J.M. Guss, P.R. Harrowell, M. Murata, V.A. Norris and H.C. Freeman, *J.Mol.Biol.*, **192**, 361 (1986).
2.   C.A. Collyer, J.M. Guss, Y. Sugimura, F. Yoshizaki and H.C. Freeman, *J.Biol.Chem.*, **211**, 617 (1990).
3.   A.G. Sykes, *Structure and Bonding*, **75**, 175-224 (1991).
4.   A.G. Sykes, *Adv.Inorg.Chem.*, **36** 377-408 (1991).
5.   M. Nordling, T. Olausson and L.G. Lundberg, *FEBS Lett.*, **276**, 98 (1990).
6.   F.A. Armstrong, J.N. Butt, J. McGinnis, R. Powls and A.G. Sykes, *Inorg.Chem.*, **29**, 4858 (1990).
7.   D.G.A.H. de Silva, D. Beoku-Betts, P. Kyritsis, K. Govindaraju, R. Powls, N.P. Tomkinson and A.G. Sykes, *J.C.S. Dalton Trans.* 2145 (1992).
8.   J.C. Gray, *Eur.J.Biochem.*, **82** 133 (1978).
9.   M.G. Segal and A.G. Sykes, *J.Am.Chem.Soc.*, **100**, 4585 (1978).
10.   J.D. Sinclair-Day and A.G. Sykes, *J.C.S., Dalton Trans.*, **2069**, (1986).
11.   J. McGinnis, J.D. Sinclair-Day, A.G. Sykes, R. Powls, J. Moore and P.E. Wright, *Inorg.Chem.*, **27**, 2306 (1988).
12.   M.P. Jackman, J.D. Sinclair-Day, M.J. Sisley, A.G. Sykes, L.A. Denys and P.E. Wright, *J.Am.Chem.Soc.*, **109**, 6443 (1987).
13.   G.C. King, R.A. Binstead and P.E. Wright, *Biochem.Biophys.Acta*, **806**, 262 (1985) and G.C. King, Ph.D. Thesis, University of Sydney (1984).
14.   M.P. Jackman, J. McGinnis, A.G. Sykes, C.A. Collyer, M. Murata and H.C. Freeman, *J.C.S.Dalton Trans.*, 2573 (1987

15. J. McGinnis, J.D. Sinclair-Day and A.G. Sykes, *J.C.S.Dalton Trans.*, 2011 (1986).
16. (a) D.J. Cookson, M.T. Hayes and P.E. Wright, *Biochim.Biophys.Acta*, **591**, 162 (1980), (b) M. Handford, H.A.O. Hill, R. W.-K. Lee, R.A. Henderson and A.G. Sykes, *J.Inorg.Biochem.*, **13**, 83 (1980).
17. J. McGinnis, J.D. Sinclair-Day and A.G. Sykes, *J.C.S.Dalton Trans.*, 2007 (1986).
18. M. Nordling, K. Sigfridsson, S. Young, L.G. Lundberg and O. Hansson, *FEBS Lett.*, **291**, 327, (1991).
19. O. Landt, H.-P. Grunert and U. Hahn, *Gene*, **96**, 125, (1990).
20. R. Maurer, B.J. Meyer and M. Ptashne, *J.Mol.Biol.*, **139**, 147, (1980).
21. P. Kyritsis, L.G. Lundberg, M. Nordling, T. Vänngård, S. Young, N.P. Tomkinson and A.G. Sykes, *J.C.S.Chem.Comm.*, 1441 (1991).
22. D. Beoku-Betts, S.K. Chapman, C.V. Knox and A.G. Sykes, *Inorg.Chem.*, **24**, 1677 (1985).
23. S. He, S. Modi, D.S. Bendall and J.C. Gray, *EMBO J.*, **10**, 4011 (1991).

STUDIES OF CNI COPPER COORDINATION COMPOUNDS: WHAT DETERMINES THE ELECTRON-TRANSFER RATE OF THE BLUE-COPPER PROTEINS?

Scott Flanagan,[a] Jorge A. González,[a] Joseph E. Bradshaw[a], Lon J. Wilson[a], David M. Stanbury[b], Kenneth J. Haller[c], and W. Robert Scheidt[c]

[a]Department of Chemistry and Institute for Biochemical and Genetic Engineering, Rice University, P. O. Box 1892, Houston, Texas 77251, USA.
[b]Department of Chemistry, Auburn University, Auburn, Alabama 36849, USA.
[c]Department of Chemistry and Biochemistry, University of Notre Dame, Notre Dame, Indiana 46556, USA.

Unlike heme iron and iron-sulfur electron-transfer proteins, cuproteins have no extrudable coordination complex, since the active-site structure exists only through chelation of the copper ion with protein residues.[1] Thus, the study of small-molecule copper complexes offers one of the few means to evaluate the active-site contribution to electron transfer for copper proteins. Small-molecule model compounds for copper protein electron-transfer dynamics should ideally demonstrate coordination number invariance (CNI) and an outer-sphere mechanism of electron transfer. *Synthetic copper systems rarely meet these two criteria*, and the literature documents only a few well-defined candidates.[2-6] The high kinetic lability of copper and its tendency to adopt different coordination numbers and geometries in the +1 and +2 oxidation states pose formidable obstacles in the design and synthesis of appropriate small-molecule systems. Despite such difficulties, we have obtained data for several CNI five-coordinate complexes and for one four-coordinate complex by synthesizing ligands carefully tailored to help control such problems. The synthesis and characterization of the five-coordinate complexes have been described elsewhere,[6] while the four-coordinate complex is presented here for the first time.

Both plastocyanin (Pc) and azurin (Az) have been structurally characterized in their Cu(I) and Cu(II) oxidation states.[7,8] In both instances, the major structural difference between the +1 and +2 oxidation states is a slight lengthening of the copper-ligand bonds in the reduced forms. The geometry of the coordination sites in these proteins has been described as a compromise between the preferred geometries of Cu(I) and Cu(II). The resulting sterically-imposed "entactic state" is frequently cited as the source of the "fast" self-exchange rates of these proteins, due to the (supposed) lowering of the enthalpic component of the activation energy.[8] A fact worth noting, however, is that for at least one measurement of the activation parameters for azurin a favorable entropic component of the activation energy is implicated as being far more influential than the enthalpic component.[9]

The technique of NMR line broadening has proven extremely useful for the measurement of self-exchange rate constants due largely to the conceptual and experimental simplicity of the method. $1/T_2$ can be determined from the line width ($\Delta\upsilon_{1/2}$) by using the relationship:

$$1/T_2 = \pi\Delta\upsilon_{1/2}.$$

The self-exchange rate constant, k, may then be obtained as the slope of a plot of $1/T_2$ vs. [Cu(II)L]:

From K.D. Karlin and Z. Tyeklár, Eds., *Bioinorganic Chemistry of Copper* (Chapman & Hall, New York, 1993).

$$1/T_2 = k[Cu(II)L] + 1/T_{2n}.$$

Pseudo-self-exchange kinetics, obtained by stopped-flow measurements, offer a means to verify the NMR line-broadening results. These measurements also may be used to check agreement with Marcus theory in order to help verify outer-sphere electron transfer.[2a]

Figure 1 displays the ligands for the four CNI complexes for which self-exchange rate constant data are available and X-ray structural data have been determined for both the

Figure 1. Ligands for the CNI Copper Compounds

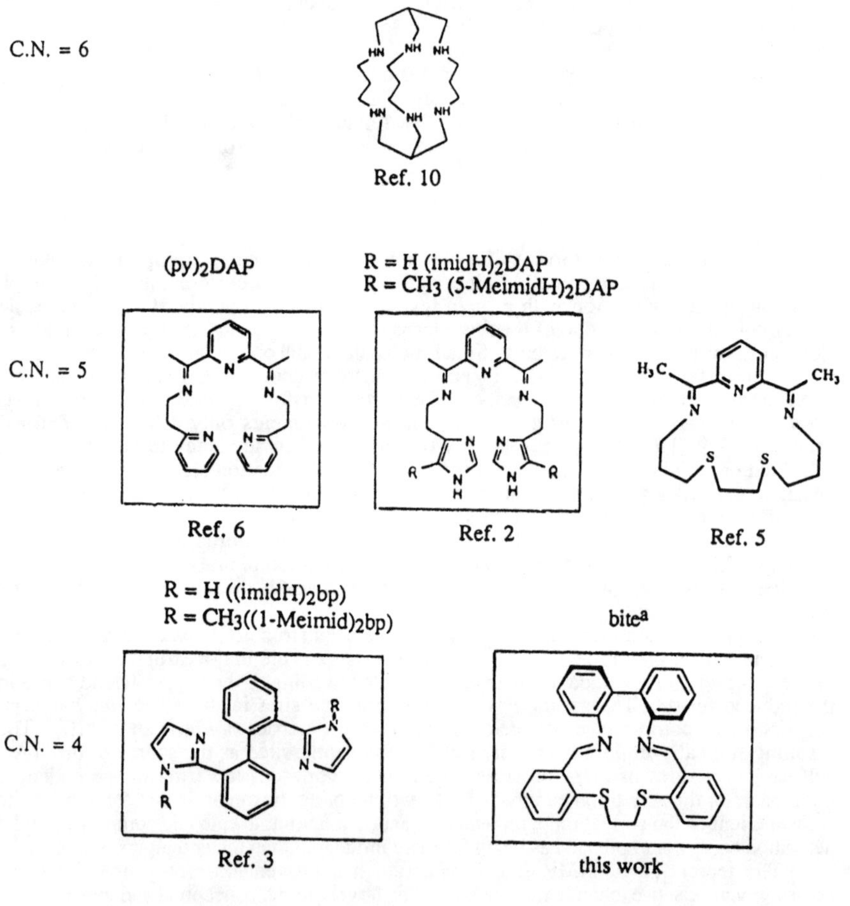

C.N. = 6

Ref. 10

(py)₂DAP

R = H (imidH)₂DAP
R = CH₃ (5-MeimidH)₂DAP

C.N. = 5

Ref. 6

Ref. 2

Ref. 5

R = H ((imidH)₂bp)
R = CH₃((1-Meimid)₂bp)

bite[a]

C.N. = 4

Ref. 3

this work

[a] bite = biphenyldiiminodithioether

+1 and +2 oxidation states (see structures in the squares). The two other ligands in the figure are also under investigation, but complete structural and kinetic determinations are not yet in hand for their copper complexes.[5,10]

Table I shows structural data for the two proteins, as well as for three of the CNI

92

Table I. X-ray Crystal Structure Data for the Coordination Spheres of Selected [$Cu^{I,II}(L)$]$^{2+,+}$ Couples.

	Cu(I)	Cu(II)	Δ (Å)
[$Cu^{I,II}(imidH)_2DAP$](BF_4)$_{1,2}$[a]			
Cu-N$_{central\ py}$	1.895 (33)	1.93 (1)	0.035
Cu-N$_{terminal\ imidH}$	1.887 (29)	1.99 (1)	0.103
	1.933 (31)	2.20 (2)	0.267
Cu-N$_{imino}$	2.282 (31)	2.05 (1)	0.232
	2.534 (29)	2.07 (1)	<u>0.464</u>
			avg = 0.220
[$Cu^{I,II}(py)_2DAP$](BF_4)$_{1,2}$[b]			
Cu-N$_{central\ py}$	2.094 (14)	1.920 (2)	0.174
Cu-N$_{terminal\ py}$	2.032 (12)	2.033 (2)	0.001
	2.083 (12)	2.129 (2)	0.046
Cu-N$_{imino}$	2.273 (14)	2.010 (2)	0.263
	2.240 (14)	2.026 (2)	<u>0.214</u>
			avg = 0.140
[$Cu^{I,II}(Az)$][c]			
Cu-O(Gly 45)	3.25	3.16	0.09
	[3.19]	[3.09]	[0.10]
Cu-N(His46)	2.17	2.08	0.09
	[2.09]	[2.09]	[0.00]
Cu-S(Cys112)	2.22	2.12	0.10
	[2.31]	[2.17]	[0.14]
Cu-N(His117)	2.05	2.01	0.04
	[2.05]	[1.99]	[0.06]
Cu-S(Met121)	3.21	3.12	0.09
	[3.25]	[3.10]	[<u>0.15</u>]
			avg = 0.08
			[avg = 0.09]
[$Cu^{I,II}((imidH)_2bp)_2$](ClO_4)$_{1,2}$[d]			
Cu-N$_{imidH}$ (1)	2.029 (6)	1.950 (8)	0.079
	[2.031 (6)]	[1.977 (8)]	[0.054]
Cu-N$_{imidH}$ (3)	2.035 (6)	1.962 (8)	0.073
	[2.053 (6)]	[1.941 (9)]	[<u>0.112</u>]
			avg = 0.076
			[avg = 0.083]
[$Cu^{I,II}(Pc)$](Cu(I) at pH 7.8)[e]			
Cu-N(His37)	2.12	2.04	0.08
Cu-N(His87)	2.25	2.10	0.15
Cu-S(Cys 84)	2.11	2.13	0.02
Cu-S(Met92)	2.90	2.90	<u>0.00</u>
			avg = 0.06

[a] Cu(I) values from ref. 6a. Cu(II) values supersede those from ref. 2b. [b] From ref. 6b. [c] From ref. 8. Standard deviations of the Cu^{II}-ligand bond distances are estimated to be ~0.05 Å (ref. 8). Standard deviations of the Cu^I-ligand bond distances are estimated to be ~0.05 - 0.10 Å (ref. 8). [d] From ref. 3. [e] From ref. 7. Standard deviations of the $Cu^{I/II}$-ligand bond distances are estimated to be ~0.05 Å (ref. 7).
Brackets indicate values for crystallographically distinct molecules in the same cell unit.

complexes for which rate constant data are also available (or can be reasonably well inferred from related complexes). If one considers reorganization of the coordination spheres in terms of metal-ligand bond length changes (admittedly, ignoring angular contributions), apparently only one of the small-molecule compounds has as small a structural change with oxidation state change as do the blue copper proteins. In fact, the average bond length change of $[Cu^{I,II}((imidH)_2bp)_2]^{2+/+}$ lies between the average bond length changes of Pc and Az. The average bond length changes of the two, five-coordinate complexes are considerably larger than found for the protein active sites.

Table II summarizes the self-exchange rate constants of the various $[Cu^{I,II}(L)]^{2+/+}$ couples in Table I. If the large self-exchange rate constant of Az is due to minimal reorga-

Table II. Electron Self-Exchange Rate Constants of Selected $[Cu^{I,II}(L)]^{2+,+}$ Couples.

Complex	k_{ex}, M^{-1} s^{-1}	Solvent	Ref.	T, K
$[Cu^{I,II}(imidH)_2DAP](BF_4)_{1,2}$	$1.31(0.16) \times 10^4$	CD_3CN	6c	298
$[Cu^{I,II}(py)_2DAP](BF_4)_{1,2}$	$1.76(0.16) \times 10^3$	CD_3CN	6b	298
$[Cu^{I,II}(Az)]$	$2.4(1.0) \times 10^6$	H_2O	9c	298
$[Cu^{I,II}((1\text{-}Meimid)_2bp)](BF_4)_{1,2}$	$< 1 \times 10^2$	CD_3CN	3c	253 to 293
$[Cu^{I,II}(Pc)]$	$\ll 2 \times 10^4$	D_2O	11	323

nization energy at the active site, one might expect that small-molecule compounds having equally small reorganization processes would also exhibit rapid self exchange. A quick examination of the data in Tables I and II shows this view to be overly simplistic. What, then, controls the rate of electron transfer? As previously implied, for a protein environment a large entropic factor due to solvent release from the protein surface may lead to an enhanced electron-transfer rate. Obviously, for small-molecule compounds, this factor is not as important. Knapp, et al., have suggested that sufficient overlap between the metal-ligand orbitals is necessary for efficient outer-sphere electron transfer.[3c] They also suggest that such delocalization may be a factor at the Pc active site. Thus, in this view, rapid electron transfer for the protein occurs because the protein presents favorable pathways for effective electronic coupling.

Paralleling our studies of electron-transfer rates, we have also investigated the ligand dynamics of small-molecule Cu(I) complexes by variable-temperature NMR studies.[2,6] Table III shows the results of a variable-temperature NMR study of $[Cu^I(py)_2DAP]^+$ and

Table III. Self-Exchange Rate Constants for the CNI $Cu^{I,II}$ Couples and Coalescence Temperatures (T_c) for Their Cu^I Compounds.

Molecule	C.N.	T_c (CH_2Cl_2)	k_{ex} (M^{-1} s^{-1}) in CD_3CN	Ref.
$[Cu^{I,II}(py)_2DAP]^{2+/+}$	5	~ 253 K	1.76×10^3 @ 298 K	6c
$[Cu^{I,II}(imidH)_2DAP]^{2+/+}$	5	—	1.31×10^4 @ 298 K	
$[Cu^{I,II}(5\text{-}MeimidH)_2DAP]^{2+/+}$	5	~ 203 K	3.5×10^4 @ 293 K	2a

$[Cu^I(5\text{-}MeimidH)_2DAP]^+$, which seems to indicate a greater degree of ligand flexibility in the latter with the lower coalescence temperature of 203 K. Thus, for the only two CNI copper couples for which electron-transfer data, structural data, and ligand mobility data have all been obtained, the electron-transfer rates parallel ligand flexibility, while opposing conventional wisdom concerning the importance of the coordination sphere reorganizational processes. Why should greater ligand mobility enhance the electron-transfer rate? In view of Knapp, et al.'s, arguments in favor of orbital overlap providing a pathway for rapid electron transfer, perhaps flexibility of the ligand environment allows for an increased probability of a collision occurring when the ligand-metal orbitals are properly aligned. As an alternative view, greater mobility may suggest lower force constants along the transition state vector which would lower the Franck-Condon barrier.

Vande Linde, et al., have obtained self-exchange data for a copper redox couple, $[Cu^{I,II}([15]aneS_5)]^{+,2+}$ which has been shown by X-ray crystallography to change coordination number with oxidation state in the solid.[12] The complex is pentacoordinate and square pyramidal in the Cu(II) state and tetracoordinate and pseudotetrahedral for Cu(I).[12b] Based on the large enthalpic changes expected for a bond breaking/bond forming process, one might predict the self-exchange rate for this small-molecule system to be quite slow. In fact, a measurement has shown that the self-exchange rate constant of k $\approx$ 2 x 10^5 M^{-1} s^{-1} (25 °C, D_2O) for $[Cu^{I,II}([15]aneS_5)]^{+,2+}$ exceeds those of all other small-molecule copper systems published to date.[12a] Vande Linde, et al., have suggested that, in fact, the enthalpic contribution to the electron transfer may be much smaller than expected for a bond breaking/bond forming process. Comparison with the activation parameters of $[Cu^{I,II}(5\text{-MeimidH})_2DAP]^{2+,+}$ (Table IV) reveals that the values for both enthalpy and

Table IV. Self-Exchange Rate Constants and Activation Parameters for the $Cu^{I,II}$ Couples

$Cu^{I,II}$ Couple	CNI	k_{ex} (M^{-1} s^{-1})	$\Delta H^{\ddagger}$ (kJ mol^{-1})	$\Delta S^{\ddagger}$ (JK^{-1} mol^{-1})	Ref.
$[Cu^{I,II}(5\text{-MeimidH})_2DAP]^{2+/+}$	Yes	3.5 x 10^4 @ 293 K (CH_3CN)	16.2 (3.3)	- 103 (12)	2b
$[Cu^{I,II}([15]aneS_5)]^{2+/+}$	No	2.0 x 10^5 @ 298 K (D_2O)	14.0 (4.0)	- 103 (11)	12a
azurin (*Pseudomonas aeruginosa*)	Yes	1.2 x 10^6 @ 309 K (H_2O)	71(8)	+ 96 (29)	9a

entropy of activation for the two systems (one CNI, the other non-CNI) are essentially identical.[12a] Since, the $[Cu^{I,II}(5\text{-MeimidH})_2DAP]^{2+,+}$ system undergoes no bond breaking or forming, the Cu-S* bond which breaks in $[Cu^{II}[15]aneS_5)]^{2+}$ must not introduce a large barrier to electron transfer. Further comparison with the activation parameters of azurin reveals that the rate constant for the protein appears to be dominated by a large positive entropic factor. Such an entropic factor is almost certainly not determined by the active site of the protein. Thus, caution should be exercised in making casual assessments about the active-site contribution to protein electron-transfer rates.

Clearly, much remains to be learned about the nature of copper electron-transfer reactions, in general, and about the contribution of the active site to blue-copper proteins electron transfer, in particular. Only through the collection of further information on systems with systematically-varied coordination numbers and geometries will the nature of these processes become clearer. Thus, we report here for the first time an $N_2S_2^*$ macrocyclic system which is four coordinate and nearly tetrahedral for the Cu(I) state but which "semi-coordinates " two BF_4^- counterions to form an axially elongated octahedral complex for the Cu(II) state, with a surprisingly "flattened" $N_2S_2^*$ donor atom set (see Figure 2).[13] The synthesis of these $[Cu^{I,II}(bite)](BF_4)_{1,2}$ salts is outlined in Figure 3. A saturated solution of $[Cu^{II}(bite)](BF_4)_2$ in CH_2Cl_2 (2 x 10^{-5} M) gives a molar conductivity of 69 $cm^2equiv^{-1}\Omega^{-1}$ as compared to 80 $cm^2equiv^{-1}\Omega^{-1}$ for $[Cu^I(bite)](BF_4)$ and 100 $cm^2equiv^{-1}\Omega^{-1}$ for [n-BuN](BF_4) at the same concentrations at 25 °C. These solution conductivities indicate a substantial degree of dissociation of the "semi-coordinated" BF_4^- counterions of the Cu(II) molecule in CH_2Cl_2, and it is possible that the $[Cu^{I/II}(bite)]^{2+/+}$ couple is CNI in this non-coordinating solvent. A preliminary measurement of the electron self-exchange rate constant of k = 2.4 x 10^4 M^{-1} s^{-1} at 25 °C in CH_3CN is on the same order of magnitude as the five-coordinate CNI compounds in Table III, despite the large anticipated reorganizational energy arising from the configurational differences seen in Figure 2. A variable-temperature NMR study of $[Cu^I(bite)](BF_4)$ displays a much higher coalescence temperature (~ 327 K) than for the five-coordinate CNI compounds, and the source of such rigidity is likely the biphenyl end of the molecule.

Figure 2. ORTEPs of the $[Cu^{I,II}(bite)]^{+,2+}$ Cations

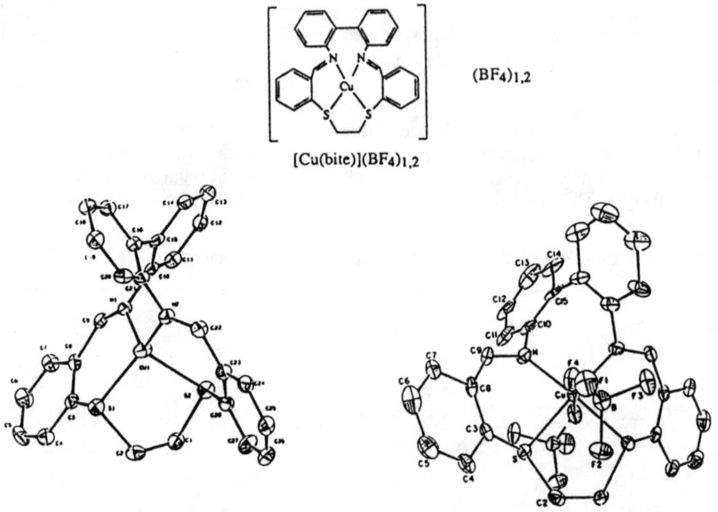

$[Cu(bite)](BF_4)_{1,2}$

$(BF_4)_{1,2}$

Cu^I-N1	1.950 (6) Å
Cu^I-N2	1.941 (6) Å
Cu^I-S1	2.198 (2) Å
Cu^I-S2	2.322 (2) Å

N1CuS1/N2CuS2
Dihedral Angle is 78°

Cu^{II}-N	1.990 (5) Å
Cu^{II}-S	2.286 (2) Å
Cu^{II}-F4	2.546 (4) Å

N1CuS1/N2CuS2
Dihedral Angle is 3°

Figure 3. Synthesis of the $[Cu^{I,II}(bite)](BF_4)_{1,2}$ Salts.

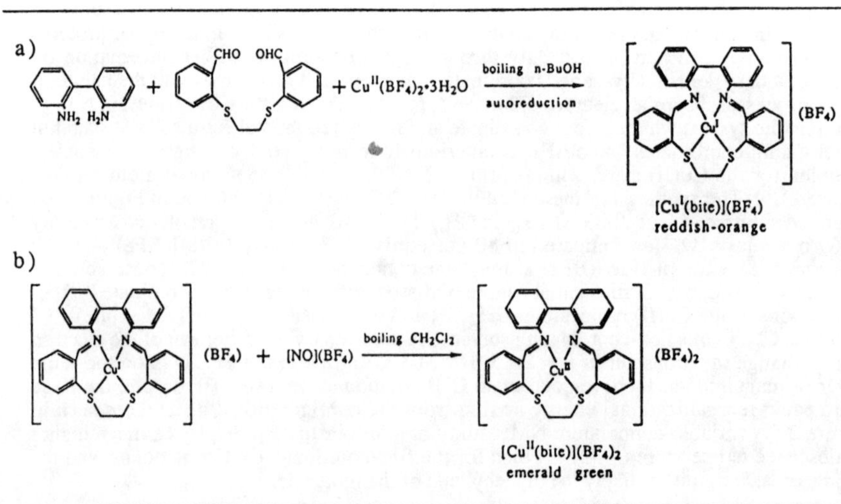

a)

$[Cu^I(bite)](BF_4)$
reddish-orange

b)

$[Cu^{II}(bite)](BF_4)_2$
emerald green

Taken together, the emerging data tends to suggest that, at least for these particular small-molecule compounds, small reorganizational processes at the copper site do not necessarily promote "fast" electron-transfer reactions, particularly when one considers the $[Cu^{I,II}((imidH)_2bp)_2]^{2+/+}$ system which has a small structural reorganization process and yet an unmeasurably small (by NMR $< 10^2$ M^{-1} s^{-1}) electron self-exchange rate.[3] Furthermore, the $[Cu^{I,II}[15]aneS_5)]^{2+/+}$ couple clearly demonstrates that a cuproprotein active site need not be CNI in order to promote a "fast" electron-transfer reaction. As a corollary to this latter observation, the blue-copper active sites could exhibit larger reorganization and still be "fast". Finally, and perhaps most importantly, these studies of CNI (and non-CNI) copper coordination compounds have emphasized that one must exercise care in ascribing properties of the cuproprotein as a whole to the copper active site.

We thank the U. S. National Institute of Health (GM-28451) and the Robert A. Welch Foundation (C-627) for support of this work. S. F. also thanks the U.S.N.I.H. for a NIGMS Training Grant (GM-08362) at Rice University. D. M. S. is an Alfred P. Sloan Fellow.

REFERENCES

1. E.I. Solomon, in *Copper Coordination Chemistry: Biochemical & Inorganic Perspectives*, K.D. Karlin and J. Zubieta, Eds., (Adenine Press, Guilderland, New York, 1983), pp. 1-22.
2. (a) D.K. Coggin, J.A. González, A.M. Kook, D.M. Stanbury, L.J. Wilson, *Inorg. Chem.* **30**, 1115 (1991). (b) D.K. Coggin, J.A. González, A.M. Kook, C. Bergman, T.D. Brennan, W.R. Scheidt, D.M. Stanbury, L.J. Wilson, *Inorg. Chem.* **30**, 1125 (1991).
3. (a) S. Knapp, T.P. Keenan, X. Zhang, R. Fikar, J.A. Potenza, H.J. Schugar, *J. Am. Chem. Soc.*, **109**, 1882 (1987). (b) S. Knapp, T.P. Keenan, J. Liu, J.A. Potenza, H.J. Schugar, *Inorg. Chem.*, **29**, 2191 (1990). (c) S. Knapp, T.P. Keenan, X. Zhang, R. Fikar, J.A. Potenza, H.J. Schugar, *J. Am. Chem. Soc.*, **112**, 3452 (1990).
4. C.M. Groenveld, J. van Rijn, J. Reedijk, G.W. Canters, *J. Am. Chem. Soc.*, **110**, 4893 (1988).
5. (a) M.G.B. Drew, C. Cairns, S.G. McFall, S.M. Nelson, *J. Chem. Soc., Dalton Trans.*, 2020 (1980). (b) M.G.B. Drew, C. Cairns, S.M. Nelson, J. Nelson, *J. Chem. Soc., Dalton Trans.*, 942 (1981).
6. (a) J.A. Goodwin, G.A. Bodager, L.J. Wilson, D.M. Stanbury, W.R. Scheidt, *Inorg. Chem.*, **28**, 35 (1989). (b) J.A. Goodwin, D.M. Stanbury, L.J. Wilson, C.W. Eigenbrot, W.R. Scheidt, *Inorg. Chem.*, **109**, 2979 (1987). (c) J.A. Goodwin, L.J. Wilson, D.M. Stanbury, R.A. Scott, *Inorg. Chem.*, **28**, 42 (1989).
7. J.M. Guss, P.R. Harrowell, M. Murata, V.A. Norris, H.C. Freeman, *J. Mol. Biol.* **192**, 361 (1986), and references therein.
8. W.E.B. Shepard, B.F. Anderson, D.A. Lewandoski, G.E. Norris, E.N. Baker, *J. Am. Chem. Soc.*, **112**, 7817 (1990), and references therein.
9. (a) C.M. Groenveld, G.W. Canters, *Eur. J. Biochem.* **153**, 559 (1985). (b) C.M. Groenveld, G.W. Canters, *Rev. Port. Quim.*, **27**, 145 (1985). (c) C.M. Groenveld, S. Dahlin, B. Reinhammar, G.W. Canters, *J. Am. Chem. Soc.* **109**, 3247 (1987).
10. A. M. Sargeson, personal comunication.
11. J.K. Beattie, D.J. Fensom, H.C. Freeman, E. Woodcock, H.A.O. Hill, A.M. Stokes, *Biochim. Biophys. Acta*, **405**, 109 (1975).
12. (a) A.M.Q. Vande Linde, K.L. Juntunen, O. Mols, M.B. Ksebati, L.A. Ochrymowycz, D.B. Rorabacher, *Inorg. Chem.*, **30**, 5037 (1991). (b) P.W.R. Corfield, C. Ceccarelli, M.D. Glick, I.W. Moy, L.A. Ochrymowycz, D.B. Rorabacher, *J. Am. Chem. Soc.*, **107**, 2399 (1985).
13. Full details of the synthesis and crystal structure determinations will be published elsewhere.

Natural and Synthetic Regulation of Gene Expression

CHEMICAL AND GENETIC STUDIES OF COPPER RESISTANCE IN *E. coli*

James W. Bryson,[a] Thomas V. O'Halloran,[a] Duncan A. Rouch,[b] Nigel L. Brown,[b] Jim Camakaris,[c] and Barry T. O. Lee[c]

[a]Department of Chemistry; Northwestern University, Evanston, IL 60208 USA,
[b]School of Biological Sciences, University of Birmingham, Birmingham B15 2TT, UK,
[c]Department of Genetics, University of Melbourne, Parkville, Victoria 3052, Australia

INTRODUCTION

Copper is an essential trace element required for bacterial growth, but is toxic at high levels of free ions. Bacteria are thus presented with the complex problem of obtaining and storing sufficient quantities of copper for normal function of several enzymes while, on the other hand, being able to survive when confronted with concentrations of copper that exceed a toxic threshold. The molecular mechanisms of metal ion detoxification are well understood only for a few systems, most notably for mercury resistance.[1,2] For recent reviews of a number of bacterial metal resistance systems, see *Plasmid*, Vol. 27(1), 1992 and reference [3]. The discovery of plasmid-borne copper resistance in bacteria has provided accessible systems for genetic and phenomenological study of copper metabolism. Study of these extrachromosomal systems has recently provided the impetus and the methods for identification of chromosomally-encoded copper homeostasis systems in *Pseudomonas syringae* pv. *tomato* (*P. syringae*)[4,5,6] and in *Echerichia coli* (*E. coli*).[7,8,9] The two best characterized copper resistance systems are the plasmid-encoded systems *cop* in *P. syringae* and *pco* in *E. coli*. These two determinants are remarkably similar genetically, but as will be shown they differ fundamentally in the mechanism of copper resistance. This paper will focus first on the genetics and regulation of copper resistance in the *cop* and *pco* systems, then on the chemical mechanisms of copper resistance.

GENETICS OF COPPER RESISTANCE

Copper resistance in *E. coli* (*pco*) has been localized to an approximately six kilobase fragment of the originally isolated plasmid pRJ1004.[10] Preliminary genetic analysis of this region suggested four complementation groups named *pcoABRC*.[9,10] More detailed sequence data and analysis now indicates seven open reading frames, renamed *pcoABCDRSE*.[11,12] Copper resistance in *P. syringae* has been localized to approximately 4.5 kilobases of the plasmid pPR23D, encoding at least four genes labeled *copABCD*.[4,13]

From K.D. Karlin and Z. Tyeklár, Eds., *Bioinorganic Chemistry of Copper*
(Chapman & Hall, New York, 1993).

Regulation

Both the *pco* and *cop* resistance systems are inducible by copper. The four *cop* genes are under the control of a single copper-inducible promoter that has been sub-cloned, but not yet precisely defined.[14] Copper-inducible expression of the *cop* genes requires a trans-acting element encoded on the plasmid or the chromosome.[5,6] The trans-acting element(s) have not yet been identified.

A trans-acting element has been identified in the *pco* system from *E. coli*, namely *pcoR*.[15] The predicted amino acid sequence of PcoR shows strong homology to a number of regulatory proteins from the family of bacterial two-component regulatory systems.[9,16] The two-component systems consist of a sensor protein and a regulatory protein. The sensor protein becomes phosphorylated by an environmental signal, and in turn phosphorylates and activates the regulator (i.e. OmpR/EnvZ or NtrB/NtrC).[17,18,19,20] It is likely that the *pco* genes are regulated in a similar manner, and the next open reading frame, *pcoS*, has been identified as the putative sensor component by sequence homology.[11] As the *pco* genes are still regulated in a *pcoR* mutant, it is presumed that other plasmid or chromosomal factors are also involved in the regulation of the resistance determinant and perhaps *vice versa*.[9] Figure 1 presents a model for the regulation of the *E. coli pco* copper resistance determinant based on genetic studies.

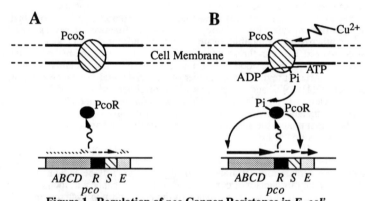

Figure 1. Regulation of *pco* Copper Resistance in *E. coli*

Model for the regulation of *E. coli* copper resistance determinant. (A) Low copper concentration. (B) High copper concentration. Dashed arrows indicate low levels of transcription, while solid arrows indicate high levels of trascription.

Resistance Proteins

The predicted amino acid sequences of the first four proteins of the *pco* determinant show extensive identity to the four proteins of the *cop* plasmid-mediated copper resistance determinant from *P. syringae* (see Figure 2).[13] Until the full sequence of the *pco* genes is completed, it will be valuable to examine the strongly analogous proteins in the *P. syringae* system. Interestingly, the copper resistance mechanism of the *cop* system operates by accumulation/sequestration of copper rather than enhanced export as is observed for the *pco* system (more details regarding the mechanism of copper resistance for the *pco* system are discussed below). Cooksey and co-workers have demonstrated that the copper resistant strains of *P. syringae* accumulate significantly more copper than the parental strains.[21] The accumulated copper is enough to turn resistant bacterial colonies blue in color.

CopA and CopC are periplasmic proteins, and have been purified. Initial results indicate that they bind ~11 and ~1 atom of copper per protein, respectively.[22] The accumulation of copper in the cells cannot, however, be attributed to these proteins alone as their concentration in the periplasm does not increase with media or cellular copper concentrations.

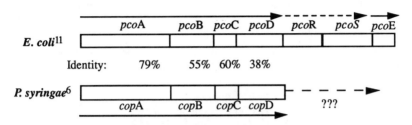

Figure 2. Comparison of the *pco* and *cop* determinants.
Percent identity in predicted amino acid sequences for *cop* and *pco* proteins are shown.

It has been noted that CopA contains sequences with strong homologies to the type 1 copper site in multicopper oxidases.[23] The multicopper oxidase family of proteins, including ascorbate oxidases, laccases and ceruloplasmins, contain one type 1, one type 2 and two type 3 copper ions. Closer examination shows that CopA contains identities or strong homologies to all four of the copper binding regions of the multicopper oxidases. Ligands to all four of the oxidase coppers are completely conserved in CopA. Partial alignments of CopA with the multicopper oxidases are shown in Figure 3. It will be interesting to see if CopA and the analogous PcoA have oxidase activity.

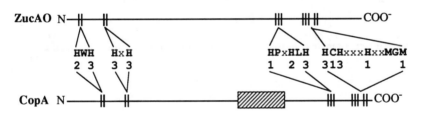

Figure 3. Identities Between Ascorbate Oxidase and CopA/PcoA
Comparison of the four highly conserved copper binding regions of the multicopper oxidases with CopA. Amino acid sequences are shown for zucchini ascorbate oxidase (**ZucAO**, total 552 a.a., from Messerschmidt et al.)[24,25] and *P. syringae* pv. *tomato* CopA (**CopA**, total 609 a.a., from Mellano and Cooksey)[13] with x representing conservative changes. Beneath each set of sequences, numbers indicate the residues that ligate the three different types of copper, Types 1, 2 and 3, as demonstrated in the crystal structure of zucchini ascorbate oxidase.[24,25] Shaded box in CopA represents the location of repeated putative copper binding sequences (see below).

In addition to the conserved multicopper oxidase copper binding regions, CopA contains several repeats of a putative copper binding sequence, AspHisXaaXaaMetXaaXaaMet (see Figure 4, below).[13] Similar repeated sequences are not observed in any of the other multicopper oxidases and this sequence appears only once in all other proteins in the common databases. It has been postulated that this repeated motif is responsible for binding the remaining copper ions observed in purified

CopA.[22] It might be postulated that CopA and CopC serve as the principal reservoirs for accumulation and sequestration of copper ions in the periplasm of resistant cells. This does not appear to be the case, however, as concentrations of CopA and CopC in resistant cells did not increase continuously with exposure to increasing concentrations of copper. In fact, the percentage of accumulated copper that could be accounted for by binding of 11 copper ions to CopA and one to CopC actually decreased dramatically from ~20% to 7% at high media copper concentrations and thus high concentrations of accumulated copper.[22] Perhaps instead CopA serves as an initial copper binding reservoir in the periplasm, and to deliver copper to other sites, perhaps on the membranes, where copper may be permanently accumulated and sequestered.[22]

> Asp His Gly Ser Met Asp Gly Met
> Asp His Ser Lys Met Ser Thr Met
> Asp His Gly Ala Met Ser Gly Met
> Asp His Gly Ala Met Gly Gly Met

Figure 4. Putative copper binding motifs in CopA.

THE MECHANISM OF COPPER RESISTANCE

In light of the high degree of identity observed between the sequences of *copABCD* and *pcoABCD*, one would expect the mechanisms of the two resistance systems to be similar as well. As we have just discussed, the mechanism of copper resistance in *P. syringae* is one of accumulation and sequestration of copper. So much copper is accumulated that colonies of the resistant strain of *P. syringae* are visibly blue when grown on agar plates supplemented with copper.[6] On the other hand, Rouch, Lee and Camakaris suggest that the mechanism of copper resistance in *E. coli* bearing the copper resistance plasmid pRJ1004 involves enhanced *export* of copper and not reduced uptake or increased accumulation, based on ^{64}Cu uptake studies.[10,15,26] It has been postulated that the copper resistance system involves export of copper in some "modified" form that is unavailable to the uptake system.[9,10,16] This was originally postulated on the basis of the dark brown appearance of colonies of the resistant strain of *E. coli* growing on agar plates supplemented with copper. Interestingly, copper resistant strains bearing a subclone of the *pco* copper resistance determinant (ED8739/pPA173)[15] rather than the large plasmid originally isolated from piggery effluent (pRJ1004)[27] do not appear as brown colonies on high copper plates, and look more like the parental strain.

Accumulation versus Export
The accumulation of copper by parental and copper resistant strains of *E. coli* was determined for cultures of both strains grown at copper concentrations just sub-toxic to the parental strain. Cells were harvested and dried, the mass of dry cells was determined, cells were dissolved in concentrated acid and copper concentrations determined by atomic absorption. The results shown in Figure 5 clearly show that the copper resistant strain accumulates significantly less copper than the parental strain. This is in dramatic contrast to the resistance mechanism of the genetically similar *cop* operon in *P. syringae* in which the copper resistant cells accumulate significantly <u>more</u> copper than the parental strain, as demonstrated by similar experiments published by Cooksey and Azad for *P. syringae* which are also pictured in Figure 5.[21] Thus copper resistance encoded by the *cop* determinant in *P. syringae* operates by accumulation or sequestration of copper, whereas that encoded by the *pco* determinant in *E. coli* operates by enhanced export of copper. This fact is particularly interesting and puzzling considering the large degree of genetic homology between these two plasmid-borne resistance systems.

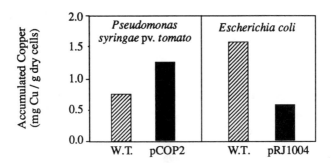

Figure 5. Accumulation vs. Export

Accumulated copper in cells grown at copper concentrations just sub-toxic to the non-resistant wild-type strains (W.T.) were measured by atomic absorption and are indicated by cross hatched bars for wild-type strains, solid bars for copper resistant strains. *E. coli* media: A-media + 100 μM each amino acid + 0.3 mM $CuSO_4$. *P. syringae* data from Cooksey and Azad,[21] media: MGY + 0.3 mM $CuSO_4$.[21]

Precipitation of CuS

In order to examine more closely the mechanism of resistance and to determine if a "modified" complex of copper is indeed exported from the copper resistant cells, we have undertaken the examination of copper complexes produced in liquid cultures by copper resistant *E. coli*. In the rich culture medium LB (10 g/l tryptone, 5 g/l yeast extract, 10 g/l NaCl)[28], the plasmid pRJ1004 confers resistance to up to 20 mM $CuSO_4$, as compared to tolerance up to 8 mM $CuSO_4$ for the parental strain.[29] At copper sulfate concentrations of greater than 4 mM $CuSO_4$, the resistant strain (harboring pRJ1004) produces a small amount of a black precipitate that co-sediments with the cells upon centrifugation of the culture. The parental strain does not produce any of the black precipitate. Although not thought to be a direct product of the copper resistance mechanism, the black precipitate has been isolated and characterized as microcrystalline CuS, the mineral covellite, by elemental analysis, EPR and X-ray powder diffraction (see Figure 6, below). Elemental analysis (18.37% C, 2.48% H, 4.43% N, 19.88% S) also suggests that some peptide or peptides are associated with the microcrystalline precipitate. Amino acid analysis indicates the presence of Glu/Gln, Gly, Cys, Thr, Ala, and Pro. The predominant presence of Glu/Gln, Gly and Cys suggests that glutathione might be one component of the peptide/CuS precipitate. Glutathione does not, however, play a crucial role in copper resistance (*vide infra*).

Similar precipitation of copper sulfide has been observed in a copper resistant strain of *Mycobacterium scrofulaceum*.[30] In that case the resistance mechanism has been attributed to the increased production of H_2S leading to the precipitation copper sulfide. In the case of the *pco* determinant, however, copper resistance cannot be attributed to this mechanism. No significant increase in H_2S has been observed, and the formation of CuS is not observed in minimal media. Copper resistant cells growing in high concentrations (concentrations toxic to the parental strain) of copper sulfate in various minimal media (M9, A media, or either supplemented with amino acids)[31] do not produce any of the copper sulfide precipitate, although they are obviously expressing the copper resistance system. Finally, the amount of copper precipitated as CuS is only a very small fraction of the total amount of copper present in media. The production of CuS may be an interesting result of copper resistance in rich media, but it is not the principle mechanism of copper resistance encoded by the *pco* determinant.

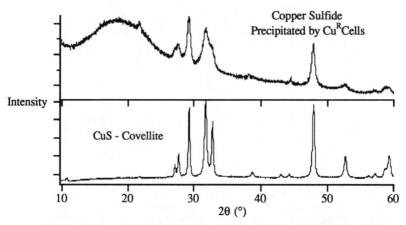

Figure 6. X-ray Powder Diffraction Patterns: Covellite (CuS) and Precipitate from Copper Resistant Cell Cultures in LB + 16 mM CuSO$_4$

In Search Of A Modified Exported Copper Complex

Spent media supernatants from cultures of the resistant strain and the parental strain were examined for a soluble copper complex unique to the resistant strain. Media supernatants were fractionated by gel filtration and ion exchange chromatography, before or after concentration, and copper species were monitored by atomic absorption. Gel filtration (Sephadex G75, Pharmacia) in the non-chelating buffer HEPES revealed a soluble copper complex of ~ 5 kilodaltons in the supernatants of the resistant strain, and was initially thought to be unique to that strain. Closer examination, however, demonstrated that this species is also present in the supernatants of the parental strain lacking *pco* grown at just below the toxic threshold copper concentration, and thus it is not unique to the resistant strain. See Figure 7 below.

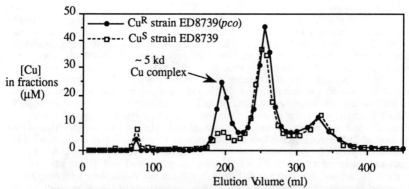

Figure 7. Sephadex G75 Fractionation of Culture Supernatants

Copper concentration in fractions of Sephadex G75 (Pharmacia) fractionation of concentrated (YM1, Amicon) media supernatants from cultures of parental (ED8739) and copper resistant (ED8739/pRJ1004) strains of *E. coli* in M9 medium[28] supplemented with 100 μM each amino acid plus 100 μM CuSO$_4$ is plotted versus elution volume. Eluent is 10mM HEPES, 50 mM NaCl pH 7.5.

Higher resolution gel filtration was attempted using Sephacryl S-100 (Pharmacia). Due to copper binding to the resin, however, it was necessary to use a the potentially chelating (Tris-HCl, pH 8.5) buffer and higher ionic strength (200 mM NaCl). Under these conditions no difference was observed in the UV or copper concentration profiles of the parental and copper resistant strains (see Figure 8, below). Thus we have been unable to identify a copper complex from media supernatants that is unique to the resistance mechanism. While the ~ 5 kd. complex observed in G75 fractionations does appear to be present in higher concentrations in resistant cultures, it is not unique and is not stable to isolation in the presence of even a mildly chelating buffer.

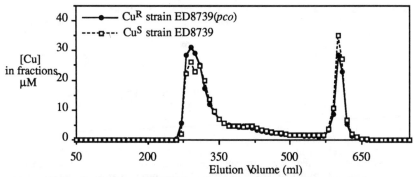

Figure 8. Sephacryl S-100 Fractionation of Culture Supernatants
Copper concentration in fractions of Sephacryl S-100 (Pharmacia) gel filtration fractionation of media supernatants from cultures of parental (ED8739) and copper resistant (ED8739/pRJ1004) strains of *E. coli* in A-media[31] supplemented with 100 μM each amino acid plus 350 μM CuSO4. Eluent is 20 mM Tris-HCl, 200 mM NaCl pH 8.5.

Another possible means for isolation of a unique copper complex could be extraction of media supernatants with organic solvents. Such a strategy has proven effective in the isolation of bacterial iron chelating agents, the siderophores.[32] Supernatants of parental and resistant strain cultures at just sub-toxic copper concentrations were extracted with a variety of solvents: ethyl acetate, chloroform, ether, 1:1 phenol:chloroform, and benzyl alcohol. While ethyl acetate and phenol/chloroform did extract some copper species, no differences in extractable copper were observed between the parental and resistant strains.

Glutathione Depletion Does Not Inhibit Copper Resistance
Glutathione is a likely and abundant ligand for copper *in vivo*, and might be likely to play a role in an exported putative "modified" form of copper. The role of glutathione in copper resistance was therefore tested by inhibition of glutathione synthesis. Two common glutathione synthase inhibitors were employed, L-buthione-S,R-sulfoximine (BSO) and dethylmaleate (DEM).[33,34,35] Both compounds slowed but did not disrupt growth of both the parental and resistant strains at low or just sub-toxic concentrations of copper. Similarly, both compounds slowed, but did not inhibit growth of the resistant strain in high concentrations of copper. While cellular concentrations of glutathione were not monitored directly in these experiments, they do suggest that glutathione is not a critical component of the copper resistance mechanism.

Media Copper is Not Rendered Harmless by Resistant Strain

It was necessary to test directly the hypothesis that copper resistance functions by export of a modified copper complex that is unable to re-enter the cell via the normal uptake pathways. If this hypothesis is true, then the copper in media supernatants of the resistant strain should be partially detoxified, i.e. a portion of the media copper should be non-toxic. It follows that if media supernatants from the resistant strain are used as the copper source for the preparation of new media (in place of CuSO4), the parental strain should be able to survive higher total concentrations of copper since part of the added copper would be non-toxic.

Supernatants from cultures of the resistant strain in copper concentrations just above the toxic threshold for the parental strain were sterilized by filtration and re-inoculated directly with either the parental or resistant strain. While the resistant strain was able to grow, the parental strain was not. Concentrated supernatants from resistant strain cultures were used to supplement fresh media to various concentrations. No change was observed in the copper tolerance levels of the parental strain. The copper in resistant strain supernatants is not significantly detoxified. This strongly suggests that the mechanism of copper resistance is not detoxification of the media copper, but perhaps is just simple export of copper ions to keep internal concentrations low.

Conclusions

Despite the extensive genetic similarities and identities in predicted protein sequence between these two plasmid-encoded bacterial copper resistance systems, there remains a fundamental difference in the observed mechanisms of resistance. In *P. syringae* pv. *tomato*, the *cop* determinant provides resistance to copper via accumulation and sequestration of copper in the periplasm. In *E. coli*, the *pco* determinant provides resistance to copper via enhanced export of copper from the cell. Reconciliation of the basic difference in strategy with the extensive similarity of the proteins involved will have to wait until more detailed examinations of those proteins and their functions are completed. These and other investigations will continue to contribute to our growing understanding of the genetics and molecular mechanisms of bacterial metal resistance and metal ion homeostasis.

ACKNOWLEDGMENTS

JWB is a Howard Hughes Medical Institute Predoctoral Fellow. This work was supported by the National Institutes of Health grant GM45972 to TVO, by Medical Research Council grant G.9025236CB and Agricultural and Food Research Council grant AG6/537 to NLB, and by Australian Research Council grant A09030665 and the Murdoch Institute to BTOL and JC. We thank J. Realme for assistance, D. Ralston and G. Munson for helpful advise and D. Cooksey for helpful discussions and for communicating results prior to publication.

REFERENCES

1. C. T. Walsh, M. D. Distefano, M. J. Moore, L. M. Shewchuk, G. L. Verdine, *FASEB J.*, **2**, 124-130 (1988).
2. T. K. Misra, *Plasmid*, **27**, 4-16 (1992).
3. S. Silver, M. Walderhaug, *Microbiol. Rev.*, **56**, 195-228 (1992).
4. C. L. Bender, D. A. Cooksey, *J. Bacteriol.*, **169**, 470-474 (1987).
5. D. A. Cooksey, H. R. Azad, J.-S. Cha, C.-K. Lim, *Appl. Environ. Microbiol.*, **56**, 431-435 (1990).
6. D. A. Cooksey, *Annu. Rev. Phytopathol.*, **28**, 201-219 (1990).
7. D. Rouch, J. Camakaris, B. T. O. Lee, in *Metal Ion Homeostasis: Molecular Biology and Chemistry, UCLA Symposium on Molecular and Cellular Biology, New Series*, D. Winge, D. Hamer, Eds. (Alan R. Liss Inc., New York, 1989),

vol. 98, pp. 469-477.
8.   S. D. Rogers, *J. Bacteriol.*, **173**, 6742-6748 (1991).
9.   N. L. Brown, D. A. Rouch, B. T. O. Lee, *Plasmid*, **27**, 41-51 (1992).
10. D. Rouch, B. T. O. Lee, J. Camakaris, in *Metal Ion Homeostasis: Molecular Biology and Chemistry, UCLA Symposium on Molecular and Cellular Biology, New Series*, D. Winge, D. Hamer, Eds. (Alan R. Liss Inc., New York, 1989), vol. 98, pp. 439-446.
11. S. Silver, B. T. O. Lee, N. L. Brown, D. A. Cooksey, in *Chemistry of the Copper and Zinc Triads*, A. J. Welch, Ed. (Royal Society of Chemistry, 1992).
12. N. L. Brown, B. T. O. Lee, D. A. Rouch, S. R. Barrett, manuscript in preparation.
13. M. A. Mellano, D. A. Cooksey, *J. Bacteriol.*, **170**, 2879-2883 (1988).
14. M. A. Mellano, D. A. Cooksey, *J. Bacteriol.*, **170**, 4399-4401 (1988).
15. D. A. Rouch, Doctoral Thesis, University of Melbourne (1986).
16. N. L. Brown, J. Camakaris, B. T. O. Lee, T. Williams, A. P. Morby, J. Parkhill, D. A. Rouch, *J. Cellular Biochem.*, **46**, 106-114 (1991).
17. C. W. Ronson, B. T. Nixon, F. M. Ausubel, *Cell*, **49**, 579-581 (1987).
18. B. Magasanik, *Trends Biochem. Sci.*, **13**, 475-479 (1988).
19. J. B. Stock, A. M. Stock, J. Mottonen, *Nature*, **344**, 345-400 (1990).
20. J. S. Parkinson, E. C. Kofoid, *Annu. Rev. Genet.*, **26**, in press (1992).
21. D. A. Cooksey, H. R. Azad, *Appl. Environ. Microbiol.*, **58**, 274-278 (1992).
22. J.-S. Cha, D. A. Cooksey, *Proc. Natl. Acad. Sci. USA*, **88**, 8915-8919 (1991).
23. C. Ouzounis, C. Sander, *FEBS Lett.*, **279**, 73-78 (1991).
24. A. Messerschmidt, A. Rossi, R. Ladenstein, R. Huber, M. Bolognesi, G. Gatti, A. Marchesini, R. Petruzzelli, A. Finazzi-Agro, *J. Mol. Biol.*, **206**, 513-529 (1989).
25. A. Messerschmidt, R. Huber, *Eur. J. Biochem.*, **187**, 341-352 (1990).
26. D. Rouch, J. Camakaris, B. T. O. Lee, R. K. J. Luke, *J. Gen. Microbiol.*, **131**, 939-943 (1985).
27. T. J. Tetaz, R. K. Luke, *J. Bacteriol.*, **154**, 1263-1268 (1983).
28. J. Sambrook, E. F. Fritsch, T. Maniatis, *Molecular Cloning: A Laboratory Manual* (Cold Spring Harbor Laboratory Press, Cold Spring Harbor, NY, 1989).
29. B. T. O. Lee, N. L. Brown, S. Rogers, A. Bergemann, J. Camakaris, D. A. Rouch, in *Metal Speciation in the Environment*, J. A. C. Broekaert, S. Cucer, F. Adams, Eds. (Springer-Verlag, Berlin, 1990), vol. G23, pp. 625-632.
30. F. X. Erardi, M. L. Failla, J. O. Falkinham, *Appl. Environ. Microbiol.*, **53**, 1951-1954 (1987).
31. F. M. Ausubel, R. Brent, R. E. Kingston, D. D. Moore, J. G. Seidman, J. A. Smith, K. Struhl, Eds., *Current Protocols in Molecular Biology* (John WIley & Sons, New York, 1987).
32. J. B. Neilands, in *Siderophores from Microorganisms and Plants* (Springer-Verlag, Berlin, 1984), vol. 58, pp. 1-23.
33. O. W. Griffith, A. Meister, *J. Biol. Chem.*, **254**, 7558-7560 (1979).
34. J. H. Freedman, M. R. Ciriolo, J. Peisach, *J. Biol. Chem.*, **264**, 5598-5605 (1989).
35. W. R. Moore, M. E. Anderson, A. Meister, K. Murata, A. Murata, *Proc. Natl. Acad. Sci. USA*, **86**, 1461-1464 (1989).

CUPROUS-THIOLATE POLYMETALLIC CLUSTERS IN BIOLOGY

D.R. Winge[a], C.T. Dameron[a], G.N. George[b], I.J. Pickering[b] and I.G. Dance[c]

[a] Departments of Medicine and Biochemistry, University of Utah Medical Center, Salt Lake City, UT 84132
[b] Stanford Synchrotron Radiation Laboratory, P.O. Box 4349 Bin 69, Stanford University, Stanford, CA 94309
[c] School of Chemistry, University of New South Wales, Kensington, N.S.W. 2033, Australia

Copper-sulfur (CuS) multinuclear clusters exist in biological macromolecules that function in a variety of cellular processes ranging from copper ion buffering, signal transduction to copper ion storage. The sulfur ligands are typically provided exclusively by cysteinyl thiolates and proteins and peptides containing CuS clusters have an abundance of cysteine residues. A common sequence motif in these polypeptides is Cys-Xaa-Cys or Cys-Xaa-Xaa-Cys in which Xaa represents any other amino acid.

Copper-sulfur clusters exist in a family of metallothionein (MT) proteins. These cysteine-rich polypeptides function in part in the buffering the cellular cytoplasm of copper ions to minimize any Cu-induced toxicity. Cells respond to changes in the copper uptake flux by modulating the intracellular concentration of MT through expression of MT genes.[1-3] The only apparent phenotype of genetic disruption of the MT encoding gene(s) in yeast is exquisite sensitivity to copper-induced cytotoxicity.[4,5] In MT's role in cellular copper resistance, the CuS cluster in MT is the sequestration form of copper within the cell. No information exists on the subsequent fate of MT-bound Cu ions in yeast. MT is also postulated to function in metal homeostasis in animal cells, although direct roles remain to be established.[6] CuMT may also participate in intracellular Cu(I) channeling and/or storage within cells. There is a paucity of information on the mechanism of copper ion channeling within cells to provide copper ions for the myriad of biological functions. CuMT has been suggested to function in this pathway.[7] Fetal livers of most animal species contain high concentrations of CuMT that may be a storage form of copper for subsequent development.

From K.D. Karlin and Z. Tyeklár, Eds., *Bioinorganic Chemistry of Copper*
(Chapman & Hall, New York, 1993).

The expression of MT genes is metalloregulated at the level of transcription. Metalloregulation occurs in yeast through intracellular copper-sensor molecules that mediate the transcriptional activation of MT gene expression. Formation of a polynuclear CuS cluster in the transcriptional activator proteins, ACE1 in *Saccharomyces cerevisiae* and AMT1 in *Candida glabrata*, is required for the functioning of these proteins in the expression of MT genes.[8-12] The activation of ACE1 and AMT1 is specific for Cu(I) and Ag(I) ions.[8,9,11] Other metal ions bind to ACE1, but these metallo-complexes are inactive in transcriptional activation.

A CuS polynuclear cluster coated with glutathione-related isopeptides occurs in *Schizosaccharomyces pombe*.[13] The isopeptides, $(\gamma Glu-Cys)_n Gly$, differ from glutathione in having multiple $(\gamma Glu-Cys)$ dipeptide units ranging from 2-6.[13] As this yeast does not contain MT, the intracellular sequestration of Cu(I) ions is achieved by CuS cluster formation within isopeptide-coated particles. The sulfurs in these Cu(I) clusters are provided exclusively by cysteinyl sulfurs.[13] This is in contrast to the cadmium:sulfur clusters that form in this yeast. Sulfur ligands for Cd(II) ions are provided by both cysteinyl sulfur and acid-labile sulfide ions.[14]

In addition, a number of cysteine-rich Zn-proteins are known that undergo facile metal exchange reactions with Cu(I) in vitro. One example is the papilloma virus E7 protein that in conjunction with a second protein (E6) appear to be responsible for mammalian cell transformation by papilloma viruses.[15,16] If Cu/Zn exchange leads to the formation of Cu-proteins in vivo, the physiological function of proteins like E7 may be altered.

Structural details are not currently available for copper thiolate polymetallic clusters in proteins, but chemical precedence exists for multinuclear copper-thiolate species.[17] Structures of several synthetic CuS clusters have been elucidated.[17] The chemistry of these clusters is relevant to the chemistry of CuS clusters in biological macromolecules. The structures of CuS multinuclear clusters in biology is important as the cluster structure is the basis for function in proteins such as CuACE1. This chapter will compare the known chemistry of synthetic CuS polymetallic clusters to the chemistry of CuS polymetallic clusters in biological macromolecules.

Synthetic CuS Clusters:

Copper thiolate compounds share the richness that exists in the inorganic literature on small metal thiolate compounds.[18] Monometallic copper complexes with only thiolate ligands are rare, and Cu(I) thiolate clusters $[Cu_x(SR)_y]^{x-y}$ dominate this chemistry. More than 10 cluster types have been synthesized and structurally characterized.[17,19-33] Polymetallic coordination clusters are commonly observed in reactions of cupric salts with limited excesses of monothiolate ligands in aprotic solvents, in reactions generalized in the following equation:

$$xCu(II) + (x+y)RS^- \rightarrow [Cu_x(SR)_y]^{x-y} + x/2(RSSR)$$

Reactions of thiols or thiolates with Cu(II) causes reduction to Cu(I) except under very specialized circumstances.[34-37] Cu(I) thiolate compounds $[CuSR]_n$ are commonly non-molecular in structure and insoluble, except in cases where the substituent R is bulky.[38] However, the presence of excess thiolate usually causes $[CuSR]_n$ compounds to redissolve with the formation of anionic complexes $[Cu_x(SR)_y]^{x-y}$. Where there is a large excess of thiolate the complexes are monometallic, $[Cu(SR)_y]^{1-y}$, where y is usually 2.[19] The anionic clusters occur in the intermediate range of excess thiolate.

Polymetallic CuS clusters have been characterized with the following stoichiometries: (references to these clusters are listed in parentheses)

$[Cu_3(SRS)_3]^{3-}$ chelating dithiolate and tiangular Cu_3 (29)
$[Cu_4(SR)_4]$ approximately square Cu_4 (39)
$[Cu_4(SR)_6]^{2-}$ tetrahedral Cu_4 (40)
$[Cu_5S_6]^-$ trigonal bipyramidal Cu_5 (22)

$[Cu_5S_7]^{2-}$ edge-bridged tetrahedral Cu_5 (23)
$[(Cu/Au)_6(SR)_8]^{2-}$ (30,45)
$[Cu_7(SRS)_4(SR)]^{2-}$
$[Cu_8(SR)_8]$ Cu_8 forms an approximate cube (39)
$[Cu_8(SR)_{12}]^{4-}$ Cu_8 cube (24-26)
$[Cu_{12}(SR)_{12}]$

In recent years arylthiolate ligands with sterically encumbering substituents in the ortho positions have been introduced into metal thiolate chemistry. The consequences of this encumbrance are reduced bridging of metal atoms, clusters that are restricted in size by the ligand volume, and in some cases reduced coordination numbers. These compounds have lesser biological relevance and the following descriptions emphasize clusters with ligands that are sterically and electronically akin to cysteine.

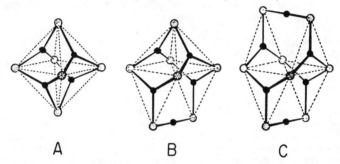

Figure 1. CuS polymetallic clusters are shown with the following stoichiometries: Cluster A is the tetracopper $Cu_4(SR)_6$ cage, cluster B is the pentacopper $Cu_5(SR)_7$ cluster and cluster C is the $Cu_6(SR)_8$ cluster.

$Cu_4(\mu\text{-}SR)_6$ Clusters: This is the most commonly formed structure type (Fig. 1A). The alternative tetrametallic, $Cu_4(\mu\text{-}SR)_4$ in approximately square array occurs with hindered thiolate ligands,[39] or with additional coordination by heteroligands,[41,42] and is less relevant as a model for biological clusters. The $Cu_4(SR)_6$ clusters are formed with alkyl or aryl monothiolate or chelating thiolate ligands.[19-21,27,33] This structure type is also formed with thioketone ligand types,[43] but these have little biological relevance. The cluster polyhedron consists of an octahedron of bridging thiolates intersected by a tetrahedron of Cu(I) atoms. The alternate description is that each edge of the Cu_4 tetrahedron is doubly-bridged by thiolate. The coordination around each Cu(I) atom is approximately trigonal planar.

    The Cu-S-Cu angles in this idealized geometry would be very acute, and are relieved to a mean of 73° by displacement of the Cu(I) atoms slightly outwards by 0.04 to 0.13 Å from the plane defined by three bonded sulfur neighbors. The mean Cu-S bond distance in a variety of Cu_4S_6 clusters is 2.28 ± 0.02 Å and the Cu-Cu separation averages 2.75 Å.[27] Idealized Cu-Cu distances of 2.64 Å were calculated for a Cu_4S_6 cluster with trigonal-planar coordination and Cu-S bond lengths of 2.29 Å.[19] The observed elongation corresponds to the expansion of the Cu-S-Cu angles. The question of whether Cu-Cu bonding is significant at this distance is addressed below. The coordination at Cu is not exactly trigonal, with a variability of S-Cu-S angles from 101 to 142° and a range of Cu-S distances from 2.24 to 2.35 Å. The reason for this derives from the orientation of the substituents R, which must have a symmetry lower than that of the Cu_4S_6 core. Interactions between the substituents affect the S-Cu-S angles, and in trigonal $M(SR)_3$ coordination there is a strong correlation between the M-S distance and the opposite S-M-S angle.[17] Should the $Cu_4(S\text{-}Cys)_6$ cluster occur in copper proteins, similar distortions due to cluster topology may occur as well as distortions imposed by the protein structure.

$Cu_5(\mu\text{-SR})_6$ Clusters: This cluster contains a trigonal bipyramid of Cu atoms with thiolate ligands doubly bridging each of the axial-equatorial edges.[22] Consequently, the two axial Cu atoms are trigonally coordinated, and the three equatorial Cu atoms are digonally coordinated. The S_6 polyhedron can twist about its threefold axis, with little effect on the copper coordination, and is partway between a trigonal prism and an octahedron in $[Cu_5(SBu^t)_6]^-$. The orientations of the thiolate substituents need not reduce the core symmetry in this structure type. In a comparable $Ag_5(SR)_6$ structure with a solubilized alkyl thiolate ligand $(SCH_2CH_2CH_2NHMe_2{}^+)$ the S_6 polyhedron is an undistorted trigonal prism.[44] Mean values for the $Cu^{dig}\text{-}S$ bonds and $Cu^{trig}\text{-}S$ bonds are 2.163 ± 0.004 Å and 2.273 ± 0.005 Å, respectively.[22] The $S\text{-}Cu^{dig}\text{-}S$ angle is 171° with the bending moving the Cu^{dig} atoms towards the center of the cluster. However the $Cu^{dig}\text{-}Cu^{dig}$ distances, 3.23 ± 0.04 Å are larger than Cu-Cu bonding distances.

$Cu_5(\mu\text{-SR})_7$ Clusters: This structure shown in Fig. 1B is regarded as an expansion of $Cu_4(SR)_6$, by insertion of CuSR at one vertex, generating a digonal site.[23] It occurs as $[Cu_5(SR)_7]^{2-}$, and as the silver homolog. The orientation of the thiolate substitutents necessarily destroy most of the symmetry possible for the core of this cluster, and consequently there are distortions of the core. S-Cu-S angles range from 105 to 138°, two of the Cu^{trig} atoms are within 0.06 Å of their coordination plane but the other two are displaced by 0.2 Å. The distances between bridged Cu atoms range from 2.65 to 3.13 Å. The $S\text{-}Cu^{dig}\text{-}S$ angle is 175°.

$Cu_6(\mu\text{-SR})_8$ Cluster: Additional application of the expansion process that generated $[Cu_5(SR)_7]^{2-}$ from $[Cu_4(SR)_6]^{2-}$, forms the structure $[Cu_6(SR)_8]^{2-}$(Fig. 1C).[45] This cluster type with only Cu(I) ions has not been synthesized, but a corresponding cluster with Cu(I) in the trigonal sites and Au in the digonal sites has been described.[30]

$Cu_8(\mu\text{-SR})_{12}$ Clusters: Cages of C_8S_{12} have been described with various chelating thiolate ligands.[24-26] These cages consist of a cube of Cu(I) atoms inside an icosahedron of sulfur atoms. Each Cu(I) is coordinated by three sulfur atoms and each sulfur atom coordinates two Cu(I) atoms on an edge of the cube.[24] The cage sizes vary with the ligand. The Cu_8 cube is larger in the D-penicillamine complex (mean Cu-Cu distances of 3.2 Å) than in the eight copper cube formed with dithiolate complexes (mean Cu-Cu distances of 2.84 Å).[24,25] There is negligible net bonding between Cu atoms.[24,46]

$Cu_{12}(SR)_{12}$ Clusters: The structure of $Cu_{12}(SC_6H_4\text{-}2\text{-}SiMe_3)_{12}$ is particularly significant in defining variations in metrical data due to variations in the coordination numbers of Cu and S. Ther structure contains six Cu atoms with distorted trigonal planar coordination, three Cu with almost linear digonal coordination, three Cu with slightly bent (163°) digonal coordination, six triply-bridging and six doubly-bridging arylthiolate ligands. Overall the cluster is like a paddlewheel with threefold symmetry about the axle. There are two parallel planes linked by the six doubly-bridging Cu atoms; each plane with composition Cu_3S_6 contains a triangle of Cu, bridged by thiolates to form an approximate hexagon, and with terminal thiolates to complete the threefold coordination at each Cu. Using superscripts to denote the coordination numbers, ther following bond lengths classes are differentiated: $Cu^2\text{-}S^2$ 2.142-2.169 Å, $Cu^2\text{-}S^3$ 2.189-2.212 Å, $Cu^3\text{-}S^2$ 2.198-2.220 Å, $Cu^3\text{-}S^3$ opposite smaller angle 2.243-2.263 Å, and $Cu^3\text{-}S^3$ opposite larger angle 2.301-2.315 Å.

The Cu-Cu distances in polycopper thiolate clusters range upwards from 2.65 Å. The question of possible Cu-Cu bonding, and its control of the Cu-Cu distances, arises in many clusters. The Cu-Cu distances are often found to be about the same as the separation 2.55 Å between Cu atoms in the metal. However, calculations indicate that even at Cu-Cu distances of 2.4 Å the binding energy is relatively small: Cu-Cu interactions are relatively soft.[46,47] In making comparisons with the Cu-Cu distances in copper metal it is necessary to recognize that the coordination number in the metal (12) is much higher than in the clusters, and therefore, that the cohesive energy per Cu-Cu interaction is much reduced. In Cu(I) clusters the interactions with the bridging ligands are generally considered to detemine the geometry.[47] In clusters such as $Cu_4(SR)_6$ and $Cu_5(SR)_6$ it is observed that there is systematic movement of the Cu atoms relative to the centroids of their donor atoms, and these have been interpreted as indicators of Cu-Cu interactions. However, it is more probable that the requirements of the bonding with the bridging ligands are determining these geometries. Therefore in biological copper cysteinate clusters it is not possible to prescribe Cu-Cu distances: the expectation is that these will be determined by the protein structure.

STRUCTURAL PRINCIPLES FOR POLYCOPPER THIOLATE CLUSTERS:

We can assemble the following generalizations for polycopper thiolate clusters as they occur in biochemical systems:
1. The clusters are held together by the bridging thiolate ligands, which are commonly doubly-bridging. There are no examples of polycopper thiolate clusters containing terminal rather than bridging thiolate ligands.
2. The Cu(I) coordination numbers can be two (digonal) or three (trigonal), and there is no evidence of preference between them. Both can occur in one cluster. Coordination numbers higher than three occur only with uncharged ligands, and it is likely that such coordination would be secondary with elongated bonds.
3. With limited connectivity due to the low coordination numbers, and with distortable coordination geometry, the $Cu_x(SR)_y$ clusters are not rigid, and can be subject to substantial distortion.
4. Doubly bridging thiolate ligands are pyramidal at sulfur, and necessarily reduce the symmetry of the Cu-S(R)-Cu bridge. This in turn can reduce the symmetry of the full cluster. Because the clusters are not rigid, when substituent configurations destroy the symmetry at trigonal copper sites there are substantial distortions of the S-Cu-S angles and the Cu-S distances.
5. Cu-S bond distances are dependent on the coordination numbers. With doubly-bridging thiolate ligands Cu^{trig}-S is about 2.27 Å and Cu^{dig}-S is about 2.16 Å.
6. Trigonal $Cu(I)(SR)_3$ coordination can be distorted both within and out of the S_3 plane. S-Cu-S angles can range from 100-140°, and there is a well-defined correlation between the Cu-S distances and the S-Cu-S angles in trigonally coordinated Cu.[48] Cu atoms can be displaced by up to 0.15 Å from the S_3 plane.
7. Digonal $Cu(I)(SR)_2$ coordination can be distorted by up to 10° from linearity.
8. Cu-Cu bonding is a minor energetic factor in these clusters, and has little influence on overall geometry.
9. Structures of CuS clusters in proteins will be dominated by protein conformation factors and less influenced by cluster bonding requirements. The identity of the R substituent has negligible effect. In the $[Cu_4(SR)_6]^{2-}$ clusters with R as methyl the mean Cu-S distance was 2.278 ± 0.02 Å, compared to Cu-S distances of 2.284 ± 0.031 Å and 2.298 ± 0.028 Å for two clusters with R as phenyl.[29]

To date, there is no definitive structural evidence on biological CuS clusters; therefore X-ray absorption spectroscopy has been used to provide insight on this problem.

X-RAY ABSORPTION SPECTROSCOPY OF SYNTHETIC CuS CLUSTERS:

X-ray absorption near-edge fine structure (or XANES) contains structure due to bound-state transitions, and to more complex phenomena such as continuum resonances. XANES spectra provide a sensitive probe of electronic structure, although they are often difficult to interpret in a quantitative manner. The XANES spectra of a number of synthetic Cu(I) compounds have been studied by Kau et al.[49] This work revealed that both trigonally and digonally coordinated Cu(I) XANES possessed a low energy feature near 8984 eV assigned as a 1s→4p bound state transition.[49] The intensity of this edge feature was shown to be indicative of whether the coordination is digonal or trigonal.[49] The intensity of the edge feature was shown to be indicative of whether the coordination is digonal or trigonal.[49] For digonally coordinated compounds the edge feature is a well defined peak, with an intensity close to 100% of the maximal absorbance. The feature in trigonally coordinated compounds is reduced to a shoulder with an intensity of approximately 60% of the maximal absorbance. We measured XANES spectra for a series of synthetic Cu(I)-thiolate clusters. The spectra of clusters having stoichiometries Cu_4S_6, Cu_5S_6 and Cu_5S_7 are shown in Fig. 2. These compounds contain different proportions of digonal and trigonal Cu(I); the tetracopper cluster $[Cu_4(SPh)_6]^{2-}$ contains only trigonally coordinated Cu(I), Cu_5S_6 contains three digonal Cu(I) ions, and Cu_5S_7 contains one digonal Cu(I). Based on the work of Kau et al.[49] we might expect that the amplitude of the 8984 eV pre-edge feature to increase with the fraction of digonal Cu(I). It is immediately apparent from Fig. 2 that this is not the case. The 8984 eV feature is of comparable intensity for all three compounds, despite the change in average coordination of the copper.

Three possible causes for this apparent discrepancy will be considered. Blackburn et al.[50] reported that the intensity of the 8984 eV feature decreases with increased doming of trigonally coordinated Cu-N compounds (see also Fig. 1 in reference 56). Likewise, deviation from linearity for digonally coordinated Cu(I) ions may also attenuate the intensity of the edge feature. For the Cu(I)-thiolate clusters examined, there is slight deviation from linearity of digonal Cu(I), with S-Cu-S angles in the range of 171-175°.[22,23] Additionally, there is a slight doming of the trigonal coppers. Taken alone, it seems unlikely that these small distortions would account for the lack of observed effects of the Cu(I) coordination number. Previous studies of digonal Cu(I) complexes have been restricted to oxygen and nitrogen ligands.[49,50] It is likely that the increased covalency of the Cu-S bond (relative to Cu-N or Cu-O) willdiminish the 4p character of the excited state, thereby attenuating the 1s→4p edge feature. Lastly, it should be remembered that each Cu(I) site in the cluster has subtly different bond lengths and bond angles, resulting in small changes in electronic structure and corresponding changes in the XANES. Since the measured XANES represents an average of the spectra from individual coppers, then subtle differences in peak position may serve to wash out the intensity of certain features.

One or all three of these possibilities may contribute to the lack of notable changes in the XANES spectra of Fig. 2. In any case, it is clear that the copper K-edge XANES cannot be used to accurately predict the presence of digonal Cu(I) in Cu(I)-thiolate clusters.

Extended X-ray absorption fine structure (or EXAFS) spectroscopy is sensitive to the local atomic environrment of the absorber (in this case, Cu). In contrast to XANES, it is readily analyzed in a quantitative manner and can provide information on the type, number and mean bond distance of ligands in the first few coordination shells. It is important to note that the absolute acccuracy (as opposed to the precision) of EXAFS bond length determination is generally not limited by the data, but rather by the transferrability of the EXAFS phase and amplitude functions used in the curve fitting analysis. In general, the bond distances are determined to an accuracy of better than 0.02 Å, whereas coordination numbers are determined to about ± 25%. EXAFS of polymetallic metal clusters will be simply a mixture of EXAFS from individual metal sites and will provide an average picture of the coordination environment of the metal. The average Cu-S coordination number cannot be determined accurately enough by EXAFS to yield useful information about the fraction of digonal copper. On the other hand, the higher accuracy of the Cu-S bond length determination can be used to determine the fraction of digonal copper, as will be shown below.

The EXAFS of the three CuS model compounds is dominated by the first shell

Cu-S scattering, with weaker contributions from Cu-Cu outer shells, which are diagnostic of a cluster. Curve-fitting of the EXAFS was carried out with fixed coordination numbers based upon the crystal structures. Bond lengths determined by our fitting procedure (the EXAFS analysis will be described in full in a future publication) were found to be highly accurate, for example the mean Cu-S distance of $[Cu_4(SPh)_6]^{2-}$, was determined by EXAFS to be 2.282 Å, and this can be favorably compared with the mean distance from X-ray crystallography of 2.287 ± 0.027 Å. The outer shell Cu-Cu distances observed by EXAFS was also similar to the Cu-Cu distances observed from cyrstallography.

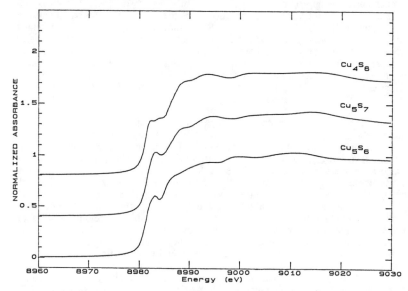

Figure 2. XANES of CuS Polymetallic Clusters. The stoichiometries of the clusters are listed.

EXAFS analysis for the three copper thiolate clusters reveals a well-defined correlation between the mean Cu-S bond distances and the fraction of digonal Cu(I) ions within the cluster (Fig. 3). The solid line, running through the points in Fig. 3, is the expected mean Cu-S bond length, based upon typical crystallographpic values of 2.28 Å and 2.16 Å for trigonal and digonal Cu-S bond distances, respectively.[22,23] The curve can be used to clearly estimate the fraction of digonal copper based upon EXAFS-derived bond lengths. This method can also be used for protein-bound copper thiolate clusters, assuming that the protein does not impose severe structural constraints at the copper sites.

If the protein does not impose constraints and angular distortions at trigonal sites in a CuS cluster, we conclude that the mean Cu-S distance may be an indicator of the presence of both digonal and trigonal coordination within the same cluster.

The outer shell Cu-Cu distances observed by EXAFS of the synthetic clusters were similar to Cu-Cu distances observed from the crystal structures. The mean Cu-Cu distance in the tetracopper cluster was found to be 2.740 Å by EXAFS and 2.74 ± 0.06 Å by crystallography.[29] In the Cu_5S_6 cluster Cu-Cu distances of 2.725 and 3.16 Å were found by EXAFS analysis, while mean Cu^{dig}-Cu^{trig} and Cu^{dig}-Cu^{dig} distances of 2.74 and 3.28 Å occur in the crystal structure.[22]

CuS CLUSTERS IN METALLOTHIONEINS:

NMR and crystal structures have been deduced for Cd(II) and Zn(II) polynuclear

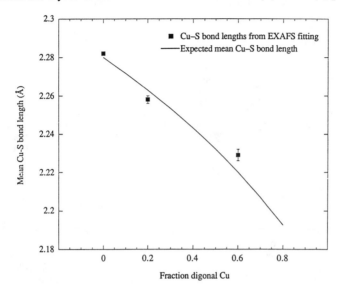

Figure 3. Correlation of the mean Cu-S bond distance and the fraction of digonal Cu(I) in a polymetallic CuS cluster. The solid line is the expected mean Cu-S bond lengths, based on crystallographic data.

clusters in mammalian metallothioneins.[51-54] A paucity of information is available on Cu(I) clusters in MTs. The metal binding stoichiometries and coordination geometries differ for Cu(I) and Cd(II) ions in MT, so the available structures do not provide a clear picture of the structure of CuMT. Spectroscopic analyses have been carried out on CuMTs with the following cluster stoichiometries: Cu_6S_9 from the β domain of rat liver, Cu_6S_7 from *Neurospora crassa*, Cu_7S_{10} from *S. cerevisiae* (formerly reported as a Cu_8 cage) and two incompletely defined CuMTs from *C. glabrata*.[55-59] Chemical properties of CuS clusters associated with these diverse MTs are similar. The copper ions are exclusively present in the cuprous state. Absorption bands in the ultraviolet are consistent with S→Cu charge transfer transitions. The presence of Cu(I) ions with thiolate ligands results in luminescent complexes. Photoexcitation of proteins containing Cu(I)-thiolate clusters with ultraviolet radiation gives rise to luminescence.[60] The emission is Stokes shifted to energies between 580-620 nm and appears to arise largely from a $3d^94s^1$ excited triplet state centered on Cu(I) as reported for synthetic CuS clusters.[60-66] There is no feature in the absorption spectrum corresponding to the excitation band. This is consistent with the expected low extinction coefficient of a spin-forbidden $3d^94s^1$ triplet → $3d^{10}$ singlet transition.[61] Therefore, metal centered transitions are likely to account for the observed emission in proteins containing CuS clusters, whereas a S→Cu ligand-metal charge transfer transition is responsible for the intense absorption in the ultraviolet. Unlike synthetic CuS clusters the emission in CuS-containing proteins is observed at room temperature in aqueous buffers. This is consistent with the Cu(I) thiolate cluster enfolded within a compact protein structure. Mononuclear CuS clusters are also emissive, so the observed luminescence is not a property of cluster formation.[65] At present, the emission properties of CuMT have a limited information content on the structure of the CuS cluster.

X-ray absorption edge spectroscopy of proteins containing CuS clusters indicates a pre-edge peak near 8984 eV for CuMTs from rat liver, dog hepatic lysosomes, *S. cerevisiae* and *N. crassa*.[55,56,67,68] The intensities of the pre-edge feature resembles the edge feature of synthetic CuS clusters with predominantly trigonal Cu(I) coordination. Prior to the XANES of the CuS polynuclear clusters described here, XANES analyses appeared to clearly indicate trigonal Cu(I) coordination in all CuMTs.[1] The well-studied CuMT from *S. cerevisiae* was predicted to contain a Cu_8S_{12} cluster with exclusive trigonal Cu(I) ion coordination.[56]

Heteronuclear (^1H-^{109}Ag) multiple quantum coherence NMR studies performed on AgMT from *S. cerevisiae* established the connectivities of ten cysteinyl residues with seven bound ^{109}Ag(I) ions.[57] As Ag(I) is isoelectronic to Cu(I), AgMT is expected to be isostructural to CuMT. This study suggested that the original proposal of the CuS cluster stoichiometry as Cu_8S_{12} may be wrong and furthermore that mixed coordination numbers may exist in CuMT. Based on the NMR results, a new model of M_7S_{10} emerged having two of the seven metal ions in digonal coordination with the remainder trigonally ligated.[57] Two digonal Ag(I) ions are consistent with the large chemical shift dispersion for Ag(I) nuclei in AgMT.[57]

The suggestion of possible digonal Ag(I) ions in yeast AgMT by NMR prompted a re-evaluation of EXAFS data of proteins with CuS clusters. There was no obvious indication from the x-ray edge spectrum of CuMT from *S. cerevisiae* that digonal Cu(I) existed in yeast MT. Yet, as shown above with synthetic CuS clusters, XANES appears insensitive to detect the presence of digonal Cu(I) ions in CuS clusters with mixed trigonal and digonal coordination.

We previously reported copper K edge and sulfur K edge EXAFS analyses of yeast CuMT yielding a mean Cu-S distance of 2.230 Å.[56] Copper K-edge EXAFS of a a truncated mutant of CuMT (CuT48) gave a very similar bond distance of 2.236 Å.[56] Re-analysis of the original EXAFS data using new phase and amplitude functions gave a mean Cu-S bond distance of 2.242 Å and a mean coordination number of 2.6. EXAFS data on a second yeast CuMT isolate gave a Cu-S distance of 2.250 Å. Comparison of these mean Cu-S distances with the synthetic Cu-S clusters as described above is suggestive of the presence of digonal Cu(I) ions in *S. cerevisiae* CuMT. From Fig. 3 we estimate that CuMT contains a fraction of digonal Cu(I) between 0.3 and 0.4. The fraction of digonal Cu(I) by HMQC NMR measurements of AgMT is likely to be 0.3. One Ag(I) resonance with a chemical shift range consistent with trigonal Ag(I) was observed, yet ligation to only two cysteinyl residues was seen. From chemical shift arguments that Ag(I) resonance is expected to be trigonal. A conclusion of digonal Cu(I) ions based on Cu-S distances is based on the premise that the protein backbone does not constrain the CuS cluster to alter Cu-S bond distances. This assumption may be valid as the MT protein structure seems to be dictated by the CuS cluster.[1]

EXAFS analyses have been carried out on the Cuβ domain of rat MT, Cu,ZnMT from pig liver, lysosomal CuMT from dog liver, and CuMT from *N. crassa* (Table I).[55,67,69] A mean Cu-S distances of 2.27 Å was reported for the CuMT β domain.[55] The stoichiometry of this latter cluster is Cu_6S_9. The Cu-S distance was based on a best fit of EXAFS data occurring with a coordination number between 2-3. The Cu_6S_7 cluster in CuMT from *N. crassa* was predicted from EXAFS to consist of 3-4 sulfur atoms at 2.20 Å and perhaps two sulfurs in an outer shell at 2.72 Å.[68] The best fit of the EXAFS data of lysosomal CuMT was reported to be four sulfur atoms at 2.27 Å.[67] It is not clear whether lower coordination numbers were tried in the curve fitting of the lysosomal CuMT. The mean Cu-S distance and XANES spectrum are both more consistent with predominantly trigonal coordination. Copper EXAFS data on Cu,ZnMT from pig liver did not distinguish between a model of trigonal Cu(I) coordination (mean Cu-S distance of 2.25 Å) or a model of trigonal Cu(I) ions with one Cu-S bond at 2.17 Å and two Cu-S bonds at 2.29 Å.[69] Curve fitting with models of lower coordination number was stated to be unsatisfactory.

Based on the correlation of Cu-S mean distances and fraction of digonal Cu(I), the pig Cu,ZnMT and *N. crassa* CuMT may be expected to contain a fraction of digonal Cu(I) ions. The *N. crassa* CuMT may be expected to have a significant portion of digonal Cu(I) based on the mean Cu-S distance of 2.20 Å in the first shell. If the mean Cu-S bond distance correlates with the fraction of digonal Cu(I), the *N. crassa* CuMT would be an excellent test case. HMQC NMR of AgMT from *N. crassa* would be a revealing study. In contrast, the Cu_6S_9 cluster in the β domain of rat MT is likely to contain predominantly trigonal coordination geometry. The Cu,ZnMT from pig liver is homologous to the MT from rat liver, so one may expect the same coordination geometry in both proteins. From the reported copper stoichiometry of 3-5 mol equivalency in the pig Cu,ZnMT, it is conceivable that the coordination geometry in CuMT may be influenced by the stoichiometry of the cluster. There is no evidence that Cu(I) ions and Zn(II) can co-exist within the same cluster.

Cu-Cu interactions were seen in each CuMT. The mean Cu-Cu distances are near 2.7 Å for each protein. This value is comparable with the mean Cu-Cu distances observed in tetracopper clusters (2.74 $\pm$ 0.06 Å).[17] The presence of a detectable Cu-Cu interaction is the best indicator that each CuMT contains a polynuclear Cu(I)-thiolate cluster.

CuS CLUSTERS IN ACE1:

Eleven of the twelve cysteinyl residues are critical for the functioning of CuACE1 in *S. cerevisiae*.[70] Physical studies have been carried out with the N-terminal half of the protein expressed in *E. coli*.[11,71] This segment of the ACE1 polypeptide, containing the eleven essential Cys residues, is responsible for specific DNA binding.[8,9] DNA binding requires the presence of bound Cu(I) or Ag(I) ions.[8,11] Cu(I) binding appears maximal between 6-7 mol equiv. Cu(I).[11] At present it is unclear whether the Cu(I) ions are clustered in a single or multiple centers. The Cu-protein exhibits S-Cu charge transfer bands in the ultraviolet and is luminescent with emission occurring between 580-620 nm.[11,72] The LMCT bands and emission are abolished upon acidification to below pH 2. These optical properties are similar to those of CuMT.[11]

X-ray absorption spectroscopy has been performed on *E. coli* isolates of CuACE1 as well as CuACE1 prepared by in vitro reconstitution protocols. The XANES spectrum is dominated by an edge feature near 8983 eV as seen in CuMT. As the CuS polynuclear cluster(s) contains 6-7 Cu(I) ions, the interpretation of the XANES edge feature is ambiguous. Based on the earlier model compound data of Kau et al.[49], the conclusion was reached that the Cu(I) ions were predominantly trigonal.[11,71]

TABLE I

Features of CuS Polymetallic Clusters in CuMTs

CuMT	Stoichiometry	Cu-S Å	Predicted Fraction of Digonal Cu(I)	(Ref.)
Rat CuMT-β	Cu_6S_9	2.27	0.1 (0)	55
Dog CuMT		2.27	0.1 (0)	67
Pig CuMT	Cu_5	2.25	0.3 (2)	69
Yeast CuMT	Cu_7S_{10}	2.24, 2.25	0.3 (2-3)	56
N.crassa CuMT	Cu_5S_7	2.2	0.8 (5)	68

Notes: The cluster stoichiometries have been reported in detail for only the rat β domain MT, *S. cerevisiae* CuMT and *N. crassa* CuMT. In the pig hepatic CuMT the sample contained both Cu and Zn and the Cu content was reported to be 5. The numbers in parentheses are the predicted number of digonal Cu(I) in each molecule. It is important to note that for the dog, pig and *N. crassa* MTs different EXAFS phase and amplitude functions have been used which may add to some uncertainty in the digonal Cu(I) estimate for these samples.

EXAFS of CuACE1 revealed a major scatterer in the first shell that fit well with sulfur atoms. A similar mean Cu-S bond distance of 2.26 was observed by two groups independently.[11,71] The data set of Nakagawa et al.[71] was best fit with a coordination number of 2.34, the best fit of the CuACE1 samples reported by Dameron et al.[11] had N values of 2.9 and 3.0. The accuracy of Cu-S bond distances was $\pm$ 0.02 Å but only $\pm$ 20% for N values.[11] The mean Cu-S distance of 2.26 Å is clearly consistent with predominant trigonal Cu(I) coordination. From the synthetic CuS cluster correlation in Fig. 3, one may conclude that if CuACE1 contains any digonal Cu(I) ions the contribution may be only a single ion. The stoichiometry of CuACE1 may therefore by 5-6 Cu^{trig} and 0-1 Cu^{dig}.

Cluster formation in CuACE1 was indicated by the observed Cu-Cu interactions apparent in the EXAFS. These Cu-Cu interactions occurring at 2.72 Å are equivalent to the short Cu-Cu distances observed in CuMTs and synthetic trigonally coordinated CuS cages. The mean number of scatters per absorbing copper atom was 2 for both ACE1 reconstituted with Cu(I) in the presence of 1 mM glutathione and native yeast CuMT. A lower number of scatterers was observed in CuACE1 prepared by reconstitution procedures in the absence of glutathione.[11]

In summary, a polynuclear CuS cluster forms in CuACE1. The formation of the cluster activates the protein for specific DNA binding. Although it is not clear whether the 6-7 Cu(I) ions are present within a single cluster, the Cu(I) ions are bound in a compact protein structure shielded from solvent interactions. Mixed coordination geometries may exist in CuACE1 as in CuMT although trigonal coordination is clearly dominant.

OTHER BIOLOGICAL CuS CLUSTERS:

Spectroscopic studies of other Cu-binding, Cys-rich peptides and proteins have been performed. We have analyzed the in vivo produced $Cu-(\gamma EC)_nG$ peptide complexes from *S. pombe* and the in vitro prepared CuE7 protein complex from the papilloma virus.[13,16] The stoichiometry of the Cu-isopeptide complex was 2.3 g atoms Cu(I) per peptide.[13] The complex is an oligomer of undefined M_r, so the Cu(I) stoichiometry per complex remains unresolved. The likely Cu(I) stoichiometry per cluster is between 4-6 mol eq. The isopeptides within the complex are heterogeneous in length with an average number of dipeptide repeats of 3.6. Thus, the mean Cu:S ratio in the isopeptide complexes is 0.6. This is compared to ratios of 0.5 to 0.8 in CuMTs. The *S. pombe* Cu clusters are luminescent which implies the presence of Cu(I). Cu(I)-S coordination is also indicated by EXAFS analysis. A mean Cu-S bond distance of 2.271 $\pm$ 0.0041 Å was determined. Trigonal coordination yielded the best fit of EXAFS data and is consistent with the corelation of Cu-S bond distances and synthetic complexes. A Cu-Cu interaction was apparent in the outer shell at 2.746 Å. This Cu-Cu scatterer peak is indicative of a polynuclear CuS cluster.

The E7 protein from papilloma viruses is a metalloprotein. The identity of the metal ion populating the binding site in virally infected cells is unclear, but expression of E7 genes from human HPV16 papilloma and rabbit CRPV viruses in an *E. coli* expression system yields E7 protein as a ZnE7 protein complex. The bacterially expressed human and rabbit E7 proteins bind maximally 1 and 2 mol eq. of divalent ions.[16] The Zn(II) ions in the E7 proteins undergo a rapid metal exchange reaction with Cu(I) ions yielded CuE7 proteins with stoichiometries of 2 and 3 mol eq. for HPV16 and CRPV CuE7s, respectively.[16] The CuE7 proteins exhibit absorption bands in the ultraviolet characteristic of Cu-S ligation. The CuE7 proteins are luminescent with a similar Stokes shift to the emission of CuMTs. This is indicative of Cu(I) ligation. X-ray absorption spectroscopy on CRPV CuE7 revealed a mean Cu-S distance of 2.265 Å and the presence of a Cu-Cu scatter peak indicative of a multinuclear CuS cluster. As CuE7 contains only 3 mol eq. Cu(I), any digonally coordinated Cu(I) ions would be expected to be more clearly evident in both the edge absorption and in the mean Cu-S distance compared to CuS clusters of higher nuclearity. The edge feature at 8983 eV and mean Cu-S distance are most consistent with exclusive trigonal Cu(I) coordination.

The oncogenic E7 protein is capable of forming a polynuclear CuS cluster. It is not clear whether the physiological form of the E7 protein is as ZnE7 or CuE7. If E7 does exist as ZnE7, it is conceivable that an in vivo metal exchange reaction may result in CuE7. As the physiological activity of E7 in viral transformation is unresolved, there is no information on whether the function of E7 is dependent on a specific metal ion.

SUMMARY:

A variety of proteins form CuS polynuclear clusters that mimic the CuS cages characterized as synthetic model compounds. There is a tendency for multinuclear cluster formation with thiolate ligands, so the similarities in cluster chemistry between biological molecules and synthetic model compounds may not be surprising. The tendency for cysteine-rich polypeptides to form Cu(I) clusters may relate to the strength of the Cu-S bond. The strength of this bond is illustrated by the observation that mercaptide sulfur is the only ligand capable of displacing acetonitrile from a $[Cu(CH_3CN)_4]^+$ complex in 50% acetonitrile.[73] All other ligands require a substantial dilutiton of acetonitrile before complexation is observed.[73] In most proteins containing CuS clusters, the polypeptide fold is largely dictated by CuS cluster formation. This is clearly the case in CuMTs. The function of ACE1 as a transcriptional activator protein is dependent on the formation of the CuS cluster. ACE1 binds Cu(I) and Cd(II) ions in different conformers and with different coordination chemistry. It therefore appears that cluster coordination chemistry is the basis for Cu(I)-specific metalloregulation seen in ACE1 and likely AMT1.

The biological CuS polynuclear clusters appear to be as diverse as the synthetic CuS model compounds. Based on synthetic CuS model compounds the nuclearity of clusters may expand from Cu_4 to Cu_8. A wider range of CuS polynuclear clusters may be observed with biological molecules. It is expected that CuS clusters with nuclearities below 4 will be seen in proteins such as CuE7. Clusters with exclusive trigonal Cu(I) coordination exist and it now appears that CuS cages with mixed digonal and trigonal coordination geometry may also occur.

The identification of the coordination geometry within a given CuS multinuclear cluster is difficult. The combination of XANES and careful analysis of Cu-S bond distances from EXAFS appear to be the best spectroscopic indicator in the absence of x-ray crystallographic and/or NMR structural data. From the Cu(I)-thiolate synthetic model compounds discussed here, it appears that the mean Cu-S bond distance may be a better indicator than the nature of the 8983 eV edge feature in XANES.

Acknowledgements: We thank Dr. G.S.H. Lee for the preparation of the synthetic CuS cluster samples. We acknowledge the support of the National Institutes of Health (ES 03817) for D.R.W. and the Australian Research Council for I.D.

REFERENCE:

1. D.R. Winge, C.T. Dameron, and G.N. George, *Advances Inorg. Biochem.* **10**, in press (1992).
2. D.H. Hamer, *Annual Rev. Biochem.* **55**, 913 (1986).

3. J.H.R. Kagi and A. Schaffer, *Biochemistry*, **27**, 8509 (1988).
4. D.H. Hamer, D.J. Thiele, and J.E. Lemontt, *Science*, **228**, 685 (1985).
5. R.K. Mehra, J.L. Thorvaldsen, I.G. Macreadie, and D.R. Winge, *Gene*, **114**, 75 (1992).
6. I. Bremner, *Experientia Suppl.*, **52**, 81 (1987).
7. I. Bremner, *J. Nutr.*, **117**, 19 (1987).
8. P. Furst, S. Hu, R. Hackett and D.H. Hamer, *Cell*, **55**, 705 (1988).
9. C. Buchman, P. Skroch, J. Welch, S. Fogel and M. Karin, *Mol. Cell.Biol.* **9**, 4091 (1989).
10. P. Zhou and D.J. Thiele, *Proc. Natl. Acad. Sci. USA*, **88**, 6112 (1991).
11. C.T. Dameron, D.R. Winge, G.N. George, M. Sansone, S. Hu, and D. Hamer, *Proc. Natl. Acad. Sci. USA*, **88**, 6127 (1991).
12. D.J. Thiele, *Nuc. Acid Res.*, **20**, 1183 (1992).
13. R.N. Reese, R.K. Mehra, E.B. Tarbet and D.R. Winge, *J. Biol. Chem.*, **263**, 4186 (1988).
14. C.T. Dameron, R.N. Reese, R.K. Mehra, A.R. Kortan, P.J. Carroll, M.L.Steigerwald, L.E. Brus and D.R. Winge, *Nature*, **338**, 596 (1989).
15. P. Hawley-Nelson, K.H. Vousden, N.L. Hubbert, D.R. Lowy, and J.T. Schiller, *EMBO J.*, **8**, 3905 (1989).
16. E.J. Roth, B. Kurz, L. Liang, C.L. Hansen, C.T. Dameron, D.R. Winge and D. Smotkin, *J. Biol. Chem.*, **267**, 16390 (1992).
17. I.G. Dance, *Polyhedron*, **5**, 1037 (1986).
18. I.G. Dance, K.J. Fisher and G.S.H. Lee in *Metallothioneins*, Eds. M.J. Stillman, C.F. Shaw, K.T. Suzuki, VCH, New York, p. 284 (1992).
19. D. Coucouvanis, C.N. Murphy, and S.K. Kanodia, *Inorg. Chem.*, **19**, 2993 (1980).
20. I.G. Dance and J.C. Calabrese, *Inorg. Chim. Acta*, **19**, L41 (1976).
21. J.R. Nicholson, I.L. Aabrahams, W. Clegg, and C.D. Garner, *Inorg. Chem.*, **24**, 1092 (1985).
22. G.A. Bowmaker, G.R. Clark, J.K. Seadon, and I.G. Dance, *Polyhedron*, **3**, 535 (1984).
23. I.G. Dance, *Aust. J. Chem.*, **31**, 2195 (1978).
24. F.J. Hollander and D. Coucouvanis, *J. Am. Chem. Soc.*, **96**, 5646 (1974).
25. P.J.M.L. Birker and H.C. Freeman, *J. Am. Chem. Soc.*, **99**, 6890 (1977).
26. L.E. McCandlish, E.C. Bissell, D. Coucouvanis, J.P. Fackler, and K. Knox, *J. Am. Chem. Soc.*, **90**, 7357.
27. I.G. Dance, G.A. Bowmaker, G.R. Clark, and J.K. Seadon, *Polyhedron*, **2**, 1031 (1983).
28. I.G. Dance, L.J. Fitzpatrick, M.L. Scudder, *J. Chem. Soc. Commun.*, 546 (1983).
29. C.P. Rao, J.R. Dorfman, R.H. Holm, *Inorg. Chem.*, **25**, 428 (1986).
30. G. Henkel, B. Krebs, P. Betz, H. Fietz, and K. Saatkamp, *Angew. Chem. Int. Ed. Engl.*, **27**, 1326 (1988).
31. E. Block, M. Gernon, H. Kang, G. Ofori-Okai, and J. Zubieta, *Inorg. Chem.*, **28**, 1263 (1989). 32. G. Henkel, B. Krebs, P. Betz, H. Fietz, and K. Saatkamp, *Angew. Chem.* **100**, 1373 (1988).
33. M. Baumgartner, H. Schmalle, and E. Dubler, *Polyhedron*, **9**, 1155 (1990).
34. N. Aoi, Y. Takano, H. Ogino, G.E. Matsubayashi, and T. Tanaka, *J. Chem. Soc. Comm.*, 703 (1985).
35. N. Aoi, G.E. Matsubayashi, and T. Tanaka, *J. Chem. Soc. Dalton Trans.* 241 (1987).
36. A. W. Addison, L.L. Borer, E. Sinn, in *Metallothioneins*, Eds. M.J. Stillman, C.F. Shaw, K.T. Suzuki, VCH, New York, p. 387 (1992).
37. I.G. Dance and J.C. Calabrese, *J. Chem. Soc. Chem. Commun.*, 762 (1975).
38. E. Block, H. Kang, G. Ofori-Okai, and J. Zubieta, *Inorg. Chim. Acta*, **167**, 147 (1990).

39. I. Schroter-Schmid and J. Strahle, *Z. Naturforsch.*, **45b**, 1537 (1990).
40. I.G. Dance, G.A. Bowmaker, R.J.H. Clark, J.K. Seadon, *Polyhedron*, **2**, 1031 (1983).
41. I.G. Dance, L.J. Fitzpatrick, D.C. Craig, and M.L. Scudder, *Inorg. Chem.*, **28**, 1853 (1989).
42. M.A. Khan, R. Kumar, and D.G. Tuck, *Polyhedron*, **7**, 49 (1988).
43. E.H. Griffith, G.W. Hurst, and E.L. Amma, *J.C.S. Chem. Comm.* 432 (1976).
44. P. Gonzalez-Duarte, J. Sola, J. Vives, and X. Solans, *J. Chem. Soc. Chem. Comm.*, 1641 (1987).
45. I. Dance, *Inorg. Chem.*, **20**, 1487 (1981).
46. A. Avdeef and J.P. Fackler, *Inorg. Chem.*, **17**, 2182 (1978).
47. P.K. Mehrotra, and R. Hoffmann, *Inorg. Chem.*, **17**, 2187 (1978).
48. I.G. Dance, *Polyhedron*, **5**, 1037 (1986).
49. L.S. Kau, D.J. Spira-Solomon, J.E. Penner-Hahn, K.O. Hodgson, and E.I. Solomon, *J. Am. Chem. Soc.*, **109**, 6433 (1987).
50. N.J. Blackburn, R.W. Stange, J. Reedijk, A. Volbeda, A. Farooq, K.D. Karlin, and J. Zubieta, *Inorg. Chem.*, **28**, 1349 (1989).
51. W. Braun, G. Wagner, E. Worgotter, M. Vasak, J.H.R. Kagi, and K. Wuthrich, *J. Mol. Biol.*, **187**, 125 (1986).
52. B.A. Messerle, A. Schaffer, M. Vasak, J.H.R. Kagi, and K. Wuthrich, *J. Mol. Biol.*, **214**, 765 (1990).
53. A.H. Robbins, D.E. McRee, M. Williamson, S.A. Collett, N.H. Xuong, W.F. Furey, B.C. Wang, and C.D. Stout, *J. Mol. Biol.*, **221**, 1269 (1991).
54. B.A. Messerle, A. Schaffer, M. Vasak, J.H.R. Kagi, and K. Wuthrich, *J. Mol. Biol.*, **225**, 433 (1992).
55. G.N. George, D. Winge, C.D. Stout, and S.P. Cramer, *J. Inorg. Biochem.*, **27**, 213 (1986).
56. G.N. George, J. Byrd, and D.R. Winge, *J. Biol. Chem.*, **263**, 8199 (1988).
57. S.S. Narula, R.K. Mehra, D.R. Winge, and I.M. Armitage, *J. Amer. Chem. Soc.*, **113**, 9354 (1991).
58. R.K. Mehra, E.B. Tarbet, W.R. Gray, and D.R. Winge, *Proc. Natl. Acad. Sci. USA*, **85**, 8815 (1988).
59. D.R. Winge, K.B. Nielson, W.R. Gray, and D.H. Hamer, *J. Biol. Chem.*, **260**, 14464 (1985).
60. M. Beltramini, G.M. Giacometti, B. Salvato, G. Giacometti, K. Munger, and K. Lerch, *Biochem. J.* **260**, 189 (1989).
61. A. Vogler and H. Kunkely, *J. Am. Chem. Soc.*, **108**, 7211 (1986).
62. M. Henary and J.I. Zink, *J. Am. Chem. Soc.*, **111**, 7407 (1989).
63. C.K. Ryu, K.R. Kyle, and P.C. Ford, *Inorg. Chem.*, **30**, 3982 (1991).
64. K.R. Kyle, C.K. Ryu, J.A. DiBenedetto, and P.C. Ford, *J. Am. Chem. Soc.*, **113**, 2954 (1991).
65. H.D. Hardt and A. Pierre, *Inorg. Chim. Acta*, **25**, L59 (1977).
66. D.R. McMillin, J.R. Kirchhoff, and D.V. Goodwin, *Coor. Chem. Rev.*, **64**, 83 (1985).
67. J.H. Freedman, L. Powers, and J. Peisach, *Biochemistry*, **25**, 2342 (1986).
68. T.A. Smith, K. Lerch, and K.O. Hodgson, *Inorg. Chem.*, **25**, 4677 (1986).
69. I.L. Abrahams, I. Bremner, G.P. Diakun, C.D. Garner, S.S. Hasnain, I. Ross, and M. Vasak, *Biochem. J.* **236**, 585 (1986).
70. S. Hu, P. Furst, and D. Hamer, *New Biologist*, **2**, 544 (1990).
71. K.H. Nakagawa, C. Inouye, B. Hedman, M. Karin, T.D. Tullius, and K.O. Hodgson, *J. Am. Chem. Soc.*, **113**, 3621 (1991).
72. J.R. Casas-Finet, S. Hu, D. Hamer, and R.L. Karpel, *FEBS Lett.*, **281**, 205 (1991).
73. V. Vortisch, P. Kroneck, and P. Hemmerich, *J. Am. Chem. Soc.*, **98**, 2821 (1976).

MECHANISMS OF COPPER ION HOMEOSTASIS IN YEAST

Valeria Cizewski Culotta, Paula Lapinskas, and Xiu Fen Liu

Division of Toxicological Sciences, The Johns Hopkins University School of Hygiene and Public Health, Baltimore, MD 21205, USA

INTRODUCTION

Since the mid 1980's, the field of metal ion chemistry has rapidly expanded to include the molecular biologists intrigued by the involvement of heavy metals in gene function and regulation. In particular, metals such as copper that are both toxic and essential for life represent an interesting dichotomy to metal ion biologists; much research currently focuses on identifying the cellular factors controlling accumulation, distribution and detoxification of these ions. The bakers yeast S. cerevisiae provides an ideal system in which to conduct these studies. The application of yeast molecular genetics should facilitate the isolation of key genes controlling metal ion homeostasis and should provide the molecular tools for probing structure and function of the encoded proteins.

The molecular mechanisms by which S.cerevisiae accumulate and detoxify Cu ions will be the subject of this report.

FACTORS CONTROLLING CELLULAR CU UPTAKE

The factors mediating cellular uptake and efflux of Cu ions most certainly play a key role in homeostasis of the metal. Although the kinetics of cellular Cu transport have been well-characterized for a variety of eukaryotic organisms, the relevant cellular proteins have not been identified, perhaps to the limitations of traditional biochemical approaches. Our laboratory has thus applied a genetic approach in yeast as a means for isolating metal uptake and efflux factors. In principle, the goal of this research has been to isolate key genes controlling cellular Cu transport and then to identify the encoded protein through gene structure and function analyses. Isolation of the genes involved in Cu transport would also allow for the mass production of the encoded polypeptides, eventually facilitating high resolution structural analyses of the putative Cu protein(s).

From K.D. Karlin and Z. Tyeklár, Eds., *Bioinorganic Chemistry of Copper* (Chapman & Hall, New York, 1993).

Dual Mechanisms of Cu Uptake in Yeast

The kinetics of Cu uptake in S. cerevisiae have been examined by Kosman's laboratory[1]. When cells are grown to a mid-logarithmic state, Cu transport appears to occur through a single saturable carrier, and the uptake parameters of V_{max} and k_m obtained in yeast were quite similar to that reported for mammalian cells [2,3]. However, mammalian cells exhibit both saturable and non-saturable Cu uptake [4,5], and in yeast, iron uptake occurs through saturable and non-saturable mechanisms as well [6]. It was thus conceivable that Cu likewise enters yeast through multiple uptake pathways.

    To explore the possibility of redundant Cu transporters in yeast, kinetics of Cu uptake were examined as a function of growth state. As shown in Fig. 1A, Cu transport in rapidly dividing cells exhibited saturation kinetics, confirming previous observations[1]. However, a quite different pattern of uptake predominated in resting cells that reached stationary phase upon nutrient starvation. In this case, transport appeared biphasic with an apparent high affinity saturable transporter evident at nM concentrations of the metal, followed by a non-saturable mechanism at higher levels (Fig. 1B).

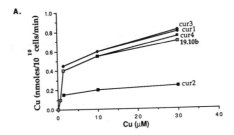

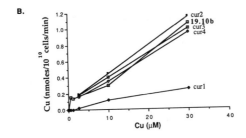

Figure 1: Kinetics of Cu uptake in S. cerevisiae: The indicated strains of yeast were grown in minimal media to either (A) mid-logarithmic state (optical density of 1.0 at 600 nm) or (B) stationary phase (optical density 3.0-4.0) and were incubated with the given concentrations of $CuSO_4$ for 5 min. Cells were rapidly harvested, washed and analyzed for Cu content through atomic absorption spectrophotometry as previously described [1].

The underlying basis for the non-saturable transport is unclear but may represent either simple diffusion or a non-interacting channel.

Overall, these biochemical studies indicated that Cu can enter the yeast cell through multiple pathways. The question then remained: do these pathways involve identical or distinct cellular factors? A genetic approach was then developed to identify the determinants of saturable and non-saturable Cu transport in yeast.

Generation of Yeast Mutants Defective for Cu Uptake

Our general approach to cloning Cu transport genes has been to isolate mutants of yeast defective for Cu uptake or efflux and then to use these mutants to isolate the corresponding gene through complementation. This strategy involved the generation of Cu-resistant (cur) mutants of yeast. Even with multiple transporters, mutations in a single transport gene would be expected to have a pronounced effect on Cu accumulation and should alter the cell's tolerance towards Cu toxicity.

Isolation of Cu-resistant strains involved the use of 19.10B, a Cu-sensitive yeast strain that lacks the large *CUP1* metallothionein gene repeat and fails to grow on Cu concentrations > 50 μM^7. Cu-resistant derivatives of this yeast were isolated by plating onto media containing 100 μM Cu and by isolating viable colonies. Approximately 20 Cu resistant mutants isolated in this manner were subject to genetic and biochemical analyses.

Genetic analyses demonstrated that in all but one of the 20 original *cur* isolates, Cu-resistance was due to a single nuclear gene mutation. Moreover, these 20 mutants could be classified further into 4 complementation or allelic groups that we have designated "cur1- cur4". In general, these mutants appeared specific for homeostasis of Cu ions. None of the mutants are unusually resistant to silver, and only one (*cur4*) exhibits a striking co-resistance to cadmium toxicity (data not shown).

To determine which if any of the *cur* mutants were defective for uptake or efflux of Cu, metal transport studies were conducted. For these experiments, the four classes of *cur* mutants and the 19.10B parent from which they were derived were cultured to either mid-logarithmic or stationary phase, and initial rates of Cu uptake were measured. In rapidly dividing yeast (Fig. 1A), only one mutant exhibited abnormal Cu uptake. Mutant *cur2* notably displayed a 60% decrease in initial velocity of uptake. However when cells were cultured to stationary phase, *cur2* exhibited normal Cu uptake while the *cur1* mutant showed significant impairments in both the apparent high affinity transporter as well as in the non-saturable uptake (Fig. 1B).

Overall, these studies demonstrated that Cu-resistance in both *cur1* and *cur2* results from decreases in Cu uptake, although distinct components of metal transport are impaired. Mutant *cur2* is affected in the saturable transport characteristic of dividing cells and *cur1* is impaired in the non-saturable uptake that predominates in resting yeast. Since *cur1* is also a Cu-resistant cell, the non-saturable transport most certainly contributes to intracellular Cu accumulation. The basis for Cu-resistance in *cur3* and *cur4* is not clear; other experiments including Cu efflux studies (data not shown) suggested that these strains are not mutants of Cu transport.

Cloning the Cu transport genes

The *CUR1* gene has recently been cloned through complementation. This gene was isolated by transforming the *cur1* mutant with a library of wild type yeast gene sequences and by screening for transformants that became Cu-sensitive. (A cell that contains both a functional and mutant copy of the *CUR1* gene was expected to demonstrate normal levels of Cu uptake and hence, Cu-sensitivity). The putative *CUR1* gene was then isolated from the yeast and was demonstrated to reproducibly convert a Cu-resistant *cur1* mutant back to a Cu-sensitive cell. Restriction analyses demonstrated that the putative *CUR1* gene is contained on a 7.0 kb segment of yeast DNA. Current studies are directed towards identifying the encoded protein through gene sequence analyses and gene structure/function studies.

Overview

Our recent studies have indicated that Cu transport in S. cerevisiae is quite complex, consisting of one or more saturable systems in combination with linear mechanisms of uptake. The *cur1* and *cur2* mutants described here should prove valuable for resolving the various components of Cu transport and for understanding some of the underlying mechanisms. Although our original genetic screen revealed only two complementation groups involved in Cu accumulation, the involvement of others is likely. We have therefore embarked on a second genetic screen and have recently isolated mutants of *cur1* that have become hyper-resistant to Cu toxicity due to mutations in additional Cu homeostasis genes. Identification of these additional Cu homeostasis genes through the aforementioned combination of biochemical, genetic and molecular approaches, together with an understanding of the CUR1-CUR4 factors, should provide a clearer picture of the mechanisms of Cu ion homeostasis in eukaryotes.

COPPER INDUCTION OF METAL DETOXIFICATION SYSTEMS

Although trace levels of Cu are vital to all life forms, non-physiological concentrations of the metal can be damaging to biomolecules. The basis for Cu toxicity appears to involve several mechanisms. This metal may inactivate essential enzymes by inappropriately complexing enzyme thiol groups[8]. Furthermore, there is an increasing body of evidence supporting the Cu-catalyzed production of toxic oxygen radicals [9-13]. To protect against Cu-related toxicity, organisms have evolved metal detoxification systems that are typically induced under conditions of metal ion stress. In most instances, this induction occurs at the level of gene transcription.

One of the best understood metal detoxification system involves the metallothioneins (MTs). These proteins are found in virtually all eukaryotes and act to protect cells against metal poisoning by binding to and sequestering metals such as Cu, Cd, and Hg[14]. Metallothioneins are induced at the level of gene transcription by many of the same metals that bind to the protein. This curious method of gene control is best understood in the yeast S. cerevisiae where Cu-induced expression of MT or the *CUP1* locus is mediated by a Cu and DNA binding transcription factor, ACE1[15].

Copper in yeast is also reported to induce the activities of superoxide dismutase (SOD) and catalase, enzymes that act to scavenge cellular O_2^- and H_2O_2 respectively[16, 17]. Indeed, these enzymes have been shown to reduce Cu-induced biological damage, presumably by circumventing the metal-catalyzed production of reactive oxygen [12,13]. Hence, induction of SOD and catalase in response to toxic Cu may represent a second line of cellular defense against metal-related damage.

Recent studies have indicated that the *SOD1* gene encoding Cu/Zn SOD in yeast is induced by Cu at the level of gene transcription and that this induction involves ACE1, the same Cu sensor factor that regulates *CUP1* [18,19]. The mechanism by which Cu regulates catalase activity however, was less clear. S. cerevisiae contains two forms of catalase: a peroxisomal catalase A encoded by the *CTA1* gene[20] and a soluble or cytosolic catalase T encoded by the *CTT1* gene[21]. To further understand metal regulation of these enzymes, our laboratory has investigated *CTT1* and *CTA1* gene expression as a function of Cu treatment.

Northern analyses of catalase gene expression in yeast

Northern blot analysis was employed to quantitate catalase gene expression in response to Cu. In these experiments, total RNA was isolated from yeast grown in the presence or absence of Cu and was resolved by denaturing gel electrophoresis. The RNA was then transferred to a nylon membrane, and reacted with a radio-labelled DNA probe consisting of *CTA1* or *CTT1* gene sequences.

Catalase A mRNA levels are not induced by copper. Catalase A mRNA was visualized by hybridization with the *CTA1* probe and as shown in Fig. 2A (top), *CTA1* mRNA levels did not increase with Cu. In our other studies, Cu failed to induce the *CTA1* gene under a wide variety of Cu concentrations and growth conditions tested (not shown). Thus, if Cu activates catalase A, induction occurs at either the translational or post-translational level.

Induction of catalase T by copper. The role of Cu in catalase T gene expression was investigated similarly using a radio-labelled probe bearing *CTT1* gene sequences. In contrast to the peroxisomal catalase, *CTT1* mRNA levels were in fact induced approximately 3 fold upon a 1 hour incubation with 100 μM Cu (Fig. 2A, bottom). Moreover, this level of induction closely paralleled that observed with the *SOD1* gene upon Cu treatment [16,19]. Substantially higher concentrations of the metal failed to induce *CTT1* gene expression, perhaps due to a Cu toxicity effect also observed with *CTA1* mRNA (Fig. 2A, top).

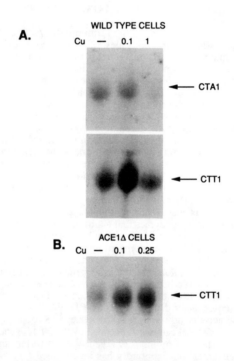

Figure 2: Regulation of Catalase mRNA levels by Copper: Yeast cells were grown to stationary phase in an enriched yeast extract based media containing 0.5% glucose and were treated with the indicated mM concentrations of $CuSO_4$. Total RNA was then subject to Northern blot analyses. Arrow indicates position of the catalase A or catalase T transcript, revealed by hybridization with the *CTA1* [20] or *CTT1* [21] gene probe, respectively. Wild type cells = strain DTY22[15]; ACE1Δ cells = DTY26 (DTY22 containing a deletion in the trans-activating domain of the ACE1 gene)[15]. NOTE: In the enriched media utilized here, 0.1-0.5 mM Cu is insufficient to induce *CUP1* (Lapinskas and Culotta, manuscript in preparation). This media presumably contains metal chelators that reduce the effective concentration of the metal.

The induction of catalase T by Cu required a restricted set of conditions. First, the *CTT1* gene is regulated in enriched media by a level of Cu that is not sufficient for *CUP1* gene induction (Fig. 2, legend). Secondly, unlike *CUP1*, Cu-regulation of catalase occurred only in cells grown to stationary phase (see ahead, Fig. 3). Other studies have demonstrated that *CTT1* gene expression is repressed in rapidly dividing yeast and becomes derepressed, or activated in stationary phase when cells are starved for essential nutrients[22]. It is thus conceivable that the relevant derepression factors cooperate with Cu regulatory factors to affect a high level of *CTT1* gene expression upon metal treatment.

Catalase gene expression does not require a functional ACE1 factor: The distinct conditions restricting Cu-induction of *CTT1* and *CUP1* suggested that separate mechanisms may govern their regulation. To assess whether ACE1 is essential for induction of catalase, mRNA from a mutant yeast containing a deletion in the *ACE1* gene was subject to Northern analyses. As shown in Fig. 2B, *CTT1* mRNA was still induced in this strain, although this mutant yeast failed to regulate *CUP1* [15]. Thus, the *CTT1* gene is induced by Cu in a manner distinct from that regulating *CUP1* and *SOD1*.

Promoter fusion studies

Since Cu-induction of catalase occurred independent of an intact ACE1 protein, Cu may have exerted control on *CTT1* mRNA stability rather than gene transcription. To test this, promoter fusions studies were conducted. These experiments utilized a yeast transformed with a fusion construct bearing the large upstream promoter of the *CTT1* gene (devoid of catalase coding sequences) fused to the coding region of the bacterial *LACZ* gene [24]. Activity of the *CTT1* promoter was then readily measured through a simple *LACZ* or ß-galactosidase assay. As shown in Fig. 3, this fusion construct was regulated by Cu in stationary, but not log phase cells and the level of induction paralleled that observed in Northern analyses (Fig. 2b). Since the *CTT1* gene promoter conferred Cu-regulation to a heterologous gene (*LACZ*), this regulation evidently occurred at the level of promoter activation and gene transcription.

The precise mechanism by which Cu induces the *CTT1* gene is still elusive. In general, regulation of this gene is quite complex and is subject to a variety of stress control signals. The *CTT1* gene is induced upon oxidative stress, nutrient starvation and heat shock [22, 24]. Cu-induction may in fact tie into one or more of these response mechanisms. One scenario explored was the possible link between metal and heat shock induction. In mammalian cells, heat shock proteins are regulated by both metals and temperature shifts, and in yeast the heat shock factor can induce *CUP1* [25, 26]. To explore the possible link between heat shock and Cu-regulation of *CTT1*, we tested a promoter fusion construct containing only the *CTT1* heat shock element fused to *LACZ*[24]. As shown in Fig. 3, this construct displayed negligible induction by Cu, indicating that if the heat shock element is involved, other regions of the *CTT1* promoter must also be necessary for efficient regulation by Cu.

Overview

In the studies presented here, yeast cells responded to Cu by inducing the transcription of the *CTT1* gene encoding cytosolic catalase T. In contrast, the *CTA1* gene encoding the peroxisomal form of the enzyme was insensitive to Cu-treatment. Unlike peroxisomal catalase, catalase T is responsive to a variety of environmental stress signals including nutrient starvation and heat shock. Copper can now be added to this list of agents that selectively induce the apparent "stress responsive" form of catalase.

The mechanism by which Cu regulates *CTT1* is not clear. Our studies indicated that this induction is distinct from the Cu-regulation of *CUP1* and *SOD1* and may not involve the *CTT1* heat shock response element. A novel Cu-sensor factor might mediate this regulation and experiments are under way to investigate this possibility.

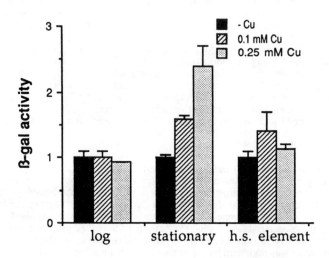

Figure 3. Cu induction of *CTT1-LACZ* fusions: Yeast strains containing an integrated copy of a *CTT1-LACZ* promoter fusion were grown as described in Fig. 2 to either logarithmic or stationary phase and were treated with the indicated concentrations of Cu for 1 hour. Cell extracts were then prepared and subject to *in vitro* analysis of ß-galactosidase activity. "log" and "stationary" = yeast strain MS903 containing *CTT1* promoter sequences -522 to -142 fused to the *LEU2* gene core promoter and the *LACZ* gene[24]. "h.s. element" = yeast strain AW1[24](as MS903 except containing only the *CTT1* heat shock element -382 to -325) grown to stationary phase.

       Overall, S. cerevisiae appears to have evolved with a highly sophisticated series of mechanisms ensuring protection against Cu toxicity. On one hand, Cu induces the *CUP1* gene encoding MT that operates to bind and sequester the metal, thereby directly protecting against Cu poisoning. As a second line of defense, Cu also induces the expression of the *SOD1* and *CTT1* genes encoding soluble superoxide dismutase and catalase. These enzymes presumably act to circumvent the Cu-catalyzed production of reactive oxygen. These detoxification mechanisms together with controlled methods of metal accumulation and distribution ensure that the homeostasis of essential and toxic Cu ions is continually maintained.

ACKNOWLEDGMENTS
       We thank H. Ruis for helpful discussion and for kindly providing the *CTT1* and *CTA1* gene probes and the AW1 and MS903 yeast strains. This work was supported by funding through the JHU NIEHS Center, grant #PO3ES-03819, through a JHU Biomedical Research grant and through a NIH research grant #ES 05794-01A1 awarded to V.C.C. P.L. is supported through an NIEHS training grant #ES-07141.

REFERENCES

1. C. M. Lin and D.J. Kosman, *J. Biol. Chem.* **265**, 9194 (1990).
2. R.C. Schmitt, H. Darwish, J. Cheney and M. J. Ettinger, *Fed. Proc.* **45**, 2800 (1986).
3. G.M. Gadd, A. Stewart, C. White and J. Mowll, *FEMS Microb. Letts*, **24**, 231 (1984).
4. F. Bronner and J. Yost, *Am. J. Physiol.* **249**, G108 (1985).
5. S. Herd, F. Camakaris, P. Wookey and D. Danks, *J. Nutr.* **12**, 1370 (1991).
6. A. Dancis, R. Klausner, A. Hinnebush and J. Barriocanal. *Mol. Cell. Biol.* **10**, 2294 (1990).
7. D. Hamer, D. Theile and J. Lemontt, *Science*, **228**, 685 (1990).
8. M. Nakamura and I. Yamazaki, *Biochim. Biophys. Acta.*, **267**, 249 (1972).
9. A. Samuni, C. Chevion, and G. Czapski. *J. Biol. Chem.* **256**, 249, (1981).
10. S. Goldstein and G. Czapski. *J. Free Radic. Biol. Med.* **2**, 3, (1986).
11. K. Yamamoto, and S. Kawanishi, *J. Biol. Chem.*, **264**, 15435 (1989).
12. L. Tkeshelashvili, T. McBride, K. Spence and L. Loeb, *J. Biol. Chem.* **266**, 6401 (1991).
13. K. Yamamoto and S. Kawanishi, *J. Biol. Chem.*, **266**, 1509 (1991).
14. D. Hamer. *Ann. Rev. Biochem.* **55**, 913 (1986).
15. D. Theile, *Mol. Cell. Biol.*, **8**, 2745 (1988).
16. F. Galiazzo, A. Schiesser and G. Rotilio. *Biochim. Biophys Acta* **965**, 46 (1988).
17. M. Greco, D. Hrab, W. Magner and D. Kosman. *J. Bacteriol*, **172**, 317, (1990).
18. M. Carri, F. Galiazzo, M. Ciriolo and G. Rotilio, *FEBS LETT*, **2278**, 263 (1991).
19. E. Gralla, D. Thiele, P. Silar, J. Valentine, *Proc. Natl. Acad. Sci. USA*, **88**, 8558 (1991).
20. G. Cohen, F. Fessl, A. Traczyk, J. Rytka and H. Ruis, *Mol. Gen. Genet*, **200**, 74 (1985).
21. W. Spevak, F. Fessl, J. Rytka, A. traczyk, M. Skoneczny and H. Ruis, *Mol. Cell. Biol.*, **3**, 1545 (1983).
22. T. Belazze, A. Wagner, R. Wieser, M. Schanz, G. Adam, A. Hartig and H. Ruis. *EMBO J.* **10**, 585 (1991).
23. R. Wieser, G. Adam, A Wagner, C. Schuller G. Marchler, H Ruis, Z. Krawiec and T. Bilinski, *J. Biol. Chem.* **266**, 12406 (1991).
24. P. Silar, G. Butler, and D. Thiele. *Mol. Cell. Biol.*, **11**, 1232 (1991).
25. W. Yang, W. Gahl and D. Hamer, *Mol. Cell. Biol.*, **11**, 3676 (1991).

RNA HYDROLYSIS BY CU(II) COMPLEXES: TOWARD SYNTHETIC RIBONUCLEASES AND RIBOZYMES

James K. Bashkin

Department of Chemistry, Campus Box 1134, Washington University, 1 Brookings Drive, St. Louis, MO 63130-4899, USA

We wish to develop reagents for the sequence-specific cleavage of RNA by non-oxidative mechanisms. The molecules we have devised may be described as synthetic ribozymes, and comprise a molecular recognition agent (such as single-stranded DNA) and an RNA hydrolysis agent. Their potential applications include antiviral or antifungal therapy, and they were designed as first-generation catalytic antisense DNA drugs for the control of gene expression. We provide a brief review of the principles of the antisense method, a discussion of the importance (for in vivo applications) of hydrolytic cleavage as an alternative to oxidative degradation of nucleic acids, and a discussion of our synthetic and mechanistic efforts toward synthetic ribozymes.

THE ANTISENSE TECHNIQUE

Infectious viral diseases such as AIDS and influenza currently pose a great threat to society. Because of the significant differences between viral and bacterial infections, no broad-spectrum pharmaceutical agents analogous to antibiotics are presently available for viral diseases. One potentially powerful, yet clinically unrealized, method for combatting viral infections is the antisense technique.[1,2] This method involves the specific interception of viral messenger RNA by a complementary nucleic acid. Successful application of antisense methods to the proper mRNA sequences would halt the production of viral coat proteins, viral proteases, reverse transcriptases, and other injurious gene products; at the same time, normal cellular protein synthesis would proceed unimpeded.

Antisense regulation of gene expression is a natural regulatory process, and it has been recognized for well over a decade that certain prokaryotic organisms synthesize antisense nucleic acids as a means of inhibiting the expression of specific genes.[3] More recently, examples of naturally-occurring antisense inhibition for eukaryotes have been discovered.[4] Furthermore, Izant and Weintraub found that antisense sequences could be

From K.D. Karlin and Z. Tyeklár, Eds., *Bioinorganic Chemistry of Copper* (Chapman & Hall, New York, 1993).

experimentally introduced into mammalian cells, and that these sequences specifically inhibited expression of the genes against which they were directed.[5]

Following the initial discovery of inhibition of gene expression by endogenously-synthesized antisense nucleic acids, a number of groups began to use chemically synthesized oligonucleotides as specific regulators of gene expression (see reference 1 for a review). Inhibition has been measured directly, by quantifying the amount of protein produced, and also indirectly, by observing the expected physiological effect when synthesis of the target protein is inhibited. Successful inhibition of viral replication was reported for Rous sarcoma virus.[6] Similar approaches have been used to inhibit replication of HIV[7] and numerous other viruses, including Vesicular Stomatitis Virus, Herpes Simplex Virus type 1, Simian Virus 40, and Influenza Virus.[1]

Although the chemical antisense method has been successful in cell culture, it has normally required high (e.g. 50μM) concentrations of antisense probe to be effective.[2] This lack of efficiency precludes any practical therapeutic applications of the technique at the present time, and may be attributed to (1) degradation of oligonucleotides by nuclease enzymes; (2) the inefficient transport of DNA across cell membranes; (3) the reversible nature of the duplex structure formed by the antisense probe and target sequence. Clearly, fundamental breakthroughs are required before the therapeutic promise of the technique is realized.

CATALYTIC ANTISENSE DRUGS (SYNTHETIC RIBOZYMES)

In 1986, we began working on the preparation and testing of synthetic, catalytic antisense probes, as illustrated schematically in Figure 1.[8,9] The principle of this method is to increase the efficiency of the chemical antisense process, because each antisense molecule

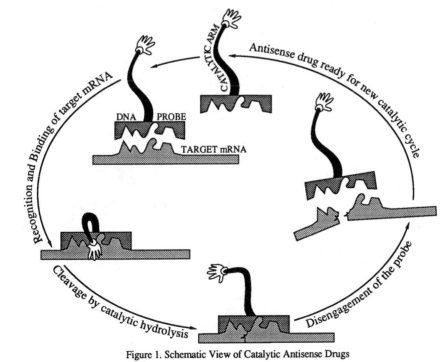

Figure 1. Schematic View of Catalytic Antisense Drugs

133

becomes a catalyst capable of destroying many copies of the target mRNA. Stein and Cohen have independently proposed that an oligonucleotide with a single pendant imidazole might function as a catalytic antisense agent.[2b] With a catalytic drug, it should no longer be necessary to introduce large amounts of antisense probe into cells. Ultimately, we would want to couple the catalytic agents with the best advances in drug delivery and nuclease-resistant DNA analogues to yield the most therapeutically promising molecule.

A different approach to catalytic antisense inhibition of gene expression was proposed in 1988 by Haselhoff and Gerlach[10a] and Cech.[10b] Their approach is based on ribozymes, the catalytic RNA molecules discovered by Cech and Altman.[10c-d] Ribozymes can be designed to bind and cleave target RNA sequences of interest, and the construction of ribozymes which attack viral mRNA that may be a viable approach to antiviral therapy.[10e] There are, however, a number of advantages that reagents based on DNA or DNA analogues may offer. For example, (1) DNA is much more stable to hydrolysis than RNA, and RNA is readily degraded by ubiquitous ribonuclease enzymes. (2) With our catalytic DNA molecules, the binding and cleavage functions are completely independent. This gives us the possibility of infinite variation of the catalyst during optimization studies, without disrupting the selective binding. (3) Since ribozymes perform both binding and cleavage with RNA residues in a complex, interdependent fashion, it may not be possible to optimize cleavage rates sufficiently for effective therapy. (4) Rates of ribozyme reactions are highly dependent on pH and the concentration of divalent cations such as Mg^{2+}, and they are optimal under nonphysiological conditions.[10a] (5) DNA probes with pendant RNA cleavage catalysts *can* be designed to operate optimally at pH 7, and they can be designed to be independent of free metal ion concentration, as will be described.

REDOX CLEAVAGE AND HYDROLYSIS OF NUCLEIC ACIDS

There has been a great deal of recent interest in the cleavage of nucleic acids by reagents termed "chemical nucleases". Current fascination with this field derives from the importance of nucleic acid processing to living cells, and from information that cleavage experiments may provide about nucleic acid structure and the nature of protein-nucleic acid interactions. Most studies on the cleavage of DNA or RNA by chemical nucleases have employed oxidative scission as the cleavage reaction. For example, in a series of novel studies, Sigman demonstrated that bis(o-phenanthroline)copper(II) cleaves nucleic acids in an oxidative fashion when O_2 and mercaptopropionic acid are present.[11] Dervan[12], Barton[13], and Tullius[14a] have reported studies of nucleic acid cleavage by synthetic redox methods. The natural product bleomycin[15], an antitumor antibiotic, has also been shown to cleave nucleic acids oxidatively, with detailed isotope labeling studies of the reaction being reported by Stubbe and Kozarich.[15b] Clearly, oxidative cleavage of nucleic acids has been a fruitful area of study, helping to develop the state-of-the-art picture of molecular recognition and nucleic acid structure. Recent developments include the guanosine-specific cleavage of DNA by Ni(II) complexes.[14b]

We have chosen to develop novel chemical nucleases that operate by *hydrolytic* rather than oxidative mechanisms.[16,17,18,19] A number of distinct advantages will result from having access to both oxidative and hydrolytic chemical nucleases, because of the inherently different chemical properties of the two reactions. Some of those properties are contrasted here in points a-d. (**a**). Oxidation tends to be destructive, especially of the sugar portion of the DNA or RNA. For example, bleomycin degrades DNA by hydrogen atom abstraction from the deoxyribose, leading to a mixture of products (i.e. base propenal) which are not simple building blocks of the parent nucleic acid.[15b] On the other hand, hydrolysis of the phosphodiester backbone of a nucleic acid yields intact nucleoside phosphate and sugar-hydroxyl groups that are good substrates for enzymatic transformations such as re-ligation. (**b**). Oxidative cleavage is not generally selective for DNA or RNA. Sequence-directed redox cleavage of RNA can be accomplished by attaching Cu-o-phen[11d], Fe(II)EDTA[12], or similar reagents to a DNA probe. These reagents usually operate by generating free hydroxyl radicals. However, the highly reactive nature of the hydroxyl radical leads to self-destruction of the DNA probe as well as destructive cleavage of the target sequence. Furthermore, there is no current strategy for controlling OH radical production in vivo, so these toxic, diffusible radicals would be produced constantly. (c). Hydrolysis provides an excellent means of chemical discrimination between RNA and DNA. RNA is considerably more susceptible to chemical hydrolysis than DNA, because the 2'-hydroxyl group of RNA promotes hydrolysis by acting as an intramolecular nucleophile. We employed these different relative reactivities in designing sequence-directed agents for the catalytic hydrolysis of RNA: catalysts that are active for RNA hydrolysis can be attached to DNA probes without danger of self-destruction, because the rate of DNA hydrolysis will be immeasurably slow. (d). Considerable cell toxicity is associated with the ancillary reagents, such as hydrogen peroxide, that are required to drive most redox cleavage. For in vivo applications, is difficult to imagine that external redox equivalents could be delivered to a cell without toxic side effects. Furthermore, the high reactivity of hydroxyl radicals renders them inherently toxic. In contrast, the water required for hydrolysis is readily available, and catalytic hydrolysis of phosphate esters at pH 7 is one of the fundamental reactions of cellular biochemistry.

Our proposed approach to nucleic acid cleavage maintains the advantage of high sequence specificity exhibited by "oxidative nucleases", but avoids the chemical selectivity and toxicity drawbacks of free radical chemistry. Hydrolysis also provides us with a sound chemical basis for the preparation of long-lived, catalytic chemical nucleases. Attaching a catalyst to a DNA probe would result in a sequence-specific ribonuclease that cannot self-destruct.

TOWARD SYNTHETIC RIBOZYMES

When our program was initiated, imidazole was known to catalyze the hydrolysis of RNA by a bifunctional mechanism.[20] It was also known that metal *ions* catalyze this reaction.[21] To develop catalytic antisense reagents, we needed to identify well-defined catalysts, either metal *complexes* or imidazole derivatives, that could be attached to DNA.

First, we developed a convenient assay for RNA hydrolysis.[16] The substrate employed in this assay, oligoadenylic acid $(A)_{12-18}$, is RNA, not a model compound. This is important because there are significant reactivity differences between RNA and models such as dinucleoside phosphates or activated phosphate esters.[22] Our assay uses anion exchange HPLC to monitor the disappearance of oligoadenylic acid with time. All of the components of the substrate are resolved by the HPLC, as are the products. The assay operates under the physiologically relevant conditions of pH 7 and 37 °C, and can serve as a rapid screen for activity or as the basis for full kinetic analyses.

As part of our program, the first examples of RNA hydrolysis by well-defined metal complexes were reported.[16] Representative complexes (1) - (3) are shown below; activities are given in Table I.

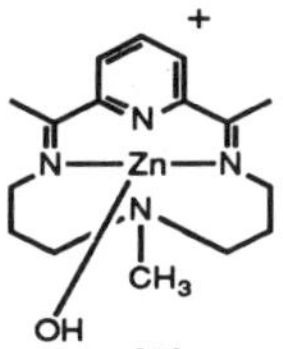

(1) 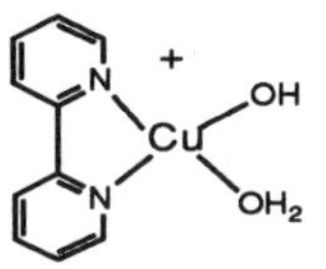   (2) 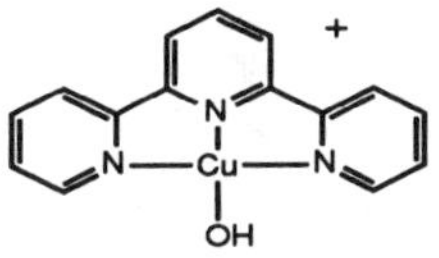   (3)

Complex	% Substrate Hydrolyzed	Complex	% Substrate Hydrolyzed
(1)	70	$ZnCl_2$	72
(2)	61	$CuCl_2$	31
(3)	97	none (control)	5

Table I. Extent of $(A)_{12-18}$ hydrolysis by transition metal complexes. Conditions: 37 °C, 20mM HEPES buffer, pH=7.1, reaction time = 20h, [complex] = 0.160 mM, [RNA] = 0.06 mM.

SEQUENCE-DIRECTED RNA CLEAVAGE AGENTS

In order to deliver an RNA hydrolysis catalyst to a specific target RNA sequence, our approach is to attach RNA hydrolysis catalysts to DNA probes using a variety of synthetic approaches. We have reported a series of Cu(II)-bipyridine/nucleoside conjugates[18] in which 2,2'-bipy derivatives were attached at the C-5 position of 2'-deoxyuridine, and via phosphodiesters at the 3'- and 5'-positions of thymidine. These conjugates were characterized by 1- and 2-D NMR studies; they were shown to bind Cu(II) through the bipyridyl group, and to be competent at RNA hydrolysis. Because of the greater thermodynamic stability of Cu(II) terpyridyl complexes than their bipyridine counterparts, and because of the activity of terpy complex (3) in RNA cleavage, we were interested in preparing DNA-terpyridine conjugates for testing as synthetic ribozymes. Several synthetic approaches are described here.

The synthesis of 4'-methyl-2,2':6',2"-terpyridine (4) was carried out according to the literature procedure.[23] As shown in **Scheme I**, this versatile starting material was then lithiated with LDA and treated with 2-(2-bromoethyl)-1,3-dioxolane at -78 °C.[24] After deprotection with 1M HCl and flash chromatography, the desired aldehyde (5) was obtained (56% isolated yield). Sodium borohydride reduction in ethanol gave (6), the corresponding alcohol (86%). The alcohol 4'-(4-hydroxybutyl)-2,2':6',2"-terpyridine is analogous to a bpy derivative that we have employed[18] as a precursor to phosphoramidite reagents, for the purpose of attaching metal complexes to DNA via a phosphodiester linkage.

To provide additional terpyridine reagents of general utility, the alcohol (6) was treated with HBr, giving the corresponding 4-bromobutyl derivative (7) in 78% yield. In a two-step procedure, 4'-(4-aminobutyl)-2,2':6',2"-terpyridine (8) was prepared via its phthalimidobutyl precursor. The purified yield of 4'-(4-phthalimidobutyl)-2,2':6',2"-terpyridine was 95%, and this was converted to the amine (9) in 89% yield on treatment with hydrazine hydrate. As shown below, we found an analogous primary amine derivative of imidazoleacetic acid to be useful for linking imidazole to nucleosides via reaction with an active ester derivative of 2'-deoxyuridine.[17]

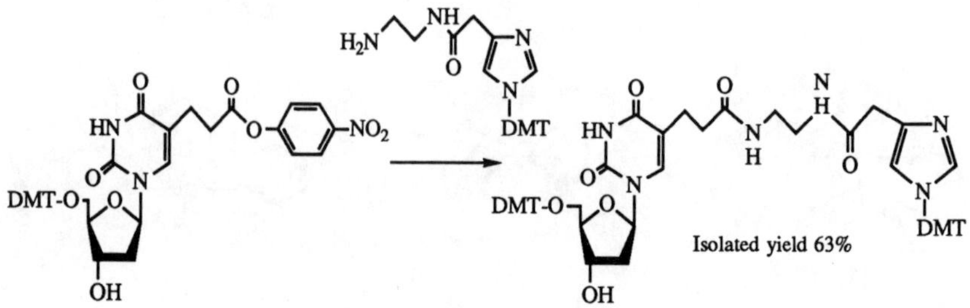

Isolated yield 63%

Compound (4) is derived from 4'-thiomethylterpyridine via a nickel-catalyzed cross-coupling reaction with methyl Grignard. Potts and Usifer reported the synthesis of the 4'-thiomethylterpyridine in a one-pot procedure from CS_2, MeI, ammonium acetate, and two equivalents of 2-acetylpyridine enolate.[23] In order to avoid the Gignard step employed prior to **Scheme I** and shorten the synthetic route to compounds similar to (9), we were

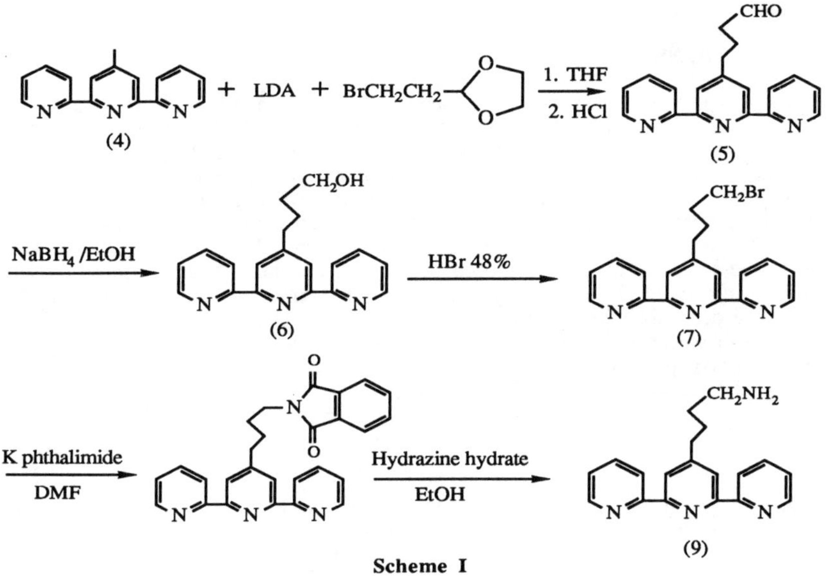

Scheme I

interested to see if alkyl halides other than MeI could be employed in the Potts/Usifer terpyridine synthesis, thereby allowing the introduction of thioether groups with useful functionality directly on the central pyridine ring. We are currently exploring the scope of this reaction, and successful examples include the use of 2-bromoethyldioxane and -dioxolane to give (10) and (11), as shown in **Scheme II**.

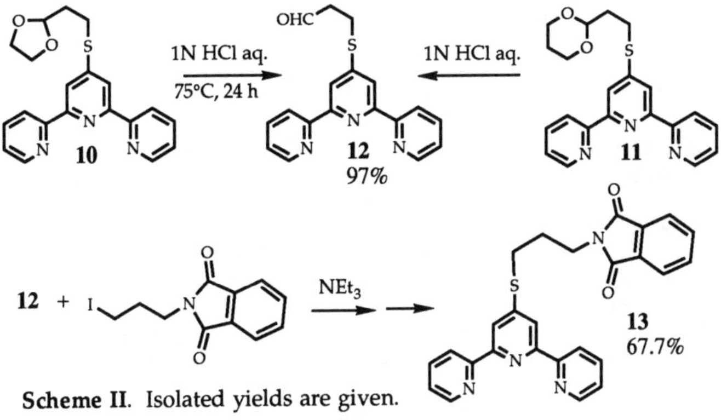

Scheme II. Isolated yields are given.

Compounds 10-12 have been characterized by 1- and 2-D ^{1}H and ^{13}C NMR, MS, HRMS, and elemental analysis. Compound (11) can be reduced to the alcohol, and cleanly undergoes ß-elimination to the corresponding thiol under basic conditions. This thiol is readily alkylated in situ, and may be our most versatile substrate for linking terpridine to nucleoside derivatives.[25] We anticipate that phthalimide (13) will convert to the corresponding amine, by analogy with the corresponding synthesis of (9) shown in **Scheme I**.

CONCLUSIONS

We wish to prepare sequence-specific chemical nucleases, or artificial ribozymes, that operate by delivering an RNA hydrolysis catalysts to a specific RNA target sequence. Molecular recognition is provided by attaching the catalyst to a DNA probe. These reagents are designed to serve as first-generation catalytic antisense reagents for the control of gene expression. We required the identification of well-defined reagents which are active in hydrolytic cleavage of RNA, and that are suitable for covalent attachment to DNA. Bi- and terpyridine derivatives of Cu(II) are among the active reagents we have found, and several strategies have been pursued for attaching these reagents to DNA probes. Synthesis of DNA-bipyridine conjugates was been described, and new routes to terpyridine derivatives have been given. Incorporation of bipyridine and terpyridine Cu(II) reagents as reactive side-chains on oligonucleotides is being pursued, as is the testing these compounds for activity as synthetic ribozymes.

Acknowledgement:- We are thankful to Midwest Center for Mass Spectrometry at the University of Nebraska-Lincoln, supported in part by the NSF Biology Division grant DIR9017262, for recording FAB MS and high resolution FAB MS. Partial financial support for this work from Grant #IN-36-32 from the American Cancer Society, the Pharmaceutical Manufacturers Association, Lucille P. Markey Center for the Molecular Biology of Human Disease, Monsanto Company, and Washington University is gratefully acknowledged.

REFERENCES

1.   E. Uhlmann and A. Peyman, *Chemical Reviews*, **90**, 544 (1990).
2.   (a) *Oligodeoxynucleotides: Antisense Inhibitors of Gene Expression,* J. S. Cohen, ed., (CRC Press, Boca Raton, (1989), (b) C.A. Stein; J.S. Cohen,*Cancer Res.*, **48**, 2659-2668 (1988).
3.   For a review, see P.J. Green et al, *Ann. Rev. Biochem.*, **55**, 569 (1986).
4.   S.M. Heywood, *Nucleic Acids Res.*, **14**, 6771 (1986).
5.   J.G. Izant and H. Weintraub, *Science,* **229**, 345 (1985).
6.   (a) P.C. Zamecnik and M.L. Stephenson, *Proc. Natl. Acad. Sci. USA*, **75**, 280 (1978); (b) M.L. Stephenson and P.C. Zamecnik, *Proc. Natl. Acad. Sci. USA*, **75**, 285 (1978).
7.   Representative examples: (a) M. Matsukura, K. Shinozuka, G. Zon, H. Mitsuya, M. Reitz, J.S. Cohen, and S. Broder, *Proc. Natl. Acad. Sci. USA*, **84**, 7706 (1987); (b)

J. Goodchild et al, *Proc. Natl. Acad. Sci. USA*, **85**, 5507 (1988).
8.   "RNA Hydrolysis", J. K. Bashkin, M. K. Stern, A. S. Modak, patent applied for, 6/90.
9.   "RNA Hydrolysis", J. K. Bashkin, A. S. Modak, M. K. Stern, patent application filed 11/1/90.
10.   (a) J. Haselhoff, W.L. Gerlach, *Nature* **334**, 585-591 (1988). (b) T. R. Cech, *J. Am. Med. Assoc.*, **260**, 3030-3034 (1988). (c) T.R. Cech, *Science*, **236**, 1532-1539 (1987). (d) S. Altman, *Angew. Chem. Int. Ed. Engl.*, **29**, 749-758 (1990). (e) for a recent overview, see L. Chrisey, J. Rossi, N. Sarver, *Antisense Res. and Dev.*, **1**, 57-63 (1991). (f) A. C. Jeffries, R.H. Symons, *Nucl. Acids Res.*, **17**, 1371-1377 (1989). (g) A.J. Zaug, J.R. Kent, T.R. Cech, *Science*, **224**, 574-578 (1984).
11.   (a) D.S. Sigman, *Acc. Chem. Res.*, **19**, 180-186 (1986). (b) C.-H.B. Chen, D.S. Sigman, *Proc. Natl. Acad. Sci. USA*, **83**, 7147-7151 (1986). (c) C.-H. B. Chen,; D.S. Sigman, *J. Am.. Chem. Soc.* **110**, 6570-6572 (1988). (d) B.C.F. Chu, L.E. Orgel, *Proc. Natl. Acad. Sci. USA*, **82**, 963-967 (1985).
12.   (a) P.B. Dervan, *Science* **232**, 464-471 (1986). (b) G.B. Dreyer, P.B. Dervan, *Proc. Natl. Acad. Sci. USA*, **82**, 968-972 (1985). (c) ibid **27**, 3635-3638 (1986).
13.   J.K. Barton, *Science*, **233**, 727-734 (1986).
14.   (a) T.D. Tullius, B.A. Dombroski, *Proc. Natl. Acad. Sci. USA*, **83**, 5469-5473 (1986). (b) X. Chen, S.E. Rokita, C.J. Burrows, *J. Am. Chem. Soc.* **113**, 5884-5886 (1991).
15.   (a) *Bleomycin: Chemical, Biochemical, and Biological Aspects*; S.M. Hecht, ed., (Springer-Verlag, New York, 1979), (b) J.W. Kozarich, L.Worth, Jr., B.L. Frank, D. Christner, D.E. Vanderwall, J. Stubbe, *Science*, **245**, 1396 (1989). (c) Y. Sugano, A. Kittaka, M. Otsuka, M. Ohno, Y. Sugiura, H. Umezawa, *Tet. Let.*, **27**, 3631-3634 (1986). (d) K. Delany, S.K. Arora, P.K. Mascharak, *Inorg. Chem.*, **27**, 705-712 (1988).
16.   M. K. Stern, J. K. Bashkin, E. D. Sall, *J. Am. Chem. Soc.*, **112**, 5357-5359 (1990).
17.   J. K. Bashkin, J. K. Gard, A. S. Modak, *J. Org. Chem.*, **55**, 5125-5132 (1990).
18.   A. S. Modak, J. K. Gard, M. C. Merriman, K. A. Winkeler, J. K. Bashkin, and M. K. Stern, *J. Am. Chem. Soc.*, **113**, 283-291 (1991).
19.   J. K. Bashkin, R. J. McBeath, A. S. Modak, K. R. Sample, W.B. Wise, *J. Org. Chem.*, **56**, 3168-3176 (1991).
20.   (a) E. Anslyn, R. Breslow, *J. Am. Chem. Soc.*, **111**, 5972-5973 (1989). (b) E. E. Anslyn, R. Breslow, *J. Am. Chem. Soc.* **111**, 4473-4482 (1989).
21.   (a) W. Farkas, *Biochim. Biophys. Acta*, **155**, 401-409 (1968). (b) J.J. Butzow, G. Eichhorn, *Biochemistry* **10**, 2019-2027 (1971). (c) J.J. Butzow, G. Eichhorn, *Nature*, **254**, 358-369 (1975). (d) H. Ikenaga, Y. Inoue, *Biochemistry*, **13**, 577-582 (1974). (e) R.S. Brown, J.C. Dewan, A. Klug, *Biochemistry*, **24**, 4785-4801 (1985). (f) L.S. Behlen, J.R. Sampson, A.B. Direnzo, O.C. Uhlenbeck, *Biochemistry*, **29**, 2515-2523 (1990). (g) K. Dimroth, H. Witzel, W. Hulsen, H. Mirbach, *Annalen*, **620**, 94-108 (1959).
22.   (a) J.M. Harrowfield, D.R. Jones, L.F. Lindoy, A.M. Sargeson, *J. Am. Chem. Soc.*, **102**, 7733-7741 (1980). (b) G. Rawji, M. Hediger, R.M. Milburn, *Inorg. Chim. Acta*, **79**, 247 (1983). (c) R.A. Kenley, R.H. Fleming, R.M. Laine, D.S. Tse, J.S. Winterle, *Inorg. Chem.*, **23**, 1870-1876 (1984). (d) D.R. Jones, L.F. Lindoy, A.M. Sargeson, *J. Am. Chem. Soc.*, **106**, 7807-7819 (1984). (e) G.H. Rawji, R.M. Milburn, *Inorg. Chim. Acta*, **150**, 227-232 (1988). (f) J. Chin, X. Zou, *J. Am. Chem. Soc.*, **110**, 223-225 (1988). (g) J.R. Morrow, W.C. Trogler, *Inorg. Chem.*, **27**, 3387-3394 (1988). (h) J.R. Morrow, W.C. Trogler, *Inorg. Chem*, **28**, 2330-2334 (1989). (i) J. Chin, M. Banaszczyk, V. Jubian, X. Zou, *J. Am. Chem. Soc.*, **111**, 187-190 (1989). (j) S.H. Gellman, R. Petter, R. Breslow, *J. Am. Chem. Soc.*, **108**, 2388-2394 (1986).
23.   K.T. Potts, D.A. Usifer, A. Guadalupe, H.D. Abruna, *J. Am. Chem. Soc.*, **109**, 3961-3967 (1987).
24.   J.K. Bashkin, A.S. Modak, unpublished results
25.   J.K. Bashkin, S.M. Touami, U. Sampath, unpublished results.

Hemocyanin and Copper Monooxygenases

THREE-DIMENSIONAL STRUCTURE OF THE OXYGENATED FORM OF THE HEMOCYANIN SUBUNIT II OF *LIMULUS POLYPHEMUS* AT ATOMIC RESOLUTION

Karen A. Magnus, Hoa Ton-That and Joan E. Carpenter

Department of Biochemistry, Case Western Reserve University, School of Medicine, 2119 Abington Road, Cleveland, Ohio 44106-4935, U.S.A.

INTRODUCTION

Hemocyanins are large multisubunit proteins that transport molecular oxygen in a variety of invertebrates. Their properties have been reviewed in references 1 and 2. Arthropod hemocyanins are copper containing proteins that occur extracellularly and they are composed of multiples of six subunits. The oxygenated form of these proteins is characterized by absorption maxima at about 340 and 580 nm, while the deoxygenated form is colorless. A major interesting question that can be addressed after determining the structure of hemocyanin using single crystal x–ray diffraction methods is: How does the protein bind molecular oxygen?

The largest known arthropod hemocyanin from the horseshoe crab found on the eastern coast of the United States, *Limulus polyphemus*, has a molecular weight of 3.3 x 10^6 Da and is composed of 48 subunits [3]. The complex may be thought of as octahexameric, and it is able to bind oxygen reversibly and cooperatively [4,5]. Each subunit is of molecular weight ~7.3 x 10^4 Da [6] and contains one dinuclear copper site. The copper ions bind one molecule of oxygen and are coordinated by six histidine residues. The subunits are of eight types with unique but related primary structures [6,7]. When hemocyanin is labelled with antibodies specific for the different subunit types and the resulting aggregates examined using electron microscopy, it is clear that each subunit type occurs only in particular locations within the 48 subunit complex [7-9].

Subunit II of *Limulus* hemocyanin occupies a potentially important location where hexamers interact and is thus believed to be a factor in regulation of cooperative ligand binding [7]. By itself, under conditions where crystallization was effected, subunit II exists as a hexamer that binds oxygen reversibly but not cooperatively, allowing crystals, for example, to be changed from oxygenated to deoxygenated while maintaining the integrity of the crystal lattice. This affords the possibility to make other interesting active site modifications such as oxidized copper ions, cobalt substituted for the copper, and azide, peroxide, carbonyl and nitrous oxide derivatives.

EXPERIMENTAL PROCEDURES

Protein Purification
The soluble protein in the hemolymph, the circulatory fluid of the *Limulus*, is approximately 85% hemocyanin. Upon dialysis into lower chloride and divalent cation concentration conditions at pH 8.9, the 48 subunit hemocyanin complex disassociates. The hemocyanin subunits can readily be separated from each other and from the other soluble proteins using anion exchange column chromatography [6] as shown in Figure 1. The

From K.D. Karlin and Z. Tyeklár, Eds., *Bioinorganic Chemistry of Copper*
(Chapman & Hall, New York, 1993).

hemocyanin subunits elute from the column in the order I, II, IIa, IIIa, IIIb, IV, V and VI. Subunit IIa, a minor component of the second hemocyanin peak, can be separated from subunit II by rerunning the column with a shallower sodium chloride gradient. Approximately 10 mg of subunit II are obtained during this procedure.

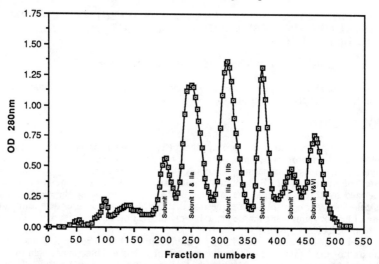

Figure 1. Elution profile showing the first step in the purification of *Limulus* subunit II for crystallization. Sodium chloride concentration varies linearly from 0.1 to 0.4 M during the elution.

Data Collection

Crystallization was effected of *Limulus* subunit II using hanging drop vapor diffusion techniques as previously described [10]. For the 2.4 Å structure determination, two sets of x-ray diffraction data with the properties summarized in Table I were collected from crystals of the oxygenated form of subunit II of *Limulus polyphemus* hemocyanin. The first set was obtained at The University of North Carolina with the assistance of Dr. C.W. Carter, Jr. and his colleagues and the second at McMaster University with the aid of Dr. D.S.-C. Yang. The crystals have the symmetry of the space group R32 with one 7.3 x 10^4 Da subunit per crystallographic asymmetric unit.

TABLE I

	UNC	McMU
Cell constants (Å)	a=b=117.20, c=285.80	a=b=117.03, c=285.20
Resolution (Å)	10.0-3.0	10.0-2.4
R-sym[a]	8.64	5.4
Unique reflections	13387	27713
Possible reflections[b](%)	86.9	71.2

[a] R-Sym = $100((S \text{ II} - <I>)/(S <I>))$
[b] Percent of theoretically observable reflections with intensities 2σ or greater that background.

Phasing
The crystallographic phase problem was solved at 4 Å resolution by molecular replacement as previously described [10]. The known structure of a subunit of the hexameric hemocyanin from *Panulirus interruptus*, now refined at 3.2 Å resolution [11], was used in a preliminary form as the test molecule for the rotation and translation function searches. The *Panulirus* hemocyanin is believed to be in a deoxygenated form [12] and has sequence identity of about 33 percent with *Limulus* II hemocyanin [13,14] (see Table III).

Model Building and Refinement
The model building work to generate atom positions for the oxygenated *Limulus* II hemocyanin structure was done using the programs FRODO [15] and O [16] and an Evans and Sutherland PS390 Picture System. Initial phases for the 2.4 Å data were calculated from the 4 Å structure which was used as the starting model.

The early models of the structure of subunit II of *Limulus* hemocyanin were refined using TNT [17] and a microVAX 3600. Subsequent work used the simulated annealing package X-PLOR [18] run at the Ohio State Supercomputer Center on a Cray Y-MP. At each cycle of refinement, an electron density map incorporating the improved phases generated from the previous cycle's atomic model was examined in detail to build a further improved model. This process was continued until the procedure had converged. Because of the importance of the oxygen binding site to the functional properties of hemocyanin, the molecular oxygen was added to the model of the subunit only after the protien molecule was highly refined and positions for the oxygen atoms clearly and reproducibly visible in difference electron density maps.

RESULTS AND DISCUSSION

The Model Parameters
The current model for the structure of the oxygenated form of subunit II of *Limulus polyphemus* hemocyanin at 2.4 Å resolution is summarized in Table II. They are as expected for a well refined structure of this size and resolution.

TABLE II

Summary of crystallographic model refinement parameters.

Resolution	6.0-2.4 Å
Number of ordered residues	573
Number of disordered residues	55
Number of non-hydrogen atoms	4656
Number of copper atoms	2
Number of solvents included	323
Number of disulfide bonds	2
Average B-value[a]	26.2 Å^2
R-value[b]	0.17
Deviations from ideality (r.m.s):	
Bond distances	0.017 Å
Bond angles	3.53°

[a] B-value-- defined by $B = 8\pi^2 u^2$, where u^2 is an atomic vibration term.
[b] R-value = $\dfrac{S \ |F(obs) - F(calc)|}{S \ F(obs)}$

The Primary and Secondary Structures

A comparison of the primary and secondary structures of *Limulus* II and *Panulirus* a hemocyanins is shown in Table III.

TABLE III[a]

```
       aaaaaaaaaa
   1   TLHDKQIRVCHLFEQLSSATVIGD...........GDKHKHSDRLKNVGK 39
       ||      ||       |    |            ||      |     |
   6   GNAQKQQDINHLLDKIYEPTKYPDLKEIAENFNPLGDTSIYNDHGAAVET 50
       aaaaaaaaaaaa          aaaaaaa              aaaaaa


                          bbbaaaaaaaaaaaaaaaaaabbbaaaaaaaaaa
  40   LQPGAI..........FSCFHPDHLEEARHLYEVFWEAGDFNDFIEIAKE 79
       |     |         | |    || |    | |   |     | |
  51   LMKELNDHRLLEQRHWYSLFNTRQRKEALMLFAVLNQCKEWYCFRSNAAY 100
       aaaaaaaa          aaaaaaaaaaaaaaaa    aaaaaaaaaa


       a      aaaaaaaaaaaaaa        aaaa       aaaaa
  80   ARTFVNEGLFAFAAEVAVLHRDDCKGLYVPPVQEIFPDKFIPSAAINEAF 129
       |   |||  ||    |  |  | |  |    |   |  | |   |
 101   FRERMNEGEFVYALYVSVIHSKLGDGIVLPPLYQITPHMFTNSEVIDKAY 150
       a           aaaaaaaaaa     bbbbbbbbb      aaaaaaaaa


                               aaaa  aaaaaaaaaaaaaa
 130   KKAHVRPEFDESPILVDVQDTGNILDPEYRLAYYREDVGINAHHWHWHLV 179
          |   ||         | |   ||  || |*   |*
 151   S.....AKMTQKQGTFNVSFTGTKKNREQRVAYFGEDIGMNIHHVTWHMD 195
       a      aaa   bbbbbbb            aaaaaaaaaaaaaa


                        aaaaaaaaaaaaaaaaaaaaaaaa
 180   YPSTWNPKYFGKKKDRKGELFYYMHQQMCARYDCERLSNGMHRMLPFNNF 229
       |   |   ||  |||||||  * |  ||  | |||||
 196   FPFWWEDSY.GYHLDRKGELFFWVHHQLTARFDFERLSNWLDPVDELHWD 244
       aa          aaaaaaaaaaaaaaaaaaaaaaa    bbbbb


                                        aaaaaaaaaaaaa
 230   DEPLAGYAPHLTHVASGKYYSPRPDGLKLRDLGDI.EISEMVRMRERILD 278
       |  || ||         |   |||      |            ||
 245   RIIREGFAP.LTSYKYGGEFPVRPDNIHFEDVDGVAHVHDLEITESRIHE 293
       bbbbbbb-bbb   bbbbbbbbbbbb         aaaaaaaaaaaa


       aaaa               aaaaaaaaa       aaaa  aaaaaa
 279   SIHLGYVISEDGSHKTLDELHGTDILGALVESSYESVNHEYYGNLHNWGH 328
       |  ||     ||          |   ||    ||| | |  ||| ||* *
 294   AIDHGYITDSDGHTIDIRQPKGIELLGDIIESSKYSSNVQYYGSLHNTAH 343
       aaaabbbbb   bbbbb    aaaaaaaa           aaaaaaa


       aaaa                      aaaaaaaaaaaaaaaaaa
 329   VTMARIHDPDGRFHEEPGVMSDTSTSLRDPIFYNWHRFIDNIFHEYKNTL 378
       |   |  ||| |      ||||    |  ||| |    *    ||||
 344   VMLGRQGDPHGKFNLPPGVMEHFQTATRDPSFFRLHKYMDNIFKKHTDSF 393
       aaaaaa                      aaaaaaaaaa aaaaaaaa
```

```
        aaaa     bbbbbbbbb     bbbbbbbbbbbb
379 KPYDHDVLNFPDIQVQDVTLHARVDNVVHFTMREQELELKHGINPGNARS 428
    || || | |       | |                        |
394 PPYTHDDLEFSGMVVDGVAIDGELITFFD.EFQYSILDAVDTGESIEDVE 442
    aaaaaa    bbbbbbbbbbbbbbbbbb bbbb                bbb

    bbbbbbbb    bbbbbbbb      bbbbbbbbbbbb        aaa
429 IKARYYHLDHEPFSYAVNVQNNSASDKHATVRIFLAPKYDELGNEIKADE 478
    | ||    |    | |      ||      || |||| |   |      ||
443 INARVHRLNHKEFTYKITMSNNNDGERLATFRIFLCPIEDNNGITLTLDE 492
    bbbbb    bbbbbbbbbbbbbbbbb    bbbbbbbbbbbbbb  bbb aaa

    aaaa bbbbbbbbb  bbbbbbbb               aaaa
479 LRRTAIELDKFKTDLHPGKNTVVRHSLDSSVTLSHQPTFEDL.......L 521
    |    ||||||    |  |  | | |||||   | |    |
493 ARWFCIELDKFFQKVPKGPETIERSSKDSSVTVPDMPSFQSLKEQADNAV 542
    aaaa bbbbbbbbb bbbbbbbbb      bbbbbb aaaaaaaaaaaaa

                                bbbbbbbbbbbbaaaa
522 HGVGLNEHKSEYCSCGWPSHLLVPKGNIKGMEYHLFVMLTDWDKDKVDGS 571
    |              ||| |   | ||    ||| |   || |||
543 NGGNHLDLSAYERSCGIPDRMLLPKSKPEGMEFNLYVAVTDGDKDTEGHN 592
              aaaaaaaa        bbbbbbbbbbbb aaaaa

            aaaa                          bbbbb
572 ESVACVDAVSYCGARDHKYPDKKPMGFPFDRPIHTEHISDFLTNNMFIKD 621
             ||      ||  |  |  |  |  |  |  |   ||
593 GGHDYGGTHAQCGVHGEAYPDNRPLGYPLERRIPDERVIDGVS...NIKH 639
             bbbbbbbb           aaaaaaaa    bbbb

    bbbbb
622 IKIKFHE. 628
    |
640 VVVKIVHH 647
    bbbbbbbb
```

a Line 1 of each set of five shows the secondary structure of the *Limulus* II amino acid residues with a representing alpha helical structure and b beta strand conformations. Line 2 is the primary structure of *Limulus* II hemocyanin [13]. A - alanine, C - cysteine, D - aspartic acid, E - glutamic acid, F - phenylalanine, G - glycine, H - histidine, I - isoleucine, K - lysine, L - leucine, M - methionine, N - aspargine, P - proline, Q - glutamine, R - arginine, S - serine, T - threonine, V - valine, W - tryptophan, Y - tyrosine, ... - gaps introduced to maximize agreement in primary and secondary structures. Line 3 has identities shown as |; the six histidine residues coordinating the two copper ions are marked *. Line 4 is the primary structure of *Panulirus* a hemocyanin [14]. Line 5 is the secondary structure for *Panulirus* hemocyanin [11].

The Tertiary Structure
Of the 628 amino acid residues in the *Limulus* II subunit, 573 are well enough ordered to be included in the model. The remaining 55 occur in regions on the surface of the molecule and not in near proximity to the oxygen binding site.

The folding of the *Limulus* II hemocyanin has the same topology as that of the *Panulirus* (deoxygenated) hemocyanin previously determined [11]. Both arthropod hemocyanins consist of three structural domains.

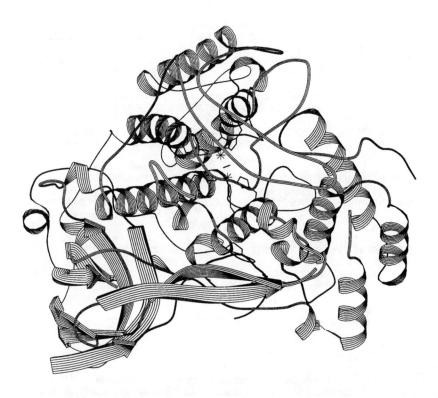

Figure 2. Diagram of the tertiary structure of the oxygenated form of *Limulus* II hemocyanin. The path of folding is shown and the side chains, copper ions and oxygen of the active site. Alpha helical regions of the protein are shown as seven stranded coils. Beta strands are represented as arrows with the tail of the arrow at the amino end of the strand and the head at the carboxyl end. The three stranded lines are well ordered areas of the protein that are neither helical nor beta structure. In the center of the molecule are drawn the six histidine side chains that coordinate the copper ions, shown as eight-pointed stars. The position of molecular oxygen is indicated by a short line segment between the two coppers. In this view, domain 1 is at the bottom of the structure, domain 2 is in the middle and on the right side of the protein and domain 3 is at the top and left.

In the *Limulus* II structure, as indicated in Figure 2, domain 1 includes amino acid residues 1 to 150 and has regions of alpha helical structure and short beta strands. Domain 2, residues 151 to 390, is composed largely of alpha helices. It includes the three histidine residues, His 173, His 177 and His 204 that coordinate the CuA ion of the oxygen binding site, as well as the residues His 324, His 328 and His 364 coordinated to the CuB ion. Domain 3 is residues 391 to 628 and has largely beta secondary structure that is, coincidentally of a topology reminiscent of super oxide dismutase [19,20].

The Oxygen Binding Site

The geometry of the oxygen binding site of *Limulus* II hemocyanin resembles the μ-η^2:η^2 peroxo dinuclear copper (II) complexes studied by Kitajima and his colleagues [21]. Molecular oxygen is bound in the hemocyanin subunit in a μ η^2:η^2 configuration with respect to the copper ions. This is shown in Figure 3.

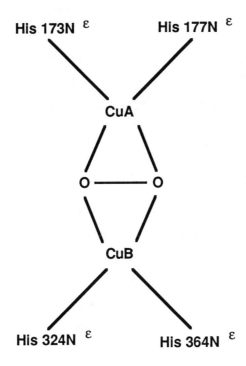

Figure 3. Schematic showing the atoms that are essentially coplanar in the oxygen binding site of Limulus II hemocyanin.

Two peaks were clearly visible in difference electron density maps corresponding to each of the oxygen atoms. These are 1.4 Å apart and symmetrically placed on either side of a line segment connection the two coppers which are 3.6 Å from each other. Molecular oxygen is essentially in the same plane as the coppers and the closer histidine ligands, His 173, His 177, His 324 and His 364. Each copper ion is in a square planar configuration with the two closest histidines and the oxygen atoms. Both coppers are coordinated more weakly to their third histidine ligand which is placed axially; His 204 for CuA and His 328 for CuB.

REFERENCES

1.   K.E. Van Holde and K.I. Miller, *Quart. Rev. Biophys.* **15**, 1 (1982).
2.   H.D. Ellerton, N.F. Ellerton, H.A. Robinson, *Prog. Biophys. Molec. Biol.* **41**, 143 (1983).
3.   M.L. Johnson and D.A. Yphantis, *Biochem.* **17**, 1448 (1978).
4.   M. Brenowitz, C. Bonaventura, J. Bonaventura, **23**, *Biochem.* 879 (1984).
5.   M. Brouwer, C. Bonaventura, J. Bonaventura, *Biochem.*, **16**, 3897 (1977).
6.   M. Brenowitz, C. Bonaventura, J. Bonaventura, E. Gianazza, *Arch. Biochem. Bioph.* **210**, 748 (1981).
7.   J. Lamy, J. Lamy, P.-Y. Sizaret, P. Billiald, P. Jolles, R.J. Feldmann, J. Bonaventura, *Biochem.* **22**, 5573 (1983).
8.   J. Lamy, P.-Y. Sizaret, J. Frank, A. Verschoor, R. Feldmann, J. Bonaventura, *Biochem.* **21**, 6825 (1982).
9.   J. Lamy, J. Lamy, J. Weill, J. Bonaventura, C. Bonaventura, M. Brenowitz, *Arch. Biochem. Biophys.* **196**, 324 (1979).
10. K.A. Magnus, E.E. Lattman, A. Volbeda, W.G.J. Hol, *Proteins* **9**, 240 (1991).
11. A. Volbeda and W.G.J. Hol, *J. Mol Biol.* **209**, 249 (1989).
12. A. Volbeda, M.C. Feiters, M.G. Vincent, E. Bouwman, Dobson, B., K.H. Kalk, J. Reedijk, W.G.J. Hol, *Eur. J. Biochem.* **181**, 669 (1989).
13. H. Nakashima, P.Q. Behrens, M.D. Moore, E. Yokota, A.F. Riggs, *J. Bio. Chem.* **261**, 10526 (1986).
14. H.J. Bak, B. Neuteboom, P.A. Jekel, N.M. Soeter, J.M. Vereijken, J.J. Beintema, *FEBS Lett.* **204**, 141 (1986).
15. J.W. Pflugrath,M.A. Saper, F.A. Quiocho, In *Methods and Applications in Crystallographic Computing*, S. Hall and T. Ashiaka, Eds., (Clarendon, Oxford, 1984), pp. 404-407.
16. T.A. Jones, J.-Y. Zou, S.W. Cowan, M. Kjeldgaar, *Acta Cryst*, **A47**, 110 (1990).
17. D. Tronrud, and L.F. Ten Eyck, In *TenEyck-Tronrud Refinement Package*, (c. Oregon State Board of Higher Education, Version 4-a., 1990).
18. A. Bruenger, *J. Mol. Biol.* **203**,803 (1988).
19. J.A. Trainer, E.D. Getzoff, K.M. Beem, J.S. Richardson, D.C. Richardson, *J. Mol. Biol.*, **160**, 181 (1982).
20. W.P.J. Gaykema, W.G.J. Hol, J.M.Vereijken, N.M. Soeter, H.J. Bak, J.J. Beintema, *Nature*, **309**, 23 (1984).
21. N. Kitajima, K. Fujisawa, C. Fujimoto, Y. Moro-oka, S. Hashimoto, T. Kitagawa, K. Toriumi, K.Tassumi, A. Nakamura, *J. Amer. Chem. Soc.*, 114 , 1277 (1992).

ACKNOWLEDGEMENTS

This work was supported grants from the National Scuence Foundation (DMB-9004561) and the Ohio Supercomputer Center.

NEW PROBES OF OXYGEN BINDING AND ACTIVATION: APPLICATION TO DOPAMINE β-MONOOXYGENASE.

Judith P. Klinman,[a]* Joseph A. Berry[b] and Gaochao Tian[a]*

[a]Department of Chemistry, University of California, Berkeley, CA 94720
[b]Department of Plant Biology, Carnegie Institution of Washington, 290 Panama St., Stanford, CA 94305

INTRODUCTION

Monooxygenases represent a class of enzymes catalyzing a reductive activation of dioxygen, linked to the insertion of an oxygen atom into a C–H bond:

$$R\text{–}H + O_2 \xrightarrow{\;2e^-,\,2H^+\;} R\text{–}OH + H_2O \tag{1}$$

Basic properties of monooxygenase-type systems which must be taken into account in the development of a reaction mechanism include (i) the splitting of the dioxygen bond, (ii) the interaction of spin unpaired (triplet) oxygen with spin paired (singlet) substrates and (iii) the large thermodynamic driving force deriving from the greater bond energy of the O–H of water relative to the C–H of substrate. In order to circumvent the spin forbidden properties of these reactions, nature has evolved a number of strategies to activate dioxygen involving the use of either organic cofactors or transition metal ions. Although the total number of enzyme systems found to use copper in this capacity is small, eukaryotic copper containing enzymes catalyze physiologically important monooxygenase reactions, e.g., those catalyzed by tyrosinase,[4] dopamine β-monooxygenase[1] and peptidyl α-amidating enzyme.[5]

In the present study we have focused on the relationship between structure and function in dopamine β-monooxygenase (DβM), with the long range goal of developing mechanistic probes that will be applicable to monooxygenases as a class. As has been reviewed,[1] the soluble form of DβM exists as active dimeric and tetrameric species. With the exception of the degree of glycosylation and small differences at the N-terminus, the structure of each enzyme subunit is identical and contains two catalytically essential copper binding sites. The ligands to copper are predominantly histidines, independent of the oxidation state of copper; additionally, a single copper atom has been proposed to be coordinated to sulfur in its reduced Cu(I) form (cf. Blackburn et al., this volume). In the absence of three dimensional structural information, the further identification of active site ligands has relied on enzyme labeling by mechanism based inhibitors. Fairly extensive

* This work supported by a grant to J.P.K. from the National Institutes of Health (GM39296) and to G.T. from the American Heart Association (91-34)

work of this nature has implicated two tyrosines at the enzyme active site: tyr-216 and tyr-357.[6,7]

The reaction catalyzed by DβM involves an insertion of oxygen into the β-carbon of the side chain of phenethylamine derivatives:

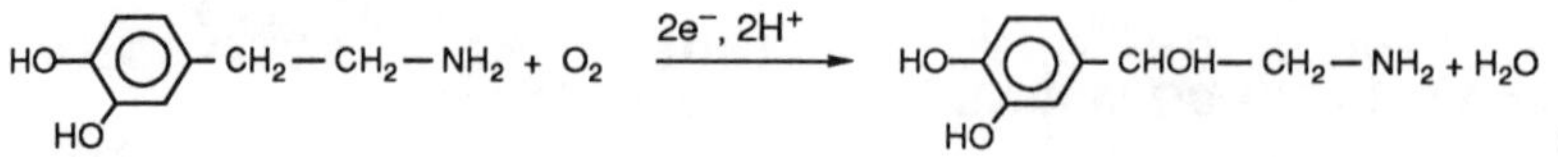

$$HO-\langle O \rangle-CH_2-CH_2-NH_2 + O_2 \xrightarrow{2e^-, 2H^+} HO-\langle O \rangle-CHOH-CH_2-NH_2 + H_2O \tag{2}$$

In vitro, the preferred 2 e⁻ donor can be demonstrated to be ascorbate. Given the presence of high concentrations of ascorbate within the DβM-containing storage vesicles, ascorbate is inferred to be the electron donor in vivo. A series of rapid mixing experiments have been carried out examining the reduction of enzyme bound copper by ascorbate and the reoxidation of enzyme bound copper by substrate and dioxygen.[8] Both coppers in free enzyme are found to be rapidly reduced by ascorbate in a single exponential process. Reoxidation of the reduced copper centers in the presence of substrate and dioxygen also occurs in a single exponential, with a rate constant which is essentially identical to that for the formation of enzyme bound product. Importantly, no evidence for spin coupling between the copper centers has been detected either in resting enzyme or in the catalytically generated enzyme product complex. This result argues that the two copper centers are at least 4 Å apart in the E and E·P complexes and hence, that catalysis is unlikely to involve a binuclear copper center of the type observed in tyrosinase and hemocyanin (cf. Solomon, this volume). Unexpectedly, the E·P complex has been found to undergo reduction by high levels of ascorbate, indicating that product and ascorbate can bind to enzyme at the same time.[8] On the basis of the available data, a model has been put forth in which the two copper centers in DβM perform different functions, that of electron transfer [Cu(II)$_A$ site] and substrate hydroxylation [Cu(II)$_B$ site],[8] Scheme 1.

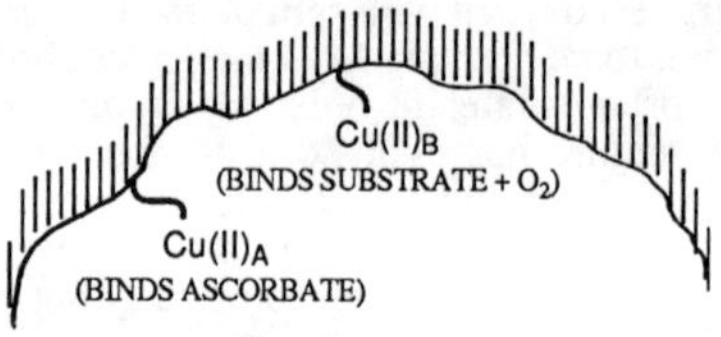

Scheme 1

The recent findings of Blackburn and co-workers, showing that carbon monoxide is competitive toward dioxygen at a single copper site per subunit, provide experimental support for this view of catalysis.[9]

At this time, a fairly detailed kinetic mechanism can be written for DβM, which involves a reduction of Cu(II) by ascorbate, followed by a random binding of substrate and dioxygen to form a ternary complex comprised of two Cu(I) centers, substrate and dioxygen, Scheme 2.

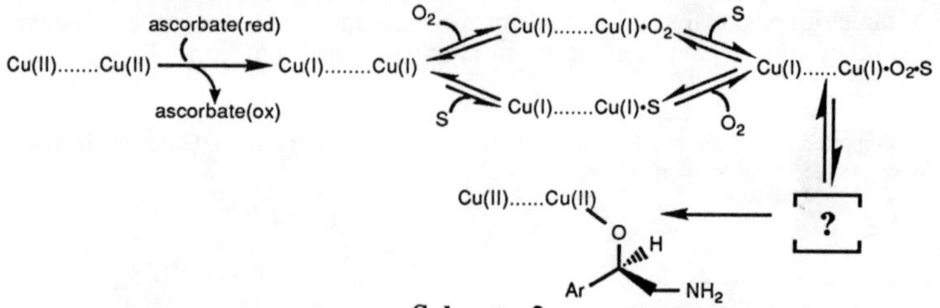

Scheme 2

152

From a chemical perspective, the challenge has been to understand the steps leading from this ternary complex to enzyme bound product, represented by the large question mark in Scheme 2. A number of previously established results bearing on the chemical mechanism include (i) the demonstration that proton uptake precedes substrate hydroxylation, ascribed to formation of a Cu(II)–OOH intermediate and (ii) the finding that activation of substrate occurs by H· abstraction, leading to the formation of a benzylic radical intermediate, R-C·H-CH$_2$-NH$_2$.[1] A major unresolved issue has been the precise nature of the oxygen species catalyzing H· abstraction, in particular whether this is Cu(II)-OOH itself or a species derived by further reductive activation of Cu(II)–OOH. Miller and Klinman originally argued that the greater energy content of the O–H bond in product water relative to the C–H bond of substrate could be used to drive a concerted process,[10] eq (3):

$$\tag{3}$$

Although the mechanism shown in eq (3) is compatible with the overall thermodynamic driving force of the DβM mechanism and provides a satisfactory working model for this enzyme as well as other non-heme metallo-oxygenases, alternative intermediates are not excluded. In particular, the possible requirement for O–O cleavage prior to substrate activation must be considered. In this instance a copper-oxygen radical is responsible for C–H activation, eq (4):

$$\tag{4}$$

Historically, the pursuit of mechanistic studies of dopamine β-monooxygenase, as well as other enzymes catalyzing C–H abstraction processes, has relied heavily on the use of hydrogen isotope effects as a probe of rate limiting steps and substrate-derived intermediates. In the case of oxygen utilizing enzymes, it has proven more difficult to establish routine protocols for the examination of dioxygen-derived species. While spectroscopy has been used in selected instances for the characterization of oxygen intermediates, such species are most often too short lived to permit their direct detection. We have initiated a program aimed at quantitating isotope effects in dioxygen dependent processes through the discrimination between O–16 and O–18. Although these isotope effects are expected to be quite small [less than 6-7% maximally[11]], they can be accurately and reproducibly measured. As we now describe, measurement of oxygen isotope effects in the DβM reaction leads to the unexpected result that Cu(II)–OOH must be reductively cleaved prior to substrate activation. Using established features of the DβM active site, a possible mechanism for the generation of a copper-oxygen radical species as the hydroxylating species is proposed.

RESULTS AND DISCUSSION

Oxygen Isotope Effects: A General Probe for Bond Order Changes at O2.
In this paper we describe the development of O_2 isotope effects as a probe for changes in bond order at O_2 in dioxygen binding and consuming reactions. A major consideration in the measurement of oxygen isotope effects is the precision with which these values can be measured. Isotope ratio mass spectrometry is the analytical method of choice, assuming first, that reactant (or product) can be quantitatively recovered from the reaction mixture and second, that the reactant can be analyzed by mass spectrometry without fragmentation. Our approach has been to isolate unreacted dioxygen from reaction mixtures in sufficient yields to allow precise measurement. Given the relatively low concentration of dissolved O_2 in buffered water solutions (ca. 0.2 mM), this requires the use of large volumes of protein solution; in general, 15-30 ml is used per O_2 analysis. Although O_2 itself could be analyzed by isotope ratio mass spectrometry, conversion to CO_2 provides a chemically inert product which is preferable for routine analysis. Each measurement involves removal of a pre-determined volume from the reaction mixture contained in the syringe attached to a high vacuum line, followed by sparging of this sample by a stream of helium, which removes O_2, N_2, CO_2 and H_2O. The latter two components are removed by liquid nitrogen traps and the former two components are trapped in molecular sieves at low temperature. Warming of the molecular sieves leads to release of N_2 and O_2, which can either be separated by gas chromatography or combusted directly to CO_2. If combustion is carried out on the mixture of gases, care must be used to assure that the presence of N_2 does not interfere with the complete conversion of O_2 to CO_2. The total concentration of CO_2 and hence, O_2 collected is quantitated by a pressure transducer, providing a measure of the fractional conversion of O_2 to H_2O in the enzymatic reaction. Finally, the CO_2 is collected and analyzed by isotope ratio mass spectrometry.

Prior to analysis of oxygen isotope effects in enzymatic reactions, it was essential to obtain background information regarding the magnitude of O-18 effects in complexes with defined structural constraints. We therefore turned to an analysis of the discrimination between O-16 and O-18 in reversible dioxygen binding to the metallo-O_2 carriers found in nature: hemocyanin (H_c), hemerythrin (H_r), myoglobin (M_b) and hemoglobin (H_b). As shown in Table 1, the equilibrium O-18 isotope effects [$^{18}(K_{eq})$] are small, but the reproducibility is excellent and experimental error is much less than the measured effects. Significantly, there is a trend as we proceed from the ancestral (H_c and H_r) to the more modern dioxygen carriers (M_b and H_b), with a high value of 1.0184 for H_c vs. 1.0039 for H_b.

Table 1. Equilibrium O-18 Isotope Effects for Reversible O_2 Binding Proteins[a]

Protein $\quad P + O_2 \underset{}{\overset{K_{eq}}{\rightleftharpoons}} P{\cdot}O_2$	$^{18}(K_{eq})$
H_c	1.0184 ± 0.0023
H_r	1.0113 ± 0.0005
M_b	1.0054 ± 0.0006
H_b	1.0039 ± 0.0002

[a] Data measured at pH 7.8, 25°C

Ideally, the values for $^{18}(K_{eq})$ summarized in Table 1 can be related to the structure of bound oxygen in each of the respective complexes. Extensive spectroscopic and x-ray investigations of the P·O_2 complexes of Table 1 indicate a partial reduction of bound O_2 by 1 to 2 e⁻, generating species which can be formulated as superoxy or peroxy intermediates. These type of reductions are summarized in eq (5), showing the attendant changes in bond order at oxygen:

$$\begin{array}{c} & {}^{1e^-} & & {}^{1e^-,\,2H^+} & & {}^{1e^-,\,1H^+} & & {}^{1e^-,\,1H^+} \\ O_2 & \longrightarrow & O_2^- & \longrightarrow & H_2O_2 & \longrightarrow & H_2O + OH\cdot & \longrightarrow & 2H_2O \end{array} \qquad (5)$$

Bond Order: 2 $\longrightarrow$ 1.5 $\longrightarrow$ 2 $\longrightarrow$ 1.5 $\longrightarrow$ 2

In an effort to correlate bond order changes with the magnitude of ${}^{18}(K_{eq})$, we have applied published computational methods to the estimation of equilibrium isotope effects[12] for the reactions in eq (5), using literature values of vibrational frequencies and/or force constants for the initial and final oxygen species.

Table 2. Calculated Values for ${}^{18}(K_{eq})$

Reaction	${}^{18}(K_{eq})$	Δ Bond Order
$O_2 \overset{1e^-}{\rightleftharpoons} O_2^-$	1.03309	-0.5
$O_2 \overset{1e^-,\,1H^+}{\rightleftharpoons} HO_2\cdot$	1.01034	0
$O_2 \overset{2e^-}{\rightleftharpoons} O_2^{2-}$	1.04962	-1
$O_2 \overset{2e^-,\,1H^+}{\rightleftharpoons} HO_2^-$	1.03430	-0.5
$O_2 \overset{2e^-,\,2H^+}{\rightleftharpoons} H_2O_2$	1.00890	0

The data summarized in Table 2 indicate a number of informative trends, which include an average value of ${}^{18}(K_{eq}) = 1.02925$ for a bond order change of -0.5 (entries 1, 3 and 4) and an average value of ${}^{18}(K_{eq}) = 1.00962$ for the conversion of O=O to O–H with constant bond order (entries 2 and 5). The value of 3% obtained for a bond order change of 0.5 is very satisfying, since previous studies indicate a value of *ca.* 6% for the complete cleavage of a C–O bond.[11] In the absence of force constants for the relevant copper (H_c) and iron (H_r, M_b, H_b) oxygen complexes, it is not possible to calculate equilibrium isotope effects which represent the individual protein oxygen complex of Table 1. However, substitution of a metal for a proton may produce a very small change, reflecting opposing effects of an increase in mass and decrease in force constant at the metal center. In support of this view, calculations of isotope effects resulting solely from an O–H *vs.* O–M bond stretching mode [where M is Cu(II) or Fe(III)] indicate virtually identical results.

In Table 3 we show the relationship of the measured values of ${}^{18}(K_{eq})$ (Table 1) to computed values (Table 2), together with structures of the deduced metal oxygen complexes of H_c, H_r and H_b. The most straightforward entry in Table 3 is H_r, which has been demonstrated to bind O_2 as a metal hydroperoxide complex, H-bonded to the bridging oxygen of the binuclear iron center. We note that the experimental value of ${}^{18}(K_{eq})$ for this complex is close to the computed value for H_2O_2. Consistent with the known H_r structure, the experimental isotope effect is slightly elevated from the computed value for H_2O_2, which we ascribe to a weakening of the O-H bond of the Fe(III)–OOH by H-bond to the bridging oxygen. In the case of H_b, structural studies provide strong support for a metal superoxy intermediate, H-bonded to a distal histidine.The reduction of the experimental value of ${}^{18}(K_{eq})$ from the value expected for protonated superoxy supports a fairly tight H-bond under the conditions of our experiment; such an interaction would increase the bonding at the terminal oxygen of the superoxy intermediate, bringing its isotope effect back down toward unity. Finally, the data for H_c can be seen to lie between H_2O_2 and HO_2^-. Given the recent evidence for an side on, μ-η^2:η^2 complex for O_2 in H_c, analogous to the structure originally reported by Kitajima in a model binuclear copper system,[13] the bonding in H_c is unlike any simple species that we have attempted to model. All one can say at the moment is that the bonding at oxygen in H_c lies somewhere between that expected for a fully protonated neutral peroxide and a monoprotonated anionic peroxide intermediate.

Table 3. Structure and Bonding in Protein Oxygen Complexes

Protein	Proposed Structure	Bonding[d]
H_c	*a*	$H_2O_2 < H_c < HO_2^-$ $1.0089 < 1.0184 < 1.0343$
H_r	*b*	$H_2O_2 \sim H_r$ $1.0089 \sim 1.0113$
H_b	*c*	$H_b < HO_2 \cdot$ $1.0039 < 1.0103$

[a] Magnus and Ton-That, this volume.
[b] Ref. 14.
[c] Ref. 15. The heme coordinated to iron is represented by a circle; HB represents the protonated distal histidine.
[d] Tables 1 and 2.

Nature of the Oxygenating Species in DβM.
With the background information available from equilibrium O–18 isotope effects, it has been possible to proceed to an analysis of kinetic O–18 effects in an enzymatic reaction, i.e. that of DβM. Earlier studies of DβM had provided both deuterium isotope effects and rate constants for the C–H bond cleavage step at the β-carbon of a range of phenethylamine derivatives:[10]

We now extend these kinetic studies to include the measurement of O–18 isotope effects, using both protonated and deuterated substrates (Table 4).

Table 4. Kinetic Isotope Effects for DβM[a]

X	Y	$^D(V/K)$	$^{18}(V/K)_H$	$^{18}(V/K)_D$
OH	OH	3.68±0.56	1.0197±0.0003	1.0256±0.0003
H	H	4.35±0.46	1.0179±0.0002	1.0238±0.0006
CH₃	H	6.50±0.51	1.0212±0.0008	1.0232±0.0009
Br	H	11.85±0.62	1.0205±0.0001	1.0214±0.0001
CF₃	H	18.50±1.40	1.0215±0.0003	1.0216±0.0003

[a] Data collected at pH 5.5, 25°C, 10 mM fumarate; these conditions lead to a kinetically ordered mechanism where substrate binds first.

One feature of the data in Table 4 that has been noted previously is the large range in the magnitude of deuterium isotope effects, reflecting an increasing rate limitation of the H-transfer step in proceeding from electron releasing to electron withdrawing substituents.[10] The effect is carried over to O-18 isotope effects, such that the magnitude of $^{18}(V/K)_H$ increases with electron withdrawing substituent. It can be seen that deuteration also increases the magnitude of $^{18}(V/K)_D$ for substrates where the H-transfer step is only partially rate limiting. Clearly, the measured O-18 isotope effects are complex, reflecting the contribution of multiple steps to the DβM reaction. Additionally, we can anticipate O-18 isotope effects on a number of processes, which include dioxygen binding, its reduction to a copper hydroperoxy intermediate and hydrogen abstraction from substrate. The general form for the expression of the O-18 isotope effect is shown in eq (6), where it can be seen that the magnitude of $^{18}(V/K)_H$ reflects a term which is the product of an equilibrium isotope effects for the formation of all reversible, oxygen dependent process, $^{18}K_{eq}$, and the O-18 isotope effect on the irreversible C-H cleavage step, ^{18}k.

$$^{18}(V/K)_H = \frac{^{18}K_{eq}{}^{18}k + C_f'}{1 + C_f} \qquad (6)$$

The terms C_f' and C_f are the commitment factors and contain ratios of rate constants for C-H cleavage relative to other steps in the reaction mechanism. We note that this equation differs from the equations for deuterium isotope effects on V/K defined by Northrop[16] in that different commitments appear in the numerator and denominator, a result of the presence of multiple O-18 isotope effects in the overall reaction. Expressions similar to eq (6) can also be written for $^{18}(V/K)_D$ and $^D(V/K)$ and these are shown in eqs (7) and (8).

$$^{18}(V/K)_D = \frac{^{18}K_{eq}{}^{18}k + C_f'/^Dk}{1 + C_f/^Dk} \qquad (7)$$

$$^D(V/K) = \frac{^Dk + C_f}{1 + C_f} \qquad (8)$$

Algebraic rearrangement of eqs (6)-(8) leads to a unique solution for the intrinsic isotope effect, $(^{18}K_{eq}{}^{18}k)$, from the experimentally accessible parameters:

$$^{18}K^{18}k = \frac{^D(V/K)^{18}(V/K)_D - {}^{18}(V/K)_H}{^D(V/K) - 1} \qquad (9)$$

The values calculated for $^{18}K^{18}k$ as a function of substrate structure are summarized in Table 5, with the attendant error propagated from experimental uncertainties in $^{18}(V/K)_H$, $^{18}(V/K)_D$ and $^D(V/K)$.

Table 5. Intrinsic O-18 Isotope Effects in the DβM Reaction.

X	Y	$k(s^{-1})^a$	$^{18}K^{18}k^c$	$^{18}k^d$
OH	OH	680	1.0281± 0.0012	1.0166
H	H	500 (0.35)^b	1.0256±0.0008	1.0141
CH$_3$	H	140 (1.86)	1.0236±0.0011	1.0122
Br	H	22 (4.09)	1.0215±0.0001	1.0101
CF$_3$	H	2 (6.89)	1.0216±0.0003	1.0102

^{a} From ref. 10.
^{b} Increase in activation energy relative to the parent compound, kcal/mol.
^{c} From eq (9) in text and the data in Table 4.
^{d} Calculated, assuming $^{18}K = 1.0113$ (Table 1).

Inspection of the calculated values for $^{18}K^{18}k$ in Table 5 indicates that these are close to $^{18}(V/K)_D$ (Table 4). This a significant result, indicating that deuteration of substrate reduces the magnitude of C_f' and C_f to the point where the intrinsic O–18 isotope effect is almost completely expressed and hence, that mechanistic deductions will be similar using either the experimental values for $^{18}(V/K)_D$ or calculated values for $^{18}K^{18}k$. It can be seen that a trend occurs, such that the magnitude of $^{18}K^{18}k$ decreases as the rate of the reaction decreases (>300-fold); this decrease in rate correlates with a change in activation energy for C–H cleavage of almost 7 kcal/mol. We have also included an estimate of values of ^{18}k, using the measured value for Fe(III)–OOH formation (Table 1) as a model for E-Cu(II)–OOH formation from DβM and dioxygen. These estimates of ^{18}k, which are based on the assumption that ^{18}K will be independent of substrate, indicate a *ca.* 60% increase in ^{18}k across the range of substrates studied. As we argue below, the relationship between $^{18}K^{18}k$ (or ^{18}k) and the rate of the reaction provides a unique probe of the oxygen species performing the C–H activation step.

Initially, a mechanism in which the active oxygen species is Cu(II)–OOH has been examined, eq (3). As shown in Scheme 3, early and late transition states for this mechanism differ in the amount of bond cleavage at the oxygen atom bound to copper, where the terminal O–H of the Cu(II)–OOH is assumed to experience very little change in bond order due to compensating effects of O–O bond cleavage and O–H bond making. According to the Hammond postulate, the slowest substrate within a series is expected to be characterized by the latest transition state.[17] We therefore predict that the decrease in bond order between the O–O bond of Cu(II)–OOH should be greatest for the *p*-CF$_3$-phenethylamine and least for the dopamine. A corollary of this prediction is that the magnitude of the O–18 isotope effect should increase as the rate of the reaction decreases. As seen in Table 5, this is in direct conflict with the experimental observations, leading to the conclusion that Cu(II)–OOH cannot be the species responsible for C–H activation in the DβM reaction.

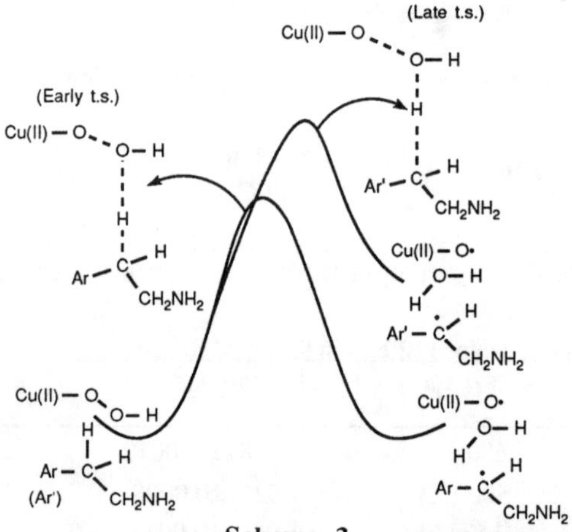

Scheme 3

As an alternative to Cu(II)–OOH, a copper oxygen radical species (Cu(II)–O·) formed by reductive cleavage of the initially formed Cu(II)–OOH has been considered [cf. eq (4)]. For this type of mechanism, hydrogen transfer from substrate to oxygen can only increase the bond order at oxygen, Scheme 4. As illustrated, proceeding from an early to a late transition state leads to an increase in bond order at oxygen and hence, a decrease in the kinetic O–18 isotope effect. As a result, the mechanism shown in Scheme 4 predicts the largest isotope effect with dopamine and the smallest isotope effect with *p*-CF$_3$-phenethylamine, in complete accord with the experimental observations.

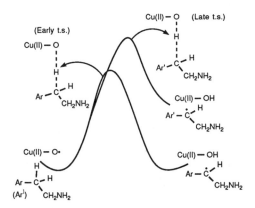

Scheme 4

We conclude that the observed changes in $^{18}K^{18}k$ with changes in substrate structure and rate implicate O–O cleavage prior to substrate activation, bringing the mechanism of DβM closer to models previously invoked for heme-iron systems such as cytochrome P-450.[18]

A Postulated Role for Protein Side Chains in Metal-Oxygen Activation.
In light of the evidence for O–O bond cleavage prior to H abstraction from substrate in a copper dependent monooxygenase, the question arises as to how this can be achieved. Unlike iron dependent monooxygenases where the metal center can exist in multiple high valence states, copper does not have ready access to valence states above +2 [for example, the redox potential for Cu(II) → Cu(III) is *ca.* 1 volt (cf. Margeram et al., this volume)]. On this basis, we eliminate a mechanism involving Cu(II) to Cu(III) conversion coupled to reductive cleavage of peroxide:

$$Cu(II){-}OOH \longrightarrow Cu(III){-}O{\cdot} + H_2O \longleftrightarrow Cu(IV){-}O^- + H_2O \qquad (9)$$

As shown in eq (9), this type of cleavage of the O–O bond produces a resonance hybrid of a Cu(III) (oxygen radical) and a Cu(IV) (oxygen anion)!
Recent studies from this laboratory on substrate analogs of the DβM reaction provide possible insight into a mechanism for Cu(II)–OOH reductive activation. As shown below, a series of compounds, **1, 3** and **4**, have been studied, which can be considered subsets of the substrates dopamine and tyramine:

Catechol	Dopamine	Phenol	p-Cresol	Tyramine
1		**3**	**4**	

Additionally, the behavior of a *p*-quinol, HO—⟨O⟩—OH (**2**), has been examined for comparison to catechol (**1**). As summarized in Table 6, compounds **1 - 4** all show inhibitory activity toward DβM. Enzyme turnover is also observed, with the exception of phenol, placing compounds **1, 2** and **4** in the category of mechanism-based inhibitors.

Table 6. Comparison of Turnover (k_{cat}) to Inactivaton (k_i) with Quinols and Phenols[a]

Compound	$k_{cat}(s^{-1})$	K_m (mM)	$k_i(s^{-1})$	K_i (mM)	$\dfrac{k_{cat}}{k_i}$
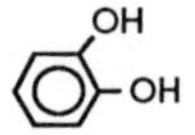 ![structure] OH–C6H4–OH (catechol)	0.34	3.0	0.016	3.5	21
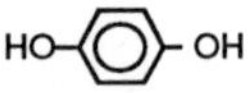 HO–C6H4–OH	0.23	1.7	0.0055	2.7	42
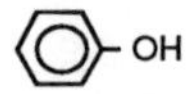 C6H5–OH	NR[b]	NR[b]	0.0017	4.3	--
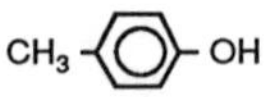 CH3–C6H4–OH [c]	3.3	1.5	0.0031	2.5	1100

[a] Ref. 3.
[b] No reaction.
[c] These values are similar to values reported earlier by Kruse and co-workers (ref. 19).

For the quinol compounds, 1 and 2, inactivation is a fairly efficient process relative to turnover, revealing low values for k_{cat}/k_i, independent of the position of the ring hydroxyl. Surprisingly, the rates of enzyme inactivation with phenol is comparable to *p*-cresol, despite the fact that phenol is incapable of supporting enzyme turnover.

In an effort to gain mechanistic insight into the processes of Table 6, a series of solvent and substrate isotope effects were undertaken. Solvent isotope effects with quinols were found to be quite small, ranging from 1.7-2.5 for both turnover and inactivation. Given the ease with which quinols can undergo 1 e$^-$ oxidation processes, the origin of these small values is most likely a combination of small (secondary) effects on the formation of radical cation, which then partitions further between a 1 e$^-$ oxidation to yield enzyme bound quinone and enzyme inactivation, Scheme 5.

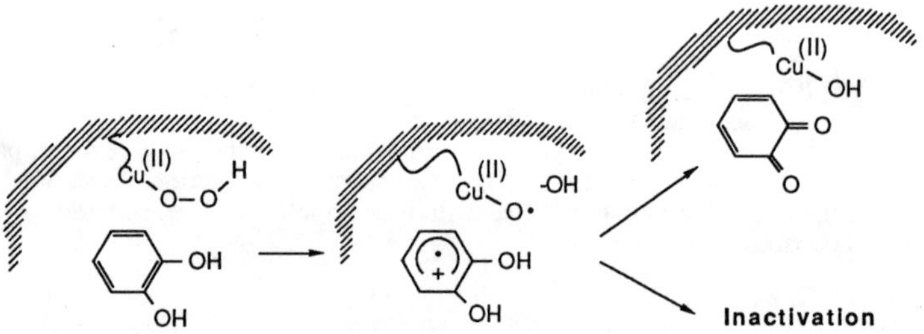

Scheme 5

The results of solvent isotope effect determinations with phenols were found to be very different from those with quinols, revealing differential effects on turnover *vs.* inactivation and extremely large solvent isotope effects on enzyme inactivation, Table 7. The magnitude of ^Dk_i is well outside the range expected for secondary effects, implicating a rate limiting abstraction of hydrogen from the solvent exchangeable, phenolic hydroxyl group. We note that these measurements provide the first experimental evidence for a DβM catalyzed O–H bond cleavage process. In contrast to ^Dk_i, the value for $^Dk_{cat}$ with cresol is close to unity. This result is not unexpected, given an expected cleavage of C–H bonds in DβM catalyzed substrate turnover.

Table 7. Solvent Isotope Effects for Phenols[a]

Compound	$D_{k_{cat}}$	D_{k_i}
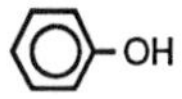 ⬡-OH	NR[b]	5.7
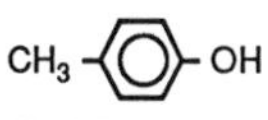 CH$_3$-⬡-OH	1.1	7.3

[a] Ref. 3.
[b] No reaction.

A comparison of substrate and solvent isotope effects in Table 8 confirms the view of different bond cleavage events for cresol turnover *vs.* enzyme inactivation.

Table 8. Comparison of Substrate and Solvent Isotope Effects with *p*-Cresol [a]

Label	$D_{k_{cat}}$	D_{k_i}
p-cresol-d7	5.2	0.95
D$_2$O	1.1	7.3

[a] Ref. 3

As summarized in Table 8, use of perdeuterated *p*-cresol reverses the trend seen in Table 7, such that a large isotope effect is now seen on substrate turnover, with no effect on enzyme inactivation.

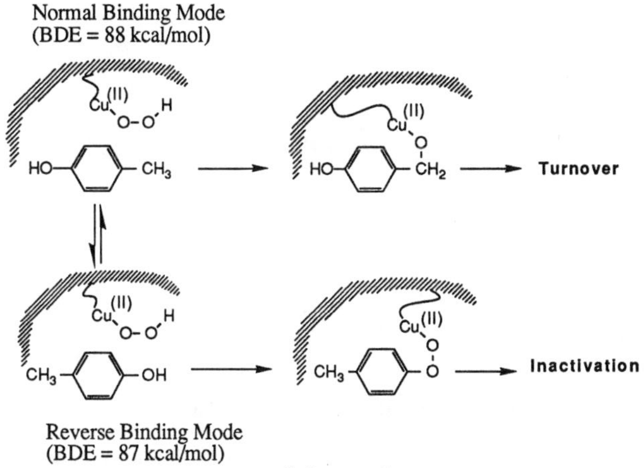

Scheme 6

A reaction mechanism comparable with the kinetic behavior of *p*-cresol is shown in Scheme 6. The small size of cresol, relative to the phenethylamine substrates, is proposed to lead to more than one orientation for *p*-cresol within the binding site. Analogous to normal amine substrates, the methyl group of cresol can bind in proximity to Cu(II)–OOH, leading to C-H activation and turnover. Alternatively, placement of the hydroxyl group of cresol near Cu(II)–OOH will produce O–H cleavage and ultimately enzyme inactivation. The much slower rate of enzyme inactivation is attributed to a lower population of the reverse binding mode, since the almost identical bond dissociation energies (BDE's) for

H· abstraction from a benzylic methyl group or phenolic hydroxyl group predict similar chemical reactivities.[3]

At this juncture, the recent evidence for a DβM catalyzed O–H abstraction[3] can be integrated with the demonstrated presence of active site tyrosine residues[6,7] to explain the trend in kinetic oxygen isotope effects observed in DβM catalysis (Table 5). A new reaction mechanism is introduced, Scheme 7, showing hydrogen transfer from the phenolic group of an active site tyrosine to Cu(II)–OOH.

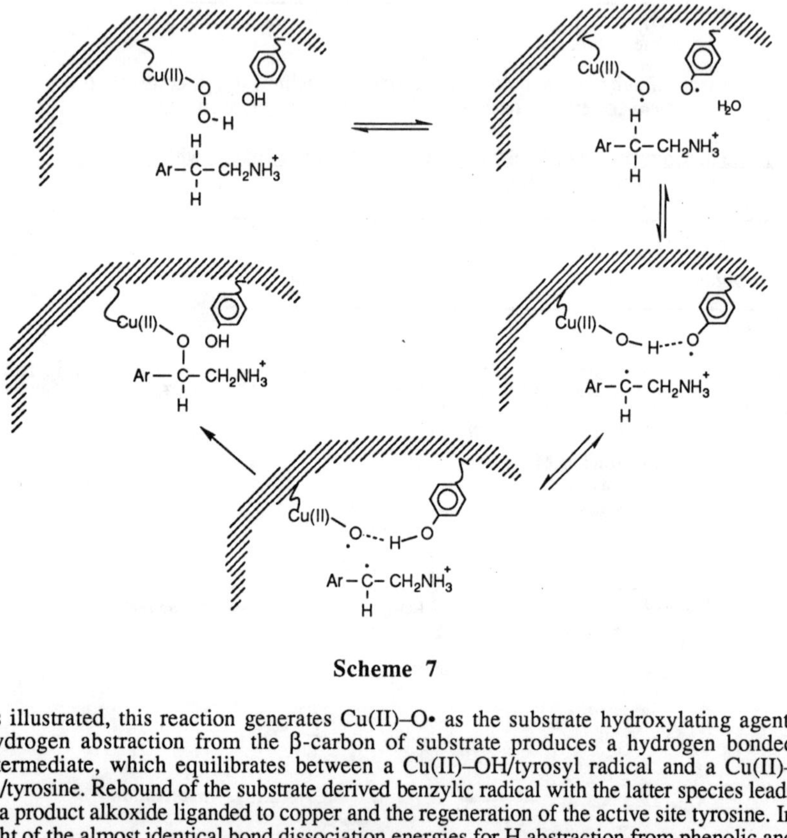

Scheme 7

As illustrated, this reaction generates Cu(II)–O• as the substrate hydroxylating agent. Hydrogen abstraction from the β-carbon of substrate produces a hydrogen bonded intermediate, which equilibrates between a Cu(II)–OH/tyrosyl radical and a Cu(II)–O•/tyrosine. Rebound of the substrate derived benzylic radical with the latter species leads to a product alkoxide liganded to copper and the regeneration of the active site tyrosine. In light of the almost identical bond dissociation energies for H abstraction from phenolic and benzylic positions, activation of the Cu(II)–OOH via a tyrosyl O–H cleavage does not offer an energetic advantage, relative to a direct C–H abstraction from phenethylamines. We propose, instead, an evolutionary advantage to the formation of protein-derived functional units for oxygen activation [Cu(II)–OOH and tyrosine in the case of DβM], providing independent reactive species with the potential to oxidize a wide range of organic substrates. It will, therefore, be very interesting to see if the type of chemistry proposed in Scheme 7 can be generalized to other non-heme, metallo-enzyme systems. In the case of DβM, our proposed mechanism can be tested by selective substitution of tyrosine residues by phenylalanines, although these experiments must await the successful expression of the cloned DβM genes.[20]

REFERENCES

1. L.C. Stewart and J.P. Klinman, *Annu. Rev. Biochem.*, **57**, 551 (1988).
2. R.P. Guy, M.F. Fogel, J.A. Berry, and T.C. Hoering, *Progr. in Photosynth. Res.*, **3**, 597 (1987).
3. S.-C. Kim and J.P. Klinman, *Biochemistry*, **30**, 8138 (1991).
4. H.S. Mason, *Annu. Rev. Biochem.*, **34**, 594 (1965).
5. a. A.F. Bradbury, M.D.A. Finnie and D.G. Smyth, *Nature,* **298**, 686 (1982).
   b. B.A. Eipper, R.E. Mains and C.G. Glembotski, *Proc. Natl. Acad. Sci USA*, **80**, 5144 (1983).
6. P.F. Fitzpatrick and J.J. Villafranca, *Arch. Biochem. Biophys.*, **257**, 231 (1987).
7. W.E. DeWolf, Jr., S.A. Carr, A. Varrichio, P.J. Goodhart, M.A. Mentzer, G.D. Roberts, C. Southan, R.E. Dolle and L.I. Kruse, *Biochemistry*, **27**, 9093 (1988).
8. M. Brenner and J.P. Klinman, *Biochemistry*, **28**, 4664 (1989).
9. N.J. Blackburn, T.M. Pettingill, K.S. Seagraves and R.T. Shigeta, *J. Biol. Chem.*, **265**, 15383 (1990).
10. S.M. Miller and J.P. Klinman, *Biochemistry*, **24**, 2114 (1985).
11. J.P. Klinman, *Adv. in Enzymology*, **46**, 415 (1978).
12. L. Melander and W.H. Saunders, *Reaction Rates of Isotopic Molecules* (Wiley, New York, 1980).
13. N. Kitajima, K. Fujisawa, Y. Moro-oka and K. Toriumi, *J. Am. Chem. Soc.*, **111**, 8975 (1989).
14. a. R.E. Stenkamp, L.C. Sieker and L.H. Jensen, *J. Am. Chem. Soc.*, **106**, 618 (1984).
    b. A.K. Shiemke, T.M. Loehr and J. Sanders-Loehr, *J. Am. Chem. Soc.*, **108**, 2437 (1986).
15. a. C.M. Barlow, J.C. Maxwell, W.J. Wallace and W.S. Caughey, *Biochem. Biophys. Res. Commun.*, **55**, 91 (1973).
    b. S.E.V. Phillips and B.P. Schoenborn, *Nature*, **292**, 81 (1981).
    c. B. Shaanan, *Nature*, **296**, 683 (1982).
16. D.B. Northrop, *Biochemistry*, **14**, 2644 (1975).
17. G.S. Hammond, *J. Am. Chem. Soc.*, **77**, 334 (1955).
18. M.J. Coon and B.E. White, in *Metal Ion Activation of Dioxygen*, T.G. Spiro, Ed., (Wiley, New York, 1980) Chapter 2, pp. 73.
19. P.J. Goodhart, W.E. DeWolf, Jr. and L.I. Kruse, *Biochemistry*, **27**, 9093 (1988).
20. a. I.A. Lamouroux, A. Vigny, N.F. Biguet, M.C. Darmon, R. Franck, J.P. Henry and J. Mallet, *EMBO J.*, **13**, 3931 (1987).
    b. J. Taljanidisz, L. Stewart, A.J. Smith and J.P. Klinman, *Biochemistry*, **28**, 10054 (1989).
    c. E.J. Lewis, S. Allison, D. Fader, V. Claflen and L. Baizer, *J. Biol. Chem.*, **265**, 1021 (1990).

CHEMICAL AND SPECTROSCOPIC STUDIES ON DOPAMINE-β-HYDROXYLASE AND OTHER COPPER MONOOXYGENASES

Ninian J. Blackburn

Department of Chemical and Biological Sciences, Oregon Graduate Institute of Science and Technology, Portland, OR 97007-1999, USA.

Introduction

The copper containing monooxygenases are vital to a number of different areas of human physiology and medical science. Dopamine-β-hydroxylase (DBH) is of central importance in the catecholamine biosynthetic pathway and catalyzes the conversion of dopamine to noradrenalin, both of which act as neurotransmitters in the sympathetic nervous system. Hence this enzyme has long been associated with affective disorders, schizophrenia, clinical depression and more recently the pathology of neuroblastoma. In addition, DBH inhibitors have been recognized as potent antihypertensive agents. Phenylalanine hydroxylase, (PAH), synthesizes tyrosine from phenylalanine, and thus occupies a central role in amino acid synthesis and metabolism. A third member of this class of proteins, peptidyl-α-amidase, catalyzes the conversion of C-terminal glycine-extended peptides to their bioactive amidated forms, and hence is responsible for the biosynthesis of essential neuropeptide hormones such as gonadotropin, vasopressin, calcitonin and oxytocin. This review summarizes recent comparative studies on the catalytic role of the active-site Cu ions in these related enzymes aimed at obtaining a deeper understanding of the reaction chemistry of copper proteins at the molecular level. Most of this work has been directed towards dopamine-β-hydroxylase, but sequence comparisons with peptydyl-α-amidase and preliminary EXAFS results on phenylalanine hydroxylase are also included.

Dopamine-β-hydroxylase (DBH) catalyzes the benzylic hydroxylation of phenylethylamines and related substrates (1-4). The enzyme contains two non-blue or type 2 (T2) copper atoms per active site (1,5,6), with a total of 8 Cu atoms per tetramer of molecular weight 290,000, and no other cofactors (7). The mechanism is known to proceed via a redox process in which the two Cu(II) centers in the catalytic unit of the resting enzyme are first reduced by ascorbate to a Cu(I) intermediate, within which dioxygen binding and substrate hydroxylation take place. The coordination chemistry of the resting Cu(II) form has been probed by a number of spectroscopic techniques. EPR (8,9) and ESEEM (10) investigations have suggested a site composed of histidine residues with solvent-accessible coordination positions on one or both of the Cu(II) centers providing binding sites for exogenous ligands such as azide and cyanide (8,11).

From K.D. Karlin and Z. Tyeklár, Eds., *Bioinorganic Chemistry of Copper* (Chapman & Hall, New York, 1993).

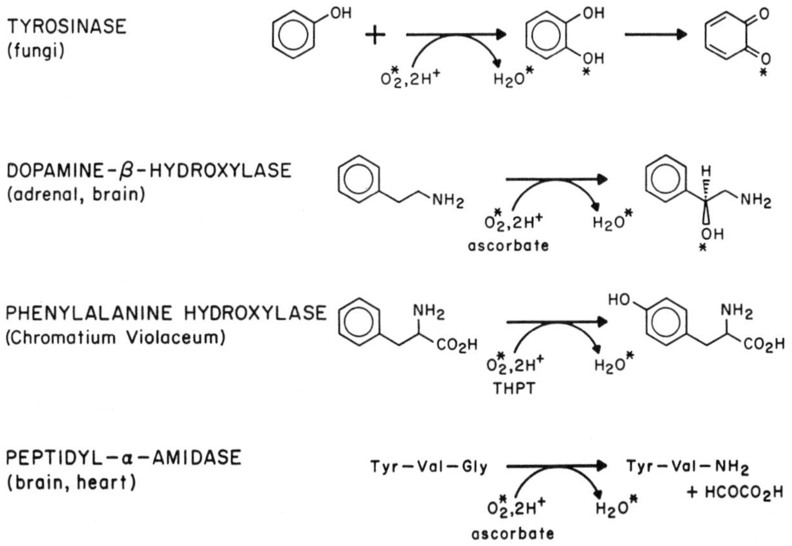

Figure 1.   *Reactions catalyzed by copper containing monooxygenases.*

Despite the requirement for two Cu centers per catalytic unit, there is no evidence for short-range magnetic interaction; hence, each Cu(II) center is believed to be mononuclear (9,12). In addition, separate binding sites have been detected kinetically for substrate and reductant, and reappearance of the Cu(II) EPR signal during single turnover of the reduced enzyme with tyramine and O_2 has been shown to occur in a catalytically competent step, making dinuclear O_2-binding chemistry extremely unlikely (13-15). This is in contrast to the situation found in the oxy-hemocyanin (16,17), oxy-tyrosinase (18,19), and inorganic Cu(II)-peroxo complexes (20-22), where dinuclear O_2 binding chemistry is the rule. Further evidence for the occurrence of an unusual Cu(I) site in DBH has come from the x-ray absorption spectroscopic data of Scott and coworkers, who inferred the presence of a S-ligand coordinated to Cu(I), not present in the oxidized Cu(II) form (23). This result is controversial, and has been challenged by Villafranca and coworkers, who detected no difference between oxidized and reduced Cu sites (24). On the basis of kinetic studies, Klinman and coworkers suggested that the copper centers were functionally inequivalent (13-15), and that one Cu center acts as a binding site for O_2 and substrate, while the other functions as an electron-transfer center. These findings offer a special challenge, because a complete description of the catalytic mechanism will require elucidation of how the intermetal connectivity influences long-range electron transfer between the copper centers (25).

The primary sequence of DBH is now known with certainty as the result of isolation of cDNAs from human (26), bovine (27-29) and rat (30). These sequences exhibit close to 90% homology to one another, and to the sequence derived from chemical methods (27,31). While the degree of homology is too high to be useful in determining conserved (active-site) amino acids, 5 his-containing regions can be identified as possible

Cu-coordination sites at positions 230 (his-his), 248 (his-his-met), 319 (his-tyr-his), 398 (his-thr-his), and 425 (his-tyr-ser-pro-his).

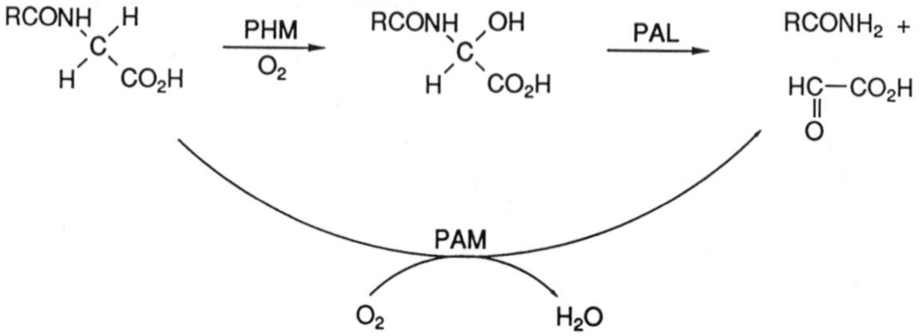

Figure 2. *Individual reactions corresponding to the PHM and PAL activities of PAM.*

Peptidyl-α-amidase (PAM) is a complex enzyme which catalyzes the conversion of C-terminal glycylpeptides to their bioactive amidated forms and glyoxalate (32,33). Strong biochemical and mechanistic similarities exist between this enzyme and DBH, to the extent that similarities in the Cu sites are likely (34,35). However, almost nothing is known about the stoichiometry or coordination chemistry of Cu in PAM. Studies on cDNAs derived from a variety of tissues have identified 3 separate domains (36-39). The NH_2-terminal domain encodes the copper/ascorbate/O_2-dependent peptidylglycine hydroxylating monooxygenase activity (PHM) which hydroxylates the α-C of the terminal glycine; the intermediate region encodes peptidyl-α-hydroxyglycine α-amidating lyase (PAL) which cleaves the α-hydroxy glycine moiety to form amidated peptide and glyoxalate; the COOH-terminal domain contains a trans-membrane domain and cytoplasmic tail. PHM and PAL have been isolated as individual activities from separate expression systems but little characterization has been done (Figure 2).

Rather different chemistry is exhibited by the pterin-dependent monooxygenases. Phenylalanine hydroxylases (PAH) catalyze the hydroxylation of phenylalanine to tyrosine by O_2. Like DBH, four electrons are transferred to O_2 in the process, two being derived from breaking the C-H bond of the substrate, and the other two from a reducing cofactor, but in these enzymes, only a single metal ion is involved in the process. Tetrahydropterin (**I**) is used as cofactor, and is oxidized by two electrons to the dihydropterin (**III**) during phenylalanine hydroxylation. The mammalian enzymes require one mole of non-heme iron per subunit for activity (40) while the bacterial enzyme from *Chromobacterium violaceum* contains a single type 2 copper atom per mole of enzyme (41). The role of the metals is not known with certainty, but reduction to the Fe(II) or Cu(I) is obligatory before catalysis can proceed (41-43). Additionally, there is no evidence that the metal ions undergo redox cycling during catalysis. This suggests that, unlike other metal-dependent monooxygenases (1,4), metal-bound reduced dioxygen intermediates exist as transitory species, if at all.

For both classes of enzyme, a 4a-OH tetrahydropterin (**II**) is produced as an intermediate (41,44,45) and it has been postulated that the active species may be the 4a-peroxy pterin. The metal is believed to serve as the initial binding site for molecular oxygen, and may subsequently facilitate transfer of O_2 to the pterin cofactor via

I   II   III

formation of a ternary Cu(I)-O_2-pterin complex. In support of this premise, an enzyme-bound O_2 intermediate has been detected by kinetics and susceptibility measurements on the Cu PAH (46). Continuous wave multifrequency EPR has provided evidence for two or three N donors to the Cu(II) centers (47), and pulsed EPR has clearly identified these as histidine residues (48). Furthermore, S-band (1.2 GHz) studies of the interaction of 5-^{14}N and 5-^{15}N labeled 6,7-dimethyl tetrahydropterin with PAH have established that the pterin is indeed able to coordinate to the Cu(II) center via the 5-N atom without displacement of a histidine ligand (48), although no information is available on the presence of a Cu(I)-pterin interaction.

Studies on the Coordination Chemistry of Oxidized and Reduced DBH

We began these studies determined to settle the controversy in the literature as to whether or not a second row scatterer was coordinated to Cu(I) in reduced DBH (23), and if so, was this S or Cl, since we considered this to be fundamental to any detailed understanding of the coordination chemistry. Metal-directed techniques for studying the coordination chemistry of Cu(I) centers are limited to x-ray absorption (EXAFS and Edge), although as will be described below, ligand-directed methodologies can also provide a wealth of information. Consequently, to address the problem, we embarked on a full reexamination of the x-ray absorption spectra of DBH. Data sets on five different samples of reduced enzyme were systematically collected over a period of a year at the NSLS (Brookhaven National Laboratory) and at the SRS (Daresbury Laboratory). In the first series of experiments, XAS data were collected on two independent samples prepared and reduced under standard conditions similar to those used in the other EXAFS studies. These both showed the presence of a second row scatterer in agreement with earlier data of Scott and coworkers (23); next we tested to see whether this was due to an exogenous Cl ligand which survived the preparative procedure. This was done by collecting data on a pair of samples prepared under rigorous exclusion of Cl ion, to one of which had been added 2 mM Br ion. In this experiment, Br would be expected to displace the Cl and leave an easily detectable EXAFS "signature" in the Cu K-absorption spectrum. Finally, data were collected on one sample which had undergone full redox cycling to eliminate the presence of any non-labile copper-bound chloride, since it was already known that no second-row scatterer was coordinated to the Cu(II) centers in the oxidized enzyme (9,23,24,50). *The EXAFS data from all five samples were identical, and clearly showed the presence of a strong wave attributable to 0.5 S per Cu.* This established beyond reasonable doubt that *at least one of the Cu(I) centers in DBH is coordinated to a protein-derived S ligand.*

167

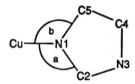

$\alpha = a - b =$ angle of rotation of the ring about an axis through the coordinated N and perpendicular to the plane of the ring.

$\beta =$ angle between the plane of the ring and an axis colinear with the Cu—N1 bond.

Figure 3.   *Atom labeling scheme and designation of angles used in the group fitting of imidazole ligation using restrained refinement multiple scattering analysis.*

EXAFS analysis of imidazole groups.

Recent progress in the simulation of imidazole ligation in copper proteins has made it possible to extend the information available from an EXAFS spectrum, and to obtain good estimates of the number of coordinated histidine residues. This "restrained refinement" allows various combinations of imidazole and O ligands to be tested with respect to their ability to simulate the raw EXAFS data. The imidazole ligands are treated as geometrically rigid units using average values of bond lengths and angles determined from the tables of crystallographic data on coordination complexes (51). The interatomic distance from Cu to any atom of the imidazole ring then depends on the Cu-N1 distance, R, and two angles, α and β (Figure 3). In practice it has been consistently found that the calculated EXAFS is relatively insensitive to variations in β; on the other hand, the majority of imidazole-containing copper protein systems that we have investigated (50,52-54), exhibit spectra that are highly sensitive to the angle α, since the simulations demand a splitting of the Cu-C2/C5 and Cu-N3/C4 into two pairs of distances centered on *ca* 3.0 and 4.1 Å respectively. Although this splitting can be rationalized by invoking protein-imposed conformational constraints as a mechanism for non-ideal sp_2 hybridization at Cu, and consequent imidazole ring tilt (55), in reality, the structural significance is less clear, since at best, the splitting represents the ring tilt averaged over all the coordinated histidine ligands. In addition, it may represent a way of compensating for unresolved differences in individual Cu-N(imid) distances and ring orientations such as are observed in many of the high-resolution crystal structures of copper proteins (56-61). A further complexity is introduced by the deformation of the ring geometry which occurs as the result of the opposing effects of σ-donation from N1 to the metal ion, π-donation from Cu(I) to imidazole π-orbitals, and partial deprotonation of N3 via H-bonding. Because of these factors, it is seldom possible to achieve satisfactory simulation of the raw data by refining R and α alone. The problem can be overcome by relaxing the geometrical constraints on the imidazole rings by the minimum amount so as to obtain a good fit to the EXAFS.

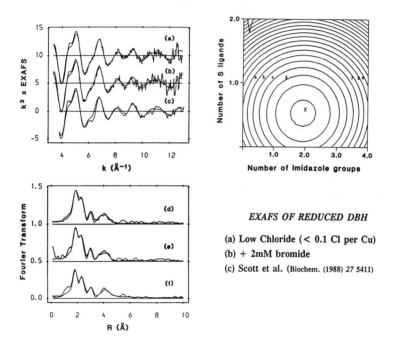

EXAFS OF REDUCED DBH

(a) Low Chloride (< 0.1 Cl per Cu)

(b) + 2mM bromide

(c) Scott et al. (Biochem. (1988) *27* 5411)

Figure 4.   *Experimental and simulated EXAFS, Fourier transforms, and least-squares minimization contour map for the various samples of reduced DBH described in the text.*

Simulation of the EXAFS of oxidized and reduced DBH.

The above methods have been applied to a detailed analysis of the coordinate structure of the Cu centers in both oxidation states of DBH (49). In the oxidized enzyme, simulations suggest 2-3 histidines and 1-2 O-donor ligands per Cu, consistent with the classification of DBH as a non-blue copper protein. The values of the first shell bond lengths (R_{av} (O/N) = 1.97 Å) are also entirely consistent with this formulation (51). No contributions from C atoms other than those derived from the imidazole rings can be detected. Although this does not rule out endogenous carboxylato or phenolato coordination as the origin of the O-donor contribution, the evidence is insufficient to propose any assignment other than O from solvent water. Upon reduction, the copper imidazole coordination number does not change (within experimental error), but the Cu-N(imid) distance decreases to 1.93 Å, and the O-donors are replaced by 0.5 S per Cu at 2.25 Å (Figure 4). The identity of the endogenous S ligand is unknown. However, DBH contains no free sulfhydryl groups, nor is there any reported spectroscopic evidence for a Cu(II)-thiol or thiolate interaction in the oxidized enzyme. Thus we propose that the S ligand is derived from a methionine. We have also found evidence for the presence of an additional weak interaction at one or both of the copper atoms, although this must be considered tenuous. The average coordination at each copper is therefore between 3 and 3.5. These conclusions have confirmed our previous suggestion based on analysis of absorption edge data (62), that Cu(I) in the reduced enzyme is close to 3-coordinate, and is further supported by comparison of the observed Cu(I)-N(imid) bond lengths with those of 3-coordinate Cu(I) model compounds.

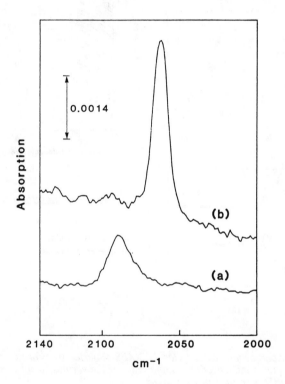

Figure 5.   *FTIR spectra of the CO complexes of (a) DBH and (b) molluscan hemocyanin.*

Carbonmonoxy dopamine-β-hydroxylase - the Cu(I) sites are inequivalent.

For some time, we and other groups working with DBH had been frustrated by the inability of the available spectroscopic probes to distinguish between the two copper centers. Since it was known with some degree of certainty that the Cu centers were mononuclear, rather than the dinuclear clusters found in tyrosinase and all inorganic Cu-dioxygen complexes, it was reasonable to suggest that the two Cu centers were catalytically distinct. Kinetic evidence had been advanced for this premise by Klinman (13), but no direct spectroscopic observation of inequivalence had been forthcoming.

The use of CO as a "ligand-directed" spectroscopic probe of the O_2-binding site, offered a novel approach to the elucidation of Cu(I) coordination chemistry since we had already demonstrated that CO was a competitive inhibitor with respect to the dioxygen substrate. Furthermore, CO offered a number of important chemical and spectroscopic properties which commended its use as a potential reporter ligand (63-70). Firstly, it was expected to form complexes only with 3-coordinate Cu(I) centers; secondly, the IR intraligand stretching frequency was known to be sensitive to the coordination environment of Cu(I), occurring in a "window" of the IR spectrum not obliterated by water absorption; thirdly, the expected linear Cu-C≡O geometry and short Cu-C bond length (1.75-80 Å) should give rise to easily identifiable multiple scattering resonances in the EXAFS, providing metrical details of ligand-binding.

Our studies on CO-DBH began by establishing that CO was a competitive inhibitor of the enzyme with respect to the oxygen substrate (71). This work strongly suggested that CO was bound at the dioxygen-binding Cu(I) center, and with this assumption, a K_i of 55 μM could be extracted from the kinetic data. Evidence that CO was indeed bound to a Cu(I) center was forthcoming from the observation that incubation of a sample of ascorbate-reduced enzyme under 1-2 atm of CO for 15 minutes at 0°C, would reproducibly produce a derivative with an IR absorption at 2089 cm^{-1}, which shifted by 46 cm^{-1} in the presence of ^{13}CO, and thus could be ascribed unambiguously to the carbonyl derivative (Figure 5). Furthermore, the CO was bound reversibly since incubation of the ^{13}CO derivative (ν_{CO} = 2043 cm^{-1}) with ^{12}CO restored the 2089 cm^{-1} band, and thus appeared to be in all respects an analogue of the dioxygen substrate (72).

The stoichiometry of CO-binding to DBH was measured using an assay system based upon quantitative transfer of CO from the copper protein to deoxy-hemoglobin (Hb), and subsequent determination of the CO-Hb so formed at 419 nm. The results showed that *reduced DBH binds 1 CO per 2 Cu.* This result was reproducible over at least four high-activity preparations, and was independent of CO concentration over a 2-fold range, implying that the carbonyl was fully formed. *Thus, the Cu(I) centers in the reduced enzyme were chemically inequivalent. Since CO competed with dioxygen, the dioxygen binding Cu(I) center was chemically and catalytically distinct from the second Cu(I) center in the catalytic unit.*

A direct and extremely desirable consequence of this was that CO could act as a *selective spectroscopic probe of the dioxygen binding site.* We were able to take advantage of this to obtain two important additional results (72). (1) ν_{CO} was found to shift to lower energy by 3 cm^{-1} in the presence of saturating levels of the substrate tyramine. (2) ν_{CO} was eliminated in the presence of the multi-substrate inhibitor 2,6-difluoro-1-hydroxybenzylimidazole-2-thione (HBIT), which is known to cross-link the substrate and O_2 binding sites (73). Thus the CO/O_2 and substrate binding sites were close enough for tyramine-binding to perturb ν_{CO}, but the substrates were not mutually exclusive. Furthermore, the downward shift in ν_{CO} implied that tyramine-binding induces a structural and/or electronic change at the CO/O_2 binding site which weakens the CO/O_2 bond and thus is equivalent to a *substrate-induced activation of CO/O_2.* On the other hand, CO binding was eliminated by HBIT, thereby directly establishing the structural contiguity of the substrate and Cu(I)-CO/O_2 binding site.

EXAFS was not expected to distinguish between the two Cu centers, but it did provide important additional information, namely an average coordination of 2-3 histidines, 0.5 S and 0.5 CO groups per Cu (72). No indication of any Cu-Cu interaction was present.

R	$\nu(CO)$
i-pr	2056
CH$_3$	2066
H	2083
Ph	2086

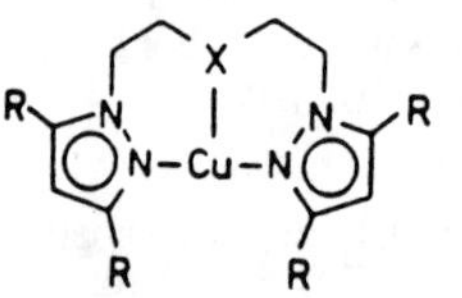

HB(3,5-R$_2$pz)$_3$Cu(I)-CO

X	$\nu(CO)$
NH	2082
O	2106
S	2123

Donor	$\nu(CO)$
(imid)$_3$	2069
(imid)$_2$X	2063
(imid)$_3$	2043
(imid)$_2$X	2089

[1,2-(CH$_3$)$_2$imid)]Cu(I)-CO

Molluscan Hc-CO

Arthropodal Hc-CO

DBH-CO

Figure 6. *Carbonyl stretching frequencies for a selection of Cu(I)-carbonyl species, illustrating the relationship between ligand basicity and Co stretching frequency. (See text for details).*

The results were consistent with CO binding to one of the two Cu centers without displacement of any other coordinated ligand. The Cu-C-O angle derived from the multiple scattering treatment of the Cu-CO unit was close to 180°, and the Cu-CO bond length was 1.78 Å, within the range found for Cu(I) carbonyls (63-70). The Cu-histidine bond lengths were essentially identical to those found for the reduced unligated enzyme. The results were entirely consistent with CO forming a reversible adduct, and supported the concept of a Cu(I) site designed for the binding of an exogenous ligand such as CO or O$_2$ with minimal perturbation of structure. Absorption edge data, and comparison to model compounds supported a 4-coordinate Cu(I)-CO entity, a premise fully in accord with the chemistry of Cu(I)-carbonyls derived from many studies in the literature.

Although these studies establish the inequivalence of the Cu(I) centers, and provide details of the average coordination at Cu(I), they do not predict the details of the individual coordination chemistry at each Cu(I) center. However, the value of the carbonyl stretching frequency can in principle, yield additional information as to the coordination around the CO-binding (O$_2$-binding) Cu center. The availability of a large literature of crystallographically characterized Cu(I)-carbonyl complexes has made it possible to obtain empirical relationships between the value of ν_{co} and the identity and

172

coordination number of the other ligands in the complex. In general, ν_{co} decreases as the basicity or coordination number of N-donor ligands increases. This is the result of the σ-donor - π-acceptor (π-acid) nature of the bonding in metal carbonyls, since the more basic the ligand, the greater the electron donation into the π^* levels on the coordinated CO.

Such trends are illustrated in Figure 6 for a selection of structurally characterized Cu(I)-carbonyls. Thus, in the series of substituted pyrazolylborate complexes, v_{co} decreases in the order phenyl, H, methyl, isopropyl, i.e. as the electron donating power of the substituent on the pyrazole ring increases (70). Similarly, ν_{co} decreases as the donor power of the coordinated atom increases from S(thioether) to N(amino) (67). Finally it can be seen that in monodentate *tris*-imidazole-ligated systems, including molluscan and arthropodal hemocyanins, ν_{co} is below 2070 cm^{-1} (74,75). Hence the value of 2089 cm^{-1} exhibited by the DBH-CO complex suggests a coordination environment less basic than the *tris*-imidazole site of hemocyanin. Two possibilities would account for this observation - a 3-coordinate site involving two histidines and one CO, or a 4-coordinate site involving two-histidines, one less-basic ligand (such as thioether S), and CO. However, with one exception, Cu(I)-carbonyls, are 4-coordinate and the $(his)_2XCu(I)$-CO coordination thus seems more likely.

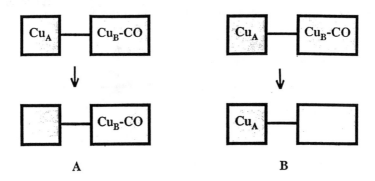

CASE A: *Binds 1 CO per Copper*

CASE B: *Does not bind CO*

Figure 7.   *Schematic representation of the two possible "half-apo" derivatives of DBH and their predicted reaction with CO.*

Preparation of Half-Apo DBH.

One way to unambiguously determine the individual coordination chemistry is to selectively remove each Cu ion, allowing the selective characterization of the other. This goal is not trivial, since the binding constants of the individual sites are similar. However, we have succeeded in selectively removing one copper via making use of the competition between CO and CN$^-$ for the Cu(I) centers. CN$^-$ interacts with both oxidized and reduced DBH removing both Cu(I) centers. Recent studies of speciation in CN-treated solutions of Cu proteins and model compounds suggests that the mechanism of copper removal involves the formation of protein-bound Cu(II) and Cu(I) cyano complexes as obligatory intermediates on route to the formation of the thermodynamically favored tricyanocuprate and tetracyanocuprate complexes (76). We reasoned therefor that since CO binds selectively to one Cu center, it should compete with cyanide for this site, and protect it from copper loss. Dialysis of ascorbate-reduced DBH with 50 μM NaCN under a CO atmosphere does indeed produce a demetallated derivative. Stoichiometry measurements indicate a CO to copper binding ratio of 1.0, indicating the formation of a genuine half-apo derivative containing the CO-binding copper center alone (Figure 7, Table 1). The carbonyl stretching frequency in the half-apo carbonyl is identical to that of the holoenzyme, providing convincing evidence that the CO/O_2-binding site is structurally unaltered by removal of the other copper.

This half-apo derivative will thus allow the coordination structure and chemistry of the dioxygen-binding site of DBH to be studied selectively. A crucial question relating to oxygen activation is whether the sulfur ligand is located at the dioxygen binding site or at the other copper center. Preliminary EXAFS data based on a single data set measured at the NSLS in January 1992, suggest that the S contribution is retained in half-apo DBH, as predicted by the value of ν_{co}. If confirmed, this result will provide the first evidence in a copper protein for a S-containing dioxygen binding site, and may provide novel insights into the electronic mechanism of oxygen activation.

Table 1.   *Stoichiometry of CO-binding to ascorbate-reduced DBH for half-apo and fully metallated derivatives.*

	Half-Apo		Fully Metallated	
Preparation	[Cu]/[E]	[CO]/[Cu]	[Cu]/[E]	[CO]/[Cu]
1	4.3	1.08	8.1	0.45
2	3.1	1.02	7.0	0.50
3	2.8	1.03	7.4	0.42
4	2.5	0.90		

Azido-dopamine-β-hydroxylase - the Cu(II) centers are also inequivalent.

At the outset of these studies, we were aware of the potential of azide to act as a powerful spectroscopic probe of both dinuclear and mononuclear Cu(II) centers in proteins as a result of the intense LMCT bands present in the 350-550 nm region (77,78). Solomon and coworkers had provided theoretical analysis which predicted two CT transitions with coincidental electronic and CD spectra in the case of terminally coordinated azide, and four transitions with non-coincidental electronic and CD spectra in the case of bridging azide (77). Thus we anticipated that a study of azide binding to DBH would provide answers to questions related to (a) the structural equivalence of the copper centers, (b) the presence or absence of N_3^--induced coupling of the Cu(II) centers, and (c) the strength of binding relative to other well-characterized systems such as Cu/Zn SOD (55,80-83). In addition to electronic spectroscopy, we planned to fully exploit the use of IR as a *ligand-directed* probe.

Table 2. *Thermodynamic data for azide-binding to fully metallated and half-apo forms of oxidized dopamine-β-hydroxylase and relevant model systems.*

Azido Cu(II)-DBH

	Fully Metallated	Half-Apo
K_D (mM)	4.2	14.1
ΔH° (kJ mol^{-1})	-14	
ΔS° (J mol^{-1} K^{-1})	-1	
Hill Constant	0.69	0.98
Cooperativity	Negative	None

K_D is defined for the reaction $EN_3 \rightarrow E + N_3^-$

Model Cu(II)-azide systems

	K_D (mM)	ΔH° (kJ mol^{-1})	ΔS° (J mol^{-1} K^{-1})	Ref
SOD + N_3^-	11.4	-4.0	28	84
Cu[dien]$^{2+}$ + N_3^-	13	-10.6	-2	84
Cu[terpy]$^{2+}$ + N_3^-		-12.1	-4.2	90

The addition of N_3^- to Cu(II)-DBH gives rise to an intense LMCT band at 380 nm, characteristic of a terminal azido complex. However, spectrophotometric titration data at this wavelength could not be interpreted by a single equilibrium, and instead exhibited negative cooperativity with a Hill coefficient of 0.69 (Table 2). Thus at least two binding sites for N_3^- must exist, both of which give rise to N_3^- to Cu(II) charge transfer. Furthermore, the negative cooperativity implies that the two sites must interact such that binding of the first azide decreases the affinity of a Cu(II) center for the second. ΔH and ΔS for binding the first azide are -14 kJ M^{-1} and -1 J K^{-1} respectively. These values contrast with the corresponding data for the azido complex of Cu/Zn SOD in which N_3^- displaces His 46, but resemble the data for Cu[dien]$^{2+}$ in which azide displaces a water molecule (84). Hence, we have interpreted this first equilibrium process of N_3^- binding to DBH by substitution of a coordinated water molecule by azide within a $Cu(II)(his)_3(H_2O)$ site.

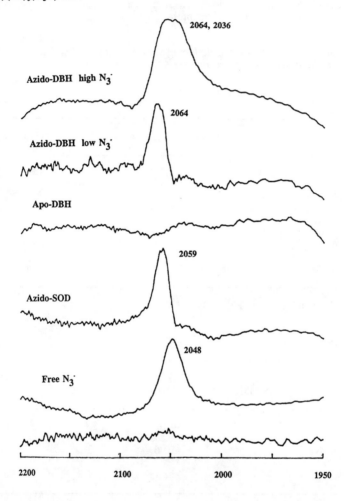

Figure 8. *FTIR spectra of azido complexes of DBH and SOD.*

The second N_3^- binding site was more elusive, since no thermodynamic data could be extracted from the non-linear part of the Scatchard plot. Significantly, azide binding to half-apo DBH is no longer negatively cooperative, implying that the second azide probably binds at the second copper site, rather than arising from *bis* complex-formation at a single copper center. Encouraged by success with CO as a ligand-directed probe, we were hopeful that IR spectroscopy might provide additional information, since the literature suggested that $\nu(N_3^-)$ was sensitive to Cu(II) ligation (77,85). FTIR titration of azide-binding showed first, the appearance of a band at 2061 cm^{-1} assignable to the intraligand stretch of coordinated N_3^- (Figure 8). Parallel studies on Cu/Zn SOD gave rise to a similar band at a similar energy (2058 cm^{-1}), suggesting an assignment to a Cu(II)(his)$_3$(N$_3$) species, and confirming the conclusions derived from the thermodynamic studies. Excess azide produced no further change in the IR spectrum of azido-SOD, but elicited the appearance of a second band at *ca* 2036 cm^{-1} in DBH, attributable to a Cu(II)-N$_3$ species, and providing direct evidence for the second binding site. The shift to lower frequency would imply (from comparison with model compounds (77)) binding of azide to a site with fewer N- ligands and more O- ligands, and we suggest that it represents N$_3^-$-binding to a Cu(II)(his)$_2$X(H$_2$O) type center. This band was found to persist in the half-apo derivative, whereas the 2061 cm^{-1} band was lost. Thus this second azide binding site must correspond to the oxidized form of the dioxygen-binding Cu center.

We believe that these results indicate that the Cu(II) centers in DBH provide distinct binding sites for azide, distinguishable by their thermodynamic binding constants and by the values of the intraligand azide stretching frequency. The energy and intensity of the LMCT band indicate terminal, equatorial coordination at mononuclear Cu(II) sites, and do not support bridging between dinuclear centers. However, the negative cooperativity does suggest some interaction between the two azide binding sites.

A model for the catalytic centers in dopamine-β-hydroxylase.

Consideration of all the data has led us to propose a model for the catalytic centers in DBH as shown in Figure 9. We believe the oxidized enzyme to contain two tetragonal Cu(II) centers coordinated to no more than 3 histidine ligands per Cu. Hence we propose that the Cu(II) sites are inequivalent, with a Cu$_A$(His)$_3$(H$_2$O).....Cu$_B$(His)$_2$X(H$_2$O) type configuration. The identity of X is unknown, available data being consistent with either a histidine or an oxygen donor ligand. Reduction by ascorbate causes loss of the bound water molecules since water is a poor ligand for Cu(I). This leaves two 3-coordinate centers, one of which, Cu$_B$, now coordinates a S from a methionine. Alternatively this S could already be present in the oxidized form as a weakly bound axial ligand, which is not detected by EXAFS. The mononuclearity of the Cu centers requires Cu$_A$ to be at some distance from Cu$_B$ and consequently must serve to shuttle electrons into the hydroxylating site as previously proposed by Klinman and coworkers (13-15).

Further support for the proposed copper ligation comes from recent sequence comparisons with peptidyl-α-amidase (PAM) (86). Because of the close biochemical similarities between DBH and PAM, it has been proposed that they are derived from a common ancestral gene, and indeed both proteins appear to have a common catalytic domain with 26% conservation of amino acid residues. Comparison of cDNA-derived amino acid sequences of frogskin PAM and human DBH indicates a total of only six conserved histidines, or a maximum of three per copper, similar to the upper limit for Cu-his coordination suggested from EXAFS. Of particular interest are the conserved histidine-rich regions V^{247}HHM, and H^{391}IFASQLHTH, both of which contain the HXH* type motifs (H*=his or met) now identified as copper-binding sequences in a number of crystallographically characterized copper proteins. The conserved met^{250} would be a strong candidate for the coordinated S-containing ligand to Cu(I), where we propose that the his^{248} and met^{250} act as ligands to one of the coppers. His^{249} would not be expected to coordinate to this Cu center, but might provide an imidazole ligand to the other copper center, resulting in a similar extended chain intermetal connectivity to that found in ascorbate oxidase (58) and nitrite reductase (59). In these proteins the two histidines of a His-Cys-His motif provide imidazole ligands to the type 2 copper while the Cys residue bridges to the type 1 center and coordinates via sulfur. By analogy, the proposed intermetal connectivity in DBH would result in a 12-Å separation between metal centers, and provide an efficient through-bond pathway for electron transfer.

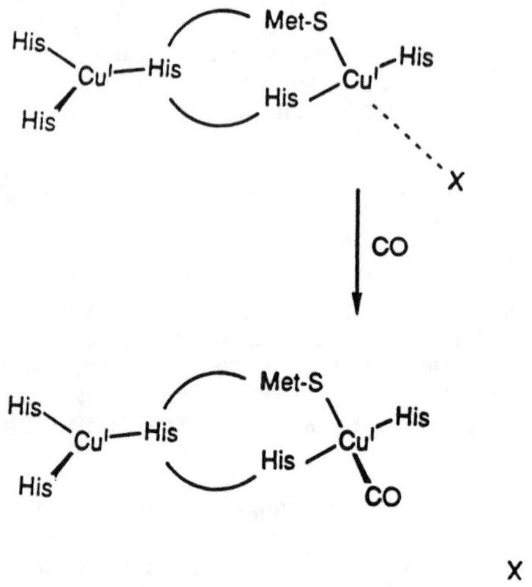

Figure 9. *A model for the catalytic Cu(I) centers in reduced DBH and their reaction with carbon monoxide.*

Phenylalanine Hydroxylase.

As part of our comparative studies on copper monooxygenases, EXAFS studies have been initiated on the Cu centers of the bacterial phenylalanine hydroxylase from *Chromobacterium violaceum*, which contains one mole of Cu and one mole of tetrahydropterin per 32 kD. The EXAFS of the Cu(II) form of the enzyme was found to resemble that of other non-blue copper proteins such as plasma amine oxidases and DBH, and was characteristic of a mixed N/O coordination shell containing both histidine and O-donor atoms. We carried out detailed simulations of the raw EXAFS data using our full curved-wave restrained refinement methodologies. The results suggest a Cu(II) coordination of two histidines and two O-donor groups (R_{av} = 1.98 Å) (87), and thus support previous spectroscopic conclusions on the oxidized form of PAH (47,48). A reasonable fit to the data could be obtained by assuming that these O atoms were derived from solvent (H_2O or OH⁻).

=The EXAFS of the dithionite-reduced enzyme showed major differences. The amplitude of the first shell in the Fourier transform was only 50% of that of the oxidized enzyme indicating a substantial reduction in coordination number. In addition, the first shell of the transform was split into two components. Adequate simulation of the reduced data could be obtained by 2 histidines at a long distance of 2.08 Å and an O ligand at a short distance of 1.88 Å, but a much better fit was produced by two histidines at a short distance of 1.90 Å and one second-row scatterer such as S or Cl at 2.20 Å. Comparison of absorption edge data on the reduced enzyme with data from

178

Cu(I) *bis*- and *tris*-(1,2-dimethylimidazole) complexes suggested a structure between 2- and 3-coordinate. The origin of the S/Cl wave is unclear, since the presence of the pterin cofactor in PAH mitigates against any strong structural analogy to DBH, but we believe the wave may originate from a S impurity present in the dithionite.

In any event, the picture which emerges is that of a highly coordinatively-unsaturated Cu(I) site. The value of the coordination number and bond length for the Cu(I)-imidazole shell both suggest a maximum of two coordinated imidazole groups. A large number of model studies have indicated that such a site would be expected to pick up one additional weakly-bound ligand so as to form a distorted 3-coordinate Cu(I) complex. Furthermore, most Cu(I) complexes which bind dioxygen are known to have distorted 3-coordinate structures (20,21), whereas two-coordinate Cu(I) complexes generally do not bind O_2 or CO (63,65,66). For example, [Cu(1,2-dimethylimidazole)$_2$]$^+$ does not bind O_2 or CO, but upon addition of one molar equivalent of (1,2-dimethylimidazole) to the *bis* complex in CH_2Cl_2, dioxygen and carbonyl adducts can be formed (87). The structure of the Cu(I) site in PAH is thus well set up for O_2-binding, although the identity of the third ligand and its relationship to the mechanism of thiol activation are unclear at this time. An attractive hypothesis is that under catalytic conditions, the pterin cofactor takes the place of this third ligand, providing a 3-coordinate site for O_2 coordination, such that the oxygen substrate is correctly oriented for reacting with the cofactor to form the 4a-peroxy pterin. Alternatively, the role of the Cu(I) center could be to provide a template for bidentate O(4),N(5)- coordination of the pterin as exemplified by the crystal structures of a number of pterin-containing model complexes (89,90), rather than to bind dioxygen. Hopefully, further details of the chemistry of Cu-pterin-dioxygen chemistry will emerge as research activity on these systems intensifies.

179

References

1. Stewart, L. C., Klinman, J. P., *Annu. Rev. Biochem.*, **1988**, *57*, 551-592.
2. Villafranca, J. J., in *Metal Ions in Biology* (Spiro, T. G., ed) **1981**, Vol. 3, pp. 263-289, Wiley, New York.
3. Klinman, J. P., Huyghe, B., Stewart, L., Taljanidisz, J., *Biol. Oxid. Syst.*, **1989**, *1*, 329-346.
4. Villafranca, J. J., Desai, P. R., *Biol. Oxid. Syst.*, **1989**, *1*, 297-327.
5. Ash, D. E., Papadopoulos, N. J., Colombo, G., Villafranca, J. J., *J. Biol. Chem.*, **1984**, *259*, 3395-3398.
6. Klinman, J. P., Krueger, M., Brenner, M., Edmondson, D. E., *J. Biol. Chem.*, **1984**, *259*, 3399-3402.
7. Klinman, J. P., Dooley, D. M., Duine, J. A., Knowles, P., Mondovi, V., Villafranca, J. J., *FEBS Letts.*, **1991**, *282*, 1-4.
8. Blackburn, N. J., Collison, D., Sutton, J., Mabbs, F. E., *Biochem. J.*, **1984**, *220*, 447-454.
9. Blackburn, N. J., Concannon, M., Khosrow Shahiyan, S., Mabbs, F. E., Collison, D., *Biochemistry*, **1988**, *27*, 6001-6008.
10. McCracken, J., Desai, P. R., Papadopoulos, N. J., Villafranca, J. J., Peisach, J., *Biochemistry*, **1988**, *27*, 4133-4137.
11. Obata, A., Tanaka, H., Kawazura, H., *Biochemistry*, **1987**, *26*, 4962-4968.
12. Walker, O. A., Kon, H., Lovenberg, W., *Biochim. Biophys. Acta,* **1977**, *482*, 309-322.
13. Stewart L. C., Klinman J. P., *Biochemistry*, **1987**, *26*, 5302-5309.
14. Brenner, M. C., Murray, C. J., Klinman, J. P., *Biochemistry,* **1989**, *28*, 4656-4664.
15. Brenner, M. C., Klinman, J. P., *Biochemistry*, **1989**, *28*, 4664-4670.
16. Volbeda, A., Hol, W. G. J., *J. Mol. Biol.*, **1989**, *209*, 249-279.
17. Solomon E. I., in *Metal Ions in Biology* **1981**, (Spiro T. G., ed.) pp. 40-108, Wiley, New York.
18. Lerch, K., in *Metal Ions Biol. Sys.*, **1981**, *13*, 143-186.
19. Robb, D. A., in *Copper Proteins and Copper Enzymes*, **1984**, (Lontie, R., ed) Vol. 2, pp. 207-241, CRC Press, Boca Raton, Florida.
20. Karlin, K. D., Gultneh, Y., *Prog. Inorg. Chem.*, **1987**, *35*, 219-327.
21. Tyeklar, Z., Karlin, K. D., *Acc. Chem. Res.*, **1989**, *22*, 241-248.
22. Kitajima, N., Fujisawa, K., Moro-oka, Y., Toriumi, K., *J. Am. Chem. Soc.*, **1989**, *111*, 8975-8976.
23. Scott, R. A., Sullivan, R. J., De Wolfe, W. E., Dolle, R. E., Kruse, L. I., *Biochemistry*, **1988**, *27*, 5411-5417.
24. Blumberg, W. E., Desai, P. R., Powers, L., Freedman, J. H., Villafranca, J. J., *J. Biol. Chem.*, **1989**, *264*, 6029-6032.
25. For a series of excellent review articles by Hoffman, Gray, Mauk and others see *Structure and Bonding*, **1991**, *Vol. 75*.
26. Lamouroux, A., Vigny, A., Faucon-Biguet, N., Darmon, M. C., Franck, R., Henry, J. P., Mallet, J., *EMBO* J., **1987**, *6*, 3931-3937.
27. Wang, N., Southan, C., De Wolfe, W. E., Wells, T. N. C., Kruse, L. I., Leatherbarrow, R. J., *Biochemistry*, **1990**, *29*, 6466-6474.
28. Taljanidisz, J., Stewart, L., Smith, A. J., Klinman, J. P., *Biochemistry*, **1989**, *28*, 10054-10061.
29. Lewis, E. J., Allison, S., Fader, D., Claflin, V., Baizer, L., *J. Biol. Chem.*, **1990**, *265*, 1021-1028.
30. Mcmahon, A., Geertman, R., Sabban, E. L., *J. Neurosci. Res.*, **1990**, *25*, 395-404.

31. Robertson, J. G., Desai, P. R., Kumar, A., Farrington, G. K., Fitzpatrick, P. F., Villafranca, J. J., *J. Biol. Chem.*, **1990**, *265*, 1029-1035.
32. Eipper, B. A., Mains, R. E., Glembotski, C. C., *Proc. Natl. Acad. Sci. U.S.A.*, **1983**, *80*, 5144-5148.
33. Murthy, A. S. N., Mains, R. E., Eipper, B. A., *J. Biol. Chem.*, **1986**, *61*, 1815-1822.
34. Young, S. D., Tambourini, P. P., *J. Am. Chem. Soc.*, **1989**, *111*, 1933-1934.
35. Katapodis, A. G., May, S. W., *Biochemistry*, **1990**, *29*, 4541-4548.
36. Stoffers, D. A., Barthel-Rosa Green, C., Eipper, B. A., *Proc. Natl. Acad. Sci. U.S.A.*, **1989**, *86*, 735-739.
37. Stoffers, D. A., Ouafik, L., Eipper, B. A., *J. Biol . Chem.*, **1991**, *266*, 1701-1707.
38. Eipper, B. A., Perkins, S. N., Husten, E. J., Johnson, R. C., Keutmann, H. T., Mains, R. E., *J. Biol. Chem.*, **1991**, *266*, 7827-7833.
39. Husten, E. J., Eipper, B. A., *J. Biol. Chem.*, **1991**, *266*, 17004-17010.
40. Gottschall, D. W., Dietrich, R. F., Benkovic, S. J., Shinman, R., *J. Biol. Chem.*, **1982**, *257*, 845-849.
41. Pember, S. O., Villafranca, J. J., Benkovic, S. J., *Biochemistry,* **1986**, *25*, 6611-6619.
42. Wallick D. E., Bloom, L. M., Gaffny, B. J., Benkovic, S. J., *Biochemistry,* **1984**, *23*, 1295-1302.
43. Marota, J. J. A., Shinman, R., *Biochemistry*, **1984**, *23*, 1303-1311.
44. Lazarus, R. A., Dietrich, R. F., Wallick, D. E., Benkovic, *Biochemistry* **1981**, *20* 6834-6841.
45. Lazarus, R. A., De Brosse, C. W., Benkovic S. J., *J. Am. Chem. Soc.* **1982**, *104*, 6869-6871.
46. Pember, S. O., Johnson, K. A., Villafranca, J. J., Benkovic, S. J., *Biochemistry*, **1989**, *28*, 2124-2133.
47. McCracken J., Pember, S. O., Benkovic, S. J., Villafranca, J. J., Miller, R. J., Peisach, J., *J. Am. Chem. Soc.*, **1988**, *110,* 1069-1074.
48. Pember, S. O., Benkovic, S. J., Villafranca, J. J., Pasenkiewicz-Gierula, M., Antholine, W. E., *Biochemistry*, **1987**, *26*, 4477-4483.
49. Blackburn, N. J., Hasnain, S. S., Pettingill, T. M., Strange, R. W., *J. Biol. Chem.*, **1991**, *266*, 23120-23127.
50. Blackburn, N. J., in *Synchrotron Radiation and Biophysics* (Hasnain, S. S., ed.), **1989**, pp 63-103, Ellis Horwood, Chichester, United Kingdom.
51. Orpen, A. G., Brammer, L., Allen, F. H., Kennard, O., Watson, D. G. Taylor, R., *J. Chem. Soc. Dalton Trans.,* **1989**, S1-S83
52. Blackburn, N. J., Strange, R. W., McFadden, L. M., Hasnain, S. S., *J. Am. Chem. Soc.*, **1987**, *109*, 7162-7170.
53. Blackburn, N. J., Strange, R. W., Farooq, A., Haka, M. S., Karlin, K. D., *J. Am. Chem. Soc.*, **1988**, *110*, 4263-4272.
54. Strange, R. W., Blackburn, N. J., Knowles, P. F., Hasnain, S. S., *J. Am. Chem. Soc.*, **1987**, *109*, 7157-7162.
55. Blackburn, N. J., Strange, R. W., McFadden, L. M., Hasnain, S. S., *J. Am. Chem. Soc.*, **1987**, *109*, 7162-7170.
56. Tainer, J. A., Getzoff, E. D., Beem, K. M., Richardson, J. S., Richardson, D. C., *J. Mol. Biol.*, **1982**, *160*, 181-217.
57. Volbeda, A., Hol, W. G. J., *J. Mol. Biol.*, **1989**, *209*, 249-279.
58. Messerschmidt, A., Rossi, A., Ladenstein, R., Huber, R., Bolognesi, M., Gatti, G., Marchesini, A., Petruzzelli, R., Finazzi-Agro, A., *J. Mol. Biol.* **1989**, *206*, 513-529.
59. Gooden, J. W., Turley, S., Teller, D. C., Adman, E. T., Liu, M. Y., Payne, W.

J., LeGall, P. J., *Science*, **1991**, *253*, 438-442.
60. Ito, N., Phillips, S. E. V., Stevens, C., Ogel, Z. B., McPherson, M. J., Keen, J. N., Yadav, K. D. S., Knowles, P. F., *Nature*, **1991**, *350*, 87-90.
61. Guss, J. M., Freeman, H. C., *J. Mol. Biol.*, **1983**, *169*, 521-563.
62. Blackburn, N. J., Strange, R. W., Reedijk, J., Volbeda, A., Farooq, A., Karlin, K. D., Zubieta, J., *Inorg. Chem.*, **1989**, *28*, 1349-1357.
63. Pasquali, M., Floriani, C. in *Copper Coordination Chemistry, Biochemical and Inorganic Perspectives*, **1984**, (Karlin, K. D., Zubieta, J., eds) pp. 311-330, Adenine Press, New York.
64. Villacorta, G. M., Lippard, S. J., *Inorg. Chem.*, **1987**, *26*, 3672-3676.
65. Sorrell, T. N., Borovick, A. S., *J. Am. Chem. Soc.*, **1986**, *108*, 2479-2481.
66. Sorrell, T. N., Borovick, A. S., *J. Am. Chem. Soc.*, **1987**, *109*, 4255-4260
67. Sorrell, T. N., Malachowski, M. R., *Inorg. Chem.*, **1983**, *22*, 1883-1887.
68. Patch, M. G., Choi, H., Chapman, D. R., Bau, R., McKee, V., Reed, C. A., *Inorg. Chem.*, **1990**, *29*, 110-119
69. Thompson, J. S., Whitney, J. F., *Inorg. Chem.*, **1984**, *23*, 2813-2819.
70. Kitajima, N., Fujisawa, K., Fujimoto, C., Moro-oka, Y., Hashimoto, S., Kitagawa, T., Toriumi, K., Tatsumi, K., Nakamura, A., *J. Am. Chem. Soc.*, **1992**, *114*, 1277-1291
71. Blackburn, N. J., Pettingill, T. M., Seagraves, K. S., Shigeta, R. T., *J. Biol. Chem.*, **1990**, *265*, 15383-15386.
72. Pettingill, T. M., Strange, R. W., Blackburn, N. J., *J. Biol. Chem.*, **1991**, *266*, 16996-17003.
73. Kruse, L. I., De Wolfe, W. E., Chambers, P. A., Goodhart, P. J., *Biochemistry*, **1986**, *25*, 7271-7278.
74. Sanyal, I., Karlin, K. D., Strange, R. W., and Blackburn, N. J., manuscript in preparation.
75. Fager, L. Y., Alben, J. O., *Biochemistry*, **1972**, *11*, 4786-4792.
76. Han, J., Blackburn, N. J., Loehr, T. M., *Inorg. Chem.*, **1992**, *31*, 3223-3229.
77. Pate, J. E., Ross, P. K., Thamann, T. J., Reed, C. A., Karlin, K. D., Sorrell, T. N., Solomon, E. I., *J. Am. Chem. Soc.*, **1989**, *111*, 5198-5209.
78. Casella, L., Gullotti, M., Pallanza, G., Buga, M., *Inorg. Chem.*, **1991**, *30*, 221-227.
79. Fee, J. A., Gaber, B. G., *J. Biol. Chem.*, **1972**, *247*, 60-65.
80. Banci, L., Bertini, I., Luchinat, C., Monnanni, R., Scozzafava, A., *Inorg. Chem.*, **1988**. *27*, 107-109.
81. Banci, L, Bertini, I., Luchinat, C., Scozzafava, A., *J. Biol. Chem.*, **1989**, *264*, 9742-9744.
82. Banci, L., Bencini, A., Bertini, I., Luchinat, C., Piccioli, M., *Inorg. Chem.*, **1990**, *29*, 4867-4873.
83. Banci, L., Bertini, I., Luchinat, C., Hallewell, R. A., *J. Am. Chem. Soc.*, **1988**, *110*, 3629-3633.
84. Dooley, D. M., Cote, C. E., *Inorg. Chem.*, **1985**, *24*, 3996-4000.
85. Agrell, I., *Acta Chem. Sc.*, **1971**, *25*, 2965-2974.
86. Southan, C., Kruse, L. I., *FEBS Lett.*, **1989**, *255*, 116-120.
87. Blackburn, N. J., Strange, R. W., Carr, R. T., Benkovic, S. J., *Biochemistry*, **1992**, *31*, 5298-5303.
88. Sanyal, I., Strange, R. W., Blackburn, N. J., Karlin, K. D., *J. Am. Chem. Soc.*, **1991**, *113*, 4692-4693.
89. Perkinson, J., Brodie, S., Yoon, K., Mosny, K., Carroll, P. J., Morgan, T. V., Burgmayer, S. J., *Inorg. Chem.*, **1991**, *30*, 719-727.

90. Holwerda, R. A., Stevens, G., Anderson, C., Wynn, M., *Biochemistry*, **1982**, *21*, 4403-4407.

THE COPPER IONS IN THE MEMBRANE-ASSOCIATED METHANE MONOOXYGENASE

Sunney I. Chan,[a] Hiep-Hoa T. Nguyen,[a] Andrew K. Shiemke,[a] and Mary E. Lidstrom[b]

[a]A. A. Noyes Laboratory of Chemical Physics and
[b]W. M. Keck Laboratory of Environmental Engineering,
California Institute of Technology, Pasadena, CA 91125 USA.

INTRODUCTION

The enzyme methane monooxygenase (MMO)[1] in methanotrophic bacteria catalyzes the conversion of methane to methanol using molecular oxygen as co-substrate [1]. Two forms of MMO differing in cellular location are known to exist [2]. The soluble methane monooxygenase (sMMO) found in the cytosolic fraction of the bacteria appears to be restricted to only *Methylococcus* and *Methylosinus* strains and is expressed only at limiting copper levels in the growth medium [3-5]. The particulate methane monooxygenase (pMMO) found in the membrane fraction of the cells appears to be expressed in all types of methanotrophic bacteria [3-6]. Only the sMMO has been purified and extensively characterized [7-11].

Progress toward understanding the pMMO has been hampered by the difficulty in maintaining the enzyme activity in vitro. However, copper ions have been shown to exert a dramatic influence on this system both during protein synthesis and in MMO activity [3-5]. For strains capable of expressing both the sMMO and the particulate form of the enzyme, only the pMMO is expressed at high levels of copper ions. Even though the copper concentration threshold to ensure pMMO expression is very low (<< 0.3 μM), high copper concentrations stabilize and enhance pMMO activity both in vivo and in vitro [4,5]. For instance, increased pMMO activity has been observed for membrane fractions obtained from *M. capsulatus* (Bath) when these cells are grown in copper-enriched media [5]. The pMMO activity is stimulated even further with the addition of copper ions during activity assay [5].

The pMMO is different from the sMMO in several aspects. The pMMO has a narrow range of substrates, capable of oxidizing only C4 hydrocarbons or smaller [4]. The pMMO is sensitive to several inhibitors which do not affect sMMO. In addition, three new polypeptides of apparent molecular weight 46 kDa, 35 kDa, and 26 kDa appear on SDS-polyacryamide gels of these membranes when cells expressing sMMO are

---

[1] The abbreviations used: MMO, methane monooxygenase; sMMO, soluble methane monooxygenase; pMMO, particulate methane monooxygenase; EPR, electron paramagnetic resonance; DCPE, direct current plasma emission; SDS-PAGE, sodium dodecyl sulfate polyacrylamide gel electrophoresis.

From K.D. Karlin and Z. Tyeklár, Eds., *Bioinorganic Chemistry of Copper*
(Chapman & Hall, New York, 1993).

switched to express pMMO [3-5]. The pMMO expression in these organisms is associated with the formation of extensive intracytoplasmic membranes. A substantial increase in the carbon conversion efficiency of methane to biomass is also associated with the change from sMMO to pMMO expression [3].

Data to date consistently point to the importance of copper to the activity of the pMMO, thus suggesting that pMMO might be a copper-containing protein. In this study, we report additional experiments that strongly support the contention that pMMO is a copper-containing enzyme system. In addition, the nature of the copper sites have also been investigated by EPR and magnetic susceptibility measurements.

THE EFFECTS OF COPPER IONS

Existing evidence suggests an important role of copper ions during pMMO protein synthesis and in MMO activity. Upon adding copper ions to copper-free medium during bacteria growth, strains capable of expressing both forms of the enzyme switch from sMMO to pMMO expression. Three bands corresponding to polypeptides of molecular weight 46 kDa, 35 kDa, and 26 kDa have been implicated for the pMMO as a result of these experiments. Recently, we have determined by SDS-PAGE analysis of membrane fractions obtained from $\underline{M. capsulatus}$ (Bath) the level of expression of these polypeptides as a function of the copper concentration in the growth medium. As illustrated in Figure 1, the levels of expression of these polypeptides do not change significantly, whereas the specific activity of pMMO in these samples varies by nearly a factor of 80 (Table 1). This result then suggests the existence of a copper-deficient, inactive pMMO in cells grown at low Cu levels, which in turn may explain the activity stimulation effect of copper ions, particularly during activity assay. The binding of copper ions to the apo-pMMO results in the activation of the enzyme; accordingly, an increase in the activity of the pMMO is observed. In addition, as evidenced in Figure 1, the pMMO appears to be over-expressed in these membranes since the 46 kDa and 26 kDa polypeptides are the most intense bands. The major band above the 46 kDa polypeptide seen in Figure 1 is the large subunit of the methanol dehydrogenase which is loosely bound to the membranes and can be removed by salt-wash.

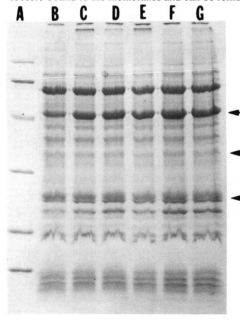

Figure 1. SDS polyacrylamide gel electrophoresis of the membrane fractions from $\underline{M. capsulatus}$ (Bath) grown under increasing Cu^{2+} concentration. A, molecular weight standards (92, 66.2, 45, 31, 21.5 and 14.4 kDa); B, no added copper; C, 1 μM Cu^{2+}; D, 2.5 μM Cu^{2+}; E, 5 μM Cu^{2+}; F, 10 μM Cu^{2+}; G, 20 μM Cu^{2+}. Lanes B-G contain 75 μg total protein. Protein concentration was determined by the Lowry method. Electrophoresis was performed on a 15% slab gel according to the Laemmli method and the gel was stained with Coomassie brilliant blue.

A   B   C   D   E   F   G

◀46 kDa

◀35 kDa

◀26 kDa

We have confirmed earlier results that membrane fractions obtained from *M. capsulatus* (Bath) exhibit greater pMMO specific activity with increasing copper concentration in the growth medium and addition of copper ions during activity assay stimulates pMMO activity even further [5]. Furthermore, we have also established the connection between the propene epoxidation activity of pMMO and the membrane-associated copper via metal analysis by direct current plasma emision (DCPE) spectroscopy. Copper analysis of *M. capsulatus* (Bath) whole cells grown with varying copper concentration shows that most of the copper in the medium is eventually taken up by the bacteria (Table I). The bulk of this intracellular copper is found to be incorporated into the membranes, and the specific activity of pMMO is directly proportional to the membrane copper content (Table I). These startling new findings strongly suggests that membrane copper is required to activate and/or stabilize the pMMO.

On the other hand, a strong correlation does not exist between the pMMO specific activity and the iron content (iron/protein ratio) of the membrane fractions. The iron content increases only by a factor of 4 over the total range of copper concentration varied despite the fact that the initial iron concentration in the growth medium was constant (0.9 μM). However, this trend in the iron/protein ratio appears not to be reproducible in recent experiments. As a result, the present data do not permit any conclusion regarding the presence of iron in the pMMO. Our results also indicate that the iron detected appears to be associated with membrane-bound c-type cytochromes. Metal analysis also reveals <u>no other metal ions</u> (including Zn, Co, Mn, Ni, Mg) aside from Fe in the membrane preparations at higher than trace levels.

EPR CHARACTERIZATION OF THE MEMBRANE-BOUND COPPER IONS

The membrane-associated copper is EPR detectable (Figure 2a). The Cu EPR spectrum is observed in the membrane fractions only when the pMMO is expressed. The intensity of the Cu EPR signal is found to correlate directly with the in vitro pMMO specific activity in isolated membrane fractions. A linear relationship is observed when the pMMO specific activity is plotted versus the EPR-detected copper concentration, i.e., the copper concentration inferred from the observed EPR intensity (Figure 3). These results also suggest that the pMMO activity is associated primarily with the membrane-bound copper ions. Interestingly, the intensity of the copper EPR observed for the membrane fractions isolated from organisms grown in high copper medium (> 5 μM) accounts for only about 1/3 of the expected intensity based on the total copper content.

The analysis of the EPR spectrum of membrane fractions obtained from *M. capsulatus* (Bath) exhibiting high in vitro activity suggests the presence of two distinct copper signals. One signal (Fig. 2a) can be attributed to type 2 copper centers on the basis of A and g values ($g_{||} = 2.25$, $A_{||} \sim 18 \times 10^{-3}$ cm^{-1}, $g_{\perp} \sim 2.058$). The other signal occurs near g ~ 2.06. This signal is broad and nearly isotropic. The relative contribution of each type of Cu^{2+} ions to the EPR spectral intensity is very sensitive to temperature. The type 2 Cu^{2+} EPR signal can be easily saturated at low temperatures (<10 K) whereas the "isotropic" signal is not, and hence can be isolated and directly observed by recording the spectrum at high microwave powers (Fig. 2b). The appearance of this signal is not a result of lineshape distortion due to high microwave power but rather is a manifestation of dramatic difference in the relaxation characteristics of the two species which give rise to these two EPR signals. In addition, the intensity of the "isotropic" signal is found to correlate directly with the in vitro pMMO activity.

These features of the "isotropic" signal suggest that it may have origin in a multinuclear copper cluster. Consistent with this idea, the unusual signal decreases in intensity and gives way to a second isotropic signal at the same g value but with resolved hyperfine features (|A| ~ 15 Gauss) upon incubation and reduction with limiting, i.e., sub-stoichiometric amounts of dithionite (Figure 2c). If the "isotropic" signal arises from a multinuclear cluster, <u>partial</u> reduction might ultimately yield a <u>mixed-valence</u> system with an unpaired-electron spin delocalized over all the Cu ions of the cluster. The observed hyperfine interaction pattern (Figure 2c & Inset) is consistent with this

Table I

Effect of Growth Medium Copper Concentration on Bacterial Copper Content and pMMO Activity

Cu Conc.[a] of Media (µM)	Total Intracellular[b] Copper (µmole)	Cu Content (µmole)[c]		Cu/Protein (µg/mg)[d]		Specific Activity (nmole/min/mg)[e]	
		Soluble	Membrane	Soluble	Membrane	As isolated	w/ Copper added
<0.3	0.7 (~100)	0.17 (25)	0.53 (75)	0.41	1.32	0.68	19.5
2.0	4.4 (~100)	0.48 (11)	3.88 (89)	1.06	7.08	9.6	27.3
5.0	8.7 (87)	0.78 (9)	7.92 (91)	1.55	13.4	17.6	27.7
10.0	13.7 (69)	1.03 (8)	12.7 (92)	2.53	22.2	26.8	30.2
20.0	22.8 (57)	2.0 (9)	20.8 (91)	4.27	37.5	53.7	22.1

[a] Copper concentration of media after supplemental copper was added. The value of 0.3 µM is the upper limit of the copper concentration in the un-supplemented media. [b] The total amount of copper found in the bacteria (whole cells), determined by DCPE. The number in parentheses is the percentage of the total copper available in the media that is ultimately found in the bacteria. [c] The copper content of the soluble and membrane fractions. The number in parenthesis is the portion of total bacteria copper found in each fraction. [d] The copper/protein ratio of the soluble and the membrane fractions. Protein concentration was determined by the Lowry method. [e] The specific activity of the pMMO in the membrane fractions (per mg of total membrane protein) assayed as isolated or with copper added to the assay buffer (150 µM).

expectation and suggests that the splitting arises from the coupling of an unpaired electron spin to three equivalent $I = 3/2$ nuclear spins. That is, the "isotropic" signal with resolved <u>ten nuclear hyperfine lines</u> (Figure 2 - Inset) arises from a mixed-valence trinuclear copper cluster wherein the electron spin of a Cu (II) is delocalized over two

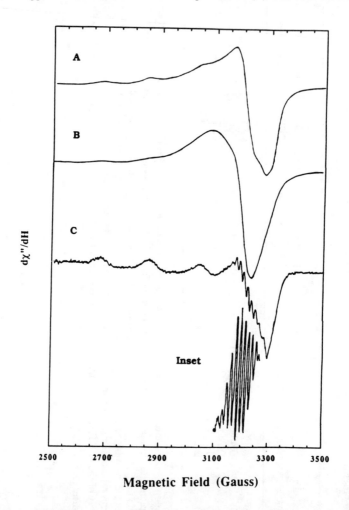

Magnetic Field (Gauss)

Figure 2. X-band EPR spectra of membrane fractions obtained from _Methylococcus capsulatus_ (Bath) cells grown with 20 μM Cu^{2+}. (A) Spectrum obtained with microwave power of 0.2 mW; (B) Spectrum obtained with microwave power of 40 mW. EPR spectra (A) and (B) were recorded at 7 K with modulation amplitude of 16 G, modulation frequency of 100 kHz, and gain of 2.5 x 10^2 for (A) and 1.6 x 10^2 for (B); (C) Spectrum obtained after 3 hrs incubation with sodium dithionite and was recorded at 8.4 K with 1 mW of microwave power, modulation amplitude of 5 G, modulation frequency of 100 kHz, time constant of 0.25 sec., and gain of 3.2 x 10^3. Inset : Second derivative of the absorption at g = 2.06.

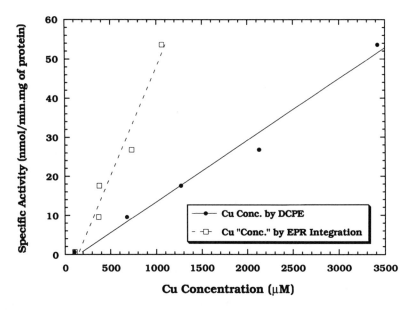

Figure 3. Specific activity versus copper content as detected by DCPE and by EPR double integration. Each sample was adjusted to a total protein concentration of 10 mg/ml. EPR intensity was calibrated with the concentration of a series of $CuSO_4$ standards. EPR spectra of the samples and $CuSO_4$ standards were obtained at 7 K at a microwave field of 9.126 GHz with 0.2 mW of power, modulation amplitude of 16 G, modulation frequency of 100 kHz, and gain of 5 x 10^3. Specific activity versus copper/protein ratio of membrane fractions obtained from *Methylococcus capsulatus* (Bath) grown at increasing Cu^{2+} concentrations. The cultures were grown in 500-ml batches in 2-l Erlenmeyer flasks using NMS medium, a 20% methane-to-air atmosphere, and continual shaking. Cultures in which the copper concentration of the medium was varied were grown in parallel with identical medium and growth conditions. The membrane fractions were isolated as described previously [5]. The MMO activity of samples was measured by the standard propene epoxidation assay [5].

additional Cu(I) ions. Furthermore, the analysis of the spectral amplitude of these hyperfine lines also supports this conclusion. The relative spectral amplitudes of the ten hyperfine lines for the aforementioned mixed-valence system in which the single unpaired electron is coupled to three equivalent I=3/2 nuclei should follow the ratio of 1:3:6:10:12:12:10:6:3:1. On the other hand, if these lines are the result of splittings from 4 equivalent nitrogen nuclei (I=1), nine lines of relative amplitude 1:4:10:16:19:16:10:4:1 are expected. A comparison of the theoretical and experimental spectral amplitudes (Figure 4) support the contention that the hyperfine pattern arises from three I=3/2 nuclei. We have also observed the EPR spectrum of the one electron-reduced trinuclear copper cluster (the so-called triplet EPR spectrum) (data not shown). As a result, it can be concluded that the "isotropic" Cu EPR signal observed for the fully oxidized enzyme arises from appropriate EPR transitions of an exchange-coupled trinuclear Cu cluster (vide infra).

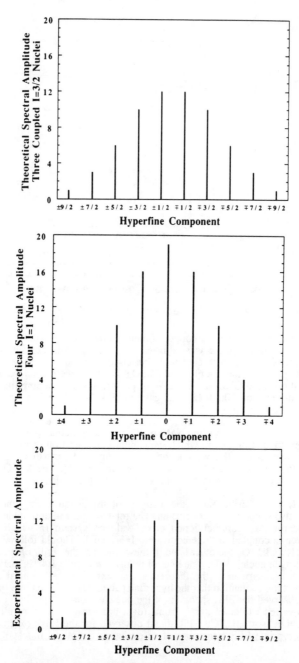

Figure 4. A comparison of theoretical hyperfine intensity distributions as predicted for three equally coupled I=3/2 nuclei and four I=1 nuclei cases with experiment.

MAGNETIC SUSCEPTIBILITY CHARACTERIZATION OF THE MEMBRANE-BOUND COPPER IONS

Although the isolated membrane fractions contain certain amounts of iron in addition to a substantial level of copper, the level of paramagnetic iron (probably low spin), however, is insignificant as compared with the amount of paramagnetic copper ions. The weak line observed at g ~ 4.3 in several EPR spectra can be attributed to non-specific iron. As a result, only paramagnetic copper ions contribute significantly to the magnetization in magnetic susceptibility measurements.

Parallel magnetic susceptibility experiments on membranes obtained from _M. capsulatus_ (Bath) confirm the existence of the above proposed exchange-coupled spin system. Coupling of the unpaired electron spins from three equivalent Cu(II) ions would give rise to a S = 3/2 state in addition to two S = 1/2 states. Since only the ground state contributes significantly to the magnetization at very low temperatures (<2 K), isothermal magnetization saturation experiments at these temperatures are expected to reveal the ground level, which can be either the quartet or the two degenerate doublets, depending on whether the exchange interactions among the Cu(II) ions are ferromagnetic or antiferromagnetic. Magnetization data measured at 1.8 K and varying field strengths (0-5.5 Tesla) fit the Brillouin function [12] with an effective spin (S_{eff}) of 1.44 ± 0.05 (Figure 5), indicating that the copper ions are ferromagnetically coupled and that the bulk of the copper ions in the membranes exists in the form of the trinuclear cluster. The value of the effective spin as deduced from the magnetic susceptibility experiment here provides strong support for our interpretation of the EPR results and establishes unequivocally the existence of a spin-coupled trinuclear copper cluster in the pMMO system.

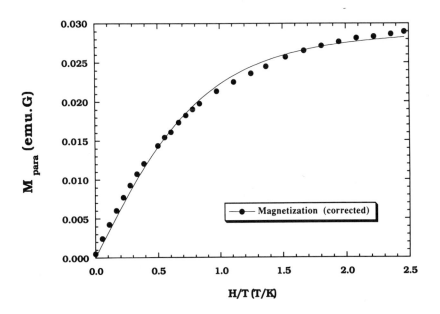

Figure 5. The Brillouin plot of the magnetization observed for membranes obtained from _Methylococcus capsulatus_ (Bath) cells grown with 20 μM Cu^{2+}. The magnetization saturation data were obtained over a range of magnetic field (0-5.5 Tesla) and at 1.8 K.

ANALYSIS OF EPR AND MAGNETIC SUSCEPTIBILITY DATA: A SPIN-COUPLING MODEL

In the case of strong exchange-coupling, the spin-coupled system consists of multiplets of total spin S. Assuming that the three copper ions in the spin-coupled trinuclear cluster are equivalent, the energy levels of the system under the application of an external magnetic field can be described by the spin Hamiltonian:

$$\hat{H} = -2J \sum_{i>j}^{3} \hat{S}_i \cdot \hat{S}_j + D[\hat{S}_z^2 - \frac{1}{3}S(S+1)] + E[\hat{S}_x^2 - \hat{S}_y^2] + \beta g\hat{S}\cdot\hat{H}$$

where J is the exchange coupling constant, D and E are the axial and rhombic zero-field splitting (ZFS) parameters, and g is the isotropic g value of the trimer. The energy levels for such a spin-coupled trinuclear copper cluster are illustrated schematically in Figure 6, where we have assumed that the exchange coupling is ferromagnetic, as indicated by the low-temperature magnetization saturation experiment described above. This scheme provides a basis for the in-depth analysis of the EPR and magnetic susceptibility data.

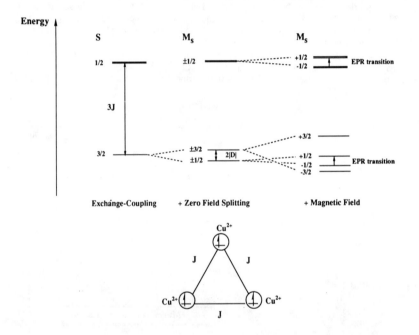

Figure 6. Energy level diagram for a ferromagnetically-coupled trinuclear copper cluster. J and D refer to the exchange-coupling constant and the axial zero-field splitting, respectively (See the equation in text). The spacing between the energy levels is not drawn to scale.

With the overwhelming presence of the coupled trinuclear copper clusters, magnetic susceptibility data can be analyzed in greater detail according to the coupling model and the derived energy levels. The magnetic saturation data can be fitted to a quartet (S=3/2) with about 5-10% of isolated type-2 Cu^{2+} ions (S=1/2) if not less. The analysis of the temperature-dependence of the magnetization also shows that the ground state is indeed a quartet with $D < 0.05$ cm^{-1} and J is in the order of about 15 - 20 cm^{-1} (data not shown). The inclusion of other species of S=1/2 (presumably isolated type 2 Cu^{2+}) into the calculations indicates that only ~5% of the total copper ions if not less are in the form the type 2 Cu^{2+} ions.

For the D_{3h} model under consideration here, EPR signals at X-band frequency are expected from the |-1/2> → |+1/2> transitions of the two degenerate doublet manifolds and from the |-1/2> → |+1/2> transition within the quartet manifold (Figure 6). At 7 K, the temperature at which the EPR measurements were made, the |-1/2> → |+1/2> transition of the quartet manifold should make the dominant contribution to the intensity of the EPR signal. In the case of |D| within the order of magnitude as estimated from the magnetic susceptibility experiments, resonance field calculations indicate that this signal is quasi-isotropic with g_{eff} ~ 2 both when the applied field is parallel or perpendicular to the principal molecular axis. This result is consistent with the observed EPR spectra of the fully oxidized copper cluster. On the other hand, for a system with large |D|, i.e. |D| >> gβH, the EPR spectrum becomes highly anisotropic, with g_{eff} ~ 2 when the magnetic field is parallel to the normal of the equilateral triangle formed by the three copper ions and g_{eff} ~ 4 when the applied field is in the plane. Obviously, this is not the case since the intense features expected at g ~ 4 are not observed. For D of that magnitude, calculations of resonance fields also indicate that the |-3/2> → |-1/2> and |+1/2> → |+3/2> transitions of the quartet manifold are highly anisotropic. As a result, only the |-1/2> → |+1/2> transition of the quartet manifold contributes to the EPR spectrum at 7 K. This would contribute to an apparent intensity anomaly when the intensity of the "isotropic" signal at g = 2.06 is used to quantify the copper concentration of the membrane fractions.

A MODEL FOR THE DIOXYGEN CHEMISTRY AND MECHANISM OF METHANE ACTIVATION

The correlation between the pMMO specific activity and the copper clusters in these membranes suggests that these clusters might be the active sites of the pMMO. An indication of this possible important catalytic role of the copper clusters is suggested by the reactivity of the partially reduced clusters toward dioxygen. Anaerobic titration of the membranes by dithionite leads to a substantial decrease in the intensity of the copper EPR signal. Upon re-exposure of the sample to air, we observed a full restoration of the copper EPR signal to its original intensity and appearance prior to electron reduction. Because the system reacts so readily with dioxygen, it seems likely that the copper clusters correspond to site(s) of oxygenase activity.

The exchange-coupled trinuclear copper clusters found in the pMMO system appear reminiscent of the interacting type 2 copper/type 3 binuclear copper cluster center in multiple blue copper oxidases in which the binuclear cluster is the dioxygen binding and reduction site. The similarity between these two prosthetic groups appears to be a result of their similar function. Conceivably, the trinuclear clusters seen here have evolved specifically for the purpose of methane oxygenation. A mechanism for dioxygen activation and methane oxidation involving these clusters can be proposed (Figure 7).

According to this scheme, the trinuclear cluster will react most readily with dioxygen at the three-electron level of reduction. Subsequently, the O-O bond is cleaved heterolytically, resulting in the formation of an oxygen "radical" species. This highly reactive intermediate will abstract a hydrogen atom from methane, followed by the addition of the hydroxyl radical into the methyl radical to yield methanol. Since the oxygenase reaction is obviously kinetically driven and the activation energy barrier is small, once the oxygen intermediate is formed, the oxygenase reaction is expected to be facile. Several examples of this type of chemistry are known.

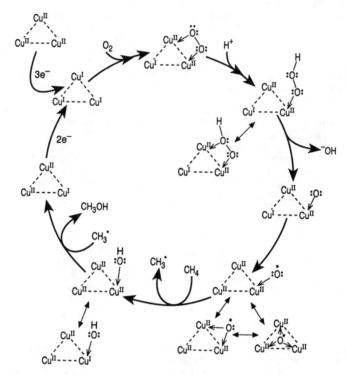

Figure 7. Dioxygen reduction and methane activation by trinuclear copper cluster. A mechanistic proposal.

PRELIMINARY ESTIMATES OF THE NUMBER OF COPPER ACTIVE SITES IN THE pMMO

EPR and magnetic susceptibility results suggest that the copper ions in the membrane fractions are highly homogenous, i.e., the bulk of the copper ions is in the form of a trinuclear copper cluster. No direct evidence has been obtained for the presence of other types of copper centers such as type 1 copper or a binuclear copper cluster. As a result, the copper/protein ratio for the pMMO can be estimated by assuming that the pMMO system is the major copper-containing protein in the membrane fraction. Results of metal analysis by DCPE and Lowry assays indicate a very high copper/protein ratio in these membranes, in the range of 12- 25 mg Cu/g protein for cells grown at 20 μM Cu^{2+}. If the pMMO is the major protein contributing to the Lowry assay and an apparent molecular weight of about 107 kDa is taken for the enzyme, these results suggest a copper content of approximately 30 Cu/protein under the current growth conditions. Since the Lowry assay almost certainly overestimates the pMMO protein concentration due to contribution from other proteins, the actual Cu/protein ratio is expected to be much higher.

CONCLUSION

Several intriguing aspects of these membranes emerge from this study. The finding that virtually all of the copper in the growth medium is incorporated into the membrane fraction suggests that these membranes, and/or the pMMO, act as a copper sponge. Our magnetic susceptibility studies indicate that the bulk of the copper ions are arranged in trinuclear clusters and that there are many such clusters per protein molecule. Since the pMMO activity correlates with the copper content of the membranes in which the enzyme resides, and the copper ions appear to exist as trinuclear clusters with well-defined magnetic and redox properties, it seems reasonable to conclude that the copper clusters are associated primarily with the pMMO. If each trinuclear copper cluster can serve as a complete catalytic unit, capable of redox and dioxygen chemistry, then the pMMO system represents a new class of monooxygenase with virtually unknown properties. Efforts are now under way to obtain more direct evidence that the pMMO is indeed a copper protein, to establish the ligand structures of the copper clusters, and to elucidate the dioxygen and methane activation chemistry that might be at work here.

Acknowledgment

This work was supported by grants GM 22432 (S.I.C.) and GM 40859 (M.E.L.) from the National Institute of General Medical Sciences, U.S. Public Health Service. Seed funding from the Caltech Chemistry and Chemical Engineering Industrial Consortium to initiate this project is also gratefully acknowledged.

REFERENCES

[1] C. Anthony, *The Biochemistry of Methylotrophs*, (Academic Press, London, 1982), Chapter 1,6, and 8.
[2] C. Anthony. *Advances in Microbial Physiology*, **27**, 113-209 (1986).
[3] S.H. Stanley, S.D. Prior, D.J. Leak, H. Dalton, *Biotechnology Letters*, **5**, 487-492 (1983).
[4] K.J. Burrow, A. Cornish, D. Scott, I.J. Higgins, *J. Gen. Microbiol.*, **130**, 3327-3333 (1984).
[5] S.D. Prior, H. Dalton, *J. Gen. Microbiol.*, **131**, 155-163 (1985).
[6] C. Bedard, R. Knowles, *Microbiol. Rev.*, **53**, 68-84 (1989).
[7] B.G. Fox, J.D. Lipscomb, *Biochem. Biophys. Res. Commum.*, **154**, 165-170 (1988).
[8] B.G. Fox, K.K. Surerus, E. Munck, J.D. Lipscomb, *J. Biol. Chem.*, **263**, 10553-10556 (1988).
[9] M.P. Woodland, H. Dalton, *J. Biol. Chem.*, **259**, 53-60 (1984).
[10] J. Green, H. Dalton, *J. Biol. Chem.*, **264**, 17698-17703 (1989).
[11] J.G. Dewitt, J.G. Bentsen, A.C. Rosenzweig, B. Hedman, J. Green, S. Pilkington, G.C. Papaefthymiou, H. Dalton, K.O. Hodgson, S.J. Lippard, *J. Am. Chem. Soc.*, **113**, 9219-9235 (1991).
[12] R.L. Carlin, *Magneto Chemistry* (Springer-Verlag, Berlin, 1986).
[13] M.D. Allendorf, D.J. Spira, E.I. Solomon, *Proc. Natl. Aca. Sci. U.S.A.*, **82**, 3063-3067 (1985).
[14] A. Messerschmidt, et al. , *J. Mol. Biol.*, **206**, 513-529 (1989).

THE ENZYMOLOGY OF PEPTIDE AMIDATION

David J. Merkler[a,1], Raviraj Kulathila[a], Stanley D. Young[a], John Freeman[b], and Joseph J. Villafranca[b,2]

[a]Analytical Protein and Organic Chemistry Group, Unigene Laboratories, Inc., 110 Little Falls Rd., Fairfield, NJ 07004, USA,
[b]Department of Chemistry, Pennsylvania State University, University Park, PA, 16802, USA

INTRODUCTION

Since the 1950's, it has been recognized that most peptide hormones contain a C-terminal amide[3-6]. In the last 40 years, over 100 amidated peptides have been identified[7,8] and structure-activity relationships have shown that the C-terminal amide is key to the activity elicited by most amidated peptides[9-12]. The amide moiety arises by the post-translational, oxidative cleavage of a C-terminal glycine-extended prohormone (peptidyl-Gly) at the α-carbon of the glycine in a reaction which requires a reducing equivalent, copper, and O_2[13,14]. Peptidylglycine α-amidating enzyme (α-AE, EC 1.14.17.3)[15,16] is the enzyme that catalyzes this reaction in vivo. In this review, we will focus on the chemical and enzymological aspects of peptide amidation. For more comprehensive treatments of the subject, two recent reviews are recommended[8,17].

DISCOVERY, PURIFICATION AND SUBSTRATE SPECIFICITY OF α-AE.

An enzyme capable of converting a glycine-extended substrate, [^{125}I]-D-Tyr-Val-Gly, to an amidated product, [^{125}I]-D-Tyr-Val-NH$_2$, was first described in 1982[13]. Using the appropriate labeled substrates, D-Tyr-Val-[^{15}N]-Gly and D-Tyr-Val-[^{14}C]-Gly, Bradbury et al.[13] showed that the nitrogen of the C-terminal glycine was retained in the amide moiety of the product and that the carbon atoms of the C-terminal glycine were incorporated into the other product of amidation, glyoxylate. Subsequently, Eipper et al.[14] showed that conversion of D-Tyr-Val-Gly to D-Tyr-Val-NH$_2$ required copper, O_2, and ascorbate (as an electron-donor). They further recognized that the cofactor requirements of α-AE were similar to those of dopamine ß-hydroxylase[18,19] (DßH) and suggested that the two enzymes were related. Later, a comparison of the DßH and α-AE cDNA sequences showed a 31% homology between the enzymes indicating that the two proteins are related[20]. α-AE has been purified from a variety

From K.D. Karlin and Z. Tyeklár, Eds., *Bioinorganic Chemistry of Copper*
(Chapman & Hall, New York, 1993).

of tissues from several organisms with molecular weights ranging from 37 kDa to 75 kDa[8,17,21,22] and there are reports for α-AE species > 100 kDa[23,24]. The substantial heterogeneity in the apparent molecular weight results from the alternate splicing of the α-AE mRNA and the subsequent processing of the α-AE proteins at paired basic amino acid sites[25-27]. A discussion of the generation of multiple forms of α-AE in the rat, cow, frog, and human is presented by Eipper et al.[8].

Despite the size heterogeneity, the substrate specificities of the enzymes that have been studied are the same. Treatment of α-AE with metal chelators abolishes amidation activity which is only restored upon copper addition[28]. The production of each mole of amidated peptide (peptidyl-NH$_2$) requires the input of two electrons[29] which are, most likely, provided in vivo by ascorbate[30]. The enzyme accepts electrons from a variety of structurally unrelated reductants[31,32], but ascorbate exhibits the highest V_{MAX}/K_M value[33]. The two electrons required per α-AE turnover are supplied in two one-electron reductions indicating that semidehydroascorbate, not dehydroascorbate, is a product of the amidation reaction[29]. Similarly, semidehydroascorbate is also found as a product for the DßH reaction[34,35]. In contrast to the wide reductant specificity, α-AE, with one exception, has a strict requirement for a C-terminal glycine. Peptides terminating in other D- or L-amino acids are not substrates or inhibitors for the enzyme[28,36,37]. The only exception is D-alanine-extended peptides which are poor substrates for α-AE[38]. The ratio of the V_{MAX}/K_M for dansyl-Tyr-Val-Gly to the V_{MAX}/K_M for dansyl-Tyr-Val-D-Ala is ~30 (D.J. Merkler, unpublished). The penultimate amino acid of the glycine-extended peptide exerts a tremendous influence on the ability of α-AE to amidate its substrate. By substituting each of the twenty natural amino acids in the penultimate position of dansyl-(Gly)$_4$-X-Gly, Tamburini et al.[39] found that the V_{MAX}/K_M for peptidyl-Gly varied 1250-fold with dansyl-(Gly)$_4$-Glu-Gly having the lowest V_{MAX}/K_M and dansyl-(Gly)$_4$-Phe-Gly with the highest V_{MAX}/K_M value. Under the conditions of their experiments, no amidation of dansyl-(Gly)$_4$-Asp-Gly was observed[39]. In general, the enzyme prefers glycine-extended substrates with a hydrophobic or a sulfur-containing amino acid in the penultimate position.

In addition to peptide amidation, α-AE will carry out other oxidative chemistry. Bradbury and Smyth showed that the enzyme catalyzed the oxidation of glyoxylate phenylhydrazone to oxalate phenylhyrazide[40]. Katopodis and May extended this observation showing that α-AE catalyzed sulfoxidation, amine N-dealkylation, and O-dealkylation[41]. The alternate oxidative reactions catalyzed by α-AE underscore the similarities between α-AE and DßH because DßH also catalyzes sulfoxidation[42] and amine N-dealkylation[43].

PEPTIDYL-α-HYDROXYGLYCINE AS AN INTERMEDIATE OF BIFUNCTIONAL α-AE.

The labeling studies of Bradbury et al.[13] demonstrated the oxidative nature of the amidation reaction and eliminated the direct amidation of the C-terminal carboxylate[44] or transamidation of the penultimate amino acid-glycine peptide bond[45] as prospective mechanisms for α-AE. Bradbury et al.[13] initially proposed an oxidative mechanism consisting of an initial dehydrogenation of the glycyl C_α-N bond to form a C-terminal N-acylimine which then hydrolyzed to the amidated peptide plus glyoxylate (Scheme 1, mechanism A). Ramer et al.[46] suggested N-hydroxylation of the C-terminal glycine as an alternate route to the N-acylimine intermediate. A second mechanism involving the direct α-hydroxylation by O_2 and the subsequent conversion of the peptidyl-α-hydroxyglycine to the amidated peptide plus glyoxylate was later proposed based on model studies[47] and the finding that α-AE oxidized glyoxylate phenylhydrazone[40] (Scheme 1, mechanism B). Data exists to support each of the mechanistic possibilities. For mechanism A, evidence comes from studies on the oxidative cleavage of peptides by Cu(III), generated from Cu(II) by Ir(IV) or electrochemical oxidation[48-50]. Such systems produce amidated peptides plus a keto product and are proposed to proceed through a N-acylimine intermediate. The formation of a N-acylimine intermediate after N-hydroxylation is consistent with the mechanism suggested for the oxidation of N-aroylglycines to N-(acetoxymethyl)benzamides and N-formylbenzamides by lead tetraacetate[51]. Support for mechanism B comes from data on the P-450 catalyzed O-dealkylation of 7-ethoxycoumarin which most likely involves direct α-hydroxylation[52], studies showing that the chemical oxidation of N-salicyloylglycine produces N-salicyloyl-α-hydroxyglycine[53] and the similarity of α-AE to DßH[20], which also hydroxylates its substrate.

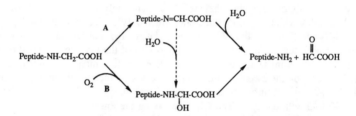

Scheme 1. Proposed mechanisms for α-AE.

Using dansyl-Tyr-Val-N-hydroxyglycine and both diastereomers of dansyl-Tyr-Val-α-hydroxyglycine, Young and Tamburini[54] found that only one diastereomer of dansyl-Tyr-Val-α-hydroxyglycine was converted to dansyl-Tyr-Val-NH_2 in a copper, ascorbate, and O_2 independent reaction[54-56]. α-AE did not metabolize either dansyl-Tyr-Val-N-hydroxyglycine nor the other diastereomer of dansyl-Tyr-Val-α-hydroxyglycine[54]. A stereochemical assignment for the diastereomers of dansyl-Tyr-Val-α-hydroxyglycine was not possible. These results strongly favor mechanism B and eliminate the N-hydroxylation pathway entirely. Additional evidence that peptidyl-α-hydroxyglycine is an intermediate in the amidation reaction (mechanism B) has come from the recent discovery that the conversion of peptidyl-Gly to peptidyl-NH_2 requires two active sites which can be separated into distinct proteins[8,33,57-61]. Peptidylglycine α-hydroxylating monooxygenase (PHM, EC.1.4.17.3) catalyzes the ascorbate, O_2, and copper-dependent formation of peptidyl-α-hydroxyglycine and peptidylamidoglycolate lyase (PAL, EC.4.3.2.5) catalyzes the O_2 and cofactor independent conversion of peptidyl-α-hydroxyglycine to peptidyl-NH_2 and glyoxylate. Separate PHM and PAL enzymes which catalyze either the formation or dealkylation of the carbinolamide intermediate as well as bifunctional enzymes (still designated as α-AE) which alone catalyze the conversion of peptidyl-Gly to peptidyl-NH_2 amidation have been described[33,57,60-63].

The report of Katopodis et al.[57] describing the isolation of bovine PAL led to the proposal that peptide amidation resulted from the sequential action of two enzymes[57,58,61,63]. This proposal brought into question reports of apparently homogenous proteins which catalyzed the conversion of peptidyl-Gly to peptidyl-NH_2[22,31,64,65]. These preparations represented either (i) homogenous α-AE encompassing both activities, (ii) purified PHM or PAL contaminated by trace levels of the missing activity, or (iii) purified PHM assayed under conditions where peptidyl-α-hydroxyglycine is spontaneously converted to peptidyl-NH_2. Peptidyl-α-hydroxyglycine is susceptible to alkaline dealkylation[59,66,67] and reports of α-AE based on amidation assays performed at pH ≥ 8.0[21,68] raise the suggestion that some of the proteins under study were actually PHM. At acidic pH, peptidyl-α-hydroxyglycine is reasonably stable ($t_{1/2}$ at pH 5.5 is ~7 days[66]). Since the amidation of peptides occurs in secretory granules which have an internal pH ~5.5[69], the non-enzymatic conversion of peptidyl-α-hydroxyglycine would not occur to any great extent.

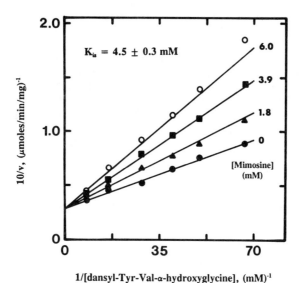

Figure 1. The inhibition of N-dealkylation of dansyl-Tyr-Val-α-hydroxyglycine by mimosine. Initial rates were measured at 37 °C in 100 mM MES pH 6.0, 30 mM KCl, 30 mM KI, 1 % (v/v) ethanol, 0.001 % (v/v) Triton X-100, and 15-124 μM dansyl-Tyr-Val-α-hydroxyglycine. The points are the experimental rates and the lines were drawn with the use of the constants obtained by computer fit to the equation for slope-linear, competitive inhibition[118].

With the availability of a soluble, 75-kDa form of rat α-AE cloned and expressed into a system with virtually no background amidation activity[25,70], we could determine if it were possible for the activities necessary to amidate glycine-extended peptides to reside on a single protein. As reported, cell culture medium conditioned by mouse C127 cells transfected with a rat α-AE cDNA converted both dansyl-Tyr-Val-Gly and dansyl-Tyr-Val-α-hydroxyglycine at levels ~1000-fold higher than medium conditioned by untransfected C127 control cells[55]. This result strongly suggested that rat 75-kDa α-AE contained both PHM and PAL activities and we proposed that bifunctional forms of the enzyme were possible. To account for reports of distinct PHM and PAL enzymes, we[55] and others[57,60] proposed that the activities could be separated from a common precursor by proteolytic digestion.

The selective inactivation of the PHM activity of rat 75-kDa α-AE mediated by ascorbate[33] and the transient accumulation of peptidyl-α-hydroxyglycine during amidation[67,71] are consistent with the proposal that this enzyme is bifunctional. Additional evidence for the bifunctionality of this form of rat α-AE comes from mimosine inhibition studies. We have previously shown that mimosine is competitive vs. ascorbate with a $K_{is} = 4.0 \pm 0.3\ \mu M$[72]. As shown in Fig. 1, mimosine is also competitive vs. dansyl-Tyr-Val-α-hydroxyglycine; however, the K_{is} is 4500 ± 300 μM. The simplest explanation for these data is that the enzyme has separate PHM and PAL active sites which have affinities for mimosine that differ ~ 1000-fold. These results taken together results strongly suggest that the rat 75-kDa α-AE is bifunctional.

Proteolytic separation of PHM and PAL from a bifunctional precursor is consistent with all the data concerning enzymes capable of amidating peptidyl-Gly or capable of catalyzing only one step of the amidation reaction. Other bifunctional enzymes are known[73-76] and, in some cases, the different activities have been separated by in vitro proteolysis[77]. For α-AE, proteolytic separation of PHM and PAL seems to be controlled by the alternate splicing of the α-AE mRNA. A schematic representation of the two α-AE cDNAs, types A and B, isolated from rat medullary thyroid carcinoma (MTC) is shown in Fig. 2. The two rat MTC cDNAs differ by

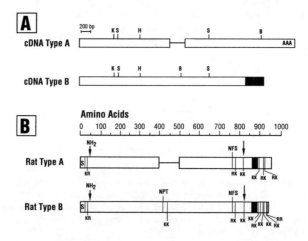

Figure 2. Two types of α-AE cDNA in rat MTC tissues. (A) Schematic representation of the two types of rat α-AE cDNA. The shaded box denotes nonidentical sequences and the letters are restriction sites: K, Kpn I; S, Sph I; H, Hinc II and B, Bam HI. (B) Schematic representation of the proteins predicted by translation of the two rat α-AE cDNAs, S, signal sequence; NH$_2$, N-terminus; shaded box, membrane spanning domain; arrow ($\downarrow$), C-terminus; paired basic amino acids representing potential cleavage sites are indicated by the single letter code below the site, and potential sites of N-linked glycosylation are given in single letter code above the site. The alternatively spliced 315 bp cDNA (A) or 105 amino acid peptide (B) is indicated by the solid line. Reprinted from Bertelsen et al.[25] by permisson from Academic Press.

an in-frame 315 bp exon (exon A, see Eipper et al.[8]) between the proposed PHM and PAL active sites that is alternatively spliced out of the type A cDNA[25]. An in-frame 315 bp exon between the PHM and PAL sites has also been identified in cDNAs from rat pituitary[26], rat atrium[78], frog[79,80], and human[81,82]. Contained within the 105 amino acids[82] coded by exon A is a paired basic amino acid site which can serve as a site for endoproteolytic processing[83]. Thus, type B α-AE DNA, which retains exon A, codes for an enzyme that can be processed into separate PHM and PAL enzymes. In type A α-AE DNA, exon A has been spliced out removing the proteolytic processing site and, thus, codes for a bifunctional enzyme. The α-AE that we have proposed to be bifunctional[33,55] is a type A enzyme[25].

^{18}O INCORPORATION INTO PEPTIDYL-α-HYDROXYGLYCINE OR PYRUVATE.

The evidence discussed above indicating that peptide amidation occurs in two steps with the initial formation of peptidyl-α-hydroxyglycine clearly favors mechanism B (Scheme 1). Mechanism B predicts O_2 as the source of the hydroxyl oxygen atom of the carbinolamide and the aldehydic oxygen of glyoxylate. Mechanism A predicts H_2O as the source of oxygen in the oxidation products. Thus, an important test for mechanism B is the analysis of glyoxylate or peptidyl-α-hydroxyglycine for ^{18}O incorporation after amidation reactions in $^{18}O_2$ or $H_2^{18}O$. Using porcine PHM, Zabriskie et al.[84] reported the partial incorporation of ^{18}O (30%) into D-Phe-Phe-α-hydroxyglycine from $^{16}O_2/^{18}O_2$ (50:50). This evidence supports mechanism B, but does not eliminate the possibility of some N-acylimine formation. More recent results of Noguchi et al.[56] showed that hydroxylation of D-Tyr-Val-Gly by rat brain PHM produced D-Tyr-Val-[α-^{18}O]-hydroxyglycine only in the presence of $^{18}O_2$. No incorporation of ^{18}O into the hydroxyl group from $H_2^{18}O$ was found.

Incorporation of ^{18}O into glyoxylate using bifunctional α-AE is not feasible because glyoxylate readily hydrates to the hemiacetal[85] and rapidly exchanges incorporated ^{18}O for ^{16}O from solvent. Attempts to trace the source of oxygen into 2-aminoacetaldehyde generated by the DβH catalyzed dealkylation of N-phenylethylenediamine failed because of the rapid exchange of the aldehydic oxygen with solvent[43]. Alternatively, the amidation of D-alanine-extended peptides produces pyruvate. Since pyruvate hydrates only slightly at pH $\geq$ 3[86], ^{18}O incorporation into the α-carbonyl oxygen of pyruvate could be used to test the mechanism of bifunctional α-AE. We have demonstrated by an ^{18}O-^{13}C NMR isotopic shift that $^{18}O_2$ and not $H_2^{18}O$ is the source of the α-carbonyl oxygen of [α-^{13}C]pyruvate generated by the amidation of dansyl-Tyr-Val-[α-^{13}C]-D-Ala[29]. This result agrees with the earlier data of Zabriskie et al.[84] and Noguchi et al.[56] and provides additional support that amidation proceeds by direct hydroxylation of the C-terminal glycine.

As indicated in Scheme 1 by the broken arrow connecting the N-acylimine intermediate to peptidyl-α-hydroxyglycine, it has been argued that production of the carbinolamide does not absolutely eliminate the possibility of N-acylimine formation[43,46,50,84,87]. These authors suggest that H_2O addition to the N-acylimine would form peptidyl-α-hydroxyglycine. Malassa and Matthies[88] have shown that water can add to an N-acylimine to form a carbinolamide. We[29] and others[56] have found no incorporation of ^{18}O from $H_2^{18}O$ into product which implies that H_2O addition to the N-acylimine cannot account for the α-AE catalyzed formation of

201

peptidyl-α-hydroxyglycine. A proposal[84] that α-AE reduces $^{18}O_2$ to $H_2^{18}O$ in the active site which then adds to the N-acylimine to form peptidyl-[α-^{18}O]-hydroxyglycine cannot be eliminated by the available experimental data. If this latter mechanism is correct, the α-AE active site must be isolated from solvent in order to be consistent with the ^{18}O labeling experiments.

REACTION STEREOCHEMISTRY AND STOICHIOMETRY

Using stereospecifically tritiated peptides, D-Tyr-Val-[α-^3H_S]-Gly and D-Tyr-Val-[α-^3H_R]-Gly, Ramer et al.[46] showed that the *pro*-S hydrogen of the C-terminal glycine was abstracted during amidation. This result was consistent with the earlier observation that α-AE amidated D-alanine-extended peptides, but not L-alanine-extended peptides[38]. It has also been determined, using purified PHM and PAL sequentially[89] and bifunctional α-AE[71], that only one diastereomer of peptidyl-α-hydroxyglycine is enzymatically formed and that this diastereomer is the substrate for PAL. The other diastereomer of peptidyl-α-hydroxyglycine is not dealkylated by PAL[54,71,89]. Because of the instability of the carbinolamide and our inability to crystalize dansyl-Tyr-Val-α-hydroxyglycine (S.D. Young, unpublished), it had not been possible to assign the absolute configuration of the diastereomers of peptidyl-α-hydroxyglycine. By reacting racemic N-cinnamoyl-α-hydroxyglycine sequentially with PAL and acylase I, Ping et al.[89] found that both enzymes utilized the same enantiomer as a substrate. Thus, the peptidyl-α-hydroxyglycine produced by PHM and that which is a substrate for PAL is of the S-configuration. Therefore, the hydroxylation of the glycyl α-carbon proceeds with retention of configuration similar to the reactions catalyzed by DβH[90-92] and other oxygenases[93-95].

In order to formulate a reasonable hypothesis for the chemical and kinetic mechanism of α-AE, it is important to define both the stereochemical course of the reaction and the stoichiometric relationships between substrates and products. When developing a fluorimetric assay for amidation, Jones et al.[96] verified that the stoichiometry of peptidyl-NH_2 produced/peptidyl-Gly consumed was approximately 1.0. Stoichiometries of ~1.0 were also defined for peptidyl-NH_2 produced/O_2 consumed[29], glyoxylate produced/4-nitrobenzamide consumed[41], glyoxylate produced/peptidyl-α-hydroxyglycine consumed[57,67], and pyruvate produced/peptidyl-D-Ala consumed[29]. As discussed earlier, the amidation of each peptidyl-Gly requires the input of two electrons which are supplied in two one-electron oxidations of ascorbate to two molecules of semidehydroascorbate[29]. The reaction catalyzed by bifunctional α-AE consistent with the data presented in this review is presented in Scheme 2.

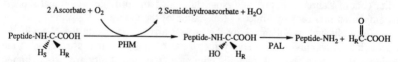

Scheme 2. The reaction catalyzed by bifunctional α-AE.

ROLE OF COPPER IN PEPTIDE AMIDATION.

A number of investigators have shown that copper is a required cofactor for peptide amidation[14,28] (Fig. 3A). Because the PAL catalyzed N-dealkylation of peptidyl-α-

hydroxyglycine was unaffected by o-phenanthroline or diethyldithiocarbamate[54,56,57,60], it was proposed that copper is only required for peptidyl-Gly hydroxylation. In agreement with the results of Eipper et al.[60], we found that EDTA ≥ 100 μM inhibited $\sim 50\%$ of the PAL activity (Fig. 3B). Since the inhibition of PAL exhibited by EDTA is observed at concentrations significantly higher than the concentrations necessary to completely inhibit amidation and Cu(II) addition does not restore PAL activity (compare Fig. 3B to Fig. 3A), it is possible that EDTA inhibition is not due to copper chelation. Colombo et al.[97] showed that DßH could bind 0.5-1.5 moles of EDTA/active site. Perhaps α-AE also binds EDTA and this binding alters the catalytic properties of PAL.

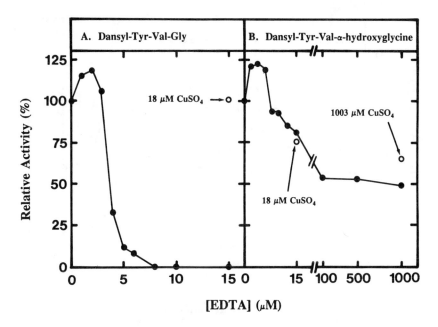

Figure 3. Effect of EDTA on the reactions catalyzed by rat MTC α-AE. Initial rates for the conversion of dansyl-Tyr-Val-Gly (A) or dansyl-Tyr-Val-α-hydroxyglycine (B) to dansyl-Tyr-Val-NH$_2$. Experiments without added CuSO$_4$ are represented by the filled circles ($\bullet$) and those with 3 μM CuSO$_4$ in excess of the EDTA concentration are represented by the open circles ($\circ$). Atomic absorption analysis indicates that our water contains sufficient trace copper (~ 0.5 μM) to support amidation.

We have applied electron paramagnetic resonance spectroscopy to α-AE in order to explore the environment of the enzyme-bound copper. EPR spectra of the enzyme associated with varying amounts of copper are shown in Fig. 4. Spectrum A (4X scale) is α-AE after exhaustive dialysis against pH 6.0 MES Buffer. The enzyme contains 0.3 Cu/active site and shows a signal consistent with an axial ligand field with $g_\perp = 2.07$ and $g_\parallel = 2.30$. This spectrum shows a symmetrical splitting pattern of 150 G, consistent with the presence of a single type 2 copper center. Addition of copper to the enzyme results in a broadening of the peaks in the $g_\parallel$ region along with the appearance of asymmetry in the $A_\parallel$ values (spectrum B, 0.9 Cu/active site, 2X scale), indicative of heterogeneity in copper binding to α-AE. This heterogeneity is present through the addition of up to 1.9 Cu/active site (spectrum C). A difference spectrum of the 0.3 Cu/active site and the 0.9 Cu/active site is shown as spectrum D (4X scale). This spectrum shows a $g_\parallel = 2.26$ with $A_\parallel = 165$ G. The heterogeneity in the signal and the ability to resolve the heterogeneity into distinct spectra suggest the presence of at least two copper binding sites in the enzyme. Preliminary data show that copper activation of the enzyme is biphasic with an inflection point at 1:1 mole ratio of copper to enzyme (data not shown). These data are consistent with the DβH data which showed that two Cu(II)/active site are required for full activity[98-100].

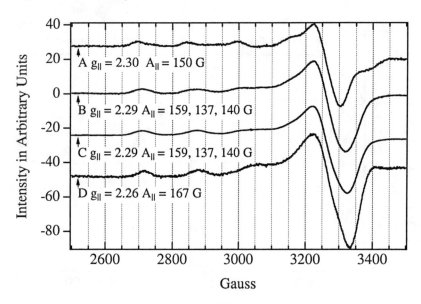

Figure 4. X band EPR spectra of rat MTC α-AE containing various amounts of copper per active site. (A), 0.3 Cu(II)/active site; (B), 0.9 Cu(II)/active site; (C), 1.9 Cu(II)/active site and (D), difference spectra of B minus A.

FORMATION OF A GLYCYL RADICAL DURING AMIDATION.

Dopamine β-hydroxylase and α-AE are remarkably similar. Alignment of their respective amino acid sequences shows regions of extensive homology and, in particular, two proposed His-X-His copper-binding motifs are conserved[20]. Both enzymes catalyze the stereospecific hydroxylation of their substrates in a reaction that requires O_2, contain two type 2 coppers/active site (the data for α-AE is not conclusive), and carry out two one-electron reductions. It has been shown that DβH forms a substrate radical during catalysis[101,102]. By analogy, α-AE may form an α-centered glycine (or alanine) radical during amidation. In the chemical literature, there is considerable support for the formation of α-centered glycyl radicals. Garrison[103] proposed an α-centered radical to account for results on the radiolytic fragmentation of glycine and alanine. In addition, numerous reports indicate the preferential reactivity of glycine in free radical reactions which has been attributed to the formation of a secondary radical at the α-carbon[104-108].

The copper and ascorbate-dependent irreversible inactivation of α-AE by D-Phe-Phe-D-vinylglycine[109] and by phenyl-substituted carboxylates possessing a double bond α or β to the carboxyl group[41,110,111] are consistent with the formation of a substrate radical during turnover. The rate or product distribution of chemical reactions that involve biradical intermediates can be modified by a magnetic field that alters intersystem crossing rates between the singlet and triplet-spin correlated states[112,113]. Recently, we have found that the V_{MAX} for amidation is decreased nearly two-fold at a magnetic field of 1000 G while the V_{MAX}/K_M is unaffected[114]. This result suggests that a spin-correlated radical intermediate is formed during amidation after the first irreversible step (most likely copper reduction by ascorbate). This is one of the first reports of a dependence of an enzymatic rate on a magnetic field and provides direct evidence for the formation of a substrate radical during catalysis.

FUTURE DIRECTIONS

The most basic questions remaining to be answered concerning α-AE are the stoichiometry of copper/active site and the kinetic mechanism. Studies addressing both of these issues are underway in our laboratory. After answering these fundamental questions, it will be possible to more fruitfully apply more sophisicated techniques to study the enzyme. EXAFS and spin-echo EPR could be used to probe the environment of the enzyme-bound copper. Isotope effects (primary deuterium, secondary deuterium, and heavy-atom effects) could be used to determine the position of the rate determining step(s) in the amidation reaction, to provide insights into the chemical mechanism, and possibly to elucidate the transition-state structure[115]. Rapid quench techniques, which would be relatively easy to apply to α-AE because of the low turnover number (~ 10 sec^{-1}), would provide a measure of the microscopic rate constants for the individual steps in the reaction (important for establishing the intrinsic isotope effects), could verify the kinetic mechanism, and determine if there is channeling of the peptidyl-α-hydroxyglycine between the PHM and PAL active sites.

Recent reports of suicide substrates for α-AE[41,109-111] offer the potential of isolating and sequencing the active site peptide. Since we have cloned and expressed the rat 75-kDa enzyme[25,72], site-directed mutagensis of the active site residues would be possible. Unlike DβH, α-AE is monomeric which means that crystalization of α-AE should be easier than the crystallization of DβH. Our goal of crystalizing α-AE is both practical and possible because we are able to produce large amounts of recombinant enzyme[116].

REFERENCES

1. To whom correspondence should be addressed.
2. Current address: Bristol-Myers Squibb Co., Pharmaceutical Research Institute, Lawrenceville, NJ 08648
3. V. du Vigneaud, C. Ressler, S. Trippet, *J. Biol. Chem.*, **205**, 949 (1953).
4. H. Tuppy and H. Michl, *Monatsh. Chem.*, **84**, 1011 (1953).
5. R. Acher and J. Chauvet, *Biochim. Biophys. Acta*, **12**, 487 (1953).

6. J.I. Harris and A.B. Lerner, *Nature (London)*, **179**, 1346 (1957).
7. D. Konopińska, G. Rosiński, W. Sobotka, *Int. J. Pept. Protein Res.*, **39**, 1 (1992).
8. B.A. Eipper, D.A. Stoffers, R.E. Mains, *Annu. Rev. Neurosci.*, **15**, 57 (1992).
9. W. Rittel, R. Maier, M. Brugger, B. Kamber, B. Riniker, P. Sieber, *Experientia*, **32**, 246 (1976).
10. A. Fournier, R. Couture, J. Magnan, M. Gendreau, D. Regoli, S. St. Pierre, *Can. J. Biochem.*, **58**, 272 (1980).
11. W. Vale, J. Spiess, C. Rivier, J. Rivier, *Science*, **213**, 1394 (1981).
12. G.E. Pratt, D.E. Farnsworth, N.R. Siegel, K.F. Fok, R. Feyereisen, *Biochem. Biophys. Res. Commun.*, **163**, 1243 (1989).
13. A.F. Bradbury, M.D.A. Finnie, D.G. Smyth, *Nature (London)*, **298**, 686 (1982).
14. B.A. Eipper, R.E. Mains, C.C. Glembotski, *Proc. Natl. Acad. Sci. USA*, **80**, 5144 (1983).
15. Abbreviations used: α-AE, peptidylglycine α-amidating enzyme; DßH, dopamine ß-hydroxylase; EPR, electron paramagnetic resonance spectroscopy; EXAFS, extended X-ray absorption fine structure spectroscopy; G, gauss; MTC, medullary thyroid carcinoma; PAL, peptidylamidoglycolate lyase; PHM, peptidyl α-hydroxylating monooxygenase
16. Another frequently used abbreviation for the enzyme is PAM, which stands for peptidylglycine α-amidating monooxygenase.
17. A.F. Bradbury and D.G. Smyth, in *Peptide Biosynthesis and Processing*, L.D. Fricker, Ed., (CRC Press, Boca Raton, 1991), pp. 231-249.
18. E.Y. Levin, B. Levenberg, S. Kaufman, *J. Biol. Chem.*, **235**, 2080 (1960).
19. S. Friedman and S. Kaufman, *J. Biol. Chem.*, **240**, 4763 (1965).
20. C. Southan and L.I. Kruse, *FEBS Lett.*, **255**, 116 (1989).
21. A.S.N. Murthy, R.E. Mains, B.A. Eipper, *J. Biol. Chem.*, **261**, 1815 (1986).
22. N.M. Mehta, J.P. Gilligan, B.N. Jones, A.H. Bertelsen, B.A. Roos, R.S. Birnbaum, *Arch. Biochem. Biophys.*, **261**, 44 (1988).
23. G.A. Beaudry and A.H. Bertelsen, *Biochem. Biophys. Res. Commun.*, **163**, 959 (1989).
24. S.N. Perkins, E.J. Husten, B.A. Eipper, *Biochem. Biophys. Res. Commun.*, **171**, 926 (1990).
25. A.H. Bertelsen, G.A. Beaudry, E.A. Galella, B.N. Jones, M.L. Ray, N.M. Mehta, *Arch. Biochem. Biophys.*, **279**, 87 (1990).
26. I. Kato, H. Yonekura, H. Yamamoto, H. Okamoto, *FEBS Lett.*, **269**, 319 (1990).
27. B.A. Eipper, C.B.-R. Green, T.A. Campbell, D.A. Stoffers, H.T. Keutmann, R.E. Mains, L.H. Ouafik, *J. Biol. Chem.*, **267**, 4008 (1992).
28. A.F. Bradbury and D.G. Smyth, in *Biogenetics of Neurohormonal Peptides*, R. Håkanson and J. Thorell, Eds., (Academic Press, New York, 1985), pp. 171-186.
29. D.J. Merkler, R. Kulathila, A.P. Consalvo, S.D. Young, D.E. Ash, *Biochemistry*, **31**, 7282 (1992).
30. U.M. Kent and P.J. Fleming, *J. Biol. Chem.*, **262**, 8174 (1987).

31. J.S. Kizer, R.C. Bateman, Jr., C.R. Miller, J. Humm, W.H. Busby, Jr., W.W. Youngblood, *Endocrinology*, **118**, 2262 (1986).

32. A.S.N. Murthy, H.T. Keutmann, B.A. Eipper, *Mol. Endocrinol.*, **1**, 290 (1987).

33. D.J. Merkler, R. Kulathila, P.P. Tamburini, S.D.Young, *Arch. Biochem. Biophys.*, **294**, 594 (1992).

34. T. Skotland and T. Ljones, *Biochim. Biophys. Acta*, **630**, 30 (1980).

35. E.J. Diliberto, Jr. and P.L. Allen, *J. Biol. Chem.*, **256**, 3385 (1981).

36. A.F. Bradbury and D.G. Smyth, *Biochem. Biophys. Res. Commun.*, **112**, 372 (1983).

37. P.P. Tamburini, B.N. Jones, A.P. Consalvo, S.D. Young, S.J. Lovato, J.P. Gilligan, L.P. Wennogle, M. Erion, A.Y. Jeng, *Arch. Biochem. Biophys.*, **267**, 623 (1988).

38. A.E.N. Landymore-Lim, A.F. Bradbury, D.G. Smyth, *Biochem. Biophys. Res. Commun.*, **117**, 289 (1983).

39. P.P. Tamburini, S.D. Young, B.N. Jones, R.A. Palmesino, A.P. Consalvo, *Int. J. Pept. Protein Res.*, **35**, 153 (1990).

40. A.F. Bradbury and D.G. Smyth, *Eur. J. Biochem.*, **169**, 579 (1987).

41. A.G. Katopodis and S.W. May, *Biochemistry*, **29**, 4541 (1990).

42. S.W. May, R.S. Phillips, P.W. Mueller, H.H. Herman, *J. Biol. Chem.*, **256**, 2258 (1981).

43. K. Wimalasena and S.W. May, *J. Am. Chem. Soc.*, **109**, 4036 (1987).

44. A.P. Scott, J.G. Ratcliff, L.H. Rees, H.P.J. Bennett, P.J. Lowry, C. McMartin, *Nature (London), New Biol.*, **244**, 65 (1973).

45. J.S. Fruton, R.B. Johnston, M. Fried, *J. Biol. Chem.*, **190**, 39 (1951).

46. S.E. Ramer, H. Cheng, M.M. Palcic, J.C. Vederas, *J. Am. Chem. Soc.*, **110**, 8526 (1988).

47. R.C. Bateman, Jr., W.W. Youngblood, W.H. Busby, Jr., J.S. Kizer, *J. Biol. Chem.*, **260**, 9088 (1985).

48. A. Levitzki, M. Anbar, A. Berger, *Biochemistry*, **6**, 3757 (1967).

49. J.S. Rybka, J.L. Kurtz, T.A. Neubecker, D.W. Margerum, *Inorg. Chem.*, **19**, 2791 (1980).

50. K.V. Reddy, S.-J. Jin, P.K. Arora, D.S. Sfeir, S.C.F. Maloney, F.L. Urbach, L.M. Sayre, *J. Am. Chem. Soc.*, **112**, 2332 (1990).

51. A.P. Gledhill, C.J. McCall, M.D. Threadgill, *J. Org. Chem.*, **51**, 3196 (1986).

52. G.T. Miwa, J.S. Walsh, A.Y.H. Lu, *J. Biol. Chem.*, **259**, 3000 (1984).

53. P. Capdevielle and M. Maumy, *Tetrahedron Lett.*, **31**, 3831 (1991).

54. S.D. Young and P.P. Tamburini, *J. Am. Chem. Soc.*, **111**, 1933 (1989).

55. D.J. Merkler and S.D. Young, *Arch. Biochem. Biophys.*, **289**, 192 (1991).

56. M. Noguchi, H. Seino, H. Kochi, H. Okamoto, T. Tanaka, M. Hirama, *Biochem. J.*, **283**, 883 (1992).

57. A.G. Katopodis, D. Ping, S.W. May, *Biochemistry*, **29**, 6115 (1990).

58. I. Kato, H. Yonekura, M. Tajima, M. Yanagi, H. Yamamoto, H. Okamoto, *Biochem. Biophys. Res. Commun.*, **172**, 197 (1990).

59. M. Tajima, T. Iida, S. Yoshida, K. Komatsu, R. Namba, M. Noguchi, H. Okamoto, *J. Biol. Chem.*, **265**, 9602 (1990).

60. B.A. Eipper, S.N. Perkins, E.J. Husten, R.C. Johnson, H.T. Keutmann, R.E. Mains, *J. Biol. Chem.*, **266**, 7827 (1991).
61. K. Takahashi, H. Okamoto, H. Seino, M. Noguchi, *Biochem. Biophys. Res. Commun.*, **169**, 524 (1990).
62. A.G. Katopodis, D. Ping, C.E. Smith, S.W. May, *Biochemistry*, **30**, 6189 (1991).
63. K. Suzuki, H. Shimoi, Y. Iwasaki, T. Kawahara, Y. Matsuura, Y. Nishikawa, *EMBO J.*, **9**, 4259 (1990).
64. J.P. Gilligan, S.J. Lovato, N.M. Mehta, A.H. Bertelsen, A.Y. Jeng, P.P. Tamburini, *Endocrinology*, **124**, 2729 (1991).
65. K. Mizuno, J. Sakata, M. Kojima, K. Kangawa, H. Matsuo, *Biochem. Biophys. Res. Commun.*, **137**, 984 (1986).
66. H. Bundgaard and K.H. Kahns, *Peptides*, **12**, 745 (1991).
67. J. Bongers, A.M. Felix, R.M. Campbell, Y. Lee, D.J. Merkler, E.P. Heimer, *Pept. Res.*, **5**, 183 (1992).
68. M. Noguchi, K. Takahashi, H. Okamoto, *Arch. Biochem. Biophys.*, **275**, 505 (1989).
69. R.E. Mains, E.I. Cullen, V. May, B.A. Eipper, *Ann. N.Y. Acad. Sci.*, **493**, 278 (1987).
70. G.A. Beaudry, N.M. Mehta, M.L. Ray, A.H. Bertelsen, *J. Biol. Chem.*, **265**, 17694 (1990).
71. A.P. Consalvo, S.D. Young, D.J. Merkler, *J. Chromatogr.*, **607**, 25 (1992).
72. D.A. Miller, K.U. Sayad, R. Kulathila, G.A. Beaudry, D.J. Merkler, A.H. Bertelsen, *Arch. Biochem. Biophys.*, **298**, 380 (1992).
73. A.J. Makoff and A. Radford, *Biochim. Biophys. Acta*, **485**, 313 (1977).
74. R. Ferone and S. Roland, *Proc. Natl. Acad. Sci. USA*, **77**, 5802 (1980).
75. M.R. El-Maghrabi, T.M. Pate, K.J. Murray, S.J. Pilkis, *J. Biol. Chem.*, **259**, 13096 (1984).
76. J. Turnbull and J.F. Morrison, *Biochemistry*, **29**, 10255 (1990).
77. H. Klenow and I. Henningsen, *Proc. Natl. Acad. Sci. USA*, **65**, 168 (1970).
78. D.A. Stoffers, C.B.-R. Green, B.A. Eipper, *Proc. Natl. Acad. Sci. USA*, **86**, 735 (1989).
79. K. Mizuno, K. Ohsuye, Y. Wada, K. Fuchimura, S. Tanaka, H. Matsuo, *Biochem. Biophys. Res. Commun.*, **148**, 546 (1987).
80. K. Ohsuye, K. Kitano, Y. Wada, K. Fuchimura, S. Tanaka, K. Mizuno, H. Matsuo, *Biochem. Biophys. Res. Commun.*, **150**, 1275 (1988).
81. J. Glauder, H. Ragg, J.W. Engels, *Biochem. Biophys. Res. Commun.*, **169**, 551 (1990).
82. Exon A for human α-AE contains 321 bp[77] and, consequently, codes for 107 amino acids.
83. R.B. Harris, *Arch. Biochem. Biophys.*, **275**, 315 (1989).
84. T.M. Zabriskie, H. Cheng, J.C. Vederas, *J. Chem. Soc., Chem. Commun.*, 571 (1991).
85. A.R. Rendina, J.D. Hermes, W.W. Cleland, *Biochemistry*, **23**, 5148 (1948).
86. Y. Pocker, J.E. Meany, B.J. Nist, C. Zadorojny, *J. Phys. Chem.*, **73**, 2879 (1969).
87. G.C. Barrett, L.A. Chowdhury, A.A. Usmani, *Tetrahedron Lett.*, **23**, 2063 (1978).

88. I. Malassa and D. Matthies, *Liebigs Ann. Chem.*, 1133 (1986).
89. D. Ping, A.G. Katopodis, S.W. May, *J. Am. Chem. Soc.*, **114**, 3998 (1992).
90. K.B. Taylor, *J. Biol. Chem.*, **249**, 454 (1974).
91. L. Bachan, C.B. Storm, J.W. Wheeler, S. Kaufman, *J. Am. Chem. Soc.*, **96**, 6799 (1974).
92. A.R. Battersby, P.W. Sheldrake, J. Stauton, D.C. Williams, *J. Chem. Soc., Perkin Trans. 1*, 1056 (1976).
93. Y. Fujita, A. Gottlieb, B. Peterkofsky, S. Udenfriend, B. Witkop, *J. Am. Chem. Soc.*, **86**, 4709 (1964).
94. S. Shapiro, J.U. Piper, E. Caspi, *J. Am. Chem. Soc.*, **104**, 2301 (1982).
95. S. Englard, J.S. Blanchard, C.F. Middelfort, *Biochemistry*, **24**, 1110 (1985).
96. B.N. Jones, P.P. Tamburini, A.P. Consalvo, S.D. Young, S.J. Lovato, J.P. Gilligan, A.Y. Jeng, L.F. Wennogle, *Anal. Biochem.*, **168**, 272 (1988).
97. G. Colombo, N.J. Papadopoulos, D.E. Ash, J.J. Villafranca, *Arch. Biochem. Biophys.*, **252**, 71 (1987).
98. D.E. Ash, N.J. Papadopoulos, G. Colombo, J.J. Villafranca, *J. Biol. Chem.*, **259**, 3395 (1984).
99. J.P. Klinman, M. Krueger, M. Brenner, D.E. Edmondson, *J. Biol. Chem.*, **259**, 3399 (1984).
100. G. Colombo, B. Rajashekhar, D.P. Giedoc, J.J. Villafranca, *Biochemistry*, **23**, 3590 (1984).
101. P.F. Fitzpatrick, D.R. Flory, Jr., J.J. Villafranca, *Biochemistry*, **24**, 2108 (1985).
102. S.M. Miller and J.P. Klinman, *Biochemistry*, **24**, 2114 (1985).
103. W.M. Garrison, *Radiat. Res., Suppl. 4*, 158 (1964).
104. J. Sperling and D. Elad, *J. Am. Chem. Soc.*, **93**, 967 (1971).
105. M. Schwarberg, J. Sperling, D. Elad, *J. Am. Chem. Soc.*, **95**, 6418 (1973).
106. C.J. Easton and M.P. Hay, *J. Chem. Soc., Chem. Commun.*, 55 (1986).
107. V.A. Burgess, C.J. Easton, M.P. Hay, *J. Am. Chem. Soc.*, **111**, 1047 (1989).
108. P. Wheelan, W.M. Kirsch, T.H. Koch, *J. Org. Chem.*, **54**, 4360 (1989).
109. T.M. Zabriskie, H. Cheng, J.C. Vederas, *J. Am. Chem. Soc.*, **114**, 2270 (1992).
110. A.F. Bradbury, J. Mistry, B.A. Roos, D.G. Smyth, *Eur. J. Biochem.*, **189**, 363 (1990).
111. A.F. Bradbury, J. Mistry, D.G. Smyth, in *Peptides 1990, Proceedings of the Twenty-First European Peptide Symposium*, E.Giralt and D. Andreu, Eds., (ESCOM Science Publishers B.V., Leiden, 1991), pp. 763-765.
112. N.J Turro, *Proc. Natl. Acad. Sci. USA*, **80**, 609 (1983).
113. U.E. Steiner and T. Ulrich, *Chem. Rev.*, **89**, 51 (1989).
114. C.B. Grissom, D.J. Merkler, L.M. Markham, A.P. Miaullis, in preparation.
115. V.L. Schramm, in *Enzyme Mechanism from Isotope Effects*, P.F. Cook, Ed., (CRC Press, Boca Raton, 1991), pp. 367-388.
116. M.V.L. Ray, P.V. Duyne, A.H. Bertelsen, D.E. Jackson-Matthews, A.M. Sturmer, D.J. Merkler, A.P. Consalvo, S.D. Young, J.P. Gilligan, P.P. Shields, Bio/Technology, **11**, (in press).
117. W.W. Cleland, *Methods Enzymol.*, **63**, 103 (1979).

Copper-Mediated Redox/Oxidative Pathways

Redox Decomposition Reactions of Copper(III) Peptide Complexes

Dale W. Margerum,* William M. Scheper, Michael R. McDonald,
Françoise C. Fredericks, Lihua Wang, and Hsiupu D. Lee

Department of Chemistry, Purdue University
West Lafayette, Indiana 47907, USA

Introduction

Coordination of deprotonated-N(peptide) groups to copper helps to stabilize the trivalent oxidation state of copper.[1] Electrode potentials ($E°$) (eq 1, where P = peptide and n = number of

$$Cu^{III}(H_{-n}P) + e \rightarrow Cu^{II}(H_{-n}P), \quad E° \qquad (1)$$

protons displaced) of tri- and tetrapeptide complexes vary from 0.37 to 1.01 V (vs NHE). In many cases, it is possible to oxidize copper(II) complex to the corresponding copper(III) complex by bulk electrolysis in a flow-through column. Copper(III) peptides with low $E°$ values are slow to undergo redox decomposition. Thus, tri-α-aminoisobutyryl amide (Aib$_3$a) forms $Cu^{III}(H_{-3}Aib_3a)$ (structure 1), a complex with an $E°$ value of 0.37 V that is indefinitely stable in neutral solutions as long as it is not exposed to light.[2] Even in 1 M acid or 1 M base the complex is stable for weeks. The $Cu^{III}(H_{-2}Aib_3)$ complex (structure 2) with $E° = 0.66$ V is sufficiently stable to permit crystals to be grown for its structure determination.[3]

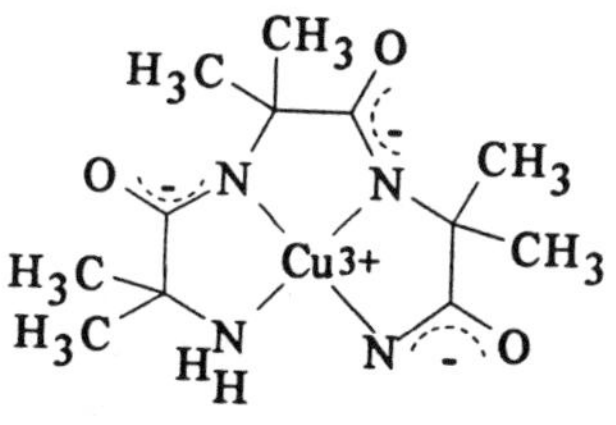

1. $Cu^{III}(H_{-3}Aib_3a)$ 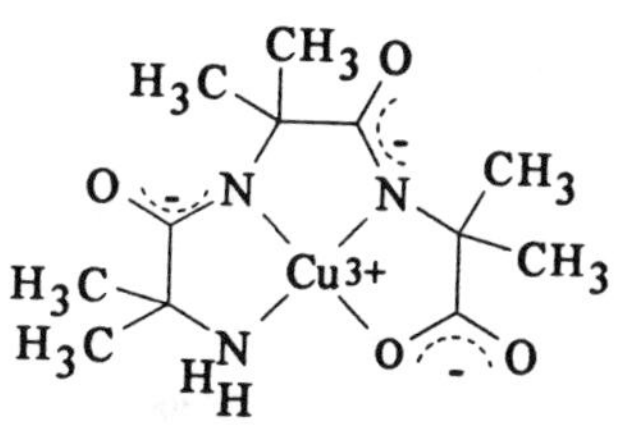   2. $Cu^{III}(H_{-2}Aib_3)$

From K.D. Karlin and Z. Tyeklár, Eds., *Bioinorganic Chemistry of Copper*
(Chapman & Hall, New York, 1993).

Glycyl, alanyl, lencyl, and α-aminoisobutyryl peptides.

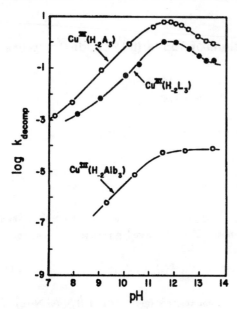

Figure 1. First-order decomposition rate constants of Cu(III) peptides.

As the $E°$ values increase for various copper(III) peptides, the complexes undergo more rapid redox decomposition. Figure 1 plots the log of the first-order decomposition rate constant (k_d, s^{-1}) against pH for three tripeptide complexes (A = L-alanine, L = L-leucine, Aib = α-aminoisobutyric acid).[4] The half-life of Cu^{III}(H$_{-2}$A$_3$) ($E°$ = 0.81 V) is only about 1 s at pH 12. Typically, the k_d values increase with pH until amine deprotonation[5] takes place. The loss of a terminal amine proton causes even stronger ligand donation to copper(III) and the resulting complex decomposes more slowly. The pK_a value for amine deprotonation (eq 2) of the trialanine complex is 11.0 and for many other tripeptide and tetrapeptide Cu(III) complexes the pK_a values fall between 11.0 and 12.5. In

$$Cu^{III}(H_{-2}A_3) \rightleftharpoons Cu^{III}(H_{-3}A_3)^- + H^+ \quad (2)$$

very high base the coordinated carboxylate group can be replaced by hydroxide ion to give an even less reactive species.

The k_d values at pH 11 for a series of tripeptides, tripeptide amides, and tetrapeptides are plotted as a function of their $E°$ values in Figure 2. The nature of the amino acid in the third residue affects the k_d values (A > G > Aib), but in all cases there is a strong correlation between the log k_d and $E°$ values. The greater the oxidizing power of the Cu(III) peptide, the faster it undergoes self-redox decomposition to give Cu(II) and oxidized peptides.

When an α-H atom is present (adjacent to the peptide nitrogen), a base decomposition pathway can be written (Scheme 1) where a Cu(II)-carbon radical is formed that is rapidly

214

oxidized by a second Cu(III) complex. Alternatively, a 2e process would lead to the same product through a Cu(I) intermediate (Scheme 2).

Figure 2. Correlation of the decomposition rate constants (s⁻¹) for Cu(III) peptides at pH 11 with their E° values. Tripeptides with 3rd residue G(●), A(▲), Aib (◆); tetrapeptides (○); Aib₃a (◊).

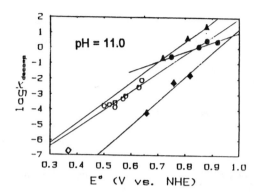

pH = 11.0

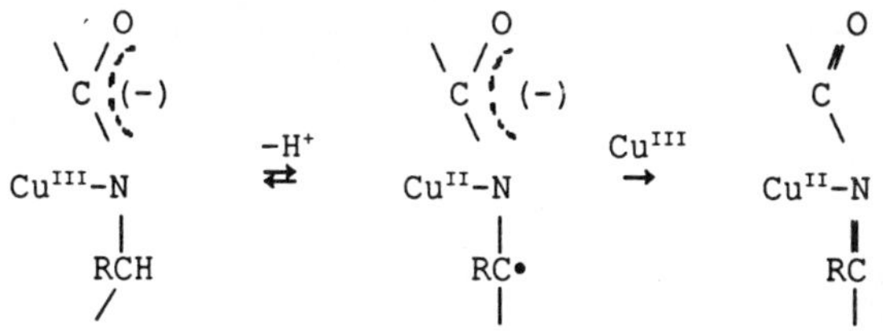

Scheme 1

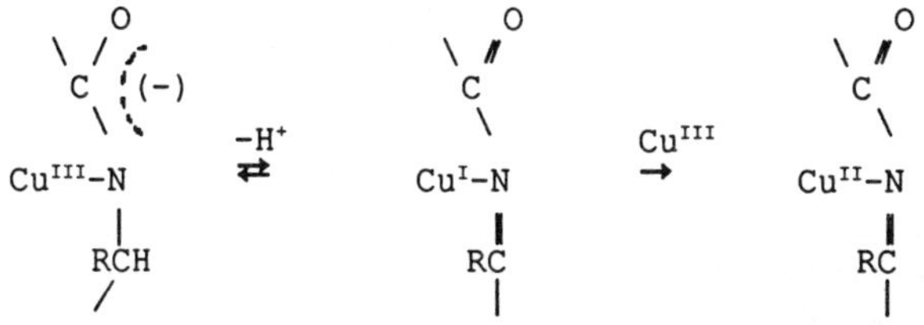

Scheme 2

The presence of Aib residues has two effects on the self-redox stability of Cu(III) peptides. First, the methyl groups enhance the donor strength of the copper-nitrogen bonds and lower the E° values. Second, the absence of an α-H atom forces a different and slower decomposition pathway. Thus, the k_d value for $Cu^{III}(H_{-3}Aib_3a)$ at pH 11 is only 2.1×10^{-7} s^{-1} ($t_{1/2}$ = 38 days).

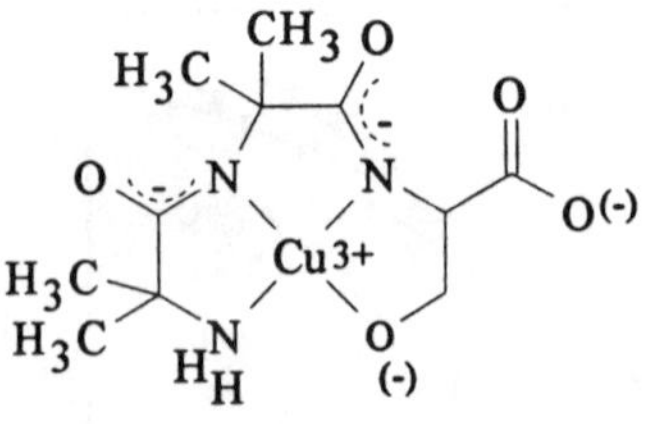

Alkoxide donors also give lower E° values,[6] for example, the $Cu^{III,II}(H_{-3}Aib_2Ser)^{-2-}$ couple (structure 3) has an E° value of 0.47 V and the k_d value for $Cu^{III}(H_{-3}Aib_2Ser)^-$ at pH 11 is 6.6×10^{-3} s^{-1}.

3. $Cu^{III}(H_{-3}Aib_2Ser)^-$

Histidine-containing Peptides.

When L-histidine is the third residue in tripeptide or tetrapeptide complexes, the E° values are much higher than for corresponding peptides with glycine in the third residue as seen in Table I. As a consequence the

Table I. Electrode Potentials (vs. NHE) for Copper(III,II) Peptide Complexes and pK$_a$ Values for Amine Deptrotonation of Cu(III) Peptides.

Peptide	E°, V vs. NHE	Amine Deprotonation pK$_a$ for Cu(III)
G$_3$	0.92	--
G$_2$A	0.88	11.1[a]
G$_2$His	0.98	8.2
G$_4$	0.63	12.1[b]
G$_2$HisG	1.01	8.8
Aib$_3$	0.66	11.6[c]
Aib$_2$His	0.81	8.7
Aib$_2$G$_2$	0.50	--
Aib$_2$HisG	0.85	8.8
G$_2$Histamine	0.97	--
G$_2$His-τ-Me	0.98	8.5
Aib$_3$amide	0.37	12.5[d]
A$_3$	0.81	11.0[a]
G$_2$-β-A	0.93	10.1

[a]Ref. 4. [b]Ref. 5. [c]D.W. Margerum *Pure and Appl. Chem.* **55**, 23 (1983). [d]Ref. 2.

decomposition rates of the histidine-containing copper(III) peptide complexes are rapid even in neutral or slightly acid conditions. For example, the half-life of $Cu^{III}(H_{-2}G_2HisG)$ is only 20 s at 25.0 °C, pH 4.7. In this case the copper(III) complex was collected rapidly from a bulk electrolysis column and its decay was immediately measured spectrophotometrically. The number of coulombs per mole of copper in the electrolysis were adjusted by the flow rate and voltage to ensure that only a one electron oxidation of Cu(II) to Cu(III) took place in the preparation of $Cu^{III}(H_{-2}G_2HisG)$. Multi-electron oxidation becomes a problem in attempts to prepare the even more reactive histidine-containing tripeptide complexes. Thus, the rate of decomposition of

4. $Cu^{III}(H_{-2}G_2HisG)$

$Cu^{III}(H_{-2}G_2His)$ is so rapid that it decomposes inside the electrolysis column and the products are also oxidized.

Rapid decomposition of the histidine-containing peptides of Cu(III) makes it difficult or impossible to measure the E° values by cyclic voltammetry. However, pulsed voltammetry techniques can be used to overcome this problem. We used the Osteryoung Square Wave Voltammetry (OSWV) technique,[7] where the sample is the stable Cu(II) peptide complex and the pulse time is sufficiently short so that decay of the Cu(III) peptide does not interfere with the E° measurement. Dervan and coworkers[8] have used G_2His linked to a DNA binding molecule and coordinated to copper in the presence of H_2O_2 and sodium ascorbate to give specific site DNA cleavage. They show that the cleavage is more specific than would be expected from the generation of hydroxyl radical. We believe that copper(III) plays an important role in these reactions. Hence the behavior of peptides with histidine in the third residue is of special interest.

Figure 3 shows the effect of pH on the electrode potential measured for the copper-Aib_2His complex. Since $Cu^{II}(H_{-2}Aib_2His)^-$ is fully formed by pH 6, the electrode potential at this pH corresponds to the E° value in eq 3. However, amine deprotonation of the Cu(III) complex (eq 4) affects the E value (eq 5). The solid line in Figure 3 corresponds to the fit for E° = 0.812 ± 0.001 V and $pK_a = 8.68 \pm 0.03$. Thus amine deprotonation for the Cu(III)

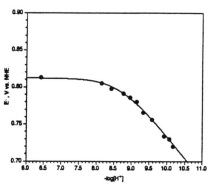

5. $Cu^{III}(H_{-2}G_2His)$

Figure 3. Determination of the amine-deprotonation pK_a for $Cu^{III}(H_{-2}Aib_2His)$.

$$Cu^{III}(H_{-2}Aib_2His) + e^- \rightleftharpoons Cu^{II}(H_{-2}Aib_2His)^-, \quad E° \qquad (3)$$

$$Cu^{III}(H_{-2}Aib_2His) \rightleftharpoons Cu^{III}(H_{-3}Aib_2His)^- + H^+, \quad K_a \qquad (4)$$

217

$$E = E^\circ - \frac{RT}{nF} \ln\left(\frac{K_a + [H^+]}{[H^+]}\right) \qquad (5)$$

complex of Aib_2His occurs much more readily than is the case for the corresponding Aib_3 complex ($pK_a = 11.6$). A similar behavior is found for G_2His ($pK_a = 8.2$) compared to G_2A ($pK_a = 11.1$). We prepared the τ-methyl derivative of G_2His and showed that its Cu(III) complex (structure 6) also has a low pK_a value for amine deprotonation ($pK_a = 8.5$). Hence, loss of a proton from the τ-nitrogen of histidine is not responsible for the low pK_a values. We also prepared the copper complex of $G_2\beta A$ (structure 7) to determine the effect of a 5,5,6-membered chelate ring sequence on these pK_a values. As shown in Table 1, the 5,5,6-chelates lower the pK_a

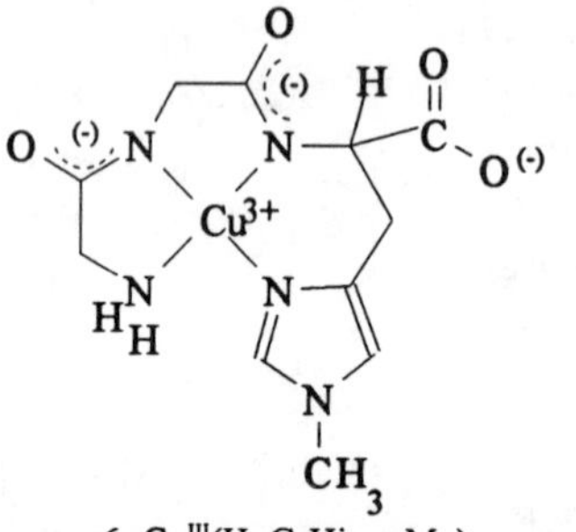

6. $Cu^{III}(H_{-2}G_2His\text{-}\tau\text{-}Me)$ 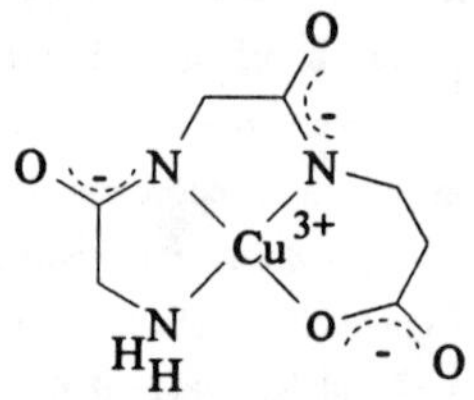   7. $Cu^{III}(H_{-2}G_2\beta A)$

value by 1.0 units for $G_2\beta A$ versus the 5,5,5-chelates in the G_2A complex. Imidazole coordination of the histidine residue to Cu(III) lowers the pK_a of amine deprotonation by another 2 pK units. The effect is unexpectedly large.

The kinetics of decomposition of Cu(III) peptides with histidine as the third residue are first order in the Cu(III) species. Rate constants at pH 4.7 - 5.1 are summarized in Table II. The

Table II. Decomposition Rate Constants for Cu(III) Peptides with Histidine or Derivatives as the Third Residue (25.0 °C).

Cu^{III}peptide	pH	E°, V vs. NHE	$k_{decomp.}$, s^{-1}
Aib_2HisG	5.0	0.87	5.5×10^{-4}
$G_2Histamine$	5.1	0.97	5.98×10^{-4}
G_2HisG	4.7	1.01	3.32×10^{-2}
G_2His	5.1	0.98	> 10

larger rate constant for G_2HisG compared to Aib_2HisG agrees with their relative $E°$ values. However, G_2His decomposes much more rapidly than G_2HisG despite their similar $E°$ values. The G_2Histamine complex (structure 8) is included for comparison. We investigated its properties in order to verify that it was not a product in the decomposition of $Cu^{III}(H_{-2}G_2His)$. The rapid decomposition of $Cu^{III}(H_{-2}G_2His)$ is associated with its fast decarboxylation. $Cu^{III}(H_{-2}Aib_2His)$ also decomposes rapidly with decarboxylation.

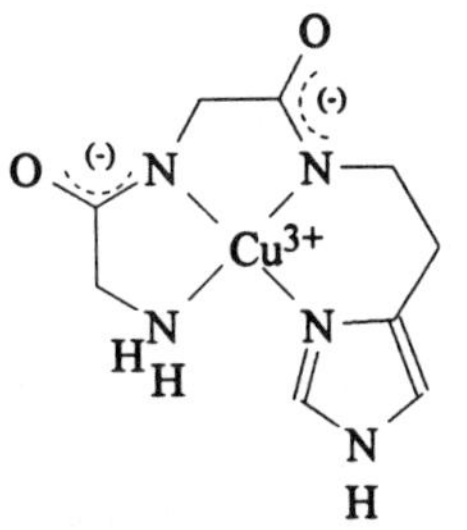

Table III summarizes the decomposition products obtained after decay of the Cu(III) complexes. Typical conditions that were used for previous one-electron bulk electrolysis of Cu(II) peptides[5] caused two or more equivalents of electrons per copper to be consumed with the histidine-containing peptides. Under these

8. $Cu^{III}(H_{-2}G_2Histamine)$

Table III. Products of Cu^{III}Peptide Decomposition

Electrolysis of Cu(II)Peptide[a]	Products
G_2His ($\geq$ 2 equiv.)	0% G_2His recovered 100% CO_2 released no circular dichroism (CD) activity no $G_2NHCH_2CH_2Im$ (histamine) main product: $G_2NHCH(OH)CH_2Im$
Aib_2His (1 equiv.)	50% Aib_2His recovered (CD) 50% CO_2 released
G_2HisG (2 equiv.)	57% G_2HisG recovered (CD)
G_2HisG (1 equiv.)	73% G_2HisG recovered (CD)

[a]Number of equivalents of electrons per copper in flow-through electrolysis is shown in parenthesis.

conditions G_2His is completely destroyed in the flow-through electrolysis procedure. The overall peptide oxidation is given by eq 6. A shorter column and faster flow rates were used with $Cu^{II}(H_{-2}Aib_2His)^-$ to give a one-equivalent oxidation to form the Cu(III) complex. In this case

$$G_2NHCH(CH_2Im)CO_2^- + OH^- \rightarrow G_2NHCH(CH_2Im)(OH) + CO_2 + 2e^- \qquad (6)$$

50% of the initial peptide was recovered. This is a typical yield when 50% of the peptide undergoes a two-electron oxidation. The decarboxylation reactions are so fast that repeated oxidation cycles use two or more equivalents and consume all of the original peptide in the case of G_2His. It is easier to generate a one-equivalent oxidation step with $Cu^{III}(H_{-2}G_2HisG)$, because it is slower to decompose. However, in this case the high recovery (73%) of the initial peptide indicates that the rest of the peptide has undergone a four-electron oxidation.

We identified the glycylglycyl-α-hydroxyhistamine product shown in eq 6 by ^1H and ^{13}C NMR. It is identical to the decarboxylated product found by Sakurai and Nakahara[9] after the reaction of $Ni^{II}(H_{-2}G_2His)^-$ with O_2. The dehydration of this peptide when coordinated to copper leads to the olefinic product whose crystal structure was reported by Meester and Hodgson,[10] *i.e.* α, β-didehydroglycylglycylhistaminatocopper(II).

The rapid decarboxylation of $Cu^{III}(H_{-2}G_2His)$ is very interesting. Pseudo-Kolbe electrolysis is the name given to anodic decarboxylations where electron transfer does not occur from the carboxylate but from a group attached to it.[11] Thus, $Cu^{III}(H_{-2}G_2His)$ might be a pseudo-Kolbe species where Cu^{III} is an electron pair acceptor that permits the loss of CO_2. We find that the rate of decarboxylation increases greatly as the acidity of the solution increases from pH 7.5 to pH 6.5. This effect cannot be due to changes in amine deprotonation, because that pK_a value is 8.2. The proton-assisted decarboxylation reaction leads us to suggest the mechanism given in Scheme 3, where protonation of the peptide nitrogen breaks its bond to copper and permits an electron pair from the α-carbon atom to form a copper-carbon bond as decarboxylation takes place. The chelate ring structure in the proposed mechanism changes from 5,5,6 for G_2His to 5,6,5 for the copper-carbon intermediate and back to 5,5,6 as the α-hydroxy

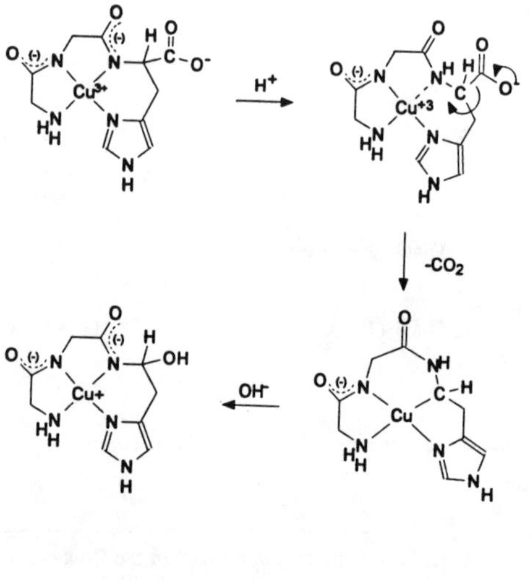

Scheme 3

220

species is formed. This flexibility may permit decarboxylation reactions that are not observed in 5,5,5-chelate ring compounds. Recent pulsed radiolysis studies by Meyerstein and coworkers[12] show that Cu^{III}-CH_3 bonds are readily formed in aqueous solution from the reaction of methyl free radicals with Cu^{II}-peptide complexes. We suggest that the copper-carbon intermediate readily undergoes hydrolytic attack at the carbon to form a Cu(I) product shown in Scheme 3 that in turn is quickly oxidized to copper(II). This proposed mechanism is highly speculative at this point, but will serve as a postulate for future tests.

Peptides with histidine in the third residue form extremely reactive copper(III) complexes that decompose rapidly and are strong oxidizing agents.

Acknowledgment. This investigation was supported by Public Health Service Grant No. GM12152 from the National Institute of General Medical Sciences.

References

1. F.P. Bossu, K.L. Chellappa, D.W. Margerum, *J. Am. Chem. Soc.* **99**, 2195 (1977).

2. J.P. Hinton, D.W. Margerum *Inorg. Chem.* **25**, 3248 (1986).

3. L.L. Diaddario, W.R. Robinson, D.W. Margerum, *Inorg. Chem.* **22**, 1021 (1983).

4. J.C. Nagy, L.L. Diaddario, H.D. Lee, D.W. Margerum. To be submitted for publication.

5. T.A. Neubecker, S.T. Kirksey, Jr., K.L. Chellappa, D.W. Margerum *Inorg. Chem.* **18**, 444 (1979).

6. L. Wang, H.D. Lee, D.W. Margerum *Inorg. Chem.* In press.

7. J.G. Osteryoung, R.A. Osteryoung *Anal.Chem.* **57**, 101A (1985).

8. D.P. Mack, B.L. Iverson, P.B. Dervan *J. Am. Chem. Soc.* **110**, 7572 (1988).

9. T. Sakurai, A. Nakahara *Inorg. Chim. Acta* **34**, L243 (1979).

10. P. de Meester, D.J. Hodgson *Inorg. Chem.* **17**, 440 (1978).

11 H.-J. Schafer, in *Topics in Current Chemistry*, E. Steckhan, Ed.; (Springer-Verlag, Berlin, 1990) vol. 152, pp. 91-150.

12. C. Mansano-Weiss, H. Cohen, D. Meyerstein *J. Inorg. Biochem.* **47**, 54 (1992).

FREE RADICAL INDUCED CLEAVAGE OF ORGANIC MOLECULES CATALYZED BY COPPER IONS - AN ALTERNATIVE PATHWAY FOR BIOLOGICAL DAMAGE

Sara Goldstein,[a] Gidon Czapski,[a] Haim Cohen,[b,c] and Dan Meyerstein[b]

[a]Department of Physical Chemistry, The Hebrew University of Jerusalem, Jerusalem 91904, ISRAEL.
[b]Department of Chemistry and R. Bloch Coal Research Center, Ben-Gurion University of the Negev, Beer-Sheva, ISRAEL.
[c]Department of Chemistry, NRCN, P.O.Box 90091, Beer-Sheva, ISRAEL.

INTRODUCTION

Free radicals are responsible for many deleterious effects in biological systems.[1-2] Many studies indicate that free radical reactions are involved in numerous diseases.[2] However, little is known on the exact mechanisms by which free radicals cause damage in biological systems. The ability of oxy-radicals and aliphatic free radicals to damage biological molecules has been demonstrated by exposure of various metabolites to high energy radiation.[3-5] Nevertheless, radiolysis is not the major source of free radicals produced in vivo, except under extreme circumstances, e.g., intentional and accidental exposure to high X-ray doses or to high level of radioactivity. Four major types of free radicals are formed in biological systems: hydroxyl (OH$\cdot$), superoxide (O$_2\cdot^-$), aliphatic (R$\cdot$) and aliphatic-peroxo (RO$_2\cdot$) radicals.

(i) Hydroxyl radicals are formed mainly via Fenton-like reactions,

$$L_mM^{n+} + H_2O_2 \longrightarrow L_mM^{(n+1)+} + OH^- + OH\cdot \qquad (1)$$

where L_mM^{n+} are low valent transition metal complexes, and by absorption of ionizing radiation.

L - Non participating ligands, which are assumed to be uncharged.
op - 1,10-phenanthroline

From K.D. Karlin and Z. Tyeklár, Eds., *Bioinorganic Chemistry of Copper* (Chapman & Hall, New York, 1993).

$$H_2O \longrightarrow OH\cdot, \ e^-_{aq}, \ H\cdot, \ H_2O_2, \ H_2, \ H_3O^+ \tag{2}$$

(ii) Superoxide radicals are generated mainly via enzymatic oxidations (e.g., xanthine/xanthine oxidase[6]), oxidation of semiquinones,[7] and reduction of oxygen by low valent transition metal ions.[8]

(iii) Aliphatic free radicals are formed mainly via H-abstraction from aliphatic compounds (RH) by OH· or $RO_2\cdot$ radicals,

$$OH\cdot/RO_2\cdot \ + \ RH \longrightarrow R\cdot \ + \ H_2O/RO_2H \tag{3}$$

by the addition of free radicals to olefins, e.g., unsaturated acids,

$$OH\cdot/R\cdot \ + \ R^1R^2C=CR^3R^4 \longrightarrow \ \cdot CR^1R^2CR^3R^4OH/\cdot CR^1R^2CR^3R^4R \tag{4}$$

and via oxidation of aliphatic compounds by transition metal complexes in high oxidation states, e.g., $M^{(n+2)+}=$ Cu(III), Fe(IV), Mn(III).

$$L_mM^{(n+2)+} \ + \ RH \longrightarrow L_mM^{(n+1)+} \ + \ R\cdot \ + \ H^+ \tag{5}$$

(iv) $RO_2\cdot$ free radicals are formed via the reaction of aliphatic free radicals with dioxygen. The latter reaction is very fast, and is expected to occur in all aerobic systems. In some cases $RO_2\cdot$ decomposes fast and O_2^- is formed.

It is well established that transition metal ions and their complexes enhance the damage caused by free radicals.[9-12] It has been suggested that the metal becomes bound to a site on the biomolecule, before or after it is reduced by a reducing agent such as superoxide, thiols, vitamin C. Thereafter, it reacts with H_2O_2 via the Fenton-like reaction (reaction 1) to produce highly oxidizing entities, which react with the biomolecule in the vicinity of the target site.[13]

However, if organic free radicals are formed in vivo in the absence or in the presence of low concentrations of oxygen, they are expected to react with transition metal complexes to form unstable intermediates with metal-carbon σ-bonds. It has been well established that a large variety of aliphatic free radicals react with first row transition metal complexes to form transients with metal-carbon σ-bonds.[14-28]

$$L_mM^{n+} + R\cdot \longrightarrow [L_mM^{(n+1)}\text{-}R]^{n+} \tag{6}$$

The mechanism of the decomposition of transient complexes with metal-carbon σ-bonds depends on the nature of the central cation, M^{n+}, the substituents on the carbon bound to the metal, and on the non-participating ligands, L. The two main reaction pathways observed for the decomposition process are:

(i) Heterolytic cleavage of the σ-bond:

$$[L_mM^{(n+1)}\text{-}R]^{n+} + H^+ \longrightarrow L_mM^{(n+1)+} + RH \tag{7}$$

$$[L_mM^{(n+1)}\text{-}R]^{n+} + H_2O \longrightarrow L_mM^{(n-1)+} + ROH + H^+ \tag{8}$$

(ii) Homolytic dissociation of the σ-bond,

$$[L_mM^{(n+1)}\text{-}R]^{n+} \rightleftharpoons L_mM^{n+} + R\cdot \tag{9}$$

followed by a variety of free radical processes.

However, when a good leaving group is bound at the β position to the carbon-centered free radical, a third pathway, namely a β elimination process, may compete with both heterolytic and homolytic cleavage of the σ-bond:

$$[L_mM^{(n+1)}\text{-}CR^1R^2CR^3R^4X]^{n+} \longrightarrow$$
$$L_mM^{(n+1)+} + R^1R^2C{=}CR^3R^4 + X^- \tag{10}$$

$$(\text{e.g., } X = OH,\ NH_3^+,\ Br,\ OPO_3^{2-},\ OC_2H_5)$$

Thus, decomposition of complexes with metal-carbon σ-bonds in many cases causes the cleavage of the organic molecule. When these reactions take place in biological systems, deleterious effects may result, especially when the decomposition process proceeds via a β-elimination reaction,[29] as in such a case the damage cannot be repaired.

We suggest an additional new pathway for biological damage initiated by free radicals. This pathway, catalyzed by transition metal complexes, emphasizes the role of metal ions in causing irreversible biological damage, particularly in anaerobic systems or in cases where the level of oxygen is very low. Under these conditions, aliphatic free radicals are expected to react with transition metal complexes to form transients with metal-carbon σ-bonds. The decomposition of these transients, especially via β-elimination processes, causes irreversible structural modifications of the aliphatic residue, which are often deleterious in biological systems.

Our recent study, which will be now reviewed, was aimed to verify the suggested mechanism. We investigated the reaction of copper ions and their complexes with various aliphatic free radicals having good leaving groups on the α and β positions to carbon-centered free radical. The study involved various aliphatic free radicals, both simple ones, e.g., derived via H-abstraction from ethanol, ethylamine, and more complicated ones derived from organic compounds with relevance to biomolecules, e.g., amino acids.

PRODUCTION OF TRANSIENTS WITH COPPER-CARBON σ-BONDS

A study of the properties of aliphatic free radicals and most transient complexes with metal-carbon σ-bonds in aqueous solutions requires the application of fast kinetic techniques as these species have very short life times. The pulse radiolysis technique is one of the most powerful tools for the study of the reactions of free radicals and the properties of short lived intermediates formed via their reactions.[30,31]

When N_2O-saturated solutions containing aliphatic compounds and copper complexes ([Cu] << [RH]) are pulse-irradiated, all the primary radicals formed via reaction 2 are converted into aliphatic free radicals through reactions 11 and 12,

$$e_{aq}^- + N_2O \xrightarrow{H_2O} N_2 + OH^- + OH^\cdot \tag{11}$$

$$H^\cdot/OH^\cdot + RH \longrightarrow R^\cdot + H_2/H_2O \tag{12}$$

which subsequently react with the copper complexes.

$$L_mCu^{2+} + R^\cdot \longrightarrow [L_mCu^{III}-R]^{2+} \tag{13}$$

$$L_mCu^+ + R^\cdot \longrightarrow [L_mCu^{II}-R]^+ \tag{14}$$

The formation and decomposition of transient complexes formed via reactions 13 and 14 can be followed (Figure 1).

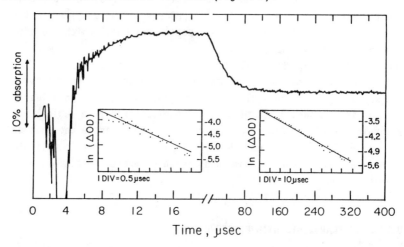

Figure 1: Typical kinetic plots of the formation and decay of the absorbance of $[Cu^{II}-CH(COO)CH_2NH_3]^+$(aq) at 380 nm. N_2O-saturated solution containing 0.1 M ß-alanine and 0.15 mM Cu^+(aq) at pH 3 was irradiated with a dose of 960 rad. The optical path length was 12.1 cm. The inset contains fits to first order rate laws (Taken from ref. 28).

The rate constants for the formation of $Cu^{II}-R^+$(aq) approach the diffusion controlled limit, and in all systems studied so far k_{14} >> k_{13}.[15-17,24-28] Solutions of cuprous ions are usually prepared either by dissolving $Cu(CH_3CN)_4PF_6$ (Aldrich) in deaerated solutions or via the reduction of Cu^{2+}(aq) by Cr^{2+}(aq) at $[Cr^{2+}(aq)]/[Cu^{2+}(aq)] \le 0.2$.[32] In both cases, particularly in the latter, the solutions contain also cupric ions. If the system contains both cuprous and cupric ions, a simultaneous formation of transient complexes with $Cu^{II}-C$ and $Cu^{III}-C$ σ-bonds may occur, although k_{13} > k_{14} by about two orders of magnitude. Under these conditions, reaction 13 can be neglected, only if the homolysis of the $Cu^{II}-C$ σ-bond is not the major route for the decomposition of the transient complex. Since homolysis of the σ-bond was found to be more prominent for $[L_mCu^{II}-R]^+$ than for $L_mCu^{III}-R^{2+}$ complexes, there are cases where $[Cu^{II}-R]^+$(aq) decomposes homolytically, and the free radical thus formed reacts with Cu^{2+}(aq).[15-17,26,33,34]

THE CASE OF CUPRIC IONS

Until recently, most of the decomposition processes of $Cu^{III}-R^{2+}(aq)$ were reported to occur via the heterolytic path,

$$[Cu^{III}-R]^{2+}(aq) \longrightarrow Cu^+(aq) + R^+ \qquad (15)$$

where R^+ is the oxidized form of the radical.[15-17,33,35-38]

We have recently demonstrated for the first time that $Cu^{2+}(aq)$ induces a β-carboxyl elimination reaction when it reacts with $\cdot CH_2CX(CH_3)COO^-$ (X = CH_3 [26]; NH_3^+ [37]) via $[Cu^{III}-CH_2C(CH_3)(X)COO]^+(aq)$ as an intermediate.

$$[Cu^{III}-CH_2CX(CH_3)COO]^+(aq) \longrightarrow Cu^+(aq) + CO_2 + CH_2=CX(CH_3) \qquad (16)$$

In the case where X = CH_3, we identified CO_2, $Cu^+(aq)$ and 2-methylpropene as the final products.[26] In the case of the amine, in addition to $Cu^+(aq)$ and CO_2, we identified acetone, formed via hydrolysis of the corresponding olefin.[34] The specific rates of formation (k_f) and decomposition (k_d) of these transients (summed up in Table I) are independent of pH and [RH].

Table I: Specific rates of formation and decomposition of $[Cu^{III}-CH_2CX(CH_3)COO]^+(aq)$

X	k_f, $M^{-1}s^{-1}$	k_d, s^{-1}
CH_3	$(3.1\pm0.3)\times10^7$	0.03 ± 0.01
NH_3^+	$(1.3\pm0.3)\times10^7$	0.15 ± 0.03

The mechanism of the decomposition of both transients was suggested to occur via cyclic intermediates, **I**, which are expected to stabilize these intermediates towards heterolysis and/or homolysis.

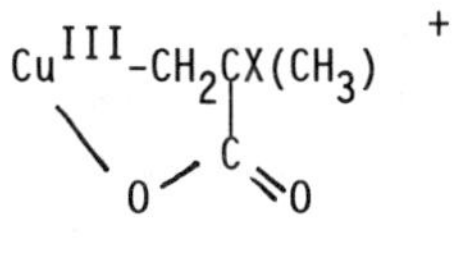

$$Cu^{III}-CH_2\underset{O}{\overset{CX(CH_3)}{C}}\overset{+}{\underset{O}{C}}$$

I

The data in Table I indicate that substitution of a methyl group by an amine group at the β-position to the carbon-centered free radical increases the rate of the β-elimination process 5-fold in these systems.

THE CASE OF CUPROUS IONS

Cuprous ions are oxidized by a large variety of aliphatic free radicals of the type $\cdot CR^1R^2CR^3R^4X$ via the formation of complexes with Cu^{II}-carbon σ-bonds as intermediates.[15,17,25,27,28,34] These transient complexes have typical absorption spectra in the visible region (Figure 2).

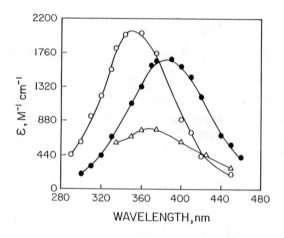

Figure 2: Absorption spectra of transient complexes obtained 100 μs after the irradiation of N_2O-saturated solutions containing 0.1 M RH, and 0.1 mM Cu^+(aq) at pH 3. (o) $[Cu^{II}-CH_2C(CH_3)(NH_3)COO]^+$(aq); (•) $[Cu^{II}-CH(CH_2NH_3)_2]^{3+}$(aq); (▵) $[Cu^{II}-CH(CH_2N(CH_3)_2H)_2]^{3+}$(aq)

The decomposition of these transient complexes via β-elimination reactions will take place unless cleavage of the σ-bonds via homolysis or heterolysis will compete efficiently with this process.

Here are described those cases where the β-elimination reaction is the major process for the decomposition of these transient complexes.

β-Hydroxyl Elimination

The observed rates of the β-hydroxyl elimination processes from a large variety of $L_mM^{(n+1)}-CR^1R^2CR^3R^4OH^{n+}$ obey equation 16.[17,18,25,27,39-41]

$$k_d = k_o + k_H[H^+] \tag{17}$$

In some cases the acid independent route (k_o) is the major process observed whereas in others the acid catalyzed process is the major one. The acid catalyzed process is suggested to proceed via the following reactions,

$$[L_mM^{(n+1)}-CR^1R^2CR^3R^4OH]^{n+} + H^+ \rightleftharpoons$$
$$[L_mM^{(n+1)}-CR^1R^2CR^3R^4(OH_2)]^{(n+1)+} \tag{18}$$

$$[L_mM^{(n+1)}-CR^1R^2CR^3R^4(OH_2)]^{(n+1)+} \longrightarrow$$
$$[L_mM^{(n+1)}-(R^1R^2C=CR^3R^4)]^{(n+1)+} + H_2O \tag{19}$$

$$[L_mM^{(n+1)}-(R^1R^2C=CR^3R^4)]^{(n+1)+} \longrightarrow$$
$$L_mM^{(n+1)+} + R^1R^2C=CR^3R^4 \tag{20}$$

where reaction 18 is a fast equilibrium process. According to this mechanism, the observed rate constant for the decomposition process is given by equation 21:

$$k_d = K_{18}k_{19}[H^+]/(1 + K_{18}[H^+]) \tag{21}$$

As $K_{18} \ll 1$ in most systems studied, the observed rate constant of the decomposition process was found to be linearly depended on $[H^+]$. Specific rates of β-hydroxyl elimination reactions for several aliphatic radicals are given in Table II.

Table II: Specific rates of β-hydroxyl elimination from $[Cu^{II}-R]^+(aq)$

R·	k_o, s^{-1}	k_H, M^{-1}s^{-1}	ref.
·CH_2CH_2OH	3.2×10^3	3.8×10^7	17
·$CH_2CH(CH_3)OH$	1.1×10^4	1.5×10^8	17
·$CH_2C(CH_3)_2OH$	5.0×10^4	8.6×10^7	25

Comparison of the specific rates of the three ß-hydroxyl elimination reactions suggests that methyl substituents at the ß-position to the carbon-centered free radical enhance the rate of the acid non-catalyzed path. The effect on the acid-catalyzed pathway seems to be more complex.

ß-Amine Elimination

We have recently shown that ß-amine aliphatic free radicals react with $Cu^+(aq)$ to form $[Cu^{II}-CR^1R^2CR^3R^4NH_3]^{2+}(aq)$.[27,28,34,42] The latter transients decompose via ß-amine elimination processes. The rates of formation and decomposition of the various organo-copper complexes studied are independent of pH, $[Cu^{2+}(aq)]$ and $[RH]$ (Table III). It is not expected that ß-amine elimination reactions would be acid-catalyzed as under all experimental conditions the amine is protonated. The observed rate constant of the acid catalyzed path, which is obtained from equation 21, equals $K_{18}k_{19}$, and the observed rate constant for the ß-amine elimination should be compared with k_{19} and not with k_H. It should be noted that $k_{19} \gg 1$.

Table III: Spectral and kinetic data for $[Cu^{II}-R]^+(aq)$

R·	max	ϵ_{max}	k_f	k_d	ref.
	nm	$M^{-1}cm^{-1}$	$M^{-1}s^{-1}$	s^{-1}	
$\cdot CH_2CH_2NH_3^+$	375	2800	9.3×10^8	2.5	42
$\cdot CH_2CH_2N(C_2H_5)_2H^+$	375	2200	6.9×10^8	190	42
$\cdot CH_2C(CH_3)_2NH_3^+$	385	2500	1.2×10^9	7.0	27
$\cdot CH(CH_2NH_3^+)_2$	355	2100	3.5×10^8	1.9	42
$\cdot CH(CH_2N(CH_3)_2H^+)_2$	370	780	3.5×10^8	200	42
$\cdot CH(COO^-)CH_2NH_3^+$	360	1500	3.4×10^9	1.1×10^4	28
$\cdot CH_2C(CH_3)(NH_3^+)COO^-$	375	2100	1.3×10^9	2.7	34

Comparison of the data in Tables II and III indicates that the ß-amine elimination is many orders of magnitude slower than that of the ß-hydroxyl elimination. The same trend was also found in the case of $[(H_2O)_5Cr^{III}-CH_2C(CH_3)_2X]^{2+}$ (X = NH_3^+; OH).[27] Since the rate-determining step of the ß-elimination process involves presumably C-N and C-O bond cleavages, and since the former bond is normally weaker than the latter,[43] it is quite puzzling to find that the ß-elimination of NH_3 is much slower than that of H_2O. A theoretical analysis suggests that protonation of the leaving group reverses the bond strength order of the C-N and the C-O bonds.[27]

It is of interest to note the very large effect of the carboxylate substituent on the carbon-centered free radical on the rate of the ß-amine elimination process. The source of this effect is under study.

Recently, it has been reported that $[(H_2O)_5Cr^{III}-CH_2CH_2OH]^{2+}$ decomposes via a ß-hydroxyl elimination reaction forming ethylene and Cr^{3+}(aq) in an acid catalyzed process with k = $(2.0 + 1.4x10^4[H^+])$ s^{-1}.[41] This rate constant should be compared with the rate constant for the decomposition of $[(H_2O)_5Cr^{III}-CH_2CH_2OC_2H_5]^{2+}$ to form Cr^{3+}(aq), ethylene and ethanol, which was reported to be $4.6x10^3[H^+]$ s^{-1}.[44] In this case the non-catalyzed term was too slow to be measured.[44] However, the data in table III indicate that the rate of the ß-amine elimination is faster when the hydrogen on the amine group is substituted by methyl or ethyl groups. Thus, O-alkylation decreases the rate of the ß-hydroxyl elimination reaction, whereas N-alkylation increases the rate of the ß-amine elimination process. The source of these opposite trends has been analyzed theoretically.[42] All the results obtained up to now are in accord with the notion that the rate of ß-elimination of a group X increases with the decrease in the $(C-X)^+$ bond strength. A mechanism of activation for the reaction was formulated based on the Shaik-Pross[45] curve crossing model, which provides the root cause for the effect of $(C-X)^+$ bond strength on the rate of the ß-elimination reactions.[42]

THE CASE OF Cu(I) COMPLEXES

1,10-Phenanthroline Ligand

The ß free radicals $\cdot CH_2CH_2OH$ and $\cdot CH_2C(CH_3)_2OH$ react with $(op)_2Cu^+$ to yield ethylene and 2-methylpropene, respectively.[24] During this process transients with copper-carbon σ-bonds have been observed. The rate constants for the formation of these transients are estimated to be ca. $1x10^{10}$ $M^{-1}s^{-1}$. These intermediates decompose via a ß-hydroxyl elimination reaction with specific rate constants of

$(1.7\pm0.4)\times10^3$ and $(1.1\pm0.3)\times10^4$ s^{-1} for $\cdot CH_2C(CH_3)_2OH$ and $\cdot CH_2CH_2OH$, respectively, in the pH range 4-10.[24] These results indicate that the acid dependent term cannot exceed 5×10^6 and 2×10^6 $M^{-1}s^{-1}$ in the case of $\cdot CH_2C(CH_3)_2OH$ and $\cdot CH_2CH_2OH$, respectively. Thus, coordination of 1,10-phenanthroline to copper affects the specific rate constants of both the acid independent and acid dependent reactions.

The ß-phospho radical, derived via H-atom abstraction from glycerol-2-phosphate, reacts with $(op)_2Cu^+$ with a specific rate constant of ca. 1×10^{10} $M^{-1}s^{-1}$ to form a transient with a metal-carbon σ-bond.[46]

$$(op)_2Cu^+ + \cdot CHOHCH(OPO_3)CH_2OH^{2-} \longrightarrow$$
$$[(op)_2Cu^{II}-CHOHCH(OPO_3)CH_2OH]^- \qquad (22)$$

We have demonstrated for the first time that the decomposition of such a transient takes place via a ß-elimination of phosphate. The rate constant for this reaction was determined to be $(8.5\pm1.5)\times10^3$ s^{-1}, independent of the acidity in the pH range 4 - 9.[46]

From the biological point of view such a ß-phosphate elimination process is very important since strand breaks in DNA are caused mainly via such reactions.[5]

π-Acid Ligands

The kinetics of the formation and decomposition of $[L(H_2O)_nCu^{II}-CH_2C(CH_3)_2OH]^+$ was studied for L = CO and C_2H_4.[47] It has been demonstrated that the decomposition takes place via a ß-hydroxyl elimination reaction.

$$[L(H_2O)_nCu^{II}-CH_2C(CH_3)_2OH]^+ \longrightarrow$$
$$Cu^{2+}(aq) + H_2C=C(CH_3)_2 + OH^- + L \qquad (23)$$

The rate constants of the decomposition process are given in Table IV. The results indicate that the π-acid ligands slow down the ß-elimination process (Table II, IV). The effect is attributed to the destabilization of the primary products of the reaction, in which the product olefin is bound to the central metal cation as a π-acid ligand or in a metallocycle, by the π-acid ligand.[46]

Table IV: Specific rates of ß-hydroxyl elimination from
$[L(H_2O)_nCu^{II}-CH_2C(CH_3)_2OH]^+$

L	k_o, s^{-1}	k_H, $M^{-1}s^{-1}$
CO	2×10^4	1.8×10^7
C_2H_4	5.3×10^3	2.9×10^7

CONCLUSIONS

Aliphatic free radicals of the type $\cdot CR^1R^2CR^3R^4X$ (X = OH, CO_2^-, NH_3^+, OPO_3^{2-}) react with both copper(II) and copper(I) complexes to form transients with copper-carbon σ-bonds. The copper(I) complexes are considerably more reactive towards aliphatic free radicals than copper(II) complexes.

The main route for the decomposition of these transient complexes occurs via ß elimination of X^-, which ultimately causes the degradation of the aliphatic residue. The specific rates of the ß-elimination processes depend on the nature of L, R and X. The results suggest that the rates are correlated to the C-X bond strength.

We suggest that analogous processes are responsible for some of the radical induced deleterious effects found in biological systems in the presence of copper complexes.

ACKNOWLEDGEMENTS

This research was supported by the GSF, Forschungszentrum fur Umwelt Gesundheit, GmbH, by The Israel Academy of Sciences and by The Israel Atomic Energy Commission. D.M. wishes to thank Irene Evans for her interest and support.

REFERENCES

1. Free Radicals in Biology and Medicine, B. Halliwell and G.M.C. Gutteridge, Eds., (Clarendon Press, Oxford, 1989).
2. D. Harman, in Free Radicals, Aging and Degenerative Diseases, J.E. Johnson, Jr., R. Walford, D. Harman and J.M. Miquel, Eds., (Alan R. Liss, Inc., New York, 1986), Vol. 8, p. 3.
3. A.J. Swallow, in Radiation Chemistry of Organic Compounds, A.J. Swallow, Ed., (Perganon Press, New York, 1960), p. 211.
4. W.M. Garrison, M.E. Tayko, W. Bennett, Radiat. Res. 16, 483 (1962).

5. The Chemical Basis of Radiation Biology, C. von Sonntag, Ed., (Taylor & Francis, London-New York-Philadelphia, 1987).
6. J.M. McCord and I. Fridovich, J. Biol. Chem. 244, 6049 (1969).
7. H. Nohl, in ref. 2, p. 77, and references cited therein.
8. S. Goldstein and G. Czapski, J. Am. Chem. Soc. 105, 7276 (1983).
9. S.D. Aust and B.C. White, Adv. Free Rad. Biol. Med. 1, 1 (1985).
10 E.R. Stadman, Free Rad. Biol. Med. 9, 315 (1990).
11. E.R. Stadman and C.N. Oliver, J. Biol. Chem. 266, 2005 (1991).
12. S. Goldstein and G. Czapski, Free Rad. Biol. Med. 2, 3, (1986) and references cited therein.
13. A. Samuni, J. Aronovitch, D. Godinger, M. Chevion, G. Czapski, Eur. J. Biochem. 137, 119 (1983).
14. K. Kelm, J. Lillie, A. Henglein, J. Chem. Soc. Faraday Trans. 1 71, 1132 (1977).
15. G.V. Buxton and R.M. Sellers, Coord. Chem. Rev. 22, 195 (1977).
16. M. Freiberg and D. Meyerstein, J. Chem. Soc. Faraday Trans. 1 76, 1825 (1980).
17. M. Freiberg, W.H. Mulac, K.H. Schmidt, D. Meyerstein, J. Chem. Soc. Faraday Trans I 76, 1838 (1980).
18. J.H. Espenson, in Advances in Inorganic and Bioinorganic Mechanisms, G. Sykes, Ed. (Academic Press, 1982), Chap. 1, p. 5.
19. D. Brault and P. Neta, J. Am. Chem. Soc. 103, 2705 (1981).
20. A. Bakac and J.H. Espenson, J. Am. Chem. Soc. 108, 719 (1986).
21. H. Cohen and D. Meyerstein, Inorg. Chem. 27, 3429 (1988).
22. A. Sauer, H. Cohen, D. Meyerstein, Inorg. Chem. 27, 4578 (1988).
23. D. Meyerstein and H.A. Schwarz, J. Chem. Soc. Faraday Trans. I 84, 2933 (1988).
24. S. Goldstein, G. Czapski, H. Cohen, D. Meyerstein, Inorg. Chem. 27, 4130 (1988).
25. H. Cohen and D. Meyerstein, J. Chem. Soc. Faraday Trans. I 84, 4157 (1988).
26. M. Masarwa, H. Cohen, J. Saar, D. Meyerstein, Isr. J. Chem. 30, 361 (1990).
27. S. Goldstein, G. Czapski, H. Cohen, D. Meyerstein, J. -K. Cho, S.S. Shaik, Inorg. Chem. 31, 798 (1992).
28. S. Goldstein, G. Czapski, H. Cohen, D. Meyerstein, Inorg. Chim. Acta 192, 87 (1992).
29. W.M. Garrison, Free Rad. Biol. Med. 6, 285 (1989).
30. Pulse Radiolysis, M. S. Matheson and L. M. Dorfman, Eds., (MIT Press, 1969).
31. D. Meyerstein, Acc. Chem. Res. 11, 43 (1978).
32. K. Shaw and J.H. Espenson, Inorg. Chem. 7, 1619 (1968).
33. H. Cohen and D. Meyerstein, Inorg. Chem. 26, 2342 (1987).
34. S. Goldstein, G. Czapski, H. Cohen, D. Meyerstein, Inorg. Chem. 31, 2439 (1992).
35. J.K. Kochi, in Free Radical, J.K. Kochi, Ed., (Wiley, New York, 1973), Vol. I, Chap. 11.
36. C. Walling, Acc. Chem. Res. 8, 125 (1974).
37. M. Freiberg and D. Meyerstein, J. Chem. Soc. Chem. Commun. 934 (1977).
38. L.J. Kirschenbaum and D. Meyerstein, Inorg. Chem. 19, 1373 (1980).
39. H. Cohen and D. Meyerstein, Inorg. Chem. 13, 2434 (1974).
40. D.A. Ryan and J.H. Espenson, Inorg. Chem. 21, 527 (1982).
41. H. Cohen, D. Feldman, R. Ish-Shalom, D. Meyerstein, J. Am. Chem. Soc. 113, 5292 (1991).
42. S. Goldstein, G. Czapski, H. Cohen, D. Meyerstein, S. Shaik, submitted.
43. Chemical Bonds and Bonds Energy, R.T. Sanderson, Ed., 2nd Edition (Academic Press, New York, 1976).

44. H. Cohen   and D. Meyerstein, <u>Angew. Chem. Int. Ed. Engl.</u> **24**, 779 (1985).
45. S.S. Shaik and A. Pross, <u>J. Am. Chem. Soc.</u> **104**, 2708 (1982).
46. S. Goldstein, G. Czapski, H. Cohen, D. Meyerstein, <u>Free rad. Biol. Med.</u> **9**, 371 (1990).
47. A. Feldman, H. Cohen, D. Meyerstein, submitted.

COPPER-MEDIATED NITROGEN LIGAND OXIDATION AND OXYGENATION

Lawrence M. Sayre,[*] Wei Tang, K. Veera Reddy, and Durgesh Nadkarni

Department of Chemistry, Case Western Reserve University, Cleveland, OH 44106, USA

INTRODUCTION

For the last several decades, bioinorganic chemists have synthesized novel mononuclear, binuclear, and more recently polynuclear ligand systems in an effort to mimic the coordination state of iron and copper in redox metalloproteins, especially those involved in the binding and activation of dioxygen. The favored utilization by nature of the histidine imidazole moiety for providing a neutral and relatively "soft" ligation of metal ions has resulted in the common use of sp^2 nitrogen (in particular pyridine and related heterocycles as well as imines) in addition to sp^3 amino nitrogen in the design of biomimetic ligands. Although sp^2 imino nitrogen is relatively resistant to oxidation, sp^3 amino nitrogen is, in general, subject to oxidation at the potentials generated in most instances of iron- and copper-O_2 interaction. Ligand dehydrogenation has been observed in some cases, usually as an unwanted side-reaction (eq 1),[1] but such instances provide important information needed for the rational design of robust ligands as well as about the nature and mechanism of metal ion-mediated amine oxidation. Based on the widespread interest in enzymatic and biomimetic monooxygenation processes, it becomes important to distinguish between authentic oxygenations (O incorporation from O_2) and those reactions where oxygen appears in the product only as a consequence of hydration of a dehydrogenated intermediate. An attempt is made in this chapter to clarify mechanistic aspects of copper-mediated oxidation and oxygenation processes pertaining to selected examples of nitrogen ligation.

$$\text{L} \overset{\frown}{\underset{Cu^m_n \cdots NH\diagdown}{\text{CH}_2}} \quad \xrightarrow{O_2} \longrightarrow \quad \text{L} \overset{\frown}{\underset{Cu^m_n \cdots N\diagdown}{\text{CH}}} \tag{1}$$

OXIDATION OF ALIPHATIC AMINES BY CuCl-O_2

The first part of this chapter deals with amine dehydrogenation chemis-

From K.D. Karlin and Z. Tyeklár, Eds., *Bioinorganic Chemistry of Copper* (Chapman & Hall, New York, 1993).

try induced by two related copper systems: (i) $CuCl-O_2$-pyridine $(Cu/O_2/PYR)$ and (ii) $CuCl-O_2-CH_3CN$ $(Cu/O_2/ACN)$.[2] The main difference between these two systems which we would predict to be manifested in the observed reactivity patterns are (i) the greater basicity of the pyridine-based system and (ii) the somewhat greater $Cu(I)$-stabilization and therefore greater copper-oxidant strength in the CH_3CN-based system. It must first be pointed out that except under quite stringent conditions,[3] $Cu(II)$ is an inadequate amine oxidant. Thus, the ready oxidation of amines in the $CuCl-O_2$ systems must be due to the generation of $Cu(III)$ or copper-oxygen species with comparable oxidant strength. Ditellurato-cuprate(III) oxidizes amines under basic conditions,[4] and the generation of $Cu(III)$ from $Cu(II)$ polyamine ligand systems either electrochemical-ly,[5] radiolytically,[6] or by basic ferricyanide[7] has been found to result in ligand dehydrogenation.

1. $Cu/O_2/PYR$

Several years ago, a brief report alleged that the $Cu/O_2/PYR$ system was capable of achieving efficient dehydrogenation of primary amines to the corresponding nitriles via intermediate imines.[8] When we explored the generality of this reaction (conditions: 1 mmol amine and 2 mmol CuCl in 30 mL pyridine, 45°C, 9 h), we observed that the high yield conversion of RCH_2NH_2 to RCN could be guaranteed in the case of benzyl (including 2- and 4-picolylamine) but not phenethyl systems, that secondary amines could be readily dehydrogenated to imines, but that tertiary amines were inert to oxidation. Since this reactivity trend, 2° ~ 1° >> 3°, is distinct from that expected for an electron-transfer oxidation mechanism, where the ease of amine oxidation is 3° > 2° > 1°,[9] we needed an oxidation mechanism which required the presence of at least one hydrogen on nitrogen. Taking into account the basic nature of the solvent and the reports that highly basic copper oxidation systems could induce imine dehydrogenation subsequent to coordination-induced N-H deprotonation (binding of amide anion to the copper),[10] we consider a mechanism (eq 2)

wherein N-H deprotonation occurs either prior to (Path A) or simultaneous with (Path B) electron transfer from nitrogen to copper. In the case of primary amines, the initially formed imine can again be dehydro-genated through a NH-deprotonation-assisted mechanism.[10] A similar mechanism probably holds for amine oxidation by ditelluratocuprate(III), which exhibits the reactivity trend 2° > 1° > 3°.

The initial report on the $Cu/O_2/PYR$ system implied that it could be used to convert any primary amine to the corresponding nitrile, and gave as an example of a non-benzylic case, the conversion of 3,4-dimethoxy-phenethylamine to 3,4-dimethoxyphenylacetonitrile.[8] We confirmed the high yield of this particular conversion, but found that it was a special case. 2-Phenethylamine itself was converted mainly to PhCOOH (54%), with smaller amounts of PhCHO (10%), PhCH$_2$CN (6%), and the amide PhCONHCH$_2$CH$_2$Ph (4%) being produced. The formation of PhCOOH and the amide can be rationalized on the basis of base/copper-mediated autoxidation of the first-formed PhCH$_2$CN to benzoyl cyanide,[11,12] which then either hydrolyzes or acylates unreacted starting amine[13] (Scheme I). The formation of PhCHO can be rationalized either on the basis of (i) an alternate course of PhCH$_2$CN autoxidation[11] (ii) or autoxidation of the initially generated PhCH$_2$CH=NH (via its enamine tautomer)[14-17] or of PhCH$_2$CH=O,[18] both well-precedented copper-mediated reactions. In comparison to 2-phenethylamine, 2-(2-pyridyl)ethylamine gave none of the corresponding nitrile and less of the chain-shortened acid (picolinic acid, 34%), but much more of the corresponding amide, N-(2-pyridyleth-yl)picolinamide (52%).

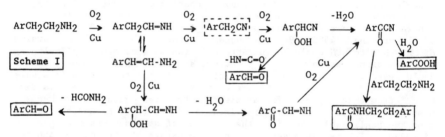

Scheme I

In both these cases, we demonstrated that the amide was arising mainly from the aroyl cyanide rather than from either (i) oxidation of the ArCHO/ArCH$_2$CH$_2$NH$_2$-based carbinolamine or (ii) direct condensation of amine with ArCOOH, by carrying out the following control study. Under the usual reaction conditions using n-propylamine as the "acceptor" in a 1:1 mixture with either ArCOOH, ArCHO, or ArCH$_2$CN, we observed formation of either no amide, a trace of amide (~ 2% yield), or a > 60% yield of amide, respectively.

Overall, the outcome for oxidation of arylethylamines can be seen to be quite sensitive to substituent electronic effects. With electron-donating substituents, enamine generation from the first-formed imine and autoxidation of the subsequently formed ArCH$_2$CN are both retarded on account of reduced C$_\alpha$ acidity, and the starting amine thus has more time to undergo the initial dehydrogenation, itself inductively accelerated. In contrast, with electron-withdrawing substituents, the initial nitrile product undergoes rapid autoxidation, generating the acylating agent ArCOCN in the presence of unconsumed starting amine, the dehydrogenation of which is inductively slowed. In the 2-(2-pyridyl)ethylamine case, persistence of starting amine could also be, in part, a reflection of chelation to copper.

For the C$_\alpha$-mono-branched primary amine 1-phenethylamine, which can be dehydrogenated only to the imine stage, we observed 51% recovered starting amine, with the major isolated product being the azine Ph(CH$_3$)C =N-N=C(CH$_3$)Ph (38%) in addition to acetophenone (5%), formed from oxidative self-coupling and hydrolysis, respectively, of the imine. In the case of the doubly C$_\alpha$-branched compound, PhCH$_2$C(CH$_3$)$_2$NH$_2$, the amine was recovered completely unchanged.

Secondary amines gave rise to products consistent with the reactivity patterns of the corresponding primary amines and the fact that only a single dehydrogenation can occur initially. Thus, for dibenzylamine, 90% of the product was the imine $PhCH=NCH_2Ph$. For N-methylbenzylamine, benzylic dehydrogenation represented the major pathway, giving $PhCH=NCH_3$ or, after workup, PhCHO in ~65% yield. Oxidative N-demethylation to $PhCH_2NH_2$ occurred to a smaller extent, as evidenced by the formation of PhCN in 8% yield. We also isolated the formamide $PhCH_2N(CH_3)CHO$ (7%), apparently arising from oxidation of the carbinolamine formed from condensation of the parent secondary amine with the HCHO released in the N-demethylation. In a similar manner, N-methyl-2-phenethylamine gave rise to the corresponding formamide $PhCH_2CH_2N(CH_3)CHO$ in 18% yield, though the major product in this case, PhCOOH (56% yield) presumably arises from the enamine autoxidation path described above (Scheme I) for phenethylamine.

A distinct course of oxidation is followed by $2-PyCH_2NHCH_3$ (2-Py = 2-pyridyl). The lack of generation of either 2-PyCN or the formamide $2-PyCH_2N(CH_3)CHO$ indicates that oxidative N-demethylation cannot compete with oxidation at the 2-pyridylmethyl group, the major products being $2-PyCH=NCH_3$ (13%), picolinic acid (34%), and N-methylpicolinamide (27%). $2-PyCH_2NH_2$ itself gave picolinamide (28%) and picolinic acid (5%) in addition to 2-cyanopyridine (46%). The production of picolinamides is curious, since no analogous benzamides were formed in the oxidation of $PhCH_2NH_2$ and $PhCH_2NHCH_3$. Also, $4-PyCH_2NH_2$ is converted to 4-PyCN without formation of $4-PyCONH_2$ (isonicotinamide). The distinct propensity of the 2-pyridyl systems to give rise to picolinamides must represent a copper coordination-mediated reaction, most easily rationalized in terms of dehydrogenation of the 2-PyCH=O-derived carbinolamines (eq 3), though copper-mediated hydrolysis of 2-cyanopyridine[19] can rationalize formation of at least some of the amide in the primary amine case.

$$R = H, CH_3 \qquad\qquad (3)$$

Another result of note is that involving (1-phenylcyclopropyl)methylamine, which undergoes only 23% conversion in the $Cu/O_2/PYR$ reaction, but yields entirely 1-phenyl-1-cyclopropanecarbonitrile. Thus, if the C_α radical $PhC(C_2H_4)CH(NH_2)\cdot$ is generated during dehydrogenation, it must be sufficiently stabilized such that further oxidation occurs faster than does ring-opening.

Capdevielle and coworkers also studied the $Cu/O_2/PYR$ system in the oxidation of primary amines.[20,21] These workers found that slow addition of CuCl at 60° resulted in improved yields of nitriles, even for simple aliphatic amines, though phenethylamine exhibited the same complications we discussed above (due to $PhCH_2CN$ being unstable under the basic oxidation conditions).[20] On the basis of a kinetic study comparing $PhCH_2NH_2$ vs. $PhCD_2NH_2$, which yielded a deuterium kinetic isotope effect of only 1.25, these workers suggested a mechanism involving rate-limiting electron transfer from amine to the copper oxidant followed by rapid deprotonation of the initially formed aminium radical.[21] In addition, an observed rate enhancement from dodecylamine to benzylamine was interpreted in terms of aryl stabilization of the aminium radical. However, the aryl group should inductively *retard* one-electron oxidation at nitrogen, and our observed inertness of tertiary amines is inconsistent with direct oxidation of the neutral amine. Our proposed mechanism (eq 2) accommodates the unreactivity of tertiary amines by invoking electron transfer only subsequent to or simultaneous with amido coordination to copper. The faster rate of $PhCH_2NH_2$ can then be explained in terms of an acid strengthening effect, in facilitating initial amido coordination, and/or in terms of benzylic stabilization of a partially rate-limiting C_α deptotonation.

Final confirmation of the role of a coordination-dependent vis-à-vis direct electron-transfer oxidation was obtained from relative rate information derived from product analyses of 1:1 co-oxidations of benzylamine/4-methylbenzylamine and benzylamine/2,4,6-trimethylbenzylamine. Whereas the ratio k_H/k_{Me} obtained from the first experiment was 1.02, the ratio k_H/k_{3Me} obtained from the second experiment was 2.78. The latter result is clearly incompatible with a direct electron-transfer oxidation, which would be aided by the electron-donating effect of the three methyl groups, and instead must represent steric retardation of a rate-limiting coordination-induced N-H deprotonation or, in part, post-electron-transfer C_α-deprotonation.

2. $Cu/O_2/ACN$

In contrast to the results discussed above in pyridine solvent, oxidation of amines (1 mmol) in CH_3CN solvent (30 mL) in the presence of CuCl (2 mmol), 45°C, 48 h, followed the reactivity order 3° > 2° > 1°, consistent with a straightforward electron-transfer oxidation mechanism.[9] In fact, simple primary amines (including benzylamine) were totally inert, and substantial conversion of secondary amines occurred only in the case of benzylic activation. Thus, whereas $PhCH_2CH_2NHCH_3$ was recovered 90% unchanged, and even $PhCH_2NHCH_3$ was only 15% oxidized (4% to PhCHO), dibenzylamine was 83% dehydrogenated, giving $PhCH=NCH_2Ph$ (80%) and traces of PhCHO and PhCOOH. The observed benzylic activation could be explained if either (i) C_α-deprotonation (resonance facilitation) or (ii) N-H deprotonation (inductive facilitation), occurring subsequent to electron-transfer oxidation at nitrogen,[22,23] is partly rate-limiting (eq 4).

Tertiary amines underwent >90% conversion to the expected products of oxidative N-dealkylation. N,N-Dimethylbenzylamine gave PhCHO (25%), $PhCH_2NHCH_3$ (33%), and $PhCH_2N(CH_3)CHO$ (37%). The latter tertiary formamide presumably arises from copper-mediated dehydrogenation, rather than dissociation, of the carbinolamine generated in the N-demethylation pathway. Since demethylation is statistically favored over debenzylation by a factor of 3, the observed product distribution, signifying $k_{CH3}/k_{CH2} = 2.8$, indicates there is only a slight reactivity preference for oxidation at the benzylic carbon. However, for the N,N-dimethyl derivatives of phenethylamine, and its "β-blocked" analog, (1-phenylcyclopropyl)methylamine, the N-demethylated secondary amines and tertiary N-methylformamides were obtained in 33-37% and 52-56% yield, respectively, with products of oxidation at the N-CH$_2$ position being limited to 8% (1-phenyl-1-cyclopropanecarboxaldehyde was obtained cleanly in the "β-blocked" case). Since the observed k_{CH3}/k_{CH2} reactivity ratio of 11 in these non-benzylic cases is nearly four times the statistical prediction, there must be a clear *intrinsic* preference for oxidation at N-methyl. Thus, by comparing the benzyl and phenethyl k_{CH3}/k_{CH2} results, it can be concluded that oxidation at N-methylene is in fact significantly enhanced by benzylic activation. The lack of observed ring opening in the tertiary cyclopropylmethylamine compound mirrors the behavior of the corresponding primary amine in the pyridine-based CuCl-O$_2$ system, again implying stabilization of any cyclopropylcarbinyl radical which might be generated as an intermediate.

In another assessment of N-methyl vs. N-benzyl reactivity, N,N-dibenzylmethylamine was found to oxidize to PhCHO (46%), N,N-dibenzylformamide (27%), and PhCH=NCH$_2$Ph (17%), the latter formed presumably from dehydrogenation of the N-demethylated product, dibenzylamine. The observed k_{CH3}/k_{CH2} in this case (0.96) is seen to be slightly *greater* than the statistically predicted value (0.75), rather than slightly smaller, as seen above for PhCH$_2$N(CH$_3$)$_2$.

In all the above cases, the conversion of R$_2$NCH$_3$ to R$_2$NCH=O was explained in terms of copper oxidant-mediated dehydrogenation of the R$_2$NCH$_2$OH moiety. It is necessary to point out that R$_2$NCH$_2$OH can arise either from hydration of R$_2$N=CH$_2^+$ (eq 5, path A) or from a HO• rebound reaction (path B), an ambiguity that exists also in the O$_2$-dependent N-dealkylation of amines by cytochrome P-450 and its chemical mimics.[24] In addition, R$_2$NCH=O can alternately arise from reaction of the intermediate R$_2$NCH$_2$• species with O$_2$ (path C). In light of the interest in biomimetic oxygenation chemistry, one might be tempted to interpret the formation of R$_2$NCH$_2$OH in our reactions in terms of an authentic copper-mediated oxygenation (by Cu(III)=O or a related species). However, in view of the lack of evidence that the CuCl-O$_2$-derived oxidant can effect oxygenation of organic compounds *in general*, in the manner accomplished by cytochrome P-450, we find no compelling reason at the present time to invoke any reaction pathway other than dehydrogenation (path A).

Under our reaction conditions, the oxygen ending up in the product

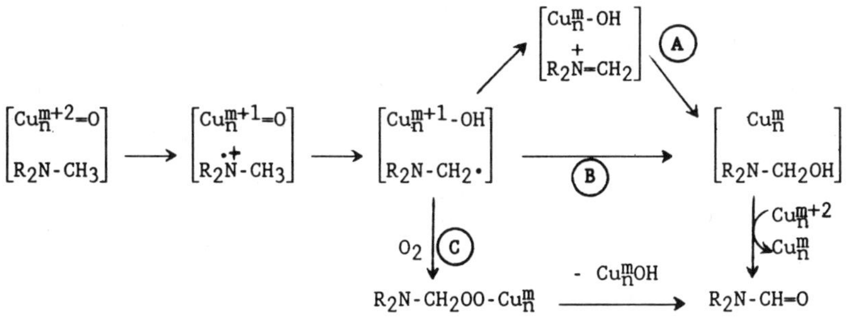

(5)

241

will be derived from O_2 regardless of which reaction pathway is followed; viz., if path A is followed, it is the H_2O formed from reduction of O_2 which adds to the iminium species. A distinction between dehydrogenation and direct monooxygenation might be obtainable by running the reaction with $^{16}O_2$ in the presence of a trace of $^{18}OH_2$, or vice versa, followed by a mass spectrometric analysis of the formamide product.

Triphenethylamine, investigated as an example of a tertiary amine which cannot undergo N-demethylation, gave rise to a multiplicity of products, including the formyl (32%), benzoyl (32%), and phenylacetyl (18%) derivative of the secondary amine (Scheme II, $R = PhCH_2CH_2$), in addition to PhCHO and PhCOOH. Generation of the formamide in the absence of a N-methyl group is surprising and requires us to invoke the enamine autoxidation pathway[14,17] discussed above (Scheme I) for the oxidation of phenethylamine in pyridine. Benzaldehyde, released from this pathway, or from copper-mediated oxygenation of $PhCH_2CHO$[18] generated in the first N-dealkylation, would be expected to recombine with the secondary amine to give the carbinolamine precursor to the observed benzamide (Scheme II).

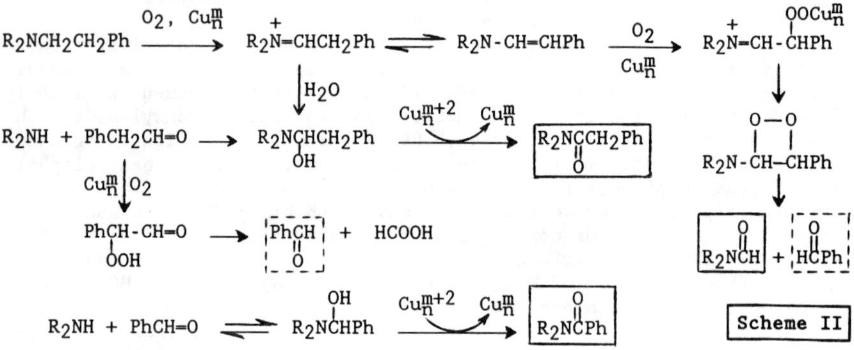

Scheme II

An interesting divergence from the "normal" reactivity pattern in CH_3CN is observed when the substrate amine contains a pyridyl group capable of *chelation* to copper (2-pyridyl but not 4-pyridyl). In such cases, the reactivity pattern switches to that observed for the $CuCl-O_2$ reactions conducted in pyridine solvent, suggesting that pyridyl chelation to copper approximates the coordinative effect that pyridine solvent has on the nature of the copper oxidant. Thus, whereas simple primary amines and 4-pyridylmethylamine were found to be inert in CH_3CN, 2-pyridylmethylamine and 2-(2-pyridyl)ethylamine were each oxidized (45-50% conversion) to the same products observed in pyridine: 2-PyCN (13%), 2-PyCONH$_2$ (13%), and 2-PyCOOH (5%) in the former case, and 2-PyCONHCH$_2$-CH$_2$Py (36%) and 2-PyCOOH (2%) in the latter case (2-Py = 2-pyridyl).

Likewise, whereas simple tertiary amines were found to be oxidized in CH_3CN but not in pyridine, N,N-dimethyl-2-pyridylmethylamine was recovered unchanged from the CH_3CN-based oxidation system. That N,N-dimethyl-4-pyridylmethylamine was also recovered unchanged, must be a reflection of the electron-withdrawing inductive effect on the one-electron oxidation at nitrogen. Switching from 2-pyridylmethyl to 2-(2-pyridyl)ethyl results in the appearance of at least marginal reactivity: 2-PyCH$_2$CH$_2$N(CH$_3$)$_2$ was 29% converted to a mixture of 2-PyCH$_2$CH$_2$N(CH$_3$)CH=O (17%) and 2-PyCOOH (12%). The partial reaction in this case appears to be a compromise between the complete oxidation of tertiary amines expected in CH$_3$CN and the inertness of tertiary amines expected in pyridine, and must be reflecting a sensitive balance between coordinative effects and attenuated electronic effects. This balance shifts back over to inertness when there are *two* 2-(2-pyridyl)ethyl groups (the third nitrogen substituent being methyl, benzyl, or phenethyl), because there is now increased pyridine-solvent-like coordination as well as a doubled inductive deactivation.

This last result is quite important in terms of the observed inertness to ligand C-N oxidation of the many bis-[2-(2-pyridyl)ethyl]-based (or related) tertiary amine ligands (whether mononuclear[25] or binuclear[26,27]) which have been investigated for copper-mediated activation of O$_2$. In contrast, tertiary amine-based ligand systems without the deactivating pyridyl coordination are expected to be more susceptible to ligand oxidation.[28]

In summary, the course of oxidation of aliphatic amines by CuCl-O$_2$ is seen to be quite sensitive to the nature of the solvent. In the more basic, but less Cu(I)-stabilizing case of pyridine, only primary and secondary amines are oxidized, a consequence of increased oxidative liability associated with N-deprotonation. In CH$_3$CN, the greater stabilization of Cu(I) apparently results in a stronger CuCl-O$_2$-based oxidant, and aliphatic amines are now oxidized in order of their ease of electron-transfer oxidation (tertiary most reactive); the inertness of primary amines in this case attests to the inability of the less basic CH$_3$CN system to achieve oxidation via N-deprotonation. Clearly, our study has so far focused on structure-reactivity aspects from the perspective of the amine substrates. Further studies will be required to clarify mechanistic details associated with O$_2$ processing and the mono- vs. binuclear nature of the CuCl-derived oxidant.

ortho-HYDROXYLATION OF PYRIDINE AND BENZOYL DERIVATIVES

In their attempts to study Cu(I)-O$_2$ chemistry in 1979, Gagne and coworkers reported a fortuitous but novel ligand oxygenation reaction, involving hydroxylation (25% yield) of an available pyridine *ortho* position in a bis-(2-pyridylimino)isoindoline-based tridentate ligand.[29] More recently, Réglier and coworkers discovered a more efficient pyridine *ortho* hydroxylation, arising when the Cu(I) complex of N-benzyl or N-phenethyl bis-[2-(2-pyridyl)ethyl)]amine is treated with 2 equiv of PhIO in dry CH$_3$CN under Ar.[24] In view of the expectations of electrophilic reactivity of a Cu(III)=O or related species presumably generated in this reaction, the observed attack at the pyridine 2-position runs contrary to the electronic bias of the pyridine ring. The mechanism we favor, which is essentially the same as that actually depicted by Réglier et al., is one involving an initial *nucleophilic* attack (the electronically preferred reaction) at the pyridine *ortho* position, followed by an internal two-electron redox (eq 6, top).

In an effort to determine the sensitivity of this reaction to variation in the ligand, we carried out a comparative study (reaction conditions as in ref 24) on 2-PyCH$_2$CH$_2$NRCH$_2$Ph, bearing as the third nitrogen substituent either 2-pyridylethyl (the prototype), 2-pyridylmethyl, 1-methylimidazol-4-ylethyl, N,N-dimethylaminoethyl, or N,N-dimethylaminopropyl. Although we could reproduce the *ortho* hydroxylation in the prototype case, all other ligands were recovered unchanged, except in the last case, where there was observed partial conversion to a multiplicity of products (not identified). The "one armed" ligand N,N-dimethyl-2-(2-pyridyl)ethylamine also failed to undergo hydroxylation.

$$(6)$$

In case the failure of reaction for the mixed 2-pyridylethyl/2-pyridylmethyl ligand reflected its inability to support generation of the active oxidant rather than reflecting a perturbed reactivity of the 2-pyridylethyl "arm", we carried out a reaction using a 1:1 mixture of the bis-(2-pyridylethyl) ligand and the mixed 2-pyridylethyl/2-pyridylmethyl ligand. Our finding of *ortho* hyroxylation only in the prototype ligand, rules out the ability of the active oxidant to effect an *inter*-molecular pyridine hydroxylation.

Our findings that the observed *ortho* hydroxylation is quite sensitive to variation of the ligand is reminiscent of the strict ligand dependence observed for O_2-dependent aryl hydroxylation of *m*-xylyl-bridged binuclear copper complexes.[26,27] In the present case, a complete understanding of mechanism will have to take into account that PhIO but not H_2O_2 or *m*-CPBA is capable of supporting the reaction. In this regard, it is worth pointing out that the generation of metal-oxo species from PhIO is not always assured,[30] and alternate mechanisms (e.g., eq 6, bottom) should also be considered.

A distinct, yet related transformation was reported by Reinaud and coworkers,[31] involving the aryl *ortho* hydroxylation of N-benzoyl α-aminoisobutyric acid (Aib) to give the corresponding N-salicyloyl derivative. The optimal reaction conditions in this case involve the use of Cu(0) (1.1 equiv) corrosion at 75°C in CH_3CN in the presence of O_2 and with excess added $Me_3N^+O^-$ (5 equiv). The *ortho* selectivity was rationalized on the basis of an intramolecular proximity effect on the action of a Cu(III)=O or related species chelated through the deprotonated amide nitrogen. The observed increase in reaction rate when electron-withdrawing aryl substituents were present, and the poor product yield obtained in the case of the 3,4,5-trimethoxybenzamide, were rationalized on the basis of increasing or decreasing the N-H acidity, respectively, thereby affecting the extent to which the copper oxidant is effectively chelated.[31b] Furthermore, the product distribution obtained for the 3-fluorobenzoyl analog was interpreted to indicate a radical (HO•-like) rather than electrophilic (HO$^+$-like) aromatic substitution.[31a]

Nonetheless, observing functionalization *ortho* to the carboxamido group would appear to be more consistent with a *nucleophilic* mechanism, and this could be the real explanation for the observed substituent effect on rate. The transformation observed by Reinaud et al. is milder than, but probably mechanistically related to, the thermal conversion of $PhCO_2Cu(II)OH$ to salicylate.[32] We believe that the key feature in both cases which insures *ortho* hydroxylation is not a proximity effect *per se*, but instead a *conjugative effect*. To illustrate the latter, we consider the mechanisms shown in Scheme III, involving nucleophilic attack with subsequent internal two-electron redox, though electronically simi-

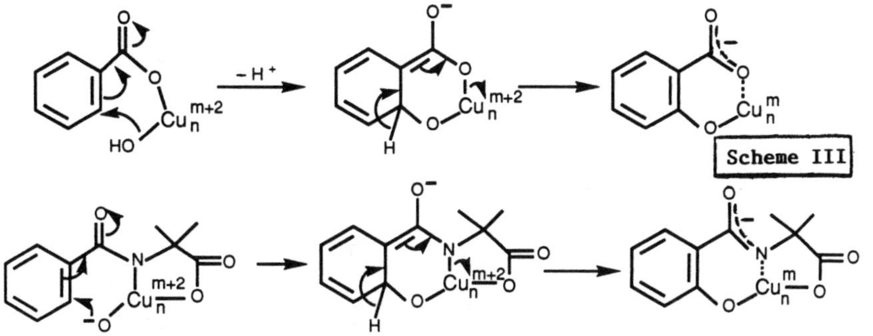

Scheme III

lar mechanisms involving radical attack with subsequent internal one-electron redox (not shown) are also possible.

Based on the presumption that the *ortho* hydroxylation is more nucleophilic than electrophilic in character, we considered that replacement of the benzene ring with a pyridine ring would preserve reactivity. Under the same conditions employed for conversion of N-benzoyl-Aib to N-salicyloyl-Aib (see above), we found that N-nicotinoyl-Aib was transformed to the mixture of hydroxylated products shown in eq 7 (isomer identities were confirmed by ^{1}H and ^{13}C NMR). The preponderance of 4-hydroxy over 2-hydroxy product (both representing *ortho* hydroxylation) is not immediately explanable, but our observation of at least some of the 6-hydroxy product requires that an *inter*molecular reaction of some type must compete with the albeit heavily favored intramolecular proximity-controlled regiochemistry. Moreover, since either 6-hydroxylation or 5-hydroxylation could theoretically result from such an intermolecular process, the generation of the former to the exclusion of the latter supports our contention that the mechanism reflects conjugative control.

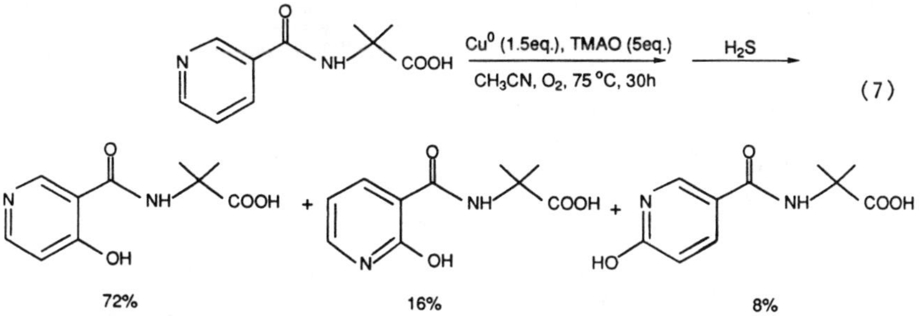

$$(7)$$

In the event that the N-benzoyl-Aib copper complex discussed above *does* generate HO•-like reactivity under the prescribed reaction conditions,[31a] we constructed some related ligand systems in hopes of observing a H•-abstraction by such species. The ligand systems we chose (eqs 8 and 9) preserve the use of one carboxylato and one amido ligand per copper center, as present in the N-benzoyl-Aib case. For the binuclear system (eq 8), the only reaction observed under several Cu(0)/Cu(I)/Cu(II)/oxidant systems was dehydrogenation α to the amido nitrogen, resulting in release of salicylamide. This reaction is not unexpected, several cases of such being observed upon generation of Cu(III) or a related strong copper oxidant.[33-35] When dehydrogenation at C_α was

245

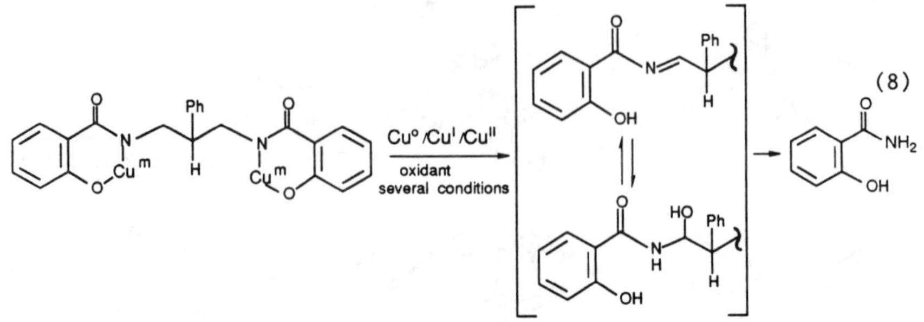

$$(8)$$

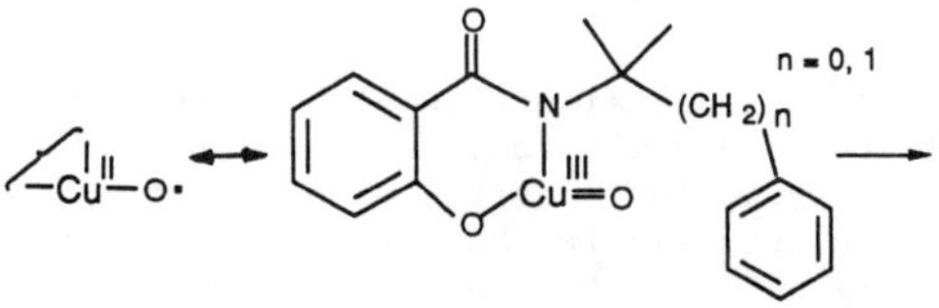

$$(9)$$

blocked (eq 9), no evidence for any type of reaction was obtained under several reaction conditions. Since only a positive result here would have permitted us to make conclusions about the nature of the copper-based hydroxylating species, complete elucidation of the mechanism of the benzamide *ortho* hydroxylation will require additional study.

OXIDATIVE N-DEALKYLATION OF N-ACYL-GLYCINE

We recently reported a Cu(III)-based system which achieves an oxidative N-dealkylation of N-acyl-glycine to carboxamides and glyoxylate, as a possible model for the peptidyl α-amidating monooxygenase (PAM) enzyme.[34] At the time, it was not yet known that the physiologic reaction is actually a two-step process, involving enzymatic C_α hydroxylation to a metastable carbinolamide, followed by an enzymatic dissociation step.[36] Since our model system formally achieved C-N dehydrogenation, with subsequent hydrolysis to the isolated products, the source of oxygen in the product carboxamide must be solvent water in our case. Thus, since the source of oxygen in the *enzymatic* product carboxamides is dioxygen (the other oxygen atom is reduced to water at the expense of the two-electron oxidation of ascorbate),[37] one could question the relevance of our Cu(III)-based model system.

However, as we pointed out,[34] if the enzyme does in fact carry out a formal C-N dehydrogenation reaction, the oxidant responsible for this would be Cu(III)=O or a related species, the two-electron reduction of which would generate a O_2-*derived water molecule* coordinated to copper, which could be selectively delivered to the RC(=O)N=CHCOOH intermediate, resulting in the observed product labeling. Thus, in a recent model study,[35] similar to ours but using O_2 itself to induce oxidation (we generated Cu(III) electrochemically or with $IrCl_6^{2-}$), the observed production of carbinolamide *does not rule out the possibility that the initial product is actually the N-acylimine*. Since the enzymatic monooxygenation is unlikely to follow a concerted oxene-insertion mechanism, the crucial question is whether, at the C_α radical intermediate stage, there is a HO• rebound step or a electron-transfer step with subsequent transfer of water from copper to C_α (eq 10). Since the N-acylimine, if produced, should be quite electrophilic, one would not expect it to survive in a model reaction in a manner which permitted its trapping during

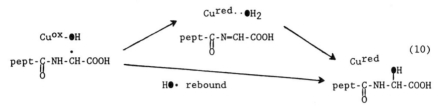

$$\text{(10)}$$

workup.[35] In studies which extend those of our published model systems,[34] we have obtained evidence for the intermediacy of a C_α radical
through use of 1-amino-1-cyclopropanecarboxylic acid (ACC) as the C-
terminal residue, but we have not yet found an uambiguous experiment to
elucidate the actual mechanism followed in eq 10.

CONCLUSION

We have here discussed several examples of distinct, yet related, nitrogen ligand oxidation/oxygenation reactions mediated by copper. These
semi-systematic studies have revealed reactivity patterns which should
help in the future design of more robust ligand systems needed in various bioinorganic studies. In addition to redox and O_2-activation roles
for copper, the strong Lewis acid character of Cu(II) and higher valent
states can result in autoxidative activation α to coordinating function-
alities such as C=O and C=N. These and related coordination-dependent
reactions must not be confused with authentic monooxygenation processes
which proceed in a "through space" manner without conjugative assistance. Lastly, it will be important to clarify how both the nature of
the copper oxidant and the mechanism of the ensuing reaction depends on
whether dioxygen itself or one of a variety of oxygen surrogates (e.g.,
PhIO, Me_3NO, H_2O_2, or m-CPBA) is employed.

Acknowledgments. This research was supported by NSF (CHE 87-06263).

REFERENCES AND NOTES

1. S. M. Nelson, *Inorg. Chim. Acta*, **62**, 39 (1982).
2. Preliminary studies have previously been reported: L. M. Sayre, F.
   Wang, and W. Tang, Abstr INOR 451, 197th National Meeting of the
   American Chemical Society, Dallas, April 9-14, 1989.
3. J. F. Weiss, G. Tollin, and J. T. Yoke III, *Inorg. Chem.*, **3**, 1344
   (1964); J. R. Clifton and J. T. Yoke III, *Inorg.Chem.*, **7**, 39 (1968);
   T. A. Lane and J. T. Yoke, *Inorg. Chem.*, **15**, 484 (1976).
4. C. P. Murthy, B. Sethuram, and T. N. Rao, *Monatsh. Chem.*, **113**, 941
   (1982).
5. D. C. Olsen and J. Vasilevskis, *Inorg. Chem.*, **10**, 463 (1971).
6. M. Anbar, R. A. Munoz, and P. Rona, *J. Phys. Chem.*, **67**, 2708 (1963).
7. F. Monacelli and G. O. Morpurgo, *Inorg. Chim. Acta*, **149**, 139 (1988).
8. T. Kametani, K. Takahashi, T. Ohsawa, and M. Ihara, *Synthesis*, 245
   (1977).
9. F. Wang and L. M. Sayre, *J. Am. Chem. Soc.*, **114**, 248 (1992).
10. A. Misono, T. Osa, and A. Koda, *Bull.Chem.Soc.Jpn.*, **40**, 912 (1967);
    41, 735 (1968).
11. H. G. Aurich, *Tetrahedron Lett.*, 657 (1964).
12. M. S. Kharasch and G. Sosnovsky, *Tetrahedron*, **3**, 97-104 (1958).
13. F. Hibbert and D. P. N. Satchell, *J. Chem. Soc. (B)*, 653 (1967); 568

(1968).

14. F. C. Schaefer and W. D. Zimmermann, *J. Org. Chem.*, **35**, 2165 (1970).

15. V. Van Rheenen, *J. Chem. Soc., Chem. Commun.*, 314 (1969).

16. T. Itoh, K. Kaneda, I. Watanabe, S. Ikeda, and S. Teranishi, *Chem. Lett.*, 227 (1976).

17. For the mechanism of product formation from enamine-derived dioxetanes, see: (a) H. H. Wasserman and S. Terao, *Tetrahedron Lett.*, 1735 (1975). (b) C. S. Foote, A. A. Dzakpasu, and J. W.-P. Lin, *Tetrahedron Lett.*, 1247 (1975).

18. S.-J. Jin, P.K. Arora, and L.M. Sayre, *J.Org.Chem.*, **55**, 3011 (1990).

19. (a) K. Watanabe, S. Komiya, and S. Suzuki, *Bull. Chem. Soc. Jpn.*, **46**, 2792 (1973). (b) P.F.B. Barnard, *J. Chem. Soc. (A)*, 2140 (1969). (c) R. Breslow and M. Schmir, *J. Am. Chem. Soc.*, **93**, 4960 (1971).

20. P. Capdevielle, A. Lavigne, and M. Maumy, *Synthesis*, 453 (1989).

21. P. Capdevielle, A. Lavigne, D. Sparfel, J. Baranne-Lafont, N. K. Cuong, and M. Maumy, *Tetrahedron*, **31**, 3305 (1990).

22. E. I. Troyanskii, V. A. Ioffe, and G. I. Nikishin, *Izv. Akad. Nauk SSSR, Ser. Khim.*, 1808 (1985).

23. The balance between N- and C_α-deprotonation is a complex issue: (a) F. D. Lewis and P. E. Correa, *J. Am. Chem. Soc.*, **103**, 7347 (1981). (b) A.S. Nazran and D. Griller, *J. Am. Chem. Soc.*, **105**, 1970 (1983).

24. (a) L. T. Burka, F. P. Guengerich, R. J. Willard, and T. L. Macdonald, *J. Am. Chem. Soc.*, **107**, 2549 (1985). (b) J. R. Lindsay Smith and D. N. Mortimer, *J. Chem. Soc., Perkin Trans. 2*, 1743 (1986).

25. M. Réglier, E. Amadei, R. Tadayoni, and B. Waegell, *J. Chem. Soc., Chem. Commun.*, 447 (1989).

26. Z. Tyeklar and K. D. Karlin, *Acc. Chem. Res.*, **22**, 241 (1989); P. P. Paul, Z. Tyeklar, R. R. Jacobson, and K. D. Karlin, *J. Am. Chem. Soc.*, **113**, 5322 (1991).

27. T.N. Sorrell, V.A. Vankai, and M.L. Garrity, *Inorg. Chem.*, **30**, 207 (1991); T.N. Sorrell and M.L. Garrity, *Inorg. Chem.*, **30**, 210 (1991).

28. Ligand autoxidation has been observed for a binuclear copper complex of N,N,N',N'-tetrakis(benzimidazolylmethyl)-1,3-diaminopropanol: M. G. Patch, V. McKee, and C. A. Reed, *Inorg. Chem.*, **26**, 776 (1987).

29. R. R. Gagne, R. S. Gall, G. C. Lisensky, R. E. Marsh, and L. M. Speltz, *Inorg. Chem.*, **18**, 771 (1979).

30. W. Nam and J. S. Valentine, *J. Am. Chem. Soc.*, **112**, 4977 (1990).

31. (a) O. Reinaud, P. Capdevielle, and M. Maumy, *J. Chem. Soc., Chem. Commun.*, 566 (1990). (b) O. Reinaud, P. Capdevielle, and M. Maumy, *Synthesis*, 612 (1990).

32. W. W. Kaeding and G. R. Collins, *J. Org. Chem.*, **30**, 3750 (1965); W. W. Kaeding and A. T. Shulgin, *J. Org. Chem.*, **27**, 3551 (1962).

33. D. W. Margerum, *Pure Appl. Chem.*, **55**, 23 (1983) and earlier references cited.

34. K.V. Reddy, S.-J. Jin, P.K. Arora, D.S. Sfeir, S.C. Feke-Maloney, F. L. Urbach, and L. M. Sayre, *J. Am. Chem. Soc.*, **112**, 2332 (1990).

35. P. Capdevielle and M. Maumy, *Tetrahedron Lett.*, **32**, 3831 (1991).

36. (a) M. Tajima, T. Iida, S. Yoshida, K. Komatsu, R. Namba, M. Yanagi, M. Noguchi, and H. Okamoto, *J. Biol. Chem.*, **265**, 9602 (1990). (b) A. G. Katopodis, D. Ping, and S. W. May, *Biochemistry*, **29**, 6115 (1990). (c) S. N. Perkins, E. J. Husten, and B. A. Eipper, *Biochem. Biophys. Res. Commun.*, **171**, 926 (1990).

37. (a) M. Noguchi, H. Seino, H. Kochi, H. Okamoto, T. Tanaka, and M. Hirama, *Biochem. J.*, **283**, 883 (1992). (b) D. J. Merkler, R. Kulathila, A. P. Consalvo, S. D. Young, and D. E. Ash, *Biochemistry*, **31**, 7282 (1992).

Dioxygen-Binding and Oxygenation Reactions

SYNTHESIS, STRUCTURE AND PROPERTIES OF μ-η²:η² PEROXO DINUCLEAR COPPER COMPLEXES MODELING THE ACTIVE SITE OF OXYHEMOCYANIN AND OXYTYROSINASE

Nobumasa Kitajima

Research Laboratory of Resources Utilization, Tokyo Institute of Technology, 4259 Nagatsuta, Midori-ku, Yokohama 227, Japan

INTRODUCTION

Hemocyanin is a ubiquitous oxygen transport protein for the arthropods and molluscs. The striking spectral features of oxyhemocyanin have attracted the attention of many chemists for quite some time.[1,2] The two coppers in oxyhemocyanin are in the cupric oxidation state, based on XAS studies, yet this state does not give an EPR signal.[3,4] Variable-temperature magnetic susceptibility studies indicate that the two cupric ions are strongly antiferromagnetically coupled, with $-2J > 600$ cm^{-1}.[5,6] Instead of the weak d-d transitions at 600-700 nm normally observed for cupric complexes, the absorption spectrum of oxyhemocyanin exhibits two characteristic bands, one at ~580 nm with ε~1000 M^{-1}cm^{-1} and an intense transition at 350 nm, ε~20000 M^{-1}cm^{-1}. Furthermore, the O-O stretching vibration, determined by resonance Raman spectroscopy and mixed labeling experiments, is unusually low (~750 cm^{-1}) while the ν(O-O) value clearly indicates that dioxygen is bound to the dicopper site as a peroxide ion in a symmetric coordination mode.[7,8] The monooxygenase tyrosinase, which is capable of oxidizing phenol to *o*-quinone, also possesses a dicopper site. The peroxo intermediate, oxytyrosinase, exhibits the characteristic properties whose features are entirely similar to those of oxyhemocyanin.[9-11]

The extremely strong antiferromagnetic coupling between the two cupric ions in oxyhemocyanin requires a superexchange pathway through a bridging ligand, because the Cu-Cu separation estimated by EXAFS is ca. 3.6 Å.[3,4] In general, the magnetic interaction through a peroxide has been believed not to be very strong, thus existence of some other endogenous ligand which is responsible for the antiferromagnetic coupling has been suggested. Originally some amino acid residue such as tyrosine was suggested to play this role, but the 3.2 Å resolution X-ray analysis of deoxyhemocyanin[12] established that this can not be the case, instead water or hydroxide is currently accepted as the most likely candidate for the endogenous bridging ligand.[13,14] The ligand displacement of the peroxide with an exogenous anion such as azide yields chemically modified hemocyanin, called met-hemocyanin. The existence of a bridging ligand mediating the strong magnetic interaction is evident in met-hemocyanin because these derivatives also exhibit strong antiferromagnetic coupling. Based on extensive spectroscopic investigations, particularly on metazido-hemocyanin, it has been suggested that this form has a μ-hydroxo-μ-1,3-azido structure.[16,17] Integration of the above results

From K.D. Karlin and Z. Tyeklár, Eds., *Bioinorganic Chemistry of Copper*
(Chapman & Hall, New York, 1993).

led to a generally accepted scenario that oxyhemocyanin binds dioxygen as a cis-μ-1,2 peroxide with an endogenous bridging ligand (X), most likely hydroxide as is shown in equation 1.

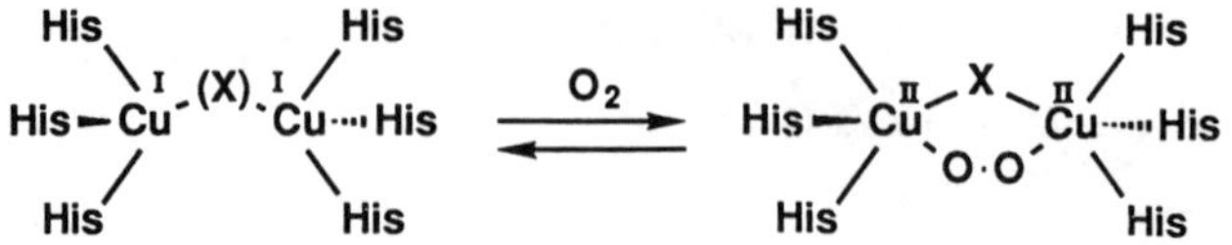

In order to provide better insight into the structure of the Cu^{2+}-O_2^{2-}-Cu^{2+} chromophore in oxyhemocyanin, extensive synthetic endeavors have been made to prepare a μ-peroxo dicopper complex which closely mimics the spectroscopic characteristics of oxyhemocyanin.[18,19] This was not easy to accomplish mainly due to the instability of such a complex. However, the series of elegant works performed by Karlin and coworkers demonstrated that by employing a proper ligand and low temperature handling, the preparation and adequate characterization of μ-peroxo complexes could be possible.[20,21] In fact, during the last decade, a considerable number of μ-peroxo complexes have been successfully prepared and characterized.[21,22] Some of these complexes mimic the spectral features of oxyhemocyanin in part yet none of them completely mimic all the characteristics of oxyhemocyanin.

By using sterically-hindered facial N_3 ligands, we succeeded in preparing dinuclear $\mu\eta^2{:}\eta^2$ peroxo copper complexes whose spectroscopic characteristics are remarkably close to those known for oxyhemocyanin.[23-26] Herein, we have described the synthesis, molecular structure and properties of the complexes and their biological relevance.

SYNTHESIS OF μ-$\eta^2{:}\eta^2$ PEROXO COMPLEXES

The ligands employed in this study are tris(pyrazolyl)borates whose structures are illustrated in Figure 1. The trigonal N_3 array provided with these ligands is reasonably similar to the ligand environment known for the copper ions in hemocyanin; X-ray analysis of deoxyhemocyanin[12] established that each copper ion is surrounded by three histidyl nitrogen atoms. In particular, the novel hindered ligand HB(3,5-iPr$_2$pz)$_3$[26,27] (hydrotris(3,5-diisopropyl-1-pyrazolyl)bor-ate) possesses striking synthetic advantages over the conventionally known tris(pyrazolyl)borates HBpz$_3$ and HB(3,5-Me$_2$pz)$_3$. Advantageous characteristics of this ligand include its high solubility in non-coordinating solvents, complete prevention of bis-ligand complex (L_2M) formation, and the highly shielding effect of the isopropyl groups to stabilize unusual coordination structures.[28]

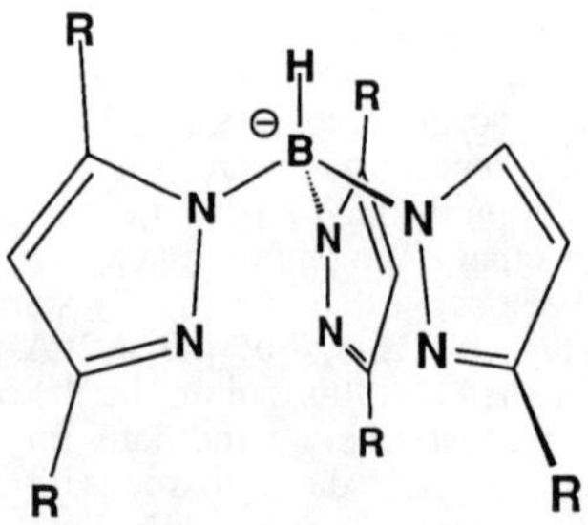

R = Me ; HB(3,5-Me$_2$pz)$_3$
R = Ph ; HB(3,5-Ph$_2$pz)$_3$
R = ^{i}Pr ; HB(3,5-^{i}Pr$_2$pz)$_3$

Figure 1. Hindered tris(pyrazolyl)borate ligands.

Cu(PPh$_3$)(HB(3,5-Me$_2$pz)$_3$), which was derived from Cu(CO)(HB(3,5-Me$_2$pz)$_3$),[29] can be converted to a μ-oxo dinuclear copper(II) complex by the reactions shown in eq. 2.

252

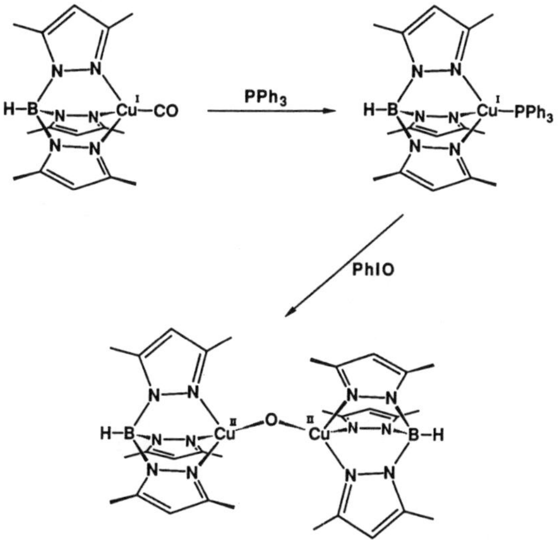

(eq. 2)

The low-temperature reaction of the μ-oxo complex $[Cu(HB(3,5-Me_2pz)_3)]_2(O)$ with hydrogen peroxide results in formation of a novel μ-peroxo complex $[Cu(HB(3,5-Me_2pz)_3)]_2(O_2)$ **(1)**, whose structure was identified with FD-MS and resonance Raman spectroscopy ($\nu(O\text{-}O)$, 731 cm^{-1} in $CHCl_3$) (eq. 3).[25]

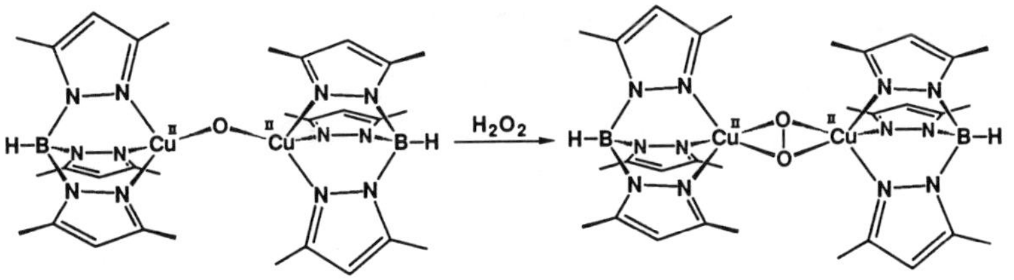

(eq. 3)

1

We also checked for the possibility of formation of this μ-peroxo complex by direct dioxygen addition to a copper(I) complex. Thus, a dinuclear copper(I) complex $[Cu(HB(3,5-Me_2pz)_3)]_2$ originally prepared by Marks et al.[29] was treated with dioxygen in a non-coordinating solvent such as CH_2Cl_2. Under certain conditions (low temperature, low concentration of dioxygen) a transient purple-colored complex, possibly a μ-peroxo complex, was observed; yet isolation of this species was not successful due to facile decomposition to a bis-ligand copper(II) complex, $Cu(HB(3,5-Me_2pz)_3)_2$.[30] Since the thermodynamic sink in these dioxygen additions appeared to be the cuprous complex $Cu(HB(3,5-Me_2pz)_3)_2$, we explored similar reactions with the more hindered tris(pyrazolyl)borate ligands $HB(3,5-iPr_2pz)_3$ and $HB(3,5-Ph_2pz)_3$ to prevent the formation of the bis-ligand complexes. In fact, both of the ligands are so sterically hindering that the corresponding bis-ligand complex is never formed. Thus, at low temperature, dioxygen reacts with the copper(I) complexes $Cu(Me_2CO)(HB(3,5-Ph_2pz)_3)$ and $Cu(HB(3,5-iPr_2pz)_3)$ to selectively give μ-peroxo complexes, $[Cu(HB(3,5-Ph_2pz)_3)]_2(O_2)$ (**2**) and $[Cu(HB(3,5-iPr_2pz)_3)]_2(O_2)$ (**3**), respectively (eq. 4).[26]

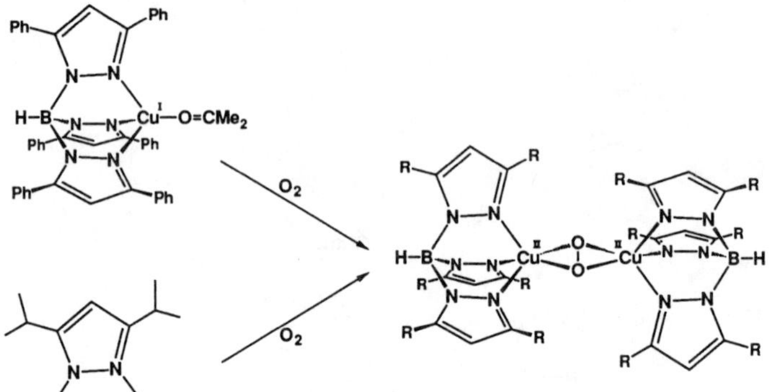

(eq. 4)

$R=Ph; \mathbf{2}$
$^iPr; \mathbf{3}$

Moreover, with $HB(3,5-iPr_2pz)_3$ a novel bis(μ-hydroxo) complex $[Cu(HB(3,5-iPr_2pz)_3)]_2(OH)_2$ was easily prepared by the straightforward method shown in eq. 5. The treatment of the bis(μ-hydroxo) complex with hydrogen peroxide again results in formation of the μ-peroxo complex **3**.[26]

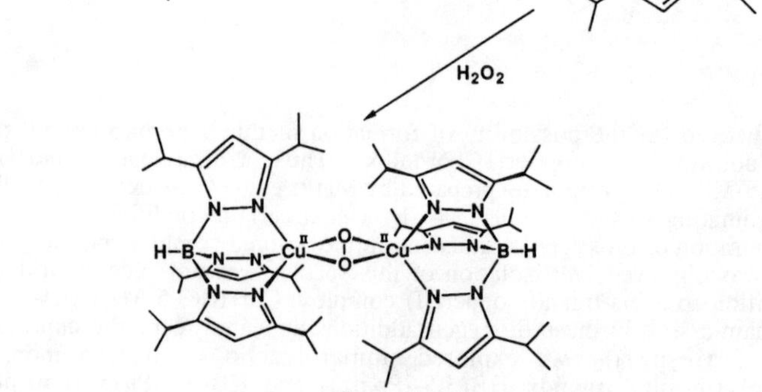

(eq. 5)

3

MOLECULAR STRUCTURE OF THE μ-η²:η² PEROXO COMPLEX

Slow recrystallization of $[Cu(HB(3,5-iPr_2pz)_3)]_2(O_2)$ (**3**) from CH_2Cl_2 gave Xray quality crystals solvated with CH_2Cl_2. The molecular structure of **3** and an expanded view of the $N_3Cu^{2+}(O_2^{2-})Cu^{2+}N_3$ moiety are shown in Figures 2 and 3, respectively.[26]

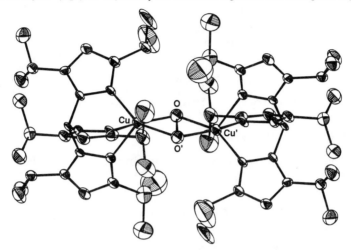

Figure 2. ORTEP view of the μ-η²:η² peroxo complex **3**.

The O-O bond distance of 1.41 Å is typical for peroxo transition metal complexes. The molecule sits on a crystallographically imposed center of symmetry, requiring that the two coppers and the peroxide ion sit on the same plane. The Cu-O (1.927 Å) and Cu'-O (1.903 Å) distances are comparable, thus the coordination mode of the peroxide ion is referred to as μ-η²:η². Although this coordination mode is known for f-block element peroxo complexes,[31] this is the first example of a structurally-characterized transition metal complex with an μ-η²:η² peroxo ligand.[24] More recently, an X-ray structure of a vanadium peroxo complex having a bent μ-η²:η² structure has been reported.[32]

Two of the Cu-N distances (2.000 and 1.993 Å) are short and almost identical to the Cu-O distances, whereas the other Cu-N bond length is significantly elongated. Therefore, the coordination geometry of the copper can

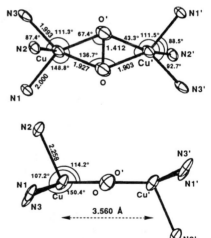

Figure 3. Expanded view of the $N_3Cu(O_2^{2-})CuN_3$ unit in **3**.

be described as square-pyramidal with the basal plane consisting of two oxygens (from the peroxide ion) and two nitrogen atoms (from the tris(pyrazolyl)borate ligand), although the structure is considerably distorted. The driving force to form an unusual μ-η^2:η^2 structure in the f-block and vanadium complexes is interpreted in terms of the strong oxophilicity of these elements, but this cannot be applied for the present complex. Rather, we ascribe the unusual μ-η^2:η^2 structure to the strong preference of cupric ion to favor five-coordination over a tetrahedral four-coordinate structure. The use of the facial trigonal ligand (a planar tetragonal structure is not accessible) is the key factor to construct this novel μ-η^2:η^2 peroxo structure. In fact, with a N_4 ligand, Karlin et al. observed the formation of a trans-μ-1,2 peroxo complex in which each copper exhibits a trigonal-bipyramidal structure with a N_4O ligand donor set.[33] It should be noted that the ligand donor sets of the copper ions in hemocyanin are N_3, not N_4.[12] The Cu-Cu separation found in **3** is 3.56 Å, which is considerably shorter than that reported for Karlin's trans-μ-1,2 peroxo complex (4.36 Å)[33] but remarkably similar to those estimated for oxyhemocyanin and oxytyrosinase by EXAFS.[3,4]

SPECTROSCOPIC PROPERTIES

The magnetic and spectral features of the μ-η^2:η^2 peroxo complexes **1-3** are summarized in Table I.[26]

Table I Physicochemical Properties of μ-η^2:η^2 Peroxo Complexes 1-3 and oxy-Hc and oxy-Tyr.

	magnetic property	absorption bands /nm (ε)	ν(O-O) (cm^{-1})	Cu···Cu (Å)
1	diamag	530(840), 338(20800)	731	---
2	diamag	542(1040), 355(18000)	759	---
3	diamag	551(790), 349(21000)	741	3.56
oxy-Hc	diamag	580(1000), 340(20000)	744-752	3.5-3.7
oxy-Tyr	diamag	600(1200), 345(18000)	755	ca. 3.6

The magnetic susceptibility of **2** was determined by the Faraday method because its thermal stability allowed the measurement at room temperature. No effective magnetic moment was observed. The diamagnetism of these complexes was also suggested by ^{1}H-NMR spectra which, for **1-3**, consist of sharp lines with no detectable paramagnetic shift. In addition, further support comes from the magnetic susceptibility measured by the Evans method.[34] Very recently, variable-temperature magnetic susceptibility measurements by SQUID were carried out for **2** and an enormously large antiferromagnetic interaction with -2J > 1000 cm^{-1} was confirmed.[35]

The absorption spectra of **1-3** exhibit two intense characteristic bands at ca. 350 and 530-550 nm, features that are qualitatively and quantitatively very similar to those known for oxyhemocyanin. The spectrum of **1** is shown in Figure 4. The 530 nm band is obviously ascribed to a $O_2^{2-} \rightarrow Cu^{2+}$ LMCT since the molar extinction coefficient of the band is too large for a d-d band, while the tailing shape may indicate overlap of a d-d band around 600 nm. There is clear enhancement of the ν(O-O) Raman band upon excitation of the $O_2^{2-} \rightarrow Cu^{2+}$ LMCT band (vide infra), which also supports the assignment.

The other unusually intense band at ca. 350 nm can also be attributed to a $O_2^{2-} \rightarrow Cu^{2+}$ LMCT band since such an intense band is not observed for other copper(II) complexes with hydrotris(pyrazolyl)borate ligands; most of the complexes exhibit no band or only a weak band with $\varepsilon < 2000$ M^{1}cm^{-1} at ~350 nm. The occurrence of a similar strong band at ~350 nm in oxyhemocyanin has been a subject of extensive studies. One possible assignment of the band is imidazole→Cu^{2+} LMCT,[36] however, another possibility which seems to be more reasonable in light of the spectra seen for **1-3** is O_2^{2-}→Cu^{2+} LMCT band. With a side-on type coordination of a peroxide, the O_2^{2-}→Cu^{2+} LMCT band is predicted to be split into two distinctive bands, designated as $\pi_\sigma^*(O_2^{2-}) \rightarrow Cu^{2+}$ and $\pi_v^*(O_2^{2-}) \rightarrow Cu^{2+}$. Recently, SCF-Xα–SW calculations were performed for a variety of coordination modes of the peroxide on cupric ions by Ross and Solomon. According to the calculations, the lower energy band (ca. 550 nm) is ascribed to the $\pi_v^*(O_2^{2-}) \rightarrow Cu^{2+}$ transition, whereas the higher energy band is ascribed to $\pi_\sigma^*(O_2^{2-}) \rightarrow Cu^{2+}$ transition. In particular, based on the double side-on structure, $\mu\eta^2:\eta^2$, the $\pi^*_\sigma(O_2^{2-})$ orbitals have the best overlap with the dx^2-y^2 orbital so that the enormous intensity of the band is predictable, with the largest splitting between the $\pi_\sigma^*(O_2^{2-}) \rightarrow Cu^{2+}$ and $\pi_v^*(O_2^{2-}) \rightarrow Cu^{2+}$ transition. This theoretical interpretation leads us to favor the assignment of the 350 nm band as the $\pi_\sigma^*(O_2^{2-}) \rightarrow Cu^{2+}$ transition. Recent Raman profile experiments on the 350 nm band further support the assignment.[35]

The μ-η^2:η^2 peroxo complexes **1-3** were studied with Raman spectroscopy, which located the v(O-O) band at 730-750 cm^{-1}. The v(O-O) stretching vibration is usually found for transition-metal peroxo complexes in the range of 800-900 cm^{-1}.[37] Thus, the values found for **1-3** are unusually low, but they are similar to those found for oxyhemocyanin and oxytyrosinase, which also exhibit low v(O-O) stretching vibrations.

Figure 5 displays the resonance Raman spectra of **3** prepared with $^{16}O_2$ (spectrum A) and a 1:2:1 mixture of $^{16}O_2/^{16}O^{18}O/^{18}O_2$ (spectrum B) in acetone at -40°[26] upon irradiation at 514.5 nm. In spectrum A there is a single band at 741 cm^{-1} due to the $v(^{16}O-^{16}O)$ while spectrum B exhibits two additional bands at 719 and 698 cm^{-1}, which are assignable to $v(^{16}O-^{18}O)$ and $v(^{18}O-^{18}O)$, respectively. Because the intensity ratio of the three peaks [$v(^{16}O-^{16}O)$, $v(^{16}O-^{18}O)$ and $v(^{18}O-^{18}O)$] is nearly 1:2:1 and they give nearly identical line widths, it is evident that only one structure contributes to the $v(^{16}O-^{18}O)$ band, indicating a

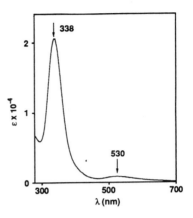

Figure 4. Electronic spectrum of the μ-η²:η² peroxo complex **1**.

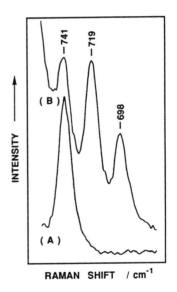

Figure 5. Resonance Raman spectrum of **3** recorded at -40°C in acetone (A) **3** prepared with $^{16}O_2$ (B) **3** prepared with a 1:2:1 mixture of $^{16}O_2$-$^{16}O^{18}O$-$^{18}O_2$.

symmetric coordination mode of the peroxide in **3**. Similar experimental results were obtained for oxyhemocyanin with mixed labeled dioxygen.[8]

The relevant physicochemical characteristics of oxyhemocyanin and oxytyrosinase are presented in Table 1 along with complexes **1-3**. The close similarities between **1-3** and the proteins are evident. Based on this remarkable resemblance, we have proposed that the peroxide is bound to the dicopper site in these proteins in the particular coordination mode seen for **1-3**, μ-η^2:η^2 (see equation 6), without the existence of an endogenous bridging ligand that was generally accepted.

$$
\begin{array}{c}
\text{His} \qquad\qquad \text{His} \\
\backslash\; _{I} \qquad _{I}\; / \\
\text{His} \!-\! \text{Cu} \qquad \text{Cu} \cdots\!\! \text{His} \\
/ \qquad\qquad \backslash \\
\text{His} \qquad\qquad \text{His}
\end{array}
\;\;\underset{\displaystyle \Longleftarrow}{\overset{O_2}{\Longrightarrow}}\;\;
\begin{array}{c}
\text{His} \qquad\qquad\quad \text{His} \\
\backslash \quad _{II}\; O\; _{II}\; / \\
\text{His} \!-\! \text{Cu} \overset{\displaystyle |}{\underset{\displaystyle O}{\diagdown\;\diagup}} \text{Cu} \cdots\!\! \text{His} \\
/ \qquad\qquad\quad \backslash \\
\text{His} \qquad\qquad\quad \text{His}
\end{array}
\qquad \textbf{(eq. 6)}
$$

REACTION CHEMISTRY OF THE μ-η^2:η^2 PEROXO COMPLEX

Although several lines of evidence imply that the coordination environment of tyrosinase is different from that of hemocyanin,[38] the extremely similar spectral features of oxytyrosinase and oxyhemocyanin indicate that the structural difference must be small and the electronic structure of the peroxide ion is essentially comparable. Hence, we suggest that the distinctive functional roles of hemocyanin (reversible dioxygen binding) and tyrosinase (monooxygenation) stem from the availability of the substrate binding site. In fact, hemocyanin from molluscs is known to be effective for phenol oxidations.[39] Since now it has become almost certain that oxytyrosinase has the μ-η^2:η^2 structure, we explored the reaction chemistry of the μ-η^2:η^2 peroxo complexes in order to shed light on the reaction mechanisms of tyrosinase, few of which have been elucidated to date.

Spontaneous Decomposition

The μ-η^2:η^2 peroxo complexes **1-3** are not very thermally stable in solution. The decomposition of **1** in CHCl$_3$ at room temperature was explored in detail.[40] The change in the electronic spectrum was followed over time. Both of the strong characteristic bands due to **1** (at 338 and 530 nm) disappeared gradually. and the intensely purple colored solution turned green within 2 h. The features of the final spectrum are identical to that of the μ-oxo complex [Cu(HB(3,5-Me$_2$pz)$_3$)]$_2$(O). The ^{1}H-NMR spectrum also suggests the quantitative formation of the μ-oxo complex from **1**. The kinetics of the decomposition were explored by following the time dependence of the absorbance of the band at 530 nm, indicating that the rate is first order with respect to the concentration of **1**. The spontaneous decomposition of μ-peroxo complexes to μ-oxo complexes is well established for iron-porphyrin compounds.[41] The decomposition rate of this reaction is also first order with respect to the concentration of the μ-peroxo ferric complex, suggesting that the rate-determining step is the homolysis of the O-O bond in the μperoxo complex. The decomposition of **1** to the μ-oxo complex should similarly proceed via homolysis of the O-O bond. The structure of the intermediate formed by the homolysis can be described as copper(III)-oxo or copper(II)-oxygen atom anion radical in its canonical form. In the homolysis of the μ-peroxo ferric complex, the intermediate species was identified as an iron(IV)-oxo[42] but a copper(III)-oxo double bond seems unlikely because of the lack of a d orbital of π-symmetry on the copper(III) ion.[43] Therefore, the copper(II)-oxygen anion radical is more reasonable as the intermediate generated by this homolysis.

In the presence of a trace amount of water, the μ-oxo complex [Cu(HB(3,5-Me$_2$pz)$_3$)]$_2$(O) is converted easily to a bis(μ-hydroxo) complex [Cu(HB(3,5-Me$_2$pz)$_3$)]$_2$(OH)$_2$, whose structure was determined by X-ray crystallography. The μ-η^2:η^2

peroxo complex **3** also decomposes spontaneously; in this case the product is solely the bis(μ-hydroxo) complex [Cu(HB(3,5-iPr$_2$pz)$_3$)]$_2$(OH)$_2$. Both bis(μ-hydroxo) complexes are strongly antiferromagnetic coupled (EPR silent) and exhibit only a weak band at ca. 650 nm due to the d-d transition. It is known that hemocyanin is converted to an inactive form over time, which can be reactivated to oxyhemocyanin with hydrogen peroxide.[44] Similarly, the reaction of the resting form of tyrosinase with hydrogen peroxide generates oxytyrosinase.[9,45] From the fact that the μ-η^2:η^2 peroxo complex spontaneously decomposes to a bis(μ-hydroxo) complex, which reacts with hydrogen peroxide to regenerate the μ-η^2:η^2 peroxo complex, we propose that the inactive form of hemocyanin and the resting form of tyrosinase have a bis(μ-hydroxo) structure.

Displacement of the Peroxide with PPh$_3$ and CO

The addition of PPh$_3$ into the solution of **1** at -20°C causes instantaneous decoloration of the purple solution along with the quantitative formation of a copper(I) complex Cu(PPh$_3$)(HB(3,5-Me$_2$pz)$_3$) with dioxygen evolution.[40] Similar reductive displacement of the peroxide with CO, to give Cu(CO)(HB(3,5-Me$_2$pz)$_3$), was observed, but this reaction requires a higher reaction temperature. Complex **3** also exhibits essentially the same reactivity toward PPh$_3$ and CO, whereas **2** does not react with PPh$_3$ presumably owing to the high steric hindrance of HB(3,5-Ph$_2$pz)$_3$.[26]

The ν(CO) for the series of tris(pyrazolyl)borate carbonyl complexes are as follows: Cu(CO)(HB(3,5-Ph$_2$pz)$_3$) (2086 cm^{-1}) > Cu(CO)(HBpz$_3$) (2083)[46] > Cu(CO)(HB(3,5-Me$_2$pz)$_3$) (2066)[29] > Cu(CO)(HB(3,5-iPr$_2$pz)$_3$) (2056). This order is in excellent accord with the electron-donating properties of the ligands; a stronger electron-donating ligand lowers the CO frequency because of increased π-back bonding from the metal center into CO π^* orbitals, which weakens the CO bond. A variety of carbonyl copper(I) complexes are known and their CO stretching vibrations are located in the range of 2080-2120 cm^{-1}, whereas the CO adduct of hemocyanin exhibits distinctively lower frequency, 2043-2063 cm^{-1}.[47] The ν(CO) of Cu(CO)(HB(3,5-iPr$_2$pz)$_3$) is the lowest among the synthetic complexes and very close to the values known for CO adducts of hemocyanin. This implies that the ligand environment provided by HB(3,5-iPr$_2$pz)$_3$ is reasonably similar to that in hemocyanin not only geometrically but also electronically.

The facile displacement of the peroxide with CO and PPh$_3$ as well as spontaneous decomposition can be explained in terms of the mechanism illustrated in Figure 6. There is an equilibrium between the μ-η^2:η^2 peroxo complex and a monomeric copper(I) complex at a relatively high temperature. PPh$_3$ or CO reacts with the copper(I) species instantaneously to give the corresponding adduct. On the other hand, the irreversible homolysis of the O-O bond generates a copper(II)-oxygen anion radical as described above. The coupling of a copper(I) species and a copper(II)-oxygen anion radical results in the formation of the μ-oxo complex, which is readily converted into a bis(μ-hydroxo) complex in the presence of water.

Reaction Aspects of μ-η^2:η^2 Peroxo Complex[40]

The spontaneous decomposition of **1** in the presence of 75 equiv of cyclohexene in CHCl$_3$ at room temperature under argon resulted in formation of ca. 10% of 2cyclohexene chloride. When the same reaction was carried out under 1 atm O$_2$, formation of oxygenated products (2-cyclohexene-1-ol (212%), 2-cyclohexene-1-one (202%), cyclohexene oxide (13%)) was noted. These results are interpreted in terms of classical type radical reactions initiated by the copper(II)-oxygen anion radical generated via homolysis of O-O bond in **1**. In fact, the labeling experiments indicated that the oxygen atoms in the oxygenated cyclohexenes originate from the dioxygen but not from the μ-η^2:η^2 peroxide. The copper(II)-oxygen anion radical should be effective for hydrogen-atom abstraction from phenol. Thus, the experimental result that anaerobic reaction of **1** and 2,6-dimethylphenol yields the corresponding diphenoquinone quantitatively is explained by a free-radical mechanism proceeding via a phenoxy radical formed by the

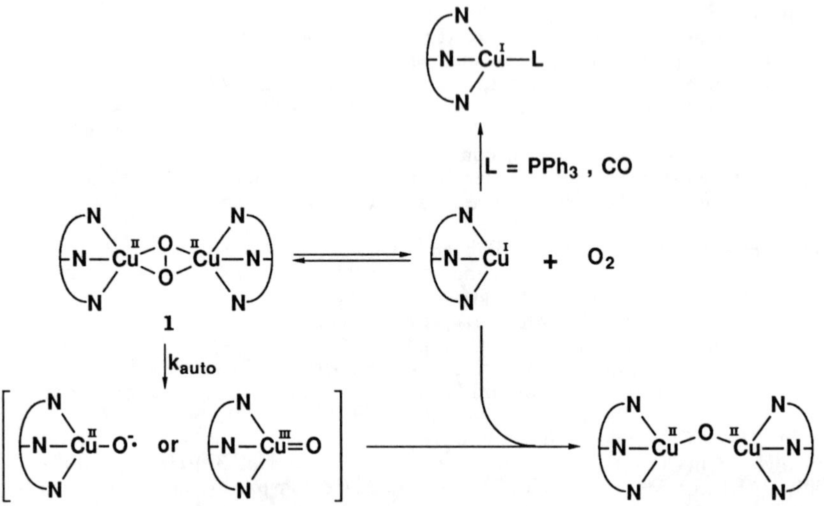

Figure 6. Reaction mechanism of spontaneous decomposition of **1**.

hydrogen abstraction. However, the overall rate of the consumption of **1** in the phenol oxidation is higher than the rate of the spontaneous decomposition of **1**. This implies that there is an alternative reaction pathway rather than the homolysis of the peroxide in the reaction between **1** and the phenol. Hence, the anaerobic oxidation of 2,6-dimethylphenol was carefully examined and the consumption rate of **1** was found to be expressed as follows (eq. 7).

$$-\frac{d[\mathbf{1}]}{dt} = (k_{auto} + k_1[DMP])[\mathbf{1}] \qquad \text{(eq. 7)}$$

Here, DMP denotes 2,6-dimethylphenol and k_{auto} is the rate constant of spontaneous decomposition of **1**. Since the second term in eq. 7 is dependent upon both the concentrations of **1** and the phenol, the reaction contributing to this term is ascribed to the formation of a phenoxy intermediate. Although this intermediate cannot be unambiguously identified in the reaction mixture of **1** and a phenol, the low-temperature reaction between **3** and a phenol such as p-fluorophenol gave a phenoxy complex, which was isolated and structurally determined by X-ray crystallography.[48] Thus, the μ-η^2:η^2 peroxo complex is nucleophilic so as to react with a phenol to cause acid/base replacement of the ligand to afford a phenoxy complex. The phenoxy copper(II) complex undergoes homolytic cleavage of the Cu-O bond to generate a phenoxy radical which is responsible for the formation of the diphenoquinone. When the reaction of **1** with 2,6-dimethylphenol was carried out in the presence of dioxygen, the consumption rate of **1** is apparently much higher and not only 3,3',5,5'-tetramethyldiphenoquinone but also 2,6-dimethylbenzoquinone was formed. In this aerobic reaction, we assume that dioxygen directly attacks the para position of 2,6-phenoxy, affording a peroxyphenoxo radical, since the para/ortho positions of a phenoxy group coordinated to a transition metal ion are accessible for the radical coupling because of the contribution of the resonance structure as indicated in eq. 8.

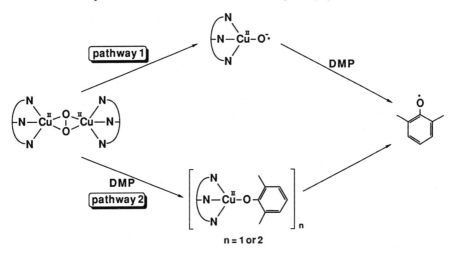

(eq. 8)

   In summary, the most likely reaction mechanism of phenol oxidation by the μ-η²:η² peroxo complex is illustrated in Figure 7. In tyrosinase, because the two copper ions are fixed by the rigid protein chains so as to reverse the homolysis of the O-O bond of the peroxide ion, the pathway 1 seems unlikely to occur. Rather it is suggested that the initial replacement of μ-η²:η² peroxide to phenoxide followed by classical type radical reactions are responsible for the oxidation reactions catalyzed by tyrosinase.

Figure 7. Mechanism of anaerobic oxidation of 2,6-dimethylphenol by the μ-η²:η² peroxo complex.

ACKNOWLEDGEMENT

   I thank Prof. Y. Moro-oka and the students, K. Fujisawa, T. Koda, C. Fujimoto, S. Hikichi, Y. Iwata and T. Katayama for their contributions to the research described herein.

REFERENCES

1.   E. I. Solomon, K. W. Penfield, D. E. Wilcox, *Struct. Bonding (Berlin)*, **53**, 1 (1983).
2.   E. I. Solomon, M. J. Baldwin, M. L. Lowery, *Chem. Rev.*, **92**, 521 (1992).
3.   J. M. Brown, L. Powers, B. Kincaid, J. A. Larrabee, T. G. Spiro, *J. Am. Chem. Soc.*, **102**, 4210 (1981).
4.   M. S. Co, K. O. Hodgson, T. K. Eccles, R. Lontie, *J. Am. Chem. Soc.*, **103**, 984 (1981).
5.   E. I. Solomon, D. M. Dooley, R. -H. Wang, H. B. Gray, M. Cerdonio, F. Mogno, G. L. Romani, *J. Am. Chem. Soc.*, **98**, 1029 (1976).
6.   D. M. Dooley, R. A. Scott, J. Ellinhaus, E. I. Solomon, H. B. Gray, *Proc. Natl. Acad. Sci., U. S. A.*, **75**, 3019 (1978).
7.   T, B, Freeman, J. S. Loehr, T. M. Loehr, *J. Am. Chem. Soc.*, **98**, 2809 (1976).

8.　T. J. Thamann, J. S. Loehr, T. M. Loehr, *J. Am. Chem. Soc.,* **99**, 4187 (1977).
9.　R. J. Jolley, Jr., L. H. Evans, N. Makino, H. S. Mason, *J. Biol. Chem.,* **249**, 335 (1974).
10.　N. C. Eickman, E. I. Solomon, J. A. Larrabee, T. G. Spiro, K. Lerch, *J. Am. Chem. Soc.,* **100**, 6529 (1978).
11.　R. S. Himmelwright, N. C. Eickman, C. D. LuBien, K. Lerch, E. I. Solomon, *J. Am. Chem. Soc.,* **102**, 7339 (1980).
12.　(a) A. Volbeda and W. G. J. Hol, *J. Mol. Biol.,* **206**, 531 (1989).　(b) A. Volbeda and W. G. J. Hol, *J. Mol. Biol.,* **209**, 249 (1989).
13.　J. Loroesch and W. Haase, *Biochemistry,* **25**, 5850 (1986).
14.　J. A. Larrabee, T. F. Baumann, S. J. Chisdes, T. J. Lyons, *Inorg. Chem.,* **31**, 3630 (1992).
15.　N. C. Eickman, R. J. Himmelwright, E. I. Solomon, *Proc. Natl. Acad. Sci., U. S. A.,* **76**, 2094 (1979).
16.　E. I. Solomon, in *Metal Clusters in Proteins*, L. Que, Jr., Ed., (American Chemical Society, Washington, D.C., 1988), pp. 116-150.
17.　J. E. Pate, P. K. Ross, T. J. Thamann, C. A. Reed, K. D. Karlin, T. N. Sorrell, E. I. Solomon, *J. Am. Chem. Soc.,* **111**, 5198 (1989).
18.　K. D. Karlin and Y. Gultneh, *Prog. Inorg. Chem.,* **35**, 219 (1987).
19.　T. N. Sorrell, *Tetrahedron,* **45**, 3 (1989).
20.　Z. Tyeklar and K. D. Karlin, *Acc. Chem. Res.,* **22**, 241 (1989)
21.　K. D. Karlin, Z. Tyeklar, A. D. Zuberbuhler, in *Bioinorganic Catalysis*, J. Reedijk, Ed., (Marcel Dekker, New York, 1992), in press.
22.　N. Kitajima, *Adv. Inorg. Chem.,* in press.
23.　N. Kitajima, T. Koda, S. Hashimoto, T. Kitagawa, Y. Moro-oka, *J. Chem. Soc., Chem. Commun.,* 151 (1988).
24.　N. Kitajima, K. Fujisawa, Y. Moro-oka, K. Toriumi, *J. Am. Chem. Soc.,* **111**, 8975 (1989).
25.　N. Kitajima, T. Koda, S. Hashimoto, T. Kitagawa, Y. Moro-oka, *J. Am. Chem. Soc.,* **113**, 5664 (1991).
26.　N. Kitajima, K. Fujisawa, C. Fujimoto, Y. Moro-oka, S. Hashimoto, T. Kitagawa, K. Toriumi, K. Tatsumi, A. Nakamura, *J. Am. Chem. Soc.,* **114**, 1277 (1992).
27.　N. Kitajima, K. Fujisawa, C. Fujimoto, Y. Moro-oka, *Chem. Lett.,* 421 (1989).
28.　(a) N. Kitajima, K. Fujisawa, Y. Moro-oka, *J. Am. Chem. Soc.,* **112**, 3210 (1990). (b) N. Kitajima, K. Fujisawa, Y. Moro-oka, *Inorg. Chem.,* **29**, 357 (1990).　(c) N. Kitajima, H. Fukui, Y. Moro-oka, Y. Mizutani, T. Kitagawa, *J. Am. Chem. Soc.,* **112**, 6402 (1990).　(d) N. Kitajima, U. P. Singh, H. Amagai, M. Osawa, Y. Moro-oka, *J. Am. Chem. Soc.,* **113**, 7757 (1991).　(e) N. Kitajima, M. Osawa, M. Tanaka, Y. Moro-oka, *J. Am. Chem. Soc.,* **113**, 8952 (1991).　(f) N. Kitajima, N. Tamura, M. Tanaka, Y. Moro-oka, *Inorg. Chem.,* **31**, 3342 (1992).　(g) N. Kitajima, K. Fujisawa, M. Tanaka, Y. Moro-oka, *J. Am. Chem. Soc.,* in press.
29.　C. Mealli, C. S. Arcus, J. L. Wilkinson, T. J. Marks, J. A. Ibers, *J. Am. Chem. Soc.,* **98**, 711 (1976).
30.　N. Kitajima, Y. Moro-oka, A. Uchida, Y. Sasada, Y. Ohashi, *Acta Cryst.,* **C44**, 1876 (1988).
31.　(a) R. Haegele and J. C. A. Boeyens, *J. Chem. Soc., Dalton Trans.,* 648 (1977).　(b) D. C. Bradley, J. S. Ghotra, F. A. Hart, M. B. Hursthouse, P. R. Raithby, *J. Chem. Soc., Dalton Trans.,* 1166 (1977).
32.　A. E. Lapshin, Y. I. Smolin, Y. F. Shepelev, P. Schwendt, D. Gyepesova, *Acta Cryst.,* **C46**, 1755 (1990).
33.　R. R. Jacobson, Z. Tyeklar, A. Farooq, K. D. Karlin, S. Liu, J. Zubieta, *J. Am. Chem. Soc.,* **110**, 3690 (1988).
34.　D. F. Evans, *J. Chem. Soc.,* **29**, 2003 (1959).
35.　M. J. Baldwin, D. E. Root, J. E. Pate, K. Fujisawa, N. Kitajima, E. I. Solomon, *J. Am. Chem. Soc.,* in press.
36.　J. A. Larrabee, T. Spiro, N. S. Ferris, W. H. Woodruff, W. A. Maltese, M. S. Kerr, *J. Am. Chem. Soc.,* **99**, 1979 (1977).

37. M. H. Gubelmann and A. F. Williams, *Struct. Bonding (Berlin)*, **55**, 1 (1983).
38. D. E. Wilcox, A. G. Porras, Y. T. Hwang, K. Lerch, M. E. Winkler, E. I. Solomon, *J. Am. Chem. Soc.,* **107**, 4015 (1985).
39. A. Nakahara, S. Suzuki, J. Kino, *Life Chem. Rep., Suppl. Ser.*, **1**, 319 (1983).
40. N. Kitajima, T. Koda, Y. Iwata, Y. Moro-oka, *J. Am. Chem. Soc.,* **112**, 8833 (1990).
41. D. -H. Chin, G. N. La Mar, A. L. Balch, *J. Am. Chem. Soc.*, **102**, 4344 (1980).
42. A. L. Balch, G. N. La Mar, L. Latos-Grazynski, M. W. Renner, V. Thanabal, *J. Am. Chem. Soc.*, **107**, 3003 (1985).
43. J. M. Mayer, *Comments Inorg. Chem.*, **8**, 125 (1988).
44. K. Heirwegh, V. Blaton, R. Lontie, *Arch. Int. Physiol. Biochem.*, **73**, 149 (1965).
45. N. Makino, P. McMahill, H. S. Mason, T. H. Moss, *J. Biol. Chem.,* **249**, 6062 (1974).
46. M. I. Bruse and A. P. P. Ostazewski, *J. Chem. Soc., Dalton Trans.*, 2433 (1973).
47. (a) L. Y. Fager and J. O. Alben, *Biochemistry*, **11**, 4786 (1972). (b) H. Van der Deen and H. Hoving, *Biophys. Chem.,* **9**, 169 (1979).
48. N. Kitajima, T. Katayama, K. Fujisawa, Y. Moro-oka, manuscript in preparation.

KINETICS AND MECHANISMS OF CU^I/O$_2$ REACTIONS

Andreas D. Zuberbühler

Department of Chemistry, Institute of Inorganic Chemistry, University of Basel, Spitalstrasse 51, CH-4056 Basel, Switzerland

INTRODUCTION

Copper proteins are found in each class of enzymes engaged in the interaction with dioxygen. Typical examples are laccase[1] as an electron transfer oxidase, tyrosinase[2] as an oxygenase and hemocyanin as an oxygen carrier[3]. Direct interaction of O$_2$ with the reduced enzyme is essential in every case. Unstable dioxygen adducts have long been postulated in enzymatic and low-molecular reaction schemes. They were, however, not really backed by direct experimental observation with the single exceptions of hemocyanin and later tyrosinase[4]. The situation has dramatically changed with the identification and characterization, first by spectroscopic methods, of pseudoreversible dioxygen adducts or peroxo complexes to low-molecular Cu^I compounds by Karlin and coworkers[5,6]. Many speculations have been put to an end upon the structural X-ray characterization of a trans-μ-peroxo bridged dicopper(II) complex[7] [Cu(L)]$_2$O$_2^{2+}$ (L = tris[(2-pyridyl)methyl]amine) and a μ-η^2:η^2 peroxo complex [Cu(L')]$_2$O$_2^8$ (L' = hydrodotris(3,5-diisopropyl-1-pyrazolyl)borate). Most recently, the η^2:η^2 binding mode also has been reported for oxyhemocyanin[9].

Reaction of low-molecular copper complexes with O$_2$ has drawn a large part of its interest from the involvement in catalytic oxidation and oxygenation of organic substrates. Here we wish to concentrate on the quasireversible interaction of copper with O$_2$ and specifically on kinetic and thermodynamic aspects of dioxygen binding to Cu^I complexes of various types. The subject is divided into four parts. We discuss the kinetics of Cu$^I_{aq}$ autoxidation with evidence for innersphere copper-oxygen interaction at room temperature[10]. We describe the kinetic and thermodynamic properties of binucleating copper complexes for which the binding of O$_2$ and the oxygenation of organic substrate can be followed spectroscopically and all relevant kinetic and thermodynamic parameters may be obtained[11,12]. Subsequent formation of 1:1 superoxo and 2:1 peroxo complexes is observed with Cu^I complexes of TMPA and some of its derivatives[13,14]. Finally, it has been shown that even Cu^I complexes with simple monodentate 1,2-dimethylimidazole may lead to a well-defined and reasonably stable peroxo species at low temperature[15]. With the analogous ligand 1-methyl-2-ethylimidazole (MEETIM) no less than 6 different absorbing intermediates have been detected by low-temperature spectrophotometry and preliminary kinetic data are presented.

From K.D. Karlin and Z. Tyeklár, Eds., *Bioinorganic Chemistry of Copper*
(Chapman & Hall, New York, 1993).

ROOM TEMPERATURE KINETIC EVIDENCE FOR DIOXYGEN ADDUCTS

The formation of an unstable dioxygen adduct is mandatory even for the autoxidation of the simple Cu^I aquo ion[16]. The system has been carefully restudied and completed by a series of measurements in the presence of a large excess of Cu^{II} [10]. A very complicated rate law has been observed. It has been analyzed in the form of concomitant one- and two-electron reduction steps of O$_2$. An unstable pseudo-reversible adduct CuO$^+_{2,aq}$ is the necessary intermediate for both types of reaction. The minimum set of mechanistic steps is given by eqns. (1)-(8). As shown, the CuO$^+_{2,aq}$ intermediate must be very versatile: it may decompose to Cu$^+_{aq}$ and O$_2$ (-1) or to Cu$^{2+}_{aq}$ and ˙O$^-_2$ (+3), react with a proton (+2) or with an additional Cu$^+_{aq}$ (+7). Many of the elementary steps indicated in (1)-(8) must be very rapid and reactions of ˙O$^-_2$ or HO$^˙_2$ with either Cu$^+_{aq}$ or Cu$^{2+}_{aq}$ have been studied independently.

$$Cu^+_{aq} + O_2 \quad \underset{k_{-1}}{\overset{k_{+1}}{\rightleftharpoons}} \quad CuO^+_2 \tag{1}$$

$$CuO^+_2 + H^+ \quad \underset{k_{-2}}{\overset{k_{+2}}{\rightleftharpoons}} \quad Cu^{2+}_{aq} + HO^˙_2 \tag{2}$$

$$CuO^+_2 \quad \underset{k_{-3}}{\overset{k_{+3}}{\rightleftharpoons}} \quad Cu^{2+}_{aq} + ˙O^-_2 \tag{3}$$

$$CuO^+_2 + OH^- \quad \overset{k_{+4}}{\rightarrow} \quad Cu^{2+} + ˙O^-_2 \; (+ \; OH^-) \tag{4}$$

$$Cu^I_{tot} + HO^˙_2 \quad \overset{k_{+5}, \, H^+}{\rightarrow} \quad Cu^{2+} + H_2O_2 \tag{5}$$

$$Cu^I_{tot} + ˙O^-_2 \quad \overset{k_{+6}, \, 2H^+}{\rightarrow} \quad Cu^{2+} + H_2O_2 \tag{6}$$

$$Cu^I_{tot} + CuO^+_2 \quad \overset{k_{+7}, \, 2H^+}{\rightarrow} \quad Cu^{2+} + H_2O_2 \tag{7}$$

$$˙O^-_2 + H^+ \quad \overset{K_8}{\rightleftharpoons} \quad HO^˙_2 \tag{8}$$

Direct formation of Cu^{2+} and superoxide in an outersphere reaction is excluded because this would imply either second order dependence on copper or make any acceleration by protons impossible, as neither Cu$^+_{aq}$ nor O$_2$ could function as a Brönsted base[17]. In fact, k_{+1} can be calculated from the experimental data, k_{+1} = 9.5·10^5 M^{-1}s^{-1} (I = 0.2 M, 20 °C)[16] or 6·10^6 M^{-1}s^{-1} (I = 0.5, 25 °C)[10] and this value has long been the only kinetic parameter directly related to quasireversible copper-dioxygen interaction.

Formation of unstable precursor dioxygen adducts has also been implied with a series of substituted imidazoles[18], vide infra. In addition, such species have long been postulated for Cu^I autoxidation in nonaqueous solvents[19]. We have most closely studied the autoxidation of Cu^I in dmso[20] which can be made the basis of an efficient and versatile catalytic cycle[21]. Here we simply mention the rate law of autoxidation, eqn. (9).

$$-d[O_2]/dt = k_{+9}[Cu^+]^2[O_2](1+k'_{+9}[H^+]) \tag{9}$$

Again, second order dependence on Cu^I as well as acceleration by protons can only be explained by the formation of one, more likely two, unstable dioxygen adducts, CuO$^+_2$ and Cu$_2$O$^{2+}_2$.

The direct relevance of Cu^I autoxidation in redox catalysis has recently been shown by establishing rate law and mechanism of the copper catalyzed oxidation of ascorbic acid in aqueous acetonitrile[22]. Under these Cu^I stabilizing conditions the rate law of ascorbate oxidation is qualitatively and quantitatively identical with that of Cu^I autoxidation in the absence of excess copper[16]. It is supplemented by one additional term describing the direct interaction of the primary dioxygen adduct CuO_2^+ with ascorbate and electron transfer within the ternary complex, eqn. (10). Obviously, eqn. (10) does not describe an elementary step and intermediacy of radicals and/or Cu^{2+} cannot be ruled out at present.

$$CuO_2^+ + AscH^- \xrightarrow{k_{,10},\ 2H^+} Cu^+ + H_2O_2 + Dha \qquad (10)$$

These earlier kinetics results can be summarized as follows: (i) In several systems, experimental data can only be explained assuming the formation of unstable pseudoreversible dioxygen adducts. Other systems are ambiguous, but there is no reason to postulate that simple outersphere electron transfer would be preferred in those cases. (ii) Competition between one- and two-electron paths may be the rule rather than the exception. (iii) In no case, a dioxygen adduct has been directly observed at room temperature and therefore, the actual properties of such species have remained unknown[23].

SPECTROSCOPICALLY CHARACTERIZED DIOXYGEN ADDUCTS

Binucleating Multidentate Ligands
Dioxygen adducts or peroxo complexes can be spectroscopically characterized and their properties studied in derivatives of the binuclear copper species 1. This work has been spurred by the low-temperature characterization of quasi-reversible O_2 binding to Cu^I complexes by Karlin[5]. All systems to be discussed henceforth have been studied in cooperation with Karlin and his coworkers who did the syntheses of the respective compounds and the primary spectroscopic investigation of O_2 binding. We are restricting ourselves to the kinetics aspects of some selected reactions.

In close analogy with tyrosinase, hydroxylation of 1 can be equally achieved starting from the cupric complex and hydrogen peroxide or from Cu^I and O_2. The H_2O_2 based reaction has been studied in 50% H_2O/dmf[24], the rate law is given by eqn. (11).

$$-d[H_2O_2]/dt = k_{+11}[Cu_2(L-H)^{4+}]^2[H_2O_2]/([H^+](1+k'_{+11}[Cu_2(L-H)^{4+}])) \qquad (11)$$

$$k_{+11} = 0.4\ M^{-1}s^{-1}\ ;\ k'_{+11} = 4500\ M^{-1}$$

term describing the direct interaction...

The second order dependence on $[Cu_2(L-H)^{4+}]$ implies a transition state containing a total of four copper ions. This is somewhat unexpected since dinuclear copper centres are sufficient to form active peroxides in enzymes like tyrosinase and in low molecular model systems. The pH dependence and the complete absence of an isotope effect for the arene proton to be replaced in hydroxylation[24] are indicating that H_2O_2 most likely binds in a 1,1-coordinated form as HO_2^-. We assume that the second molecule of complex is needed as a Lewis acid, increasing the nucleofugicity of the terminal OH group. The remaining oxene-like oxygen would be a highly reactive electrophile in line with analogous suggestions for tyrosinase and with the absence of an isotope effect in the actual hydroxylation step.

Subsequently, we are only discussing reactions starting from Cu^I complexes and O_2. All these systems were studied with a diode array stopped-flow spectrometer (512 diodes) between -100 °C and room temperature and a minimum repeat time of 8 ms. In single experiments 50 kbytes of absorbance readings are collected this way and data reduction by the method for factor analysis was indispensable. Least-squares refinement thus was achieved using a computer program which has been derived from SPECFIT[25] by replacing the routine for equilibrium models by the appropriate kinetics expression.

The reaction of **1** with O$_2$ in dichloro methane can be described by the simple mechanistic scheme (12)-(13) for all temperatures between -100 and +25°C[11]. No additional intermediates, such as a complex in which O$_2$ would be bound to only a

$$Cu_2(L-H)^{2+} + O_2 \underset{k_{-12}}{\overset{k_{+12}}{\rightleftarrows}} Cu_2(L-H)O_2^{2+} \qquad (12)$$
$$\quad \mathbf{1} \qquad\qquad\qquad\qquad\qquad \mathbf{2}$$
$$Cu_2(L-H)O_2^{2+} \overset{k_{+13}}{\rightarrow} Cu_2(L-O^-)OH^{2+} \qquad (13)$$
$$\quad \mathbf{2} \qquad\qquad\qquad\qquad \mathbf{3}$$

single copper ion, are observed. Below -50°C, formation of **2** is complete and k_{-12} as well as k_{+13} are irrelevant. From the temperature dependence, activation parameters for all three rate constants have been obtained. They are collected in Table 1, together with the analogous parameters for related complexes with a series of ligands R-Xyl-H **4** which are substituted in para position to the entering hydroxyl group[12].

Table 1. Kinetics parameters for O$_2$ interaction with R-Xyl-H Cu^I complexes.

R		NO$_2$	H	C(CH$_3$)$_3$	F
k_{+12} (M^{-1}s^{-1})	183 K	110	410	470	7.2
	223 K	280	1300	1700	270
	$\Delta H^{\ne}$ (kJmol^{-1})	6.4 ± 0.1	8.2 ± 0.1	9.1 ± 0.3	29 ± 1
	$\Delta S^{\ne}$ (JK^{-1}mol^{-1})	-167 ± 1	-146 ± 1	-140 ± 1	-66 ± 1
k_{-12} (s^{-1})	183 K	2.1·10^{-5}	1.6·10^{-5}	4.3·10^{-6}	1.5·10^{-6}
	223 K	0.027	0.076	0.094	0.025
	$\Delta H^{\ne}$ (kJmol^{-1})	59 ± 1	70 ± 1	83 ± 4	81 ± 3
	$\Delta S^{\ne}$ (JK^{-1}mol^{-1})	-8 ± 4	50 ± 6	110 ± 20	90 ± 10
k_{+13} (s^{-1})	183 K	1.6·10^{-5}	3.0·10^{-4}	6.3·10^{-3}	1.5·10^{-3}
	223 K	0.013	0.13	0.96	0.18
	$\Delta H^{\ne}$ (kJmol^{-1})	55 ± 1	50 ± 1	41 ± 2	39 ± 1
	$\Delta S^{\ne}$ (JK^{-1}mol^{-1})	-32 ± 2	-35 ± 2	-59 ± 8	-82 ± 6

The somewhat naive expectation behind this series of experiments was that all compounds might form a dioxygen adduct with similar kinetic and thermodynamic properties because of the identical copper binding sites of all ligands. On the other hand, oxygenation of the aromatic ring would be strongly influenced by the electronic effects of the substituents R in the para position. Both expectations were only partially fulfilled.

Results for R = -CN and R = -OCH$_3$ are missing in Table 1: First, the cyano

$$R = -NO_2$$
$$= -CN$$
$$= -C(CH_3)_3$$
$$= -F$$
$$= -OCH_3$$

4

compound could not be reasonably studied because of very strong photochemical interference. The electron-rich methoxy compound, for which we expected the fastest hydroxylation within this series, neither gives a dioxygen adduct nor oxygenation of the benzene ring at low temperature. This puzzling behavior could be qualitatively explained by NMR: at low temperature the copper binding arms do not freely rotate in solution, but get locked in positions which are obviously unsuitable for O_2 binding. This effect is by far most pronounced with the methoxy derivative and completely absent e.g. for the nitro compound[12]. Most likely, such hindered rotation means direct interaction of complexed copper with the benzene ring and thus critically depends on its electron density.

The results for the other four derivatives give a quite consistent, if partially unexpected picture. First, we have a remarkable tendency towards compensation for all three elementary steps: Higher enthalpies of activation are coupled with more favorable activation entropies and vice versa. Consequently, rate constants at the intermediate temperature of 223 K are relatively similar for the four compounds in each case. Activation enthalpies are generally below 10 kJmol[-1] for the formation of the peroxo complex (k_{+12}); the higher value for the fluoro compound (29 kJmol[-1]) again can be related to the hindered rotation mentioned for R = -OCH$_3$. Negative activation entropies are consistent with a more highly ordered transition state for partial oxygen binding. Trends for k_{-12} (decay into $Cu_2(R\text{-Xyl-H})^{2+}$ and O_2) and k_{+13} (hydroxylation of the aromatic ring) are weak and in opposite direction for the four ligands. The direction of the trends is as expected on the basis of simple electronic arguments. Electron rich groups with a +I or a +M effect are increasing the strength of the copper dioxygen bonds, i.e. increase $\Delta H^{\ddagger}$ for k_{-12}, but are facilitating oxygenation of the arene at the same time. Somewhat surprising, however, is the relative size of these opposed effects. The place of substitution is relatively far from the copper ions binding O_2, but must strongly influence the electronic properties of the para position to be hydroxylated. Nevertheless, the effect on the arene hydroxylation (k_{+13}) seems to be equal to or even less than the effect on quasireversible deoxygenation (k_{-12}). We feel that this behavior can only be explained by the already postulated strongly electrophilic nature of the actual hydroxylating agent, the reactivity of which depends only weakly on the properties of the substrates. Again, no isotope effect is observed when the position to be hydroxylated is marked by deuterium.

From the elementary reactions k_{+12} and k_{-12}, equilibrium parameters may be calculated for O_2 binding to the $Cu_2(R\text{-Xyl-H})^{2+}$ complexes. They are collected in Table 2, together with results for the phenoxo ligand (H-Xyl-O$^-$), the hydroxylation product of H-Xyl-H.

All systems are characterized by negative standard reaction enthalpies as well as entropies, with a marked tendency for compensation again. Binding strength correlates directly with inductive effects of the substituents, but due to the compensating entropy terms equilibrium constants are rather similar. Their relative order in fact depends on the temperature considered, as is the case for some of the kinetics parameters. This is a typical example for cautioning against conclusions concerning relative stabilities or reactivities based on data obtained at a single temperature only. Even the phenoxo compound $Cu_2(H\text{-Xyl-O}^-)^+$ fits well into the picture of the other xylidene derivatives, despite the fact that the forward reaction

Table 2. Equilibrium parameters for O$_2$ interaction with R-Xyl-H and H-Xyl-O$^-$ Cu^I complexes.

R		NO$_2$	H	C(CH$_3$)$_3$	F	H-Xyl-O$^-$
	T=183 K	$6.7 \cdot 10^6$	$2.9 \cdot 10^7$	$1.2 \cdot 10^8$	$5.0 \cdot 10^6$	$6.5 \cdot 10^8$
K$_2$ (M^{-1})	T=223 K	$1.3 \cdot 10^4$	$1.9 \cdot 10^4$	$1.9 \cdot 10^4$	$1.1 \cdot 10^4$	$2.7 \cdot 10^5$
(= k$_{+12}$/k$_{-12}$)	$\Delta_r H^{\ominus}$ a)	-53 ± 1	-62 ± 1	-74 ± 4	-52 ± 3	-66 ± 1
	$\Delta_r S^{\ominus}$ b)	-159 ± 4	-196 ± 6	-250 ± 20	-156 ± 10	-192 ± 2

a) (kJmol^{-1}); b) (JK^{-1}mol^{-1})

k$_{+12}$ is faster by more than 3 orders of magnitude[11]. This means that the additional hydroxo group does not significantly influence the oxygen binding strength and that the phenoxo compound must be in a state of analogous order before oxygen binding, the only difference being the necessary amount of reorganisation (i.e. the height of the activation barrier).

Mononuclear Cu^I Complexes Based on Multidentate Ligands
Copper dioxygen interaction can be studied in equal detail using the mononuclear tetradentate ligands TMPA (**5a**), BPQA (**5b**) and BQPA (**5c**)[12-14].

[Cu(TMPA)]$_2$O$_2^{2+}$ in fact represents the first peroxo copper complex to be fully characterized by X-ray analysis[7], making a kinetic study especially attractive. Since there are no reasonable sites for hydroxylation with this set of ligands, the dioxygen adducts are relatively stable with respect to irreversible decay and we can concentrate on the quasireversible dioxygen binding.

As we have mononuclear starting complexes we have to expect kinetics which are different from those discussed in the previous section. All three ligands are forming mononuclear complexes Cu(L)$^+$ which interact quasireversibly with O$_2$. They behave rather differently in detail, however. The bis-quinolyl derivative BQPA preferentially forms a 1:1 adduct Cu(BQPA)O$_2^+$ and only small amounts of μ-peroxo complex. The reaction with BPQA can be described by a simple one-step equilibrium $2Cu(BPQA)^+ + O_2 \rightleftharpoons [Cu(BPQA)]_2O_2^{2+}$ at all temperatures, no monomeric intermediate is detected under any conditions. With TMPA finally, both the 1:1 and 2:1 complexes are formed in sequence and all kinetic as well as thermodynamic parameters for that system have been obtained[13]. This is the first case for which the stepwise binding of O$_2$ to one and subsequently two copper moieties could be directly studied and characterized spectroscopically as well as kinetically.

The overall reaction for the complexes with TMPA, BPQA and BQPA is given by eqns. (14)-(16).

The three peroxo complexes are rather similar in their spectral properties

269

$$Cu(L)^+ + O_2 \quad \overset{k_{+14}}{\underset{k_{-14}}{\rightleftarrows}} \quad Cu(L)O_2^+ \tag{14}$$

$$Cu(L)^+ + Cu(L)O_2^+ \quad \overset{k_{+15}}{\underset{k_{-15}}{\rightleftarrows}} \quad [Cu(L)]_2O_2^{2+} \tag{15}$$

$$[Cu(L)]_2O_2^{2+} \quad \overset{k_{+16}}{\rightarrow} \quad \text{irreversible decay} \tag{16}$$

(TMPA: λ_{max} = 525 nm, ε = $1.5 \cdot 10^4$ $M^{-1}cm^{-1}$; BPQA: λ_{max} = 534 nm, ε = $0.86 \cdot 10^4$ $M^{-1}cm^{-1}$; BQPA: λ_{max} = 545 nm, ε = $0.55 \cdot 10^4$ $M^{-1}cm^{-1}$), each with a prominent shoulder around 600 nm. We thus have a slight bathochromic shift upon increasing substitution accompanied by a decrease in molar absorptivity. Undoubtedly all these species have the same trans-μ-peroxo structure established by X-ray for TMPA[7]. Also, the two spectra of the 1:1 adducts are closely related with a prominent charge transfer band near 400 nm (TMPA: λ_{max} = 410 nm, ε = $4 \cdot 10^3$ $M^{-1}cm^{-1}$; BQPA: λ_{max} = 388 nm, ε = $8 \cdot 10^3$ $M^{-1}cm^{-1}$).

With respect to kinetic and thermodynamic parameters of dioxygen interaction the three systems also show many similarities even though higher substitution and consequent steric hindrance have some specific effects. The relevant activation enthalpies and entropies are collected in Table 3, together with some rate constants calculated for 183 and 223 K. As shown in Table 3, activation enthalpies for k_{+14}, k_{-14} and k_{-15} are identical within experimental error for the systems that can be compared. Specifically, the enthalpic parts of both formation and dissociation of the 1:1 adducts with TMPA and BQPA are the same. The considerably slower rates observed with the latter are exclusively due to less favorable entropy terms for BQPA. Also, activation enthalpies of dissociation all are around 60-65 kJmol^{-1}, independent of the ligand and whether relating to 1:1 (k_{-14}) or 2:1 (k_{-15}) species i.e. to peroxo or to superoxo complexes.

Unfortunately, rate constants for the second step with BQPA could not be obtained. Despite the relative instability of $[Cu(BQPA)]_2O_2^{2+}$ this equilibrium is always established much faster than the formation of the 1:1 adduct and only equilibrium data were obtained for the μ-peroxo complex. As shown at the bottom of Table 3, the overall forward reaction to the 2:1 adducts (k_{ter}) is described by a negative activation enthalpy and a highly negative activation entropy. Consequently the rate of adduct formation decreases with increasing temperature. The negative $\Delta H^{\ddagger}$ for k_{ter} of course is due to the preequilibrium constant k_{+14}/k_{-14} which strongly decreases with increasing temperature.

The rate constants collected in Table 3 are reflecting the trends of the activation parameters just discussed. We observe that a) for TMPA and BQPA the rate constants k_{+14} and k_{-14} have a similar temperature dependence, the latter being slower by several orders of magnitude, b) dissociation always is more favored by high temperature than association, c) the rate of overall formation decreases with increasing temperature. All equilibria are shifted to the left with increasing temperature, as more directly is shown by the thermodynamic parameters which are derived from the kinetics and are collected in Table 4.

1:1 binding of dioxygen is practically identical for TMPA and BQPA, not only with respect to the reaction enthalpies but also for reaction entropies and thus the actual equilibrium constants. For BPQA, the superoxo complex $Cu(BPQA)O_2^+$ has not been detected at all. Most likely this is not due to an intrinsic instability of that species but rather to the very rapid further complexation to the peroxo compound $[Cu(BPQA)]_2O_2^{2+}$. This is in line with the extremely fast reaction observed also for BQPA. With that ligand the 1:1 complex can only be observed because of the relative instability of the μ-peroxo compound which is only seen as an 'intermediate', while $Cu(BQPA)O_2^{2+}$ is the ultimate thermodynamic sink. Despite the lacking experimental evidence we assume that the three superoxo complexes have roughly the same formation constants for TMPA, BPQA and BQPA.

Table 3. Kinetics parameters for O$_2$ interaction with TMPA, BPQA and BQPA Cu^I complexes

		TMPA	BPQA	BQPA
k_{+14} (M^{-1}s^{-1})	183 K	$2 \cdot 10^4$	-	18
	223 K	$9 \cdot 10^5$	-	$7 \cdot 10^2$
	$\Delta H^{\ddagger}$ (kJmol^{-1})	32 ± 4	-	30 ± 2
	$\Delta S^{\ddagger}$ (JK^{-1}mol^{-1})	14 ± 18	-	-53 ± 8
k_{-14} (s^{-1})	183 K	13	-	$6 \cdot 10^{-3}$
	223 K	$3 \cdot 10^4$	-	16
	$\Delta H^{\ddagger}$ (kJmol^{-1})	66 ± 4	-	65 ± 4
	$\Delta S^{\ddagger}$ (JK^{-1}mol^{-1})	137 ± 18	-	72 ± 19
k_{+15} (M^{-1}s^{-1})	183 K	$4 \cdot 10^4$	-	-
	223 K	$2.5 \cdot 10^5$	-	-
	$\Delta H^{\ddagger}$ (kJmol^{-1})	14 ± 1	-	-
	$\Delta S^{\ddagger}$ (JK^{-1}mol^{-1})	-78 ± 2	-	-
k_{-15} (s^{-1})	183 K	$1.9 \cdot 10^{-4}$	$1.9 \cdot 10^{-4}$	-
	223 K	0.3	0.3	-
	$\Delta H^{\ddagger}$ (kJmol^{-1})	61 ± 3	61 ± 1	-
	$\Delta S^{\ddagger}$ (JK^{-1}mol^{-1})	19 ± 10	21 ± 4	-
k_{ter} (M^{-2}s^{-1})	183 K	$6 \cdot 10^7$	$3 \cdot 10^6$	-
	223 K	$7 \cdot 10^6$	$1.5 \cdot 10^6$	-
	$\Delta H^{\ddagger}$ (kJmol^{-1})	-20 ± 2	-8 ± 1	-
	$\Delta S^{\ddagger}$ (JK^{-1}mol^{-1})	-201 ± 5	-160 ± 2	-

$k_{ter} = k_{+14}k_{+15}/k_{-14}$

Formation of the 2:1 complex on the other hand is significantly suppressed by steric hindrance. The binding enthalpy is continuously decreased from TMPA to BQPA and only partially compensated by less negative entropy terms. It is easily extrapolated that none of the complexes discussed here would have significant stability at room temperature.

If we compare these results with those for the xylidene derivatives discussed above, we see that all equilibria studied so far are driven by enthalpy and destabilized by highly unfavorable entropy terms. It seems questionable if room temperature stable copper peroxo complexes can be formed at all without positively influencing their entropy part.

Table 4. Equilibrium parameters for O_2 interaction with TMPA, BPQA and BQPA.

		TMPA	BPQA	BQPA
K_1	183 K	$1.5 \cdot 10^3$	-	$6 \cdot 10^3$
	223 K	27	-	80
	$\Delta_r H^\ominus$ (kJmol^{-1})	-34 ± 1	-	-36 ± 6
	$\Delta_r S^\ominus$ (JK^{-1}mol^{-1})	-123 ± 4	-	-125 ± 27
K_2	183 K	$2 \cdot 10^8$	-	2000^*
	223 K	$8.5 \cdot 10^5$	-	2000^*
	$\Delta_r H^\ominus$ (kJmol^{-1})	-47 ± 3	-	0^*
	$\Delta_r S^\ominus$ (JK^{-1}mol^{-1})	-97 ± 10	-	40^*
β_2	183 K	$3 \cdot 10^{11}$	$9 \cdot 10^9$	$2 \cdot 10^{7*}$
	223 K	$2 \cdot 10^7$	$3 \cdot 10^6$	$2 \cdot 10^{5*}$
	$\Delta_r H^\ominus$ (kJmol^{-1})	-81 ± 3	-69 ± 2	-40^*
	$\Delta_r S^\ominus$ (JK^{-1}mol^{-1})	-220 ± 11	-181 ± 5	-80^*

* relatively uncertain values; $\beta_2 = K_1 K_2$

Copper Peroxo Complexes Based on Monodentate Imidazole Derivatives
The interaction of dioxygen with copper complexes of substituted imidazoles, L, has been studied quite some time ago, at room temperature in aqueous solution[18]. A relatively complicated rate law, eqn. (17), was found, suggesting two different paths of O_2 reduction. The first one, which is not influenced by Cu^{II}, represents two-electron reduction directly to H_2O_2. The second step describes one-electron reduction through superoxide. Both paths need the formation of an unstable $CuL_2O_2^+$ intermediate, for reasons analogous to those discussed for Cu_{aq}^+. The minimum set of mechanistic steps is given by eqns. (18-23).
    The mechanism implies the formation of unstable monomeric superoxo complexes which may either decompose into free superoxide or react with a second

$$-d[O_2]/dt = [CuL_2^+]^2[O_2] \; \frac{k_a}{1+k_b[CuL_2^+]} + \frac{k_c[L]+k_d}{[CuL_2^+]+k_e[Cu_{tot}^{II}]} \tag{17}$$

metal ion under direct formation of H_2O_2. While the absolute minimum of steps consistent with the experimental rate law is given by eqns. (18)-(23), we may assume that in fact eqns. (21) and (23) are also describing inner-sphere reactions with dimeric peroxo, $[Cu(L)]_2O_2^{2+}$, and monomeric superoxo, $CuL_3O_2^+$, complexes as additional intermediates. Needless to say that none of the intermediates shown had been observed spectroscopically or by any other direct means. It was concluded[18] that competition between one- and two-electron reduction of O_2 may be the rule rather than the exception for simple copper complexes at room temperature.
    By switching from water to dichloromethane and from room temperature to

$$CuL_2^+ + O_2 \quad \underset{k_{-18}}{\overset{k_{+18}}{\rightleftharpoons}} \quad CuL_2O_2^+ \tag{18}$$

$$CuL_2O_2^+ \quad \underset{k_{-19}}{\overset{k_{+19}}{\rightleftharpoons}} \quad Cu^{II} + {}^{\bullet}O_2^- \tag{19}$$

$$CuL_2^+ + {}^{\bullet}O_2^- \quad \overset{k_{+20}, \, 2H^+}{\rightarrow} \quad Cu^{II} + H_2O_2 \tag{20}$$

$$CuL_2O_2^+ + CuL_2^+ \quad \overset{k_{+21}, \, 2H^+}{\rightarrow} \quad 2\,Cu^{II} + H_2O_2 \tag{21}$$

$$CuL_2^+ + L \quad \underset{k_{-22}}{\overset{k_{+22}}{\rightleftharpoons}} \quad CuL_3^+ \tag{22}$$

$$CuL_3^+ + O_2 \quad \overset{k_{+23}}{\rightarrow} \quad Cu^{II} + {}^{\bullet}O_2^- \tag{23}$$

183 K, the postulated intermediates could be observed only too well. The first report in this direction has been made for 1,2-dimethylimidazole[15]. An intensely colored binuclear species [Cu(L)$_3$]$_2$O$_2^{2+}$ has been postulated to possess a trans-μ-peroxo structure analogous to the one discussed above for [Cu(TMPA)]$_2$O$_2^{2+}$. The number of species and their stabilities increase considerably with increasing substitution of the imidazole ring in the 2-position, i.e. by going from 1-methylimidazole to 1,2-dimethylimidazole and 1-methyl-2-ethylimidazole. We are concentrating here on the latter ligand for which no less than 6 colored intermediates are found by low temperature spectrophotometry[26]. It has not been possible so far to derive complete sets of activation parameters for six, or, taking back reactions into account, twelve rate constants. Obviously, kinetics may get terribly complicated if simple rather than complicated ligands are used.

Here we discuss some first results, mainly based on data obtained for 15 mM copper and various ligand to metal ratios at -90 °C. Using a ligand to copper ratio of 3:1, a charge transfer spectrum develops with λ_{max} = 495 nm and a second peak at 650 nm within about 10 minutes. This spectrum closely resembles the absorbance first observed for 1,2-dimethylimidazole[26] and is assumed again to be a dimeric μ-peroxo complex [Cu(L)$_3$]$_2$O$_2^{2+}$. The final spectrum changes dramatically if the ligand to metal ratio is reduced to about 2.5. With roughly the same rate of formation we now have an intensely blue species with a maximum at 610 nm and unknown stoichiometry. With intermediate ligand ratios a mixture of the two species is formed strictly in parallel with a single relaxation. In each case some lower-intensity spectral features related to initial stages of the reaction are observed. The formation of the final high intensity products can be described by a kind of clock reaction, the time for onset being easily changed between 20 and several hundred seconds by the ligand concentration. Ligand to copper ratios of four and higher completely block the formation of final products at -90 °C. On the other hand, no reaction at all is observed for [L]:[Cu^I] $\leq$ 2, which means that CuL$_2^+$ is inert towards O$_2$ in a non-coordinating solvent.

At least three additional relaxations can be distinguished preceding the high intensity final products described above. The total of our observations has been put into a general rectivity scheme. As none of the species has been isolated, all of the stoichiometries displayed on the scheme should be considered tentative. However, those given in bold are backed by relatively strong evidence, while those in italics are not more than educated guesses.

The reactive cuprous species is CuL$_3^+$, as evidenced by the elemental analysis of the corresponding complex with 1,2-dimethylimidazole[15] and by the absence of a ligand dependence in the first reaction step for [L]:[Cu^I] $\geq$ 3 [26]. We need also unreactive CuL$_2^+$ in a rapid equilibrium as a source of free ligand which is needed for

Primary Dimers
$$[Cu(L)_4]_2O_2^{2+}, [Cu(L)_x]_2O_2^{2+}$$

$\uparrow\downarrow$

Cuprous complexes $\xrightarrow{O_2}$ Primary monomer(s) $\longrightarrow$ Non-CT Cupric complex

CuL_2^+, CuL_3^+ $CuL_3O_2^+, CuL_4O_2^+$ $CuL_4^{2+} + {}^\cdot O_2^-$

$CuL_3^+ \;\downarrow\uparrow\; O_2$

Final dimer(s)
$$[Cu(L)_3]_2O_2^{2+}, [Cu(L)_2]_2O_2^{2+}$$

General Reactivity Scheme for Cuprous Complexes with 1-methyl-2-ethylimidazole

the formation of CuL_4^{2+}. Both CuL_2^+ and CuL_3^+ are nonabsorbing at $\lambda > 350$ nm.

Starting from the cuprous complexes, an unstable primary monomer is formed first in a reaction noticeable only by a rapid increase at the lowest wavelengths (360-380 nm), with no charge-transfer band in the visible. It is formed to an estimated extent of 5-20%. The primary monomer(s) are formed with a rate constant k_{obs} of about 2 s^{-1} (30 mM Cu^I, -90 °C). There is circumstantial kinetic evidence presently for the additional formation of $CuL_4O_2^+$ besides the main species $CuL_3O_2^+$.

The primary dimers are unequivocally a mixture of two species, again in rapid ligand dependent equilibrium. The observed rate constant is about 0.2 s^{-1}, i.e. one tenth of the primary monomer, little dependent on the ligand concentration. The dimeric nature and incomplete formation are evidenced by a 4-fold increase in maximum amount and more than four-fold increase in initial rate by doubling the copper concentration. Four ligands are needed for complete formation of the main species which is designated $[Cu(L)_4]_2O_2^{2+}$ and absorbs at 425 and 550 nm. The additional species has a characteristic absorption at 650 nm, but is never obtained in pure state. Again, the formation of the primary dimers is estimated to be 5-20%, and definitely less than 50%.

At all ligand ratios, the primary monomer is transformed into a compound with a relatively weak absorption at 600 nm and no charge transfer band in the visible range. At -90 °C to -70°C this is the thermodynamic sink of the reaction and the final species for ligand to metal ratios above four. We have a single complex for all experimental conditions and it seems to be essentially fully formed. Somewhat surprisingly, the rate of its formation follows a 1/[L] inhibition by ligand in excess of a 3:1 [L]:[Cu^I] ratio, supporting our notation that $CuL_3O_2^+$ rather than $CuL_4O_2^+$ is the reactive primary monomer. Superoxide may be associated as an ion pair.

The primary monomer, which is the central relay station of the reactivity scheme, can also be transformed into the final dimers discussed first. However, these species are immediately destroyed under most experimental conditions, i.e. as long as dioxygen and free ligand are available in significant amounts. The pseudo-first order rate constant is around 0.005 s^{-1} for both final dimers. At ratios of 3:1 and slightly above $[Cu(L)_3]_2O_2^{2+}$ is fully formed, at less than 2.5 only the intensely blue $\lambda_{max} = 610$ nm species is seen. The equilibrium is significantly temperature-dependent, at higher temperatures the 610 nm species becomes more prominent.

To summarize, no less than six colored species have been detected at low temperature for the dioxygen interaction of Cu^I complexes with 1-methyl-2-ethylimidazole, four of them in pairs of rapid ligand-dependent equilibria. It seems that most if not all reactions involving the coordination of imidazole are fast relative to any of the redox reactions. There seem to be two different thermodynamic sinks in this system, a monomeric more or less classical cupric complex at high ligand to metal ratios and dimeric copper peroxo complexes at lower ones. Obviously, this system has a complexity that can be hardly compared with those discussed in the previous sections of this chapter. We still try to quantitatively analyze at least part of the kinetic steps, but perhaps we shall have to be content with partial solutions.

CONCLUSIONS AND COMPARISON WITH HEMOCYANIN

Spectroscopic and later X-ray characterization[5-8] of copper peroxo complexes has opened a new era in kinetic and thermodynamic studies of copper-dioxygen interaction. Quasireversible formation of mononuclear superoxo, $Cu(L)^+ + O_2 \rightleftharpoons Cu(L)O_2^+$, or binuclear peroxo complexes, $2Cu(L)^+ + O_2 \rightleftharpoons [Cu(L)]_2O_2^{2+}$, has long been postulated. In some cases, such species have been backed by convincing kinetic evidence, but until recently they could never be observed directly, with the exception of the copper proteins tyrosinase and hemocyanin. Formation of the 1:1 adduct of the simple aquo ion, $CuO_{2,aq}^+$, has been the only known low molecular reaction that could be described by an individual rate constant, $k = 10^6 \ M^{-1}s^{-1}$.

With quite a number of dioxygen adducts now well characterized, kinetics and mechanism of such interactions can be studied, and activation parameters for all relevant elementary steps may be obtained. Quasireversible interaction of O_2 with cuprous complexes seems to be a rather general phenomenon. Multidentate mono- and binucleating ligands such as first developed by Karlin[5-7] have made their detailed study possible at low temperature in organic solvents. All equilibria studied so far are driven by enthalpy and destabilized at room temperature by highly unfavorable entropy terms. While different in detail, binding enthalpies for O_2 are around -60 kJmol^{-1} in peroxo complexes and activation enthalpies for the formation are low or even negative.

The results with substituted imidazoles show that complicated chelating ligands are not needed to accomplish pseudo-reversible copper dioxygen interaction. Self-assembly of the corresponding complexes may be sufficient and quite effective. On the other hand the numerous rapid equilibria involved with simple monodentate ligands may render a quantitative kinetic description of such systems rather difficult.

O_2-binding to the low-molecular systems may be compared with the biological oxygen carrier, hemocyanin. Reported binding enthalpies for hemocyanin are in the range of -46 to +13 kJmol^{-1} [27], thus binding may be easily as strong or stronger than model compounds. Also, enzymatic reaction rates above $10^6 \ M^{-1}s^{-1}$ are obtained even at very low temperature at least for $Cu_2(H\text{-}Xyl\text{-}O^-)^+$ [11]. Nevertheless, we still don't have a really good mimic of hemocyanin with sufficient room temperature stability. This is exclusively due to the strongly unfavorable entropy term of the equilibrium constants of all model compounds studied so far. How to overcome this unfavorable entropy contribution without the cooperative effect of a protein remains an open question at present.

REFERENCES

1. M.D. Allendorf, D.J. Spira, E.I. Solomon, *Proc. Natl. Acad. Sci. USA*, **82**, 3063 (1985).
2. H.S. Mason, W.L. Folks, E. Peterson, *J. Am. Chem. Soc.*, **77**, 2914 (1955).
3. T.B. Freedman, J.S. Loehr, T.M. Loehr, *J. Am. Chem. Soc.*, **98**, 2809 (1976).
4. R.L. Jolley Jr., L.H. Evans, H.S. Mason, *Biochem. Biophys. Res. Commun.*, **46**, 878 (1972).
5. K.D. Karlin and Y. Gultneh, *Prog. Inorg. Chem.*, **35**, 219 (1987), and refs. therein.
6. K.D. Karlin, Y. Gultneh, J.C. Hayes, R.W. Cruse, J. McKnown, J.P. Hutchinson, J. Zubieta, *J. Am. Chem. Soc.*, **106**, 2121 (1984).
7. R.R. Jacobson, Z. Tyeklar, A. Farooq, K.D. Karlin, S. Liu, J. Zubieta, *J. Am. Chem. Soc.*, **110**, 3690 (1988).
8. N. Kitajima, K. Fujisawa, Y. Moro-oka, K. Toriumi, *J. Am. Chem. Soc.*, **111**, 8975 (1989).
9. K. Magnus, H. Ton-That, *J. Inorg. Biochem.*, **47**, 20 (1992).

10. L. Mi and A.D. Zuberbühler, *Helv. Chim. Acta*, **74**, 1679 (1991).
11. R.W. Cruse, S. Kaderli, K.D. Karlin, A.D. Zuberbühler, *J. Am. Chem. Soc.*, **110**, 6882 (1988).
12. K.D. Karlin and A.D. Zuberbühler, to be published.
13. K.D. Karlin, N. Wei, B. Jung, S. Kaderli, A.D. Zuberbühler, *J. Am. Chem. Soc.*, **113**, 5868 (1991).
14. A.D. Zuberbühler, in *Dioxygen Activation and Homogeneous Catalytic Oxidation*, L.I. Simandi, Ed., (Elsevier Science Publishers, Amsterdam, 1991), pp. 249-257.
15. I. Sanyal, R.W. Strange, N.J. Blackburn, K.D. Karlin, *J. Am. Chem. Soc.*, **113**, 4692 (1991).
16. A.D. Zuberbühler, *Helv. Chim. Acta*, **53**, 478 (1970).
17. A.D. Zuberbühler, in *Metal Ions in Biological Systems Vol. 5*, H. Sigel, Ed., (Marcel Dekker, New York, 1976), pp. 325-368.
18. M. Güntensperger and A.D. Zuberbühler, *Helv. Chim. Acta*, **60**, 2584 (1977).
19. P.M. Henry, *Inorg. Chem.*, **5**, 688 (1966).
20. G. Rainoni and A.D. Zuberbühler, *Chimia*, **28**, 67 (1974).
21. H. Gampp and A.D. Zuberbühler, *Chimia*, **32**, 54 (1987).
22. L. Mi, A.D. Zuberbühler, *Helv. Chim. Acta*, **75**, 1547 (1992).
23. A.D. Zuberbühler, in *Copper Coordination Chemistry: Biochemical & Inorganic Perspectives*, K.D. Karlin and J. Zubieta, Eds., (Adenine Press, Guilderland, 1983), pp. 237-258.
24. R.W. Cruse, S. Kaderli, C.J. Meyer, A.D. Zuberbühler, K.D. Karlin, *J. Am. Chem. Soc.*, **110**, 5020 (1988).
25. H. Gampp, M. Maeder, C.J. Meyer, A.D. Zuberbühler, *Talanta*, **32**, 257 (1985).
26. R. Bilewicz, S. Kaderli, K.D. Karlin, M. Maeder, N. Wei, A.D. Zuberbühler, to be published.
27. E. Antonini, M. Brunori, H.A. Kuiper, L. Zolla, *Biophys. Chem.*, **18**, 117 (1983); Z. Er-el, N. Shaklai, E. Daniel, *J. Mol. Biol.*, **64**, 341 (1972).

FUNCTIONAL MODELS FOR HEMOCYANIN AND COPPER MONOOXYGENASES

Zoltán Tyeklár[‡] and Kenneth D. Karlin

Department of Chemistry, The Johns Hopkins University, Charles & 34th Streets, Baltimore, Maryland 21218 USA

INTRODUCTION

The active structures and reactivity of copper protein active sites have been of longtime interest to inorganic chemists, because of the novel coordination sites and their associated physical and spectroscopic features.[1-6] These include (i) the unusually intense blue color and high, positive reduction potentials observed in the 'blue' copper electron transfer proteins or multicopper oxidases,[1,2,7] (ii) the presence of a magnetically coupled dinuclear center in the O_2-carrier hemocyanin and monooxygenase tyrosinase,[1-4,8] (iii) the trinuclear copper centers in the oxidases laccase[1,9] and ascorbate oxidase[1,9,10] or the recently suggested Cu cluster in particulate methane monooxyenase[11] and (iv) the dinuclear porphyrin-iron and copper center in cytochrome c oxidase[12] which effects the O_2-binding and reduction to water. Thus, to further define structural and spectroscopic features in 'blue' 'Type 1' sites, there is continuing interest in the synthesis of copper complexes with thiolate ligands,[2,13] as well as in generating new di- or trinuclear clusters of interest.[14]

In the last few years, there has been an increased emphasis upon functional modeling of proteins.[15] While the structural and spectroscopic modeling of metalloprotein active sites is an important and ongoing endeavor,[16] the realization that coordination chemists can and should make significant contributions to reactivity studies and mechanism is apparent. Here, we wish to summarize our own efforts in this area, primarily concerned with O_2 reactions with copper(I) complexes. Whether involved in dioxygen transport (e.g. hemocyanin), monooxygenation, or O_2-reduction, Cu(I)/O_2 interactions are fundamentally important to all the O_2-processing and utilization by copper proteins. By employing ligands and resulting complexes which allow for facile coordination and redox changes for both Cu(I) and Cu(II) structures, and by using low-temperature syntheses and manipulations, we have had particular success in identifying and characterizing discreet dioxygen-copper adducts. In most cases, the O_2-binding is reversible, and dioxygen can be recovered from the adduct; the products are usually best described as peroxo-dicopper(II) complexes. Here, we summarize the characteristics of five types of copper-dioxygen species, differentiated by the various ligands employed, and by their varying spectroscopic features and structures. In addition, a chemical system effecting arene hydroxylation, thus a copper monooxygenase mimic, is overviewed.

[‡]Current Address: METASYN, Inc., 71 Rogers Street, Cambridge, MA 02142-1118

From K.D. Karlin and Z. Tyeklár, Eds., *Bioinorganic Chemistry of Copper*
(Chapman & Hall, New York, 1993).

AN UNSYMMETRICALLY COORDINATED DIOXYGEN COMPLEX

In 1984, we first described $[Cu_2(XYL-O-)(O_2)]^+$ (**2**, PY = 2-pyridyl), one of the first Cu_2O_2 species to be authenticated through such criteria as observation of reversible O_2-binding, and determination of an O-O stretching vibration (*vide infra*). It was stabilized by low-temperature manipulation, formed in a reaction of O_2 with a X-ray structurally characterized phenoxo-bridged dicopper(I) complex $[Cu_2(XYL-O-)]^+$ (**1**). Completed chemical studies show that **1** also reversibly binds carbon monoxide.[17-19]

1

$[Cu_2^I(XYL-O-)]^+$

2

$[Cu_2^{II}(XYL-O-)(O_2)]^+$

Dioxygen adduct **2** is formed in dichloromethane solution as a deep purple compound at -80 °C by reaction of **1** (as PF_6^- or ClO_4^- salts) with O_2 (Cu:O_2 = 2:1, manometry), with strong 505 (ϵ = 6300) and 610 (ϵ = 2400) nm absorptions. A resonance enhanced Raman band[18] at 803 cm^{-1} indicates that the oxidation state of the O_2 ligand is as peroxide (O_2^-); this band shifts to 750 cm^{-1} when using $^{18}O_2$ to generate **2**. A metal-ligand copper-oxygen stretch was found at 488 cm^{-1} (464 cm^{-1} for $^{18}O_2$), and this absorption was used in the analysis of a mixed isotope experiment with $^{16}O-^{18}O$, which showed two components at 465 and 486 cm^{-1}.[18] Since the EXAFS data[19] provided a Cu$\cdots$Cu distance of 3.31 Å for **2**, a μ-phenolato-μ-1,1-peroxo double bridging coordination could be ruled out, since this should force the Cu...Cu distance to be < 3.1 Å. Thus, it was concluded that the peroxide ligates either in a Cu-O-O- terminal binding to a single copper ion or possibly in an unsymmetrical bridging Cu-O-O$\cdots$Cu mode.

Resonance Raman enhancement profile studies[18] concluded that the 505 and 610 nm bands could be assigned to $\pi_s^* \rightarrow d(x^2-y^2)$ and $\pi_v^* \rightarrow d(x^2-y^2)$ peroxide-to-Cu(II) LMCT transitions. Two such absorptions are in fact predicted for "terminal" end-on peroxo coordination (through one O-atom) to a single Cu(II) ion. Thus, the spectroscopy "requires" that the peroxo ligand coordinates in this fashion. A structure with O_2^{2-} bridging the two copper ions would thus require equatorial peroxo coordination to a second Cu(II) in trigonal bipyramidal coordination, thus rendering it spectroscopically unaffected by this other Cu ion; in fact, such phenolate-bridged copper(II) dinuclear complexes are known.

The reversible CO and O_2 binding to the dicopper(I) complex $[Cu_2(XYL-O-)]^+$ (**1**) (i.e., functional modeling) could be nicely followed by UV-Vis spectroscopic cycling experiments, shown in Figure 1.[17] Here, we started with a colorless bis(carbonyl) adduct $[Cu_2(XYL-O-)(CO)_2]^+$ having a featureless spectrum 0. Application of a vacuum removes the coordinated carbon monoxide generating $[Cu_2(XYL-O-)]^+$ (**1**) (spectrum 1, λ_{max} = 385 nm). Oxygenation at -80 °C in CH_2Cl_2 gives the dioxygen adduct $[Cu_2(XYL-O-)(O_2)]^+$ (**2**) (spectrum 2, λ_{max} = 505 nm). Saturating this solution with CO at -80 °C and warming to ~ -50 °C bleaches the purple color, regenerating $[Cu_2(XYL-O-)(CO)_2]^+$. The cycling can be repeated a number of times, as shown, with the decomposition measured by loss in intensity of the strong absorbances of **1** and **2**. While rapid oxidation to Cu(II) species occurs if **1** is oxygenated at room temperature, the strong charge transfer bands at 505 and 610 nm are not observed under these conditions, since **2** is unstable and rapidly decomposes.

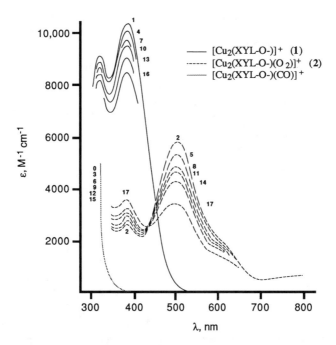

Figure 1. UV-Vis spectra demonstrating the reversible O_2 and CO binding behavior of $[Cu_2(XYL-O-)]^+$ (**1**) in CH_2Cl_2. See text for further explanation.

Complex $[Cu_2(XYL-O-)(O_2)]^+$ (**2**) exhibits quite interesting reaction chemistry, and the peroxo ligand can be readily protonated or acylated to give μ-1,1-(OOR) (R = -H or -C(O)Ar) complexes.[20,21] The latter has been structurally characterized by X-ray crystallography for the case when Ar = *meta*-chlorophenyl.[21]

Sorrell[22] has recently prepared ligands similar to that found in **1**, but possessing pyrazole or mixed pyrazole/pyridine ligands. The corresponding phenoxo-bridged dicopper(I) complexes appear to bind O_2 in the same fashion, the O_2-complexes all exhibit a characteristic purple color with strong 500-510 and 610-630 nm bands presumed also to be peroxo-to-Cu(II) LMCT transitions.

ROOM TEMPERATURE STABLE PEROXO AND HYDROPEROXO COMPLEXES

Synthetic variations of ligands and resulting modification of complexes under study is an approach we rely upon heavily, to alter structure, spectroscopy and reactivity in a way that helps us to better understand the systems at hand. One quite interesting variation has turned out to be generation of the unsymmetrical xylyl and phenoxy ligand UN-OH, found in complex $[Cu_2(UN-O^-)]^+$ (**3**), with structure indicated.[23] Peroxo species $[Cu_2(UN-O^-)(O_2)]^+$ (**4**) and hydroperoxo complexes $[Cu_2(UN-O^-)(O_2H)]^{2+}$ (**5**) were formed by oxygenating **4** or $[Cu_2(UN-OH)]^{2+}$; direct protonation of **4** gives **5**. While the closely related complex $[Cu_2(XYL-O-)(O_2)]^+$ (**2**) can only be manipulated at -80 °C in solution, room-temperature stable solids **4** and **5** were isolated; their stability is indicated by the ability to obtain analytical data, solid state UV-Vis or infra-red spectra, and to redissolve the solids obtaining UV-Vis spectra in agreement with directly low-temperature solution generated species. In addition, O_2 can be removed from solid **4** by heating under vacuum, to obtain **3** with only ~ 25 % decomposition. Solids **4** and **5** were also studied by EXAFS spectroscopy; Cu...Cu = 3.28 ± 0.07 Å and 2.95 ± 0.07 Å, respectively. These findings and others suggest that **4** has an unsymmetrical 'terminal' Cu(II)-O-O..Cu(II) ligation, similar to **2**, and **5** has a μ-1,1-hydroperoxo coordination. Supporting magnetic data and paramagnetically shifted ^{2}H NMR spectra were obtained.[23]

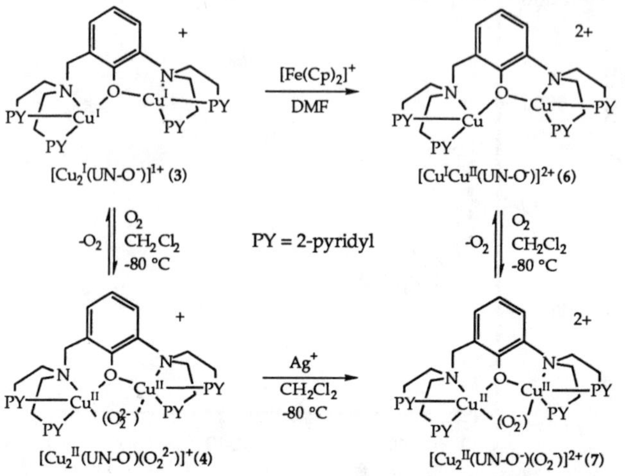

$[Cu_2^I(UN\text{-}O^-)]^{1+}$ (3) $[Cu^ICu^{II}(UN\text{-}O^-)]^{2+}$ (6)

PY = 2-pyridyl

$[Cu_2^{II}(UN\text{-}O^-)(O_2^{2-})]^+$ (4) $[Cu_2^{II}(UN\text{-}O^-)(O_2^-)]^{2+}$ (7)

A MIXED VALENCE DICOPPER COMPLEX AND ITS O$_2$ ADDUCT

The unsymmetrical nature of complexes with the UN-O- ligand prompted us to see if a dicopper complex with this ligand would support a mixed-valence compound, and we found that indeed it does.[24] A mixed-valence Cu^ICu^{II} complex **6** was synthesized via the reaction of **3** with ferrocinium ion as 1e$^-$ oxidant. $[Cu^ICu^{II}(UN\text{-}O\text{-})]^{2+}$ (**6**) exhibits a four-line EPR spectrum at 77 K, indicating a valence-trapped structure. [$g_{\parallel}$ = 2.25, $A_{\parallel}$ = 155 × 10^{-4} cm^{-1} (CH$_2$Cl$_2$/C$_7$H$_8$), μ_{eff}/Cu$_{RT}$ = 2.0 ± 0.1 B.M.]. Complex **6** reacts with O$_2$ (manometry; **6**/O$_2$ = 1.1 ± 0.1) giving [Cu$_2$(UN-O-)(O$_2$)]$^{2+}$ (**7**) (λ_{max} = 404 nm), formulated as a superoxo-dicopper(II) complex. The binding is reversible, since we can deoxygenate **7** and reoxygenate **6** for several cycles. The presence of the superoxo moiety in **7** is consistent with an EPR spin-trapping experiment using 3,3,5,5-tetramethyl-1-pyrroline N-oxide; this produces an adduct with an EPR triplet at g = 2.006 (A_N = 20 gauss). Further proof for the formulation of **7** is that it is directly related to peroxo complex **4**, by one-electron oxidation, as was shown in a spectrophotometric titration, using Ag$^+$ as stoichiometric oxidant.[24a]

A TRANS-μ-1,2-PEROXO DICOPPER(II) COMPLEX

[{(TMPA)Cu}$_2$(O$_2$)]$^{2+}$ (**9**) was the first structurally characterized Cu$_2$O$_2$ species.[25] It derives from the reaction of [(TMPA)Cu(RCN)]$^+$ (**8**) with dioxygen at -80 °C in EtCN or CH$_2$Cl$_2$ solvent (Cu:O$_2$ = 2:1, manometry) (Figure 2).[25] Complex **8** has the pentacoordinate structure shown, probably better described as a pseudotetrahedral Cu(I) species with long (~2.5 Å) Cu-N$_{alkylamine}$ interaction. Complex **9** is intensely purple in color, with strong absorptions at 440 (ε = 2000 M^{-1}cm^{-1}), 525 (11500) and 590 (sh, 7600) nm, with an additional d-d band at 1035 nm (180).

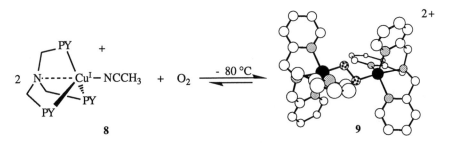

Figure 2. Reversible binding of O_2 to $[(TMPA)Cu(RCN)]^+$ (**8**) to give a structurally characterized *trans*-μ-1,2-peroxo dicopper(II) complex $[\{(TMPA)Cu\}_2(O_2)]^{2+}$ (**9**).

While strong binding occurs at reduced temperatures, the O_2 (and CO) coordination to $[(TMPA)Cu(RCN)]^+$ (**8**) is reversible. This has been confirmed in recent kinetic studies (*vide infra*) and demonstrated through chemical manipulations. Thus, when a vacuum is applied to $[\{(TMPA)Cu\}_2(O_2)]^{2+}$ (**9**) in EtCN while heating briefly, the purple solution decolorizes and **8** (R = Et) is produced. Rechilling (< -80 °C), followed by introduction of O_2 regenerates **9**. Repetitive cycling, without severe decomposition, can be followed spectrophotometrically, as shown in Figure 3.[25b]

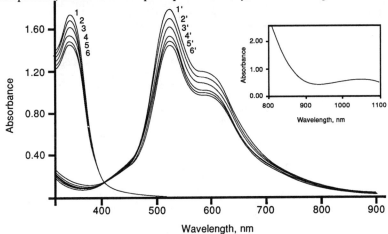

Figure 3. UV-Vis spectra demonstrating the reversible O_2-binding behavior of $[(TMPA)Cu(RCN)]^+$ (**8**) in EtCN to give the dioxygen adduct $[\{(TMPA)Cu\}_2(O_2)]^{2+}$ (**9**) (*vacuum cycling*). Reaction of **8** (λ_{max} = 345 nm) with O_2 at -80 °C produces an intensely violet solution of **9**, spectrum 1', λ_{max} = 525 nm. The inset shows the *d-d* band in the near-IR region. See text for further explanation.

Dioxygen can also be displaced from **9** by reaction with either CO or PPh_3 (in EtCN) to give the adducts $[(TMPA)Cu(CO)]^+$ or $[(TMPA)Cu(PPh_3)]^+$. In both cases, the O_2 evolved can be identified and for PPh_3, near quantitative (95 %) evolution of O_2 is observed by manometry. Carbon monoxide can be used to effect repetitive "carbonyl cycling"; in EtCN, O_2 is displaced from **9** giving $[(TMPA)Cu(CO)]^+$, the CO is removed via vacuum/Ar-purge cycles at room temperature providing $[(TMPA)Cu(RCN)]^+$ (**8**), and rechilling of the solution of **8** followed by oxygenation regenerates **9**.[25]

Thermally and hydrolytically unstable crystals of $[\{(TMPA)Cu\}_2(O_2)](PF_6)_2 \cdot 5$ Et_2O (9), were obtained at -85 °C in EtCN and X-ray data were obtained at -90 °C. The centrosymmetric complex is best described as a peroxo dicopper(II) species. As shown in Figure 2, it contains a *trans*-μ-1,2-O_2^{2-} group (derived from O_2) bridging the two Cu(II) ions. The Cu atom is pentacoordinate with a distorted trigonal bipyramidal geometry and the peroxo oxygen atoms occupy axial sites. The Cu-Cu' separation is 4.359 (1) Å and the O-O' bond length is 1.432 (6) Å which are structural parameters similarly found for peroxo-bridged dicobalt(III) complexes.[25b]

In recent spectroscopic and theoretical studies,[26] both an intraperoxide O-O stretch (832 cm^{-1}) and a Cu-O stretch (561 cm^{-1}) were identified using resonance Raman spectroscopy; the three strong visible absorptions have been assigned as peroxo-to-copper charge transfer transitions. The 832 cm^{-1} stretching vibration fixes the oxidation state of the dioxygen moiety in $[\{(TMPA)Cu\}_2(O_2)]^{2+}$ (9) as O_2^{2-}, consistent with presence of a *d-d* band at 1035 nm, expected for Cu(II) but not Cu(I). Complex 9 is EPR silent and exhibits a nearly normal ^1H NMR spectrum, suggesting that it is essentially diamagnetic. Recent magnetic susceptibility measurements performed on 9 confirm this assumption, with a finding that -2J > 700 cm^{-1}, based on $H_{ex} = -2JS_1 \cdot S_2$.[27] These results confirm that a single bridging O_2^{2-} ligand can mediate strong magnetic coupling between Cu(II) ions, even at a distance of 4.35 Å.

With A. D. Zuberbühler, a detailed kinetic study has been carried out for the reaction of $[(TMPA)Cu(RCN)]^+$ (8) with O_2 producing $[\{(TMPA)Cu\}_2(O_2)]^{2+}$ (9).[28] Details are also provided in Zuberbühler's chapter in this Volume. While $[\{(TMPA)Cu\}_2(O_2)]^{2+}$ (9) is eventually formed, the investigation revealed the intermediacy of an initially formed 1:1 adduct $[(TMPA)Cu(O_2)]^+$ (10). This was an exciting finding, since kinetic/thermodynamic parameters, along with spectral data (λ_{max} = 410 nm, ε = 4000 $M^{-1}cm^{-1}$) represent the first such information for a *primary* 1:1 copper-dioxygen interaction. In comparison with other cobalt or iron species, the formation of 10 (k ~ 10^8 $M^{-1}s^{-1}$, K_{form} ~ 0.3 M^{-1}; calculations, extrapolated to 25 °C) is faster than rates seen for most LCo(II) 1:1 oxygenation reactions, where k ~ 10^3-10^6 $M^{-1}s^{-1}$ (25 °C). For heme-proteins or porphyrin-Fe(II) model complexes, the O_2 on-rates (k ~10^6-10^9) are similar to that seen for formation of 10. However, the off-rates for iron species appear to be much smaller, giving rise to large K_{eq} values in the range of 10^4-10^6 at 20 °C.[28,29] This investigation, along with others, reveals that the primary cause for the room temperature instability of copper-dioxygen complexes is the large unfavorable entropy ($\Delta S°$) of formation from the reaction of copper(I) complexes with dioxygen.

O_2-BINDING WITH IMIDAZOLE-CONTAINING COPPER COMPLEXES

Imidazole ligation from side-chain histidine residues pervades Cu-enzyme chemistry, yet none of the synthetic model complexes described by us or others are based exclusively on imidazole nitrogen donors. However, we recently have been able to accomplish this goal by employing low-temperature techniques using a simple imidazole ligand.[30]

The two-coordinate complex $[L_2Cu]PF_6$ (11, L = 1,2-dimethylimidazole) has a linear structure {Cu-N_{im} =1.865 (8) Å, $\angle$ N-Cu-N = 179.2(7)°}, but it is unreactive towards dioxygen. However, three-coordinate $[L_3Cu]^+$ (12) reacts readily at -90 °C to give a brown solution of the peroxo-dicopper(II) complex $[\{L_3Cu\}_2(O_2)]^{2+}$ (13) [UV-Vis: λ_{max} (ε, $M^{-1}cm^{-1}$); 346 (sh, 2200), 450 (sh, 1450), 500 (1900) and 650 (600) nm]. Manometric measurements (Cu/O_2 = 2:1) confirm the stoichiometry of O_2 addition; excess imidazole does not affect the chemistry (Figure 4).

Dioxygen is not readily removed from 13 under a vacuum, presumably because of strong low temperature binding. However, O_2 appears to be reversibly bound, since addition of PPh_3 to $[\{L_3Cu\}_2(O_2)]^{2+}$ (13) liberates O_2, giving $[L_3Cu(PPh_3)]^+$ (14). The peroxidic nature of 13 is shown by its reaction with CO_2 to give a carbonato complex 7, similar to one we have previously characterized. Further evidence for the identity of 13 as a peroxo compound comes from a *peroxo transfer reaction*. Addition of the TMPA ligand to -90 °C solutions of 13 causes immediate conversion to the well-characterized complex 9, as followed by UV-Vis spectroscopy (Figure 4).[30]

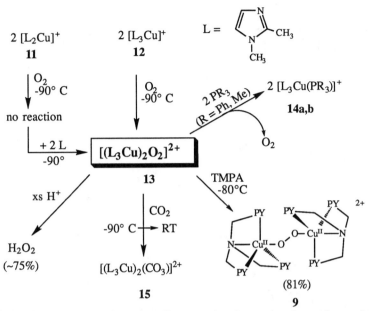

Figure 4. Scheme illustrating the copper-dioxygen chemistry using the unidentate 1.2-dimethylimidazole ligand L. At -90 °C, the three coordinate copper(I) complex [L₃Cu]⁺ (12) reacts the peroxo-dicopper(II) complex [{L₃Cu}₂(O₂)]²⁺ (13).

X-ray Absorption spectroscopy (XAS) studies (with N. J. Blackburn) on solutions of **13** (100 K) confirm a Cu(II) oxidation state assignment. Two models fit the EXAFS data for **5**, a bent μ-η^2:η^2- (**A**) or a planar *trans*-μ-1,2-peroxo (**B**) structure. Further investigations will be required to determine the actual structure.

	A	B
Cu-N (eq)	2.01 Å	2.01 Å
Cu-N(ax)	2.24 Å	
Cu-O(1)	1.92 Å	1.88 Å
Cu-O(2)	1.92 Å	2.85 Å
Cu-Cu	2.84 Å	4.30 Å

We also examined related chemistry with 1-Me-imidazole, dimethylaminopyridine (DMAP), and *t*-Bu-pyridine.[30b] Two and three-coordinate Cu(I) complexes have been prepared and studied by NMR spectroscopy and electrochemistry. An O₂-complex analogous to **13** is unstable for 1-Me-Im, but stable for DMAP; the *t*-Bu-pyridine complex is unreactive to O₂. It seems that both electronic and steric factors are involved in determining whether Cu₂O₂ complexes form.

XYLYL AND Nn DINUCLEAR COMPLEXES

Another class of compounds we have studied use dinucleating ligands where two bis[2-(2-pyridyl)ethyl]amine (PY2) units are linked through the amino nitrogens by a variable hydrocarbon spacer.[3,15,31] This can be either a xylyl group or those with a variable methylene chain $-(CH_2)_n-$ where n = 3, 4 or 5 (Figure 5). The xylyl-containing systems also bind O_2, but undergo a subsequent ligand hydroxylation reaction, to be discussed below.

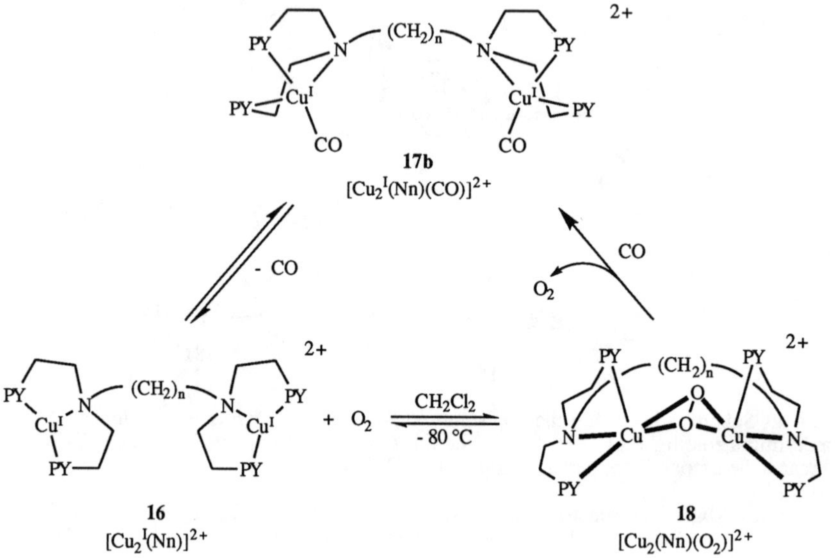

Figure 5. Reversible O_2- and CO-binding to dicopper(I) complexes **8**. Complexes **10** are suggested to have μ-η^2-η^2 peroxo coordination.

For tricoordinate complexes $[Cu_2(Nn)]^{2+}$ (**16**) or tetracoordinate nitrile adducts $[Cu_2(Nn)(CH_3CN)_2]^{2+}$ (**17a**), reversible binding can be demonstrated. These react with O_2 at -80 °C in CH_2Cl_2 solution, producing deep brown or purple species $[Cu_2(Nn)(O_2)]^{2+}$ (**18**).[3,15,31] As found for our other copper dioxygen complexes, reversible O_2 and CO binding can be followed spectrophotometrically, Figure 6. Release of intact O_2 from **18** has also been demonstrated.[31] Carbon monoxide binding to **8** is stronger than O_2, since CO can displace O_2 from **18** to give $[Cu_2(Nn)(CO)_2]^{2+}$ (**17b**). Thus, the relative binding strength of CO vs. O_2 to reduced copper ion parallels that observed for heme-proteins and porphyrin-iron(II) complexes.

$[Cu_2(Nn)(O_2)]^{2+}$ (**18**) possess striking UV-VIS properties, with multiple and strong charge-transfer absorptions. The position and relative intensities of these bands vary with the length of the polymethylene unit connecting the two PY2 donor groups,[31] reflecting subtle changes in the mode of O_2 binding. The characteristic 350-360 nm band with ϵ=16-21000 $M^{-1}cm^{-1}$ dominates; the presence of this distinctive intense absorption in part provides indications for the possible close structural relationship of $[Cu_2(Nn)(O_2)]^{2+}$ (**18**) to the Cu_2O_2 oxy-Hc chromophore, with its 345 nm (ϵ = 20,000) feature.[1-5]

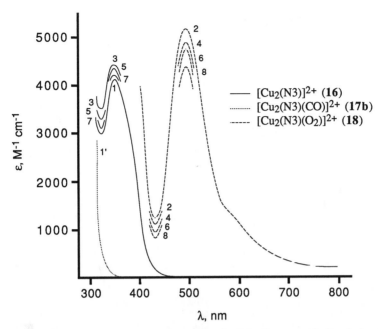

Figure 6. UV-Vis spectra demonstrating the reversible dioxygen binding behavior of $[Cu_2(N3)]^{2+}$ (**16**) in CH_2Cl_2 (*vacuum cycling*). The bis(carbonyl) complex $[Cu_2(N3)(CO)_2]^{2+}$ (**17b**, spectrum 1') is decarbonylated *in vacuo* at room temperature producing **16**, spectrum 1, (λ_{max} = 345 nm). Oxygenation at -80 °C generates the dioxygen adduct $[Cu_2(N3)(O_2)]^{2+}$ (**18**, spectrum 2, λ_{max} = 490 nm; the much stronger 365 nm absorption is not shown). Application of a vacuum to the solution and rapid warming to 100 °C removes the bound dioxygen ligand, regenerating **16**, spectrum 3. This cycle can be repeated four times, as shown.

We have yet to obtain vibrational data for complexes $[Cu_2(Nn)(O_2)]^{2+}$ (**18**), but a variety of other evidence is consistent with their peroxo-dicopper(II) formulation. These O_2-adducts (**18**) possess low-energy weak d-d absorptions, diagnostic of Cu(II) and not Cu(I) which has a filled shell d^{10} electronic configuration.   X-ray absorption measurements carried out on the N3 and N4 derivatives confirm the Cu(II) oxidation state (XANES, near-edge structure).[32] The EXAFS data also allow the determination of the Cu···Cu distances, which vary between 3.3 and 3.4 Å, depending on n. To account for EXAFS outer shell multiple scattering caused by the pyridine ligands, we proposed a μ-η^2:η^2-peroxo structure for $[Cu_2(Nn)(O_2)]^{2+}$ (**18**), shown in Figure 5. The bent 'butterfly' structure was suggested in order to also accommodate the EXAFS derived Cu...Cu distances observed (keeping reasonable Cu-O and O-O bond lengths), and is presumably caused by ligand constraints.   Species **18** are EPR silent, have normal ^{1}H NMR spectroscopic properties and exhibit solution diamagnetism,[27,31] thus they appear to provide another class of compounds where a single peroxo ligand bridges and strongly antiferromagnetically couples two Cu(II) ions. With the physical-spectroscopic evidence described above and in light of (i) the X-ray structurally described μ-η^2:η^2-peroxo dicopper(II) complex described by Kitajima and co-workers (see also his chapter in this Volume)[2,33] and (ii) a structurally characterized bent μ-η^2:η^2-peroxo divanadium(V) complex,[34] the proposed bent side-on peroxo binding proposed for **18** (Figure 5) seems justified.

A FUNCTIONAL MODEL FOR COPPER MONOOXYGENASES: THE NIH SHIFT MECHANISM IN COPPER CHEMISTRY

As mentioned in the last section, dicopper(I) complexes with a xylyl group connecting two PY2 tridentate chelates also binding dioxygen, but undergo a further hydroxylation reaction, reminiscent of copper monooxygenases, Figure 7.[3,15,35-37]

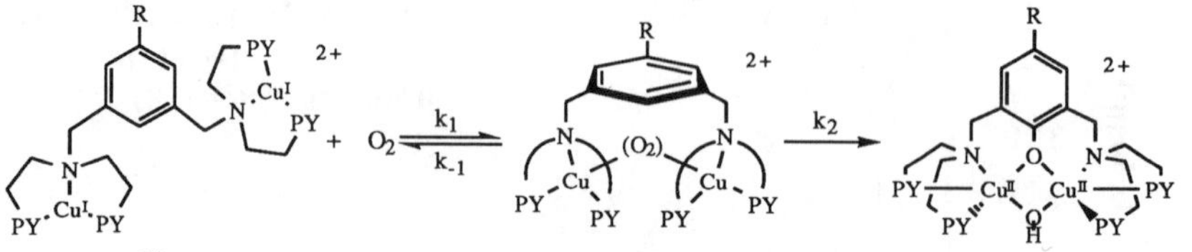

Figure 7. Reversible oxygenation of dicopper(I) complex **19** to give dioxygen adduct **20**, followed by hydroxylation yielding **21**.

Both the precursor three-coordinate dicopper(I) complex $[Cu_2(R-XYL-H)]^{2+}$ (**19**, R = H; Cu...Cu = 8.9 Å) and the hydroxylated product $[Cu_2(R-XYL-O-)(OH)]^{2+}$ (**21**, R = H; Cu$\cdots$Cu = 3.1 Å), with a phenoxo and hydroxo doubly-bridged dicopper(II) coordination have been characterized by X-ray crystallography.[35] Low-temperature stopped-flow kinetic studies[37] (also see Chapter by A. D. Zuberbühler in this Volume) provided the first evidence of O_2-binding to **19** (R = H); the process is reversible with a Cu/O_2 = 2:1 stoichiometry. Multiwavelength data analyses ($\lambda > 360$ nm) revealed distinctive features attributable to $[Cu_2(R-XYL-H)(O_2)]^{2+}$ (**20**), including a strong band in the 435-440 nm (ε ~3000-5000) range with weaker absorptions at lower energy. Supporting evidence comes from our own studies of synthetic analogs of the 'parent' xylyl system, $[Cu_2(R-XYL-H)(O_2)]^{2+}$ (**20**, R = NO$_2$, F, CN). For these derivatives, the reactions with O_2 are reasonably fast and hydroxylation still occurs, but the latter process is slowed to the point that the $[Cu_2(R-XYL-H)(O_2)]^{2+}$ (**20**) intermediates are stabilized at -80 °C and observable by usual low-temperature UV-Vis spectroscopic methods.[30b] Based on the close analogy of the ligand structure (i.e., hydrocarbon spacer between two PY2 units) and the nearly matching characteristic UV-Vis spectra, the O_2-adduct in $[Cu_2(XYL-H)(O_2)]^{2+}$ (**20**) is very similar to that found in $[Cu_2(Nn)(O_2)]^{2+}$ (**18**) (*vide supra*), suggested to have a bent μ-η^2-η^2-peroxo ligand coordinated to two Cu(II) ions. This structure is the one found in Kitajima's complex and oxy-hemocyanin.[1,2,8,33]

The reaction of **19** with O_2 in DMF or CH_2Cl_2 provides nearly quantitative yields of $[Cu_2(XYL-O-)(OH)]^{2+}$ (**21**).[35] Isotopic labeling experiments using $^{18}O_2$ have revealed that the source of oxygen atoms in this compound is dioxygen.[35] The overall reaction stoichiometry (observed (Cu:O_2=2:1, manometry) and the observed oxygen atom insertion into an aromatic carbon-hydrogen bond (e.g. **19** → **21**) are reminiscent of the action of tyrosinase, which contains a dinuclear active site essentially the same as that of hemocyanin.[1-5] Thus, we have expended considerable efforts to gain insights into the nature of intermediates and mechanism of the hydroxylation reaction, since this system represents one in which dioxygen is incorporated into an unactivated (aromatic) C-H bond, under very mild conditions, i.e. in a very rapid reaction occurring at room-temperature with 1 atm external O_2 pressure.

Additional kinetic studies[3,28b] of various 2- (site of hydroxylation) and 5-substituted (*para* to hydroxylation site) XYL-H ligands have also been revealing (see also chapter by A. D. Zuberbühler, this Volume). Important findings are that (a) when a deuterium atom is placed in the 2-position, k_2 (Figure 7) for the hydroxylation of $[Cu_2(XYL-D)]^{2+}$ is within experimental error of that seen for the -H parent compound **19**. This lack of deuterium isotope effect is consistent with electrophilic attack of the substrate arene substrate π-system, precluding C-H bond cleavage in the rate-determining step. Consistent with this notion, we also find an increase in $\Delta H^{\neq}$ of the hydroxylation step (k_2) with electron-withdrawing ability of R, when studying complexes such as $[Cu_2(R\text{-}XYL\text{-}H)]^{2+}$ (**19**, R = tBu, F, H and NO_2).

The conclusion that arene hydroxylation involves an electrophilic attack is also in accord with studies we recently completed on reactivity comparisons of three classes of peroxo-dicopper(II) complexes, $[Cu_2(XYL\text{-}O\text{-})(O_2)]^+$ (**2**), $[[\{(TMPA)Cu\}_2(O_2)]^{2+}$ (**9**), and $Cu_2(Nn)(O_2)]^{2+}$ (**18**, n = 4) (Figure 8).[38] We found that the $\mu\text{-}\eta^2\text{-}\eta^2\text{-}O_2^-$ group in $[Cu_2(N4)(O_2)]^{2+}$ (**18**) behaves as a nonbasic electrophilic peroxo ligand, in contrast to the basic or nucleophilic behavior of the two other types, which possess end-on coordination. For example, in reactions with H^+, CO_2 and PPh_3, $[Cu_2(Nn)(O_2)]^{2+}$ (**18**) does not readily protonate, it is unreactive toward CO_2 and it oxygenates PPh_3, while the others readily give H_2O_2, carbonates or liberate O_2 (respectively) in reactions with the same reagents.

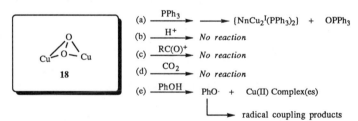

Figure 8. Reactivity comparison of end-on versus side-on bound peroxo-Cu complexes.

An rather important additional insight into the *m*-xylyl hydroxylation mechanism came from experiments in which a methyl group was placed into the 2-position of the ligand.[36] Instead of causing benzylic hydroxylation or blocking ring attack, 2-hydroxylation occurs and the methyl group undergoes a 1,2-migration. When **22** is reacted with dioxygen in CH_2Cl_2 and the resulting solution is worked up for its organic products, a phenol, MePY2, PY2 and formaldehyde (detected as a Nash adduct) are isolated or detected in excellent yield and with good material balance (Figure 9). From the initial reactions mixture, the phenol product can be isolated as a dicopper(II) complex **23**, confirming its identity. An isotope labeling experiment using $^{18}O_2$ to run the reaction also established that the source of oxygen in the phenol product was dioxygen.

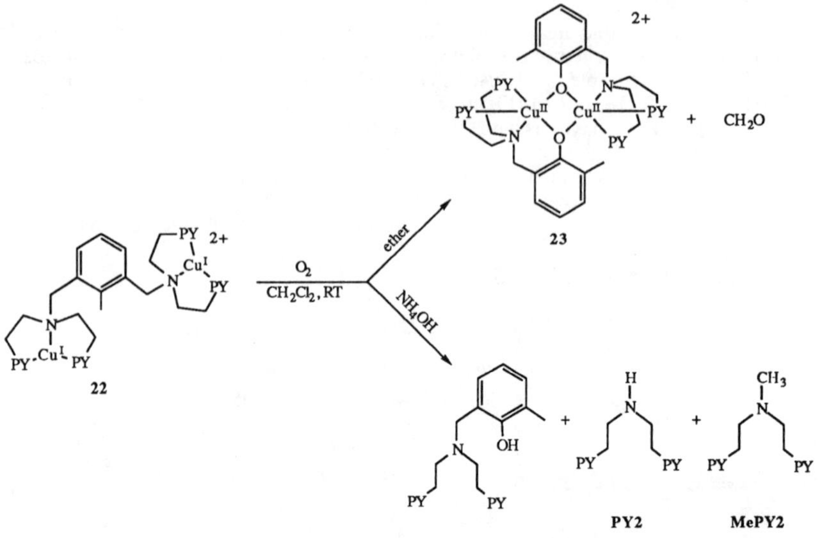

Figure 9. Oxygenation of a 2-methyl-substituted derivative $[Cu_2(XYL-CH_3)]^{2+}$ (**22**), causing an oxidatively induced 1,2-methyl migration. This reaction is reminiscent of the "NIH shift", known for certain iron monooxygenases.

The process observed is reminiscent of the "NIH shift", observed previously in iron hydroxylases,[36] where a reactive iron-oxy species (with an as yet undetermined identity) is an electrophile, attacking an arene substrate. This results in hydroxylation-induced migrations, due to the formation of carbonium ion intermediates and retention of heavier substituents in preference to -H, during rearomitization. This comparison has lead us to suggest a similar reaction pathway for the present copper system, which can account for the products observed. Thus, we have proposed a unified mechanism for the reaction of $[Cu_2(R-XYL-H)]^{2+}$ (**19**) or $[Cu_2(XYL-CH_3)]^{2+}$ (**22**) with O_2, consistent with all the data (Figure 10).[3,36a] This involves the initial formation of a dioxygen-complex intermediate, suggested to have a μ-η^2-η^2-peroxo structure capable of acting as an electrophile, also consistent with the spectroscopic and reactivity studies carried out on $[Cu_2(Nn)(O_2)]^{2+}$ (**18**) (*vide supra*). This attacks the xylyl ligand which is located in a favorable proximity. In fact, a molecular model of $[Cu_2(XYL-H)(O_2)]^{2+}$ (**20**) suggests the O-O vector is well aligned with and close to the p-π orbital of the arene carbon which is attacked, possibly an important factor in oxygen atom transfer reactions, as discussed by Sorrell.[4] This peroxo attack by the Cu_2O_2 unit generates a cationic intermediate, the methyl group in **22** undergoes a 1,2-migration, and re-aromatization with "assistance" of the amine nitrogen lone pair causes loss of an iminium ion in a retro-Mannich reaction. Under the experimental conditions employed, hydrolysis produces PY2 and the formaldehyde observed, and some reduction of the iminium salt can lead to a small amount of MePY2 observed. This mechanism can be straightforwardly applied to the parent complex $Cu_2(XYL-H)]^{2+}$ (**19**) where the 2-H atom migrates and during rearomatization, H^+ can be easily lost in preference to the ligand arm. The lack of a 2-deuterio isotope rate effect is also consistent with the proposed mechanism.

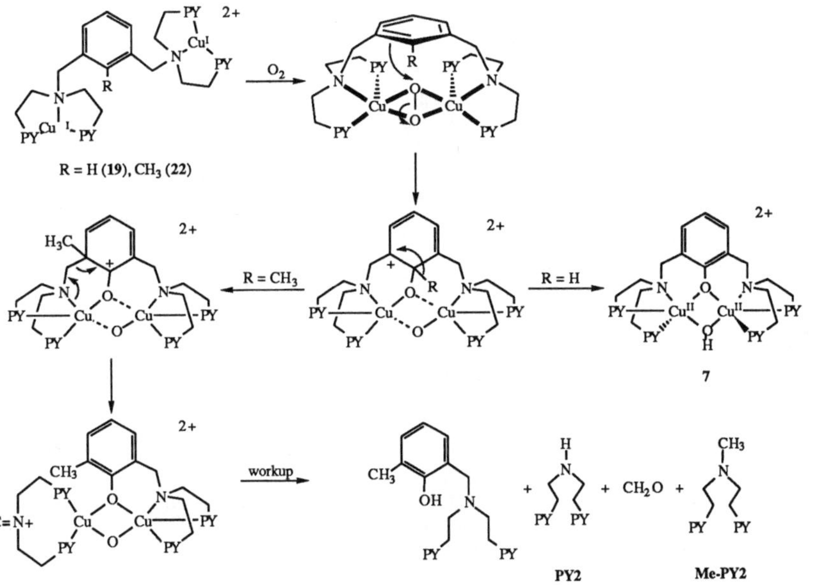

Figure 10   Unified mechanism of copper mediated arene hydroxylation and "NIH Shift" (1,2-migration) reactions. See text for further explanation.

       Other researchers[39-44] have examined alternate xylyl systems similar to [Cu$_2$(R-XYL-H)]$^{2+}$, using chelating groups other than PY2. Some of these studies are overviewed in other chapters in this Volume, e.g. by L. Casella, B. L. Feringa, A. E. Martell and T. N. Sorrell. Sorrell[39] has studied closely related complexes where 1-pyrazolyl or 2-imidazolyl donor groups fully or partially replace the 2-pyridyl ligands in XYL-H. Hydroxylation does not occur upon oxygenation of these dicopper(I) complexes and all react via four-electron reduction of O$_2$ to give bis(μ-hydroxo)copper(II) dimers. Also, if -CH$_2$PY (PY = 2-pyridyl) instead of -CH$_2$CH$_2$PY arms are used in the xylyl dinucleating ligands, only irreversible oxidation and no ligand hydroxylation takes place.[40] Sorrell has tried to correlate the tendency for hydroxylation with ligand basicity and/or Cu(II)/Cu(I) redox potential. With a somewhat limited data set, there were no clear trends. Electronic considerations should be important, but also copper chelation and peroxide proximity/orientation towards xylyl substrate have to be considered.

SUMMARY AND CONCLUSIONS

The recent developments in copper-dioxygen chemistry have been remarkable, with the synthesis of a number of well-characterized Cu$_2$O$_2$ species. The previously encountered problems of copper ion lability, peroxide disproportionation and air and/or moisture sensitivity have been overcome. It is clear that several different structural types exist. This includes a binding mode for peroxide to copper ion not previously considered, i.e. the μ-η^2:η^2-peroxo-dicopper(II) complexes suggested for species [Cu$_2$(Nn)(O$_2$)]$^{2+}$ (**18**), and structurally proven in Kitajima's complex (see his Chapter).[33] His accomplishment has entirely altered our thinking about Cu$_2$O$_2$ structure and chemistry. From the properties of these systems and the trans-μ-1,2- complex [{(TMPA)Cu}$_2$(O$_2$)]$^{2+}$ (**9**), it is clear that a single peroxide bridging ligand is capable of mediating the strong magnetic coupling observed in oxy-hemocyanin. Our own ability to generate complexes which can reversibly bind CO and O$_2$ will enable us to further understand the details concerning the ligation, structural and environmental requirements for O$_2$-binding.

The xylyl hydroxylation model system serves as a functional mimic for copper hydroxylases, in particular tyrosinase with its dinuclear copper active site. We have shown how O_2 can be activated by copper for electrophilic attack and oxygenation of an aromatic substrate, and described the first example of the NIH shift in copper chemistry. Very recent studies reveal that the bacterial copper phenylalanine hydroxylase[45] also exhibits an NIH shift[46] after its initial discovery[36b] in the chemical system described here. In summary, a Cu_2O_2 species, suggested to have a $\mu\text{-}\eta^2\text{-}\eta^2$-peroxo structure, forms in close and appropriate juxtaposition to the arene substrate. As with enzymes, this is seen to be an important factor in the xylyl chemical system, since relatively minor modification of the ligand can slow or shut down the hydroxylation reaction.

Further development of the model systems described here should be useful in answering questions concerning mechanism(s) of biological substrate oxygenation using O_2, or in developing applications to practical systems for hydrocarbon oxidation.

REFERENCES

1. E. I. Solomon, M. J. Baldwin and M. Lowery, *Chem. Rev.*, **92**, 521-542 (1992).
2. N. Kitajima, *Adv. Inorg. Chem.*, **39**, 1-77 (1992).
3. K. D. Karlin, Z. Tyeklár and A. D. Zuberbühler In *Bioinorganic Catalysis*; J. Reedijk, Ed. (Marcel Dekker, Inc., New York, 1993), Chapter 9, pp. 261-315.
4. T. N. Sorrell, *Tetrahedron*, **45**, 3-68 (1989).
5. (a) *Biological & Inorganic Copper Chemistry* , K. D. Karlin and J. Zubieta, Eds., Adenine Press: Guilderland, N.Y., Vols. 1 & 2 (1986). (b) *Copper Coordination Chemistry: Biochemical and Inorganic Perspectives* , K. D. Karlin and J. Zubieta, Eds., Adenine Press: Guilderland, N.Y. (1983).
6. E. T. Adman, *Adv. Protein Chemistry*, **42**, 145-197 (1991).
7. S. K. Chapman, *Persp. Bioinorg. Chem.*, **1**, 95-140 (1991).
8. (a) See chapter by K. Magnus, et. al., this Volume. (b) A. Volbeda and W. G. J. Hol, *J. Mol. Biol.*, **209**, 249-279 (1989).
9. See also the chapter by E. I. Solomon, et. al., in this Volume.
10. (a) A. Messerschmidt, R. Landenstein, R. Huber, M. Bolognesi, L. Avigliano, R. Petruzelli, A. Rossi and A. Finazzi-Agró, *J. Mol. Biol.*, **224**, 179-205 (1992). (b) See also the chapter by A. Messerschmidt, in this Volume.
11. (a) S. I. Chan, H. T. Nguyen, A. K. Shiemke and M. E. Lidstrom, *J. Inorg. Biochem.*, **47**, 10 (1992). (b) See also the chapter by S. I. Chan, et. al, this Volume.
12. (a) G.T. Babcock and M. Wikström, *Nature*, **356**, 301-309 (1992) and references cited therein. (b) See also the chapter by Fee et. al., this Volume.
13. (a) N. Kitajima, K. Fujisawa, M. Tanaka and Y. Moro-oka, *J. Am. Chem. Soc.*, **114**, 9233-9235 (1992). (b) N. Kitajima, K. Fumisawa and Y. Moro-oka, *J. Am. Chem. Soc.*, **112**, 3210 (1990).
14. (a) K. D. Karlin, Q.-F. Gan, A. Farooq, S. Liu and J. Zubieta, *Inorg. Chem.*, **29**, 2549-2551, (1990). (b) H. Adams, N. A. Bailey, M. J. S. Dwyer, D. E. Fenton, P. C. Hellier and P. D. Hempstead, *J. C. S. Chem. Commun.* 1297-1298 (1991). (c) P. A. Vigato, S. Tamburini and D. E. Fenton, *Coord. Chem. Rev.*, **106**, 25-170 (1990).
15. K. D. Karlin and Z. Tyeklár In *Functional Modeling of Metalloproteins*, Volume 9 of *Advances in Inorganic Biochemistry*, G. L. Eichhorn & L. G. Marzilli, Eds., (Prentice Hall, New York, 1993), pp. 123-172.
16. J. A. Ibers and R. H. Holm, Science, **209**, 223-235 (1980).
17. K. D. Karlin, R. W. Cruse, Y. Gultneh, A. Farooq, J. C. Hayes and J. Zubieta, *J. Am. Chem. Soc.*, **109**, 2668-2679 (1987).
18. J. E. Pate, R. W. Cruse, K. D. Karlin and E. I. Solomon, *J. Am. Chem. Soc.*, **109**, 2624-2630 (1987).
19. N. J. Blackburn, R. W. Strange, R. W. Cruse and K. D. Karlin, *J. Am. Chem. Soc.*, **109**, 1235-1237 (1987).
20. K. D. Karlin, P. Ghosh, R. W. Cruse, A. Farooq, Y. Gultneh, R. R. Jacobson, N. J. Blackburn, R. W. Strange and J. Zubieta, *J. Am. Chem. Soc.*, **110**, 6769-6780 (1988).

21. P. Ghosh, Z. Tyeklár, K. D. Karlin, R. R. Jacobson and J. Zubieta, *J. Am. Chem. Soc.*, **109**, 6889-6891 (1987).
22. T. N. Sorrell and V. A. Vankai, *Inorg. Chem.*, **29**, 1687-1692 (1990).
23. M. Mahroof-Tahir, N. N. Murthy, N. N., K. D. Karlin, N. J. Blackburn, S. N. Shaikh, and J. Zubieta, *Inorg. Chem.*, **31**, 3001-3003 (1992).
24. (a) M. Mahroof-Tahir and K. D. Karlin *J. Am. Chem. Soc.* **114**, 7599-7601 (1992). (b) M. S. Nasir, K. D. Karlin, D. McGowty and J. Zubieta, *J. Am. Chem. Soc.*, **113**, 698-700 (1991).
25. (a) R. R. Jacobson, Z. Tyeklár, A. Farooq, K. D. Karlin, S. Liu, and J. Zubieta, *J. Am. Chem. Soc.*, **110**, 3690-3692 (1988). (b) Z. Tyeklár, R. R. Jacobson, N. Wei, N. N. Murthy, J. Zubieta and K. D. Karlin, *J. Am. Chem. Soc.*, **115**, 0000 (1993).
26. M. J. Baldwin, P. K. Ross, J. E. Pate, Z. Tyeklár, K. D. Karlin and E. I. Solomon, *J. Am. Chem. Soc.*, **113**, 8671-8679 (1991).
27. K. D. Karlin, Z. Tyeklár, Z. Farooq, R. R. Jacobson, E. Sinn, D. W. Lee, J. E. Bradshaw and L. J. Wilson, *Inorg. Chim. Acta*, **182**, 1-3 (1991).
28. (a) K. D. Karlin, N. Wei, B. Jung, S. Kaderli and A. D. Zuberbühler, *J. Am. Chem. Soc.*, **113**, 5868-5870 (1991). (b) A. D. Zuberbühler, In *Dioxygen Activation and Homogeneous Catalytic Oxidation*, L. I. Simándi, Ed., (Elsevier: Amsterdam, 1991) pp 249-257.
29. E. C. Niederhoffer, J. H. Timmons and A. E. Martell, *Chem. Rev.*, **84**, 137-203 (1984).
30. (a) I. Sanyal, R. W. Strange, N. J. Blackburn and K. D. Karlin, *J. Am. Chem. Soc.*, **113**, 4692-4693 (1991). (b) K. D. Karlin and co-workers, to be published.
31. (a) K. D. Karlin, M. S. Haka, R. W. Cruse, G. J. Meyer, A. Farooq, Y. Gultneh, J. C. Hayes and J. Zubieta, *J. Am. Chem. Soc.*, **110**, 1196-1207 (1988). (b) K. D. Karlin, Z. Tyeklár, A. Farooq, M. S. Haka, P. Ghosh, R. W. Cruse, Y. Gultneh, J. C. Hayes, P. J. Toscano, and J. Zubieta, *Inorg. Chem.*, **31**, 1436-1451 (1992).
32. N. J. Blackburn, R. W. Strange, A. Farooq, M. S. Haka and K. D. Karlin, *J. Am. Chem. Soc.*, **110**, 4263-4272 (1988).
33. N. Kitajima, K. Fujisawa, K., C. Fujimoto, Y. Moro-oka, S. Hashimoto, T. Kitagawa, K. Toriumi, K. Tatsumi and A. Nakamura, *J. Am. Chem. Soc.*, **114**, 1277-1291 (1992).
34. A. E. Lapshin, Y. I. Smolin, Y. F. Shepelev, P. Schwendt and D. Byepesova, *Acta Cryst.* **C46**, 1753-1755 (1990).
35. K. D. Karlin, Y. Gultneh, J. C. Hayes, R. W. Cruse, J. W. McKown, J. P. Hutchinson and J. Zubieta, *J. Am. Chem. Soc.*, **106**, 2121-2128 (1984).
36. (a) M. S. Nasir, B. I. Cohen and K. D. Karlin, *J. Am. Chem. Soc.*, **114**, 2482-2494 (1992), and references cited therein. (b) K. D. Karlin, B. I. Cohen, R. R. Jacobson and J. Zubieta, *J. Am. Chem. Soc.*, **109**, 6194-6196 (1987).
37. R. W. Cruse, S. Kaderli, K. D. Karlin and A. D. Zuberbühler, *J. Am. Chem. Soc.*, **110**, 6882-6883 (1988).
38. P. P. Paul, Z. Tyeklár, R. R. Jacobson and K. D. Karlin, *J. Am. Chem. Soc.*, **113**, 5322-5332 (1991).
39. T. N. Sorrell, V. A. Vankai and M. L. Garrity, *Inorg. Chem.*, **30**, 207-210 (1991).
40. K. D. Karlin and co-workers, unpublished results.
41. L. Casella, M. Gullotti, G. Pallanza and L. Rigoni, *J. Am. Chem. Soc.*, **110**, 4221-4227 (1988).
42. (a) O. J. Gelling, F. van Bolhuis, A. Meetsma and B. L. Feringa, *J. Chem. Soc. Chem. Commun.*, 552-554 (1988). (b) O. J. Gelling and B. L. Feringa, *J. Am. Chem. Soc.*, **112**, 7599-7604 (1990).
43. R. Menif, A. E. Martell, P. J. Squattrito and A. Clearfield, *Inorg. Chem.*, **29**, 4723-2729 (1990).
44. L. Casella, M. Gullotti, M. Bartosek, G. Pallanza and E. Laurenti, *J. Chem. Soc., Chem. Commun.*, 1235-1237 (1991).
45. S. O. Pember, K. A. Johnson, J. J. Villafranca and S. J. Benkovic, *Biochem.*, **28**, 2124-2130 (1989), and references cited therein.
46. S. J. Benkovic, Penn. St. U., private communication.

DIOXYGEN ACTIVATION BY BIOMIMETIC DINUCLEAR COMPLEXES

Luigi Casella[a] and Michele Gullotti[b]

[a] Dipartimento di Chimica Generale, Università di Pavia, 27100 Pavia, Italy
[b] Dipartimento di Chimica Inorganica, Metallorganica e Analitica, Università di Milano, 20133 Milano, Italy

Metalloproteins containing dinuclear and polynuclear centers are very important in the biological reactions of dioxygen.[1] Among these, we have been mostly interested in the dicopper sites of the dioxygen carrier hemocyanin[2] and the monooxygenase tyrosinase,[3] in the polynuclear copper site of ascorbate oxidase[4] and, more recently, in the heme-copper site of cytochrome oxidase.[5] Our work with biomimetic dinuclear copper complexes started with the investigation of a series of two-coordinate copper(I)-bis(imine) complexes. The reason for this choice was that in the reaction mechanism for tyrosinase proposed by Solomon et al.[6] a dinuclear two-coordinate copper(I) site for the deoxy form of the protein was involved. Even though the structural features of this mechanism were schematic, and in fact subsequent EXAFS studies[7] and crystal structure determination of hemocyanin[2b-d] attenuated such a possibility, it was clear that the chemistry of synthetic two-coordinate copper(I) systems with ligands of potential biological relevance, in particular their reactivity with dioxygen, was little

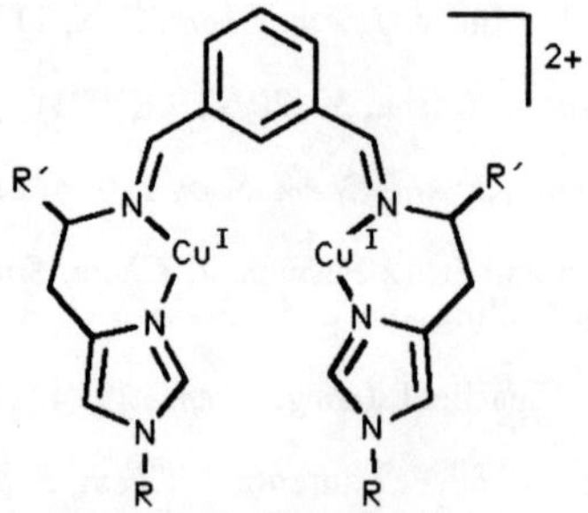

1 : R = H ,   R´ = H

2 : R = H,   R´ = COOMe

3 : R = Me,   R´ = H

4 : R = Me,   R´ = COOMe

From K.D. Karlin and Z. Tyeklár, Eds., *Bioinorganic Chemistry of Copper* (Chapman & Hall, New York, 1993).

developed. The complexes that we investigated are obtained
by template condensation of benzene-1,3-dicarboxaldehyde
and derivatives of histidine or histamine in the presence
of an excess of copper(I), 1 - 4.[8] We should also add that
the term "two-coordinate" is used to emphasize the number
of ligand donor atoms; actually, the fragments N-Cu-N
cannot achieve a perfectly linear arrangement and, at
least in solution, it is likely that the copper(I) centers
are three-coordinate, with a solvent molecule bound to
each metal site.

The complexes 1 - 4 were unable to mediate the
aromatic oxygenation of phenols, to produce catechols, by
dioxygen, i.e. the principal activity of tyrosinase, but
they exhibited still interesting reactivity with O_2, which
bears some relevance to the chemistry of this enzyme.
Compounds 1 - 4 undergo, in fact, a monooxygenase reaction
on the aromatic nucleus of the ligand according to the
stoichiometry:

$$[Cu^I_2(LH)]^{2+} + O_2 \longrightarrow [Cu^{II}_2(LO^-)(OH^-)]^{2+} \qquad (1)$$

to produce the μ-phenoxo-μ-hydroxo dicopper(II) complexes
5. However, the reaction is subject to solvent effects.
It is faster in methanol than acetonitrile and in the case
of 1 and 2 the aromatic hydroxylation is virtually stopped

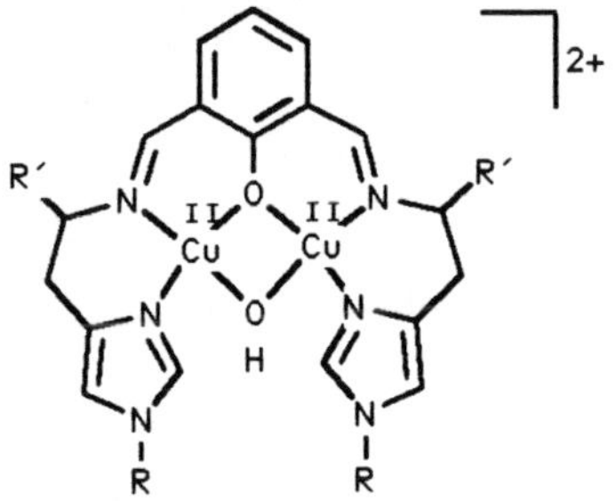

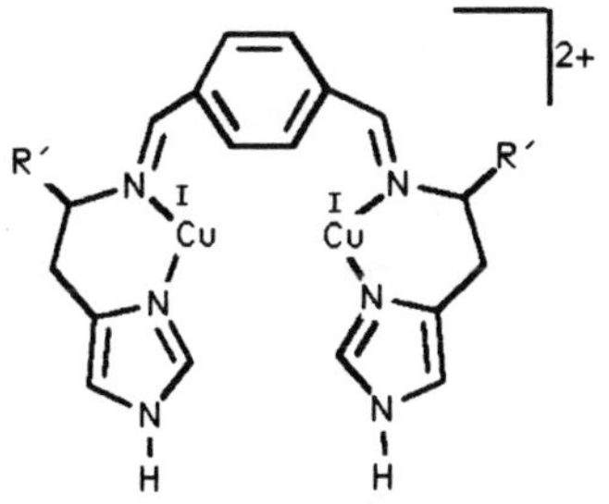

in favor of a simple oxidation of copper(I) to copper(II)
when carefully dried acetonitrile is used, according to
the reaction:

$$2 [Cu^I_2(HL)]^{2+} + O_2 \longrightarrow 2 [Cu^{II}_2(L^-)(OH^-)] \qquad (2)$$

where the protons necessary for reduction of O_2 come from
the imidazole NH groups of the ligands. In methanol
solution the amount of hydroxylated product formed by 1
and 2 is about 50%.[8b]

The monooxygenase reaction is strictly dependent on the
dinuclear structure of the complexes. Mononuclear
copper(I) analogues of 1 - 4[9] react with O_2 to give only

products of copper(I) oxidation. Steric factors are probably also very important, as shown by the failure of the aromatic nucleus of the p-substituted bis(imine) complex 6 to undergo the ligand oxygenation reaction. The larger separation between the two coppers probably prevents the formation of an intramolecular $Cu-O_2-Cu$ bridged intermediate, which is probably a key step in the monooxygenase reaction.

Although the ligand oxygenation reaction (1) had a precedent in the dicopper systems studied by Karlin et al.,[10] containing one tertiary amine and two pyridine nitrogen donors to each metal center, it is worth noting that the saturated diamino-m-xylene residue undergoing the reaction there is significantly different from the conjugated bis(imine) residues contained in 1 - 4, where the electron density on the phenyl nucleus, in particular at C-2, is clearly reduced by the electron withdrawing substituents. Another important difference is that the activity of Karlin's model system depends strictly on the presence of pyridine donors; replacement of these groups with other nitrogen heterocycles leads, in fact, to complete inhibition of the oxygen insertion reaction.[11,12] On the other hand, the monooxygenase reactivity of the bis(imine) complexes 1 - 4 can be duplicated in a range of related systems containing nitrogen and even sulfur donor atoms.

The dinuclear copper(I) complexes 7 - 9 were obtained in the same way as their imidazole-containing analogues.[13] The pyridine-containing complex 10 was also prepared by us, but it was reported independently and characterized by X-ray analysis by Feringa et al.[14]

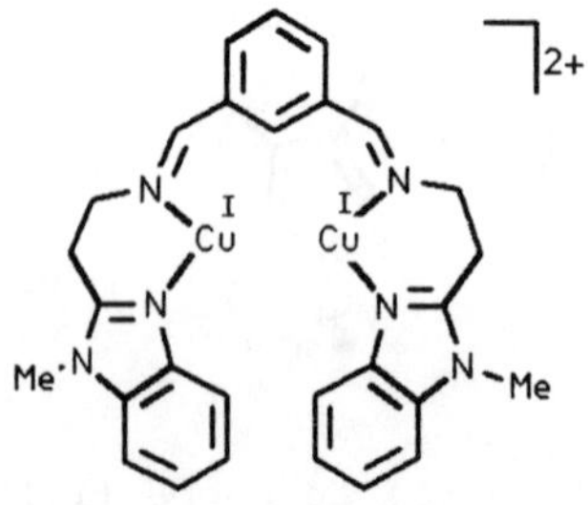

7

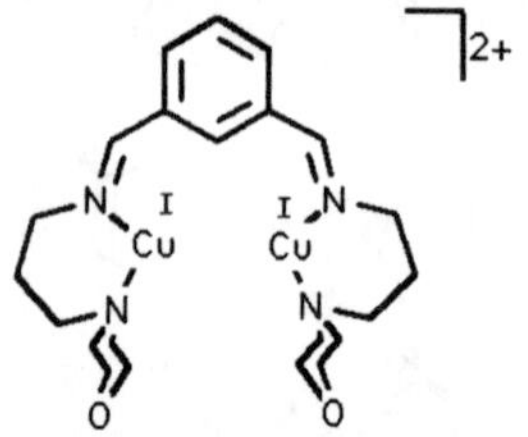

8

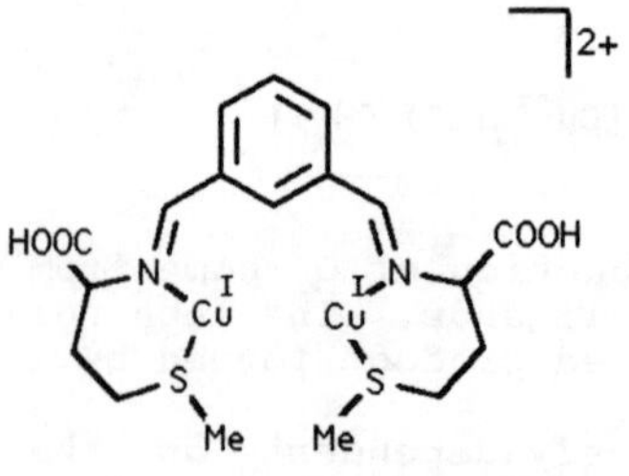

9

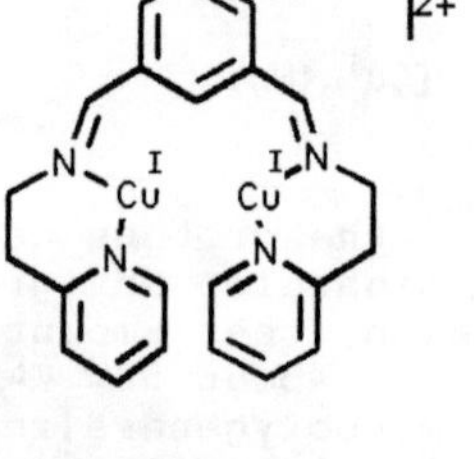

10

Compounds **7 - 10** undergo the monooxygenase reaction (1) to different extent (Table 1), as determined by the relative yield of 2-hydroxybenzene-1,3-dicarbaldehyde isolated after cleavage of the μ-phenoxo-μ-hydroxo dicopper(II) product.[13] They also react with O_2 at different rate, as it is evident from the rate of growth of the 360-nm electronic band which characterizes the 2-hydroxy-1,3-bis(imino)phenyl chromophore of the hydroxylated products

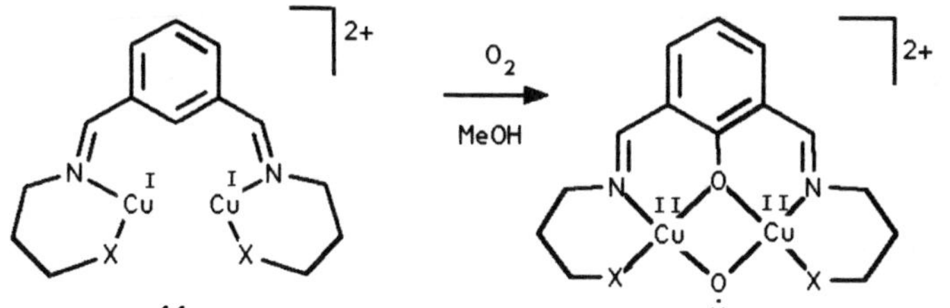

$$(3)$$

11   **12**

12.[8] Although a detailed kinetic analysis of the reaction is complicated by the competitive oxygenation and oxidation pathways, comparative experiments show that the rate of formation of **12** is inversely correlated with the basicity of the nitrogen donor group of the dicopper complex **11** (Table 2). The slower reactivity of the methionine complex **9** is probably due to the stabilization

Table 1. Yield (%) of 2-hydroxybenzene-1,3-dicarbaldehyde obtained from the oxygenation of **11** in methanol, after decomposition of **12**.

X					
%	100	88	70	50	40	50

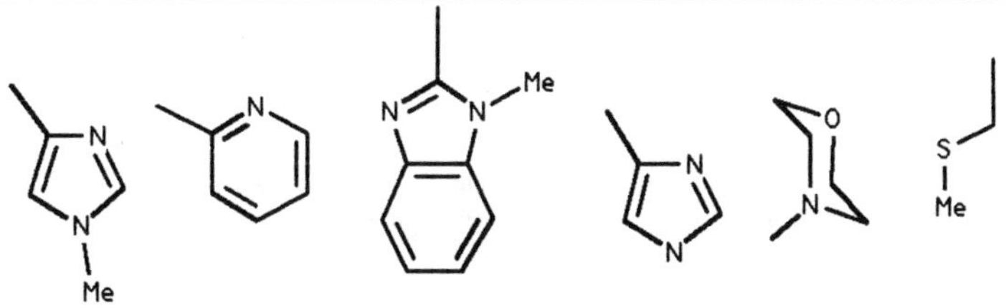

of the copper(I) state by the thioether ligand.[15] However, the monooxygenase activity of this complex has potential relevance to the sulfur-bound copper(I) site of dopamine β-hydroxylase.[16]

Also for **7 - 10** the ligand hydroxylation reaction is markedly affected by the solvent. When carefully dried acetonitrile is used the oxygenation is largely to completely depressed in favor of copper(I) oxidation. These results, together with the data in Table 2 and the effect of enhancement of the hydroxylation reaction by

Table 2. Rate of formation of **12** by reaction of **11** with air in methanol at 23 °C.

X	$[Cu^{I}_2]$, M	k_{obs}, s^{-1}
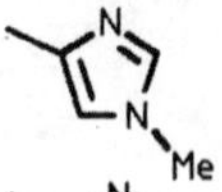 (N-methylimidazolyl)	$5.0\ 10^{-4}$	$2.5\ 10^{-2}$
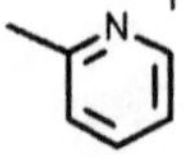 (pyridyl)	$1.4\ 10^{-4}$	$2.4\ 10^{-2}$
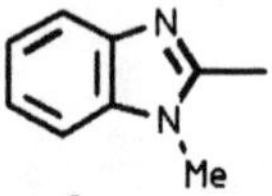 (N-methylbenzimidazolyl)	$2.5\ 10^{-4}$	$4.4\ 10^{-2}$
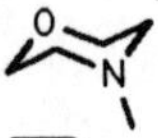 (amide)	$5.0\ 10^{-4}$	$3.1\ 10^{-3}$
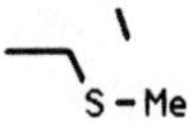 S—Me	$2.9\ 10^{-4}$	$7.8\ 10^{-4}$

k_{obs} : (pyridyl) ~ (N,N'-Me benzimidazolyl) > (N-Me imidazolyl) > (amide) > S–Me

pK_b : 8.6   8.5   6.9   ~ 4

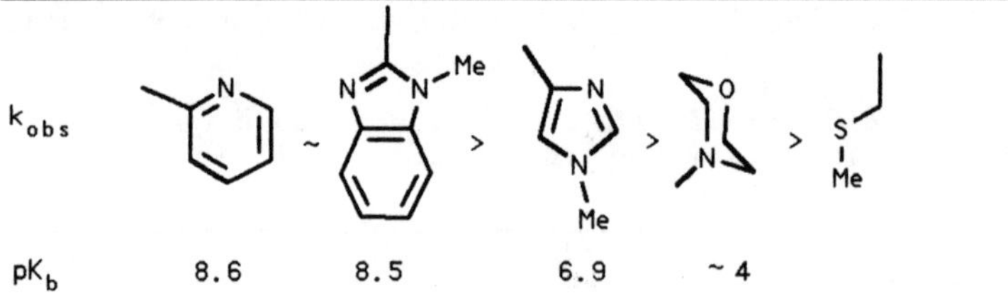

protons that we found for some of the systems,[17] suggest that an electrophilic copper-peroxo or copper-hydroperoxo complex may represent the active species in the hydroxylation process. A possible pathway for the reaction is outlined in Scheme 1. We are currently trying to characterize the intermediate by low-temperature spectral measurements.

Although the flexibility of the bis(imine) systems seems to allow further developments of this dicopper(I)/O$_2$ chemistry,[18] it was clear that modeling of tyrosinase activity, i.e. the o-hydroxylation of exogenous phenolic substrates,[8] can only be achieved in systems where the oxygenation of the endogenous ligand is prevented. We thought that a complex structurally related to that employed by Karlin, but where the pyridyl residues were replaced by benzimidazoles, which have virtually identical basicity, could be useful to this end. The dicopper(I) complex with ligand **13** was therefore synthesized. It

Scheme 1

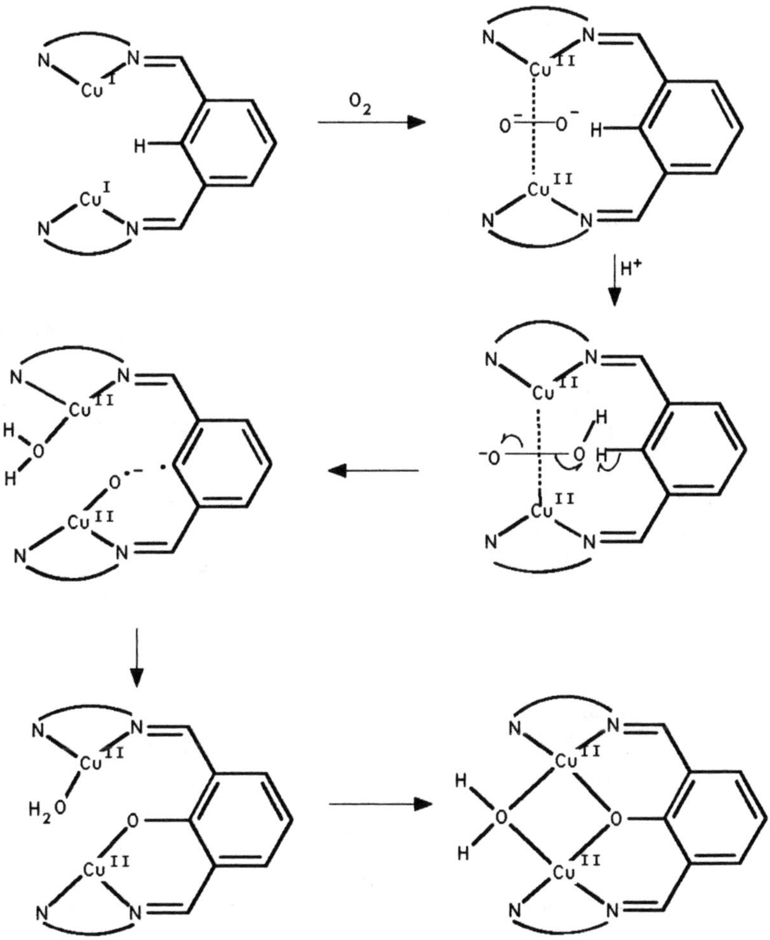

reacts with dioxygen to produce the bis(hydroxy)-dicopper(II) complex according to reaction (4),[12,19] as for all the other related dicopper(I) complexes carrying heterocyclic nitrogen donors different from pyridine in the side arms.[11] However, as expected, complex 13 is able to transfer its potential monooxygenase reactivity on exogenous phenols.[19]

The reaction occurs according to Scheme 2; dry acetonitrile was routinely employed as solvent. Binding of the phenolate salt of the substrate to the dicopper(I) complex yields the adduct [LCuI2(PhO-)]+, which is characterized by a CT band near 310 nm. Reaction of this with O2 produces the catecholate complex of copper(II) [L-

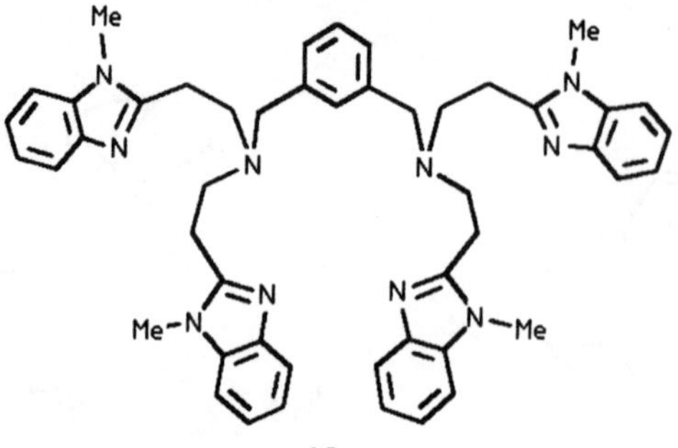

13

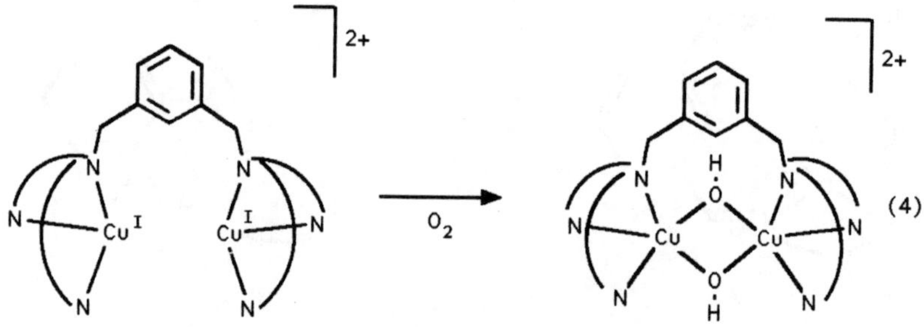

(4)

$Cu^{II}_2(Cat^{2-})(OH^-)]^+$, probably through the intermediate formation of the dioxygen adduct $[LCu_2(PhO^-)(O_2)]^+$. When the phenolate contains electron donor substituents the catecholate complex decomposes to form the quinone and dicopper(I) complex. However, the quinone undergoes further transformations to polymeric products in the reaction conditions and we therefore preferred to work with methyl 4-hydroxybenzoate, which gives a stable catecholate complex. Acid treatment of the reaction mixture and work-up, in this case, enables to isolate the o-catechol; this is the only product of the reaction, besides unreacted phenol.[19]

In order to assess how the structure of the dicopper(I) complex affects the phenol hydroxylation reaction we prepared a series of polybenzimidazole dinucleating ligands (14 -16) and the corresponding dicopper(I) complexes.[12,20] The ligands differ from 13 in the length of the alkyl chains between the tertiary amine and benzimidazole groups and this results in different sizes of the chelate rings around the copper(I) centers. In the ligand 16 a pyridine ring replaces the N-methylbenzimi- dazole group in each of the two arms of 15. The mononuclear copper(I) complex of the ligand 17 was also considered in the investigation.

298

Scheme 2

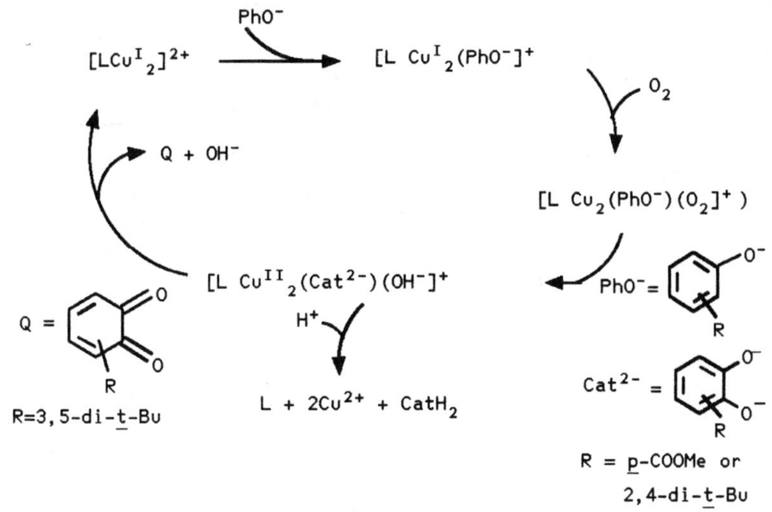

$$[LCu^I_2]^{2+} \xrightarrow{\quad PhO^- \quad} [L\ Cu^I_2(PhO^-]^+$$

$$\downarrow O_2$$

$$[L\ Cu_2(PhO^-)(O_2]^+)$$

$$Q + OH^-$$

$$[L\ Cu^{II}_2(Cat^{2-})(OH^-]^+$$

$$H^+$$

$$L + 2Cu^{2+} + CatH_2$$

Q =

R=3,5-di-t-Bu

$$PhO^- = $$

$$Cat^{2-} = $$

R = p-COOMe or
2,4-di-t-Bu

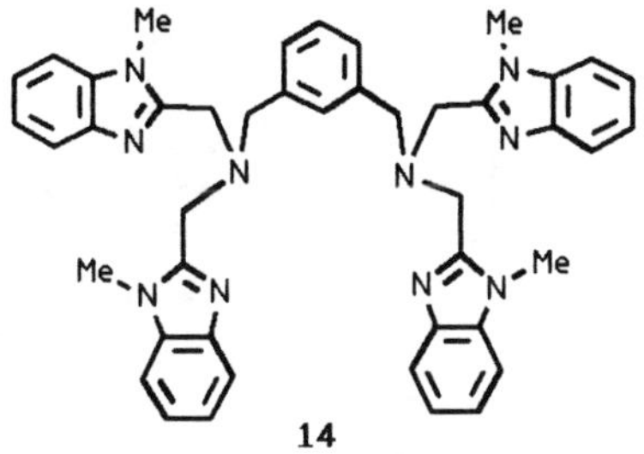

14

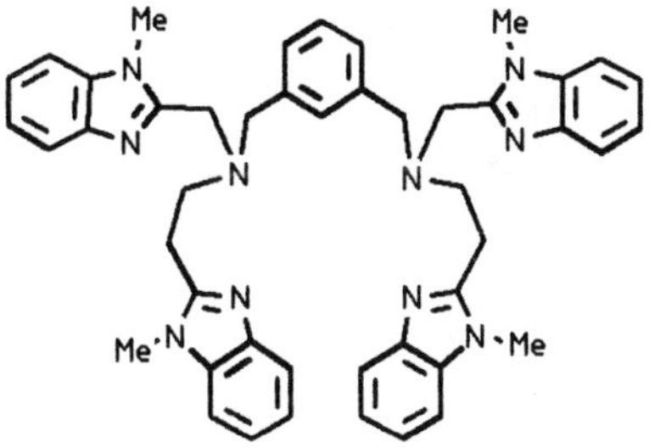

15

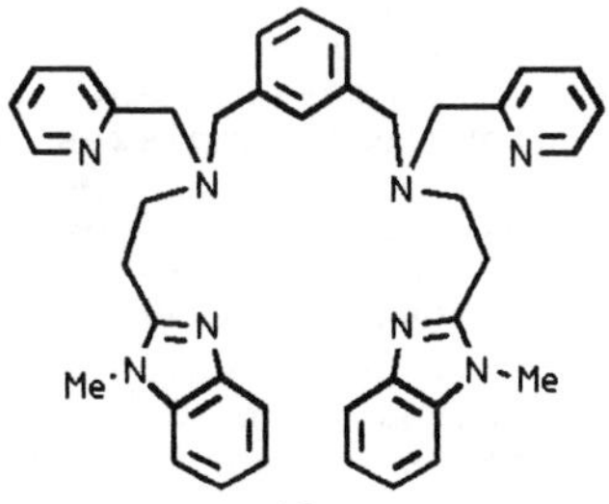

16

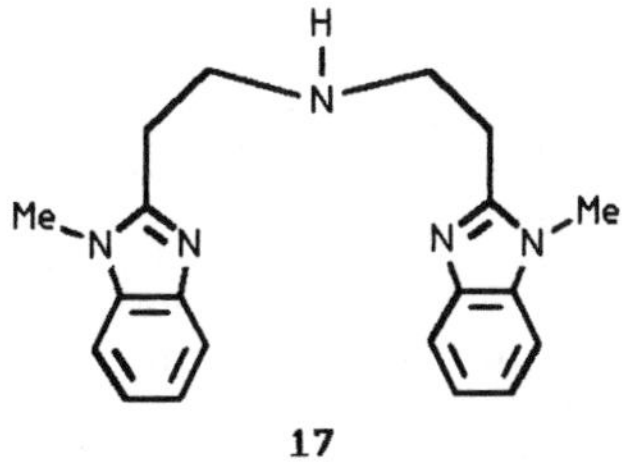

17

Comparative experiments showed that each of the dinuclear copper(I) complexes is able to mediate the

299

phenol monooxygenase reaction to some extent, but catechol formation was negligible when the mononuclear copper(I) complex was used (Table 3).[20] It is interesting that among the dinuclear complexes the amount of catechol produced in the reaction seems related to the amount of phenolate adduct formed by the dicopper(I) complex, as judged by the intensity of the 310-nm CT band. However, it is also clear that binding of the phenolate to copper(I) does not necessarily imply that the monooxygenase reaction will occur; in fact the mononuclear complex [Cu^I(17)]$^+$ produces the most intense Cu(I)-phenolate CT band ($\Delta\epsilon$=16500 M^{-1} cm^{-1}). We believe that a key point in the phenol hydroxylation is the possibility of bridging the phenolate between the two coppers in the dinuclear complexes, whereas it can only bind as a monodentate ligand to [Cu^I(17)]$^+$. Some evidence for a bridging phenolate in the adducts of [Cu$^I_2$(13)]$^{2+}$ - [Cu$^I_2$(16)]$^{2+}$ comes from their

Table 3. Formation of methyl 3,4-dihydroxybenzoic acid in the copper-mediated oxygenation of methyl 4-hydroxybenzoate.
Conditions: [Cu$^I_2$] 0.1 mmol, PhO$^-$Na$^+$ 0.1 mmol, dry MeCN 50 mL, O$_2$ 1 atm, room temperature.

Cu^I complex	Yield (%) isolated catechol
[Cu$^I_2$(13)]$^{2+}$	37
[Cu$^I_2$(14)]$^{2+}$	30
[Cu$^I_2$(15)]$^{2+}$	25
[Cu$^I_2$(16)]$^{2+}$	18
[Cu$^I_2$(17)]$^+$	trace

proton NMR spectra, since they do not show splittings of the signals for the methylene and N-methyl groups in the two arms of the ligands, that we would expect if the phenolate binds to only one of the two copper(I) centers. The difference in reactivity of the dinuclear complexes is probably due to the flexibility in the coordination sphere of the coppers. In the various steps of the reaction (Scheme 2) the metal centers need to change their coordination number, geometry and oxidation state. It is therefore conceivable that the structural changes are more easily accommodated in the complex derived from 13, which contains two flexible six-membered chelate rings to each copper.

In the attempt to further characterize the reaction system we prepared catecholate adducts of the dicopper(II) complexes of 13 and 14.[21] While we could obtain both the adducts with the monocatecholate ion 18 and the

300

dicatecholate ion **19** for $[Cu^{II}_2(14)]^{4+}$, only the latter was obtained with $[Cu^{II}_2(13)]^{4+}$. The adducts **19** are characterized by a CT band at ca. 340 nm that is absent in

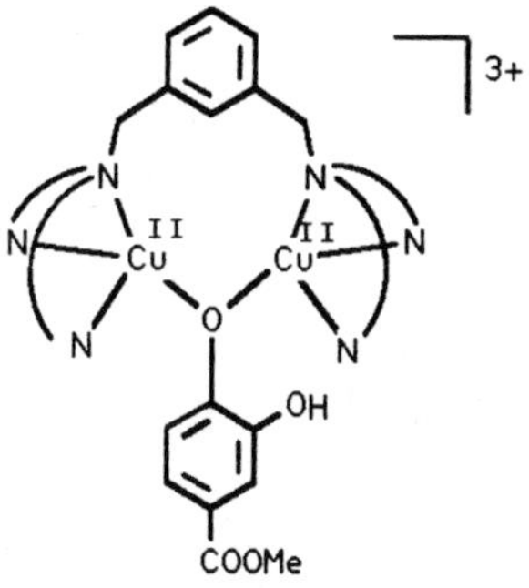

18 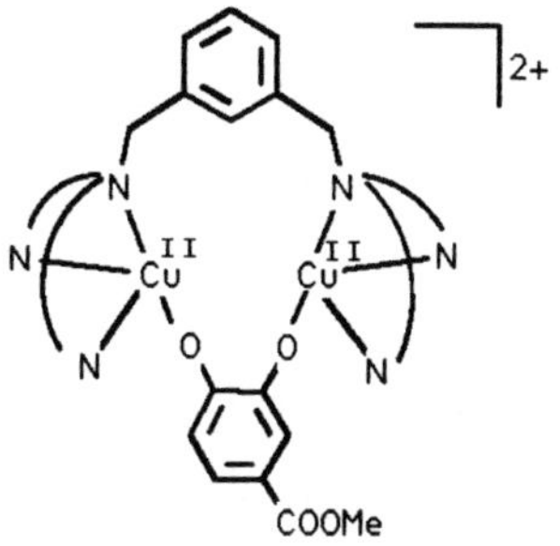   **19**

the spectrum of **18**. This band is present in the spectra of the catecholate complexes resulting from the phenol hydroxylation reaction by $[Cu_2(13)]^{2+}$ or $[Cu_2(14)]^{2+}$ and O_2. Low-temperature spectral measurements indicate that monocatecholate complexes of type **18** may also be involved as intermediates in the reaction, i.e. as the initial products resulting from the attack of the bound dioxygen molecule to the phenolate. The reactivity of substituted phenols suggests this attack is electrophilic.

The cumulative evidence from the work on tyrosinase-like reactivity of model copper complexes shows that there is requirement for a dinuclear site. The advantages of a dinuclear species, with respect to the corresponding mononuclear complex, in terms of complementarity in the number of electron transferred and proper binding and activation of the substrate, result also in a number of other, easier, oxidation reactions such as the catalytic dehydrogenation of catechols:[21]

$$2 \text{ catechol} + O_2 \longrightarrow 2 \text{ o-quinone} + 2 H_2O \qquad (5)$$

This was particularly evident in our study on oxidations catalyzed by the mononuclear complexes of **20** and **21** and the dinuclear complexes of **22** since the copper(II) coordination spheres are closely related.[22]

Our most recent interest in dinuclear complexes involved in dioxygen binding and activation turned to the field of heme-copper complexes because of their potential relevance to the dioxygen reaction site of cytochrome oxidase, the terminal protein complex of mitochondrial

20   **21**   **22**

R =   **a**   **b**   **c**   **d**

respiratory chain.[5] Several features of this protein mixed
metal center are not well understood, e.g. the nature of
the ligand that likely bridges the heme and copper centers
and is responsible for the strong magnetic coupling
observed in the oxidized resting form of the enzyme,[23] the
structure of the dioxygen adduct of the reduced enzyme,
and the mechanism of the O-O cleavage process.

23   **24**

We have prepared the mixed deuterohemin-copper(II)
complexes 23 and **24** in which an amino-polybenzimidazole
residue covalently linked to the porphyrin provides the
donor groups to copper(II).[24] The procedure for deuterohemin modification at the propionic acid side chain
followed a method we recently used for the linkage of
peptide residues to the same positions.[25] Each derivative
is obtained as a 50% mixture of the isomers containing the
propionic acid side chain modified at position 6 or 7 of
the porphyrin ring.

Binding of copper(II) to the deuterohemin-polybenzimidazole derivatives produces negligible changes in the spectral properties o the iron(III) porphyrin, indicative of the absence of interactions between the metal centers, and can only be followed by EPR spectroscopy. However, with the mixed complex 23 we observed the disappearance of the Cu(II) EPR signal (at -150 °C) on addition of 1 equiv. OH-. Measurements at liquid helium temperature are necessary to monitor the changes in the EPR signal of Fe(III) but some indication that OH- binds also to this center comes fron the proton NMR spectrum of the mixed complex, where a shift to high field of the methyl groups of deuterohemin occurs on addition of OD-. It is thus possible that a bridged Fe(III)-OH--Cu(II) species is formed in these conditions. Depletion of the Cu(II) EPR signal is observed here for the first time in a model heme-Cu system. With the mixed complex 24 the addition of OH- causes only partial decrease of the Cu(II) EPR signal, while the same perturbation of the porphyrin methyl signal is observed in the paramagnetic NMR spectrum on addition of OD-. In this system steric crowding of the three benzimidazole groups may hinder the approach of copper(II) to iron(III) which is necessary for the formation of the hydroxy bridge.

It is possible to reduce selectively Cu(II) in the mixed complexes 23 and 24, to produce the Fe(III)-Cu(I) species, using ascorbate, whereas the fully reduced species, Fe(II)-Cu(I), can be obtained by reaction with dithionite. This observation is interesting because it is known that in the electron-filling phase the copper(II) center of oxidized cytochrome oxidase is reduced first[5d] and, therefore the redox potential of the Cu(II)/Cu(I) couple of the enzyme must be higher than that of the Fe(III)/Fe(II) couple, as it is in our model complexes. The reduced Fe(II)-Cu(I) forms of 23 and 24 react rapidly with molecular oxygen. At room temperature the reaction occurs with concomitant destruction of the porphyrin, by some catalytic species produced by cleavage of the dioxygen complex, but this oxidative process is completely blocked at -50 °C where cycling between the Fe(II)-Cu(I) and Fe(III)-Cu(II) species can be repeated at will by dithionite/O_2 reactions. No intermediate species, e.g. the dioxygen adduct of the reduced complex, is accumulated in these conditions, but we believe that it will be possible to observe such intermediates at still lower temperatures, as it has been done in some cases with unhindered iron-porphyrin complexes.[26]

REFERENCES

1. (a) E.I. Solomon, M.J. Baldwin, M.D. Lowery, *Chem. Rev.*, **92**, 521 (1992). (b) L.L. Ingraham, D.L. Meyer, *Biochemistry of Dioxygen*, (Plenum Press, New York, 1985). (c) *Molecular Mechanisms of Oxygen Activation*, O. Hayaishi, Ed., (Academic Press, Inc., New York, 1974). (d) *Metal Clusters in Proteins*, L. Que Jr., Ed.,(ACS Symposium Series, **372**, American Chemical Society, Washington, D.C., 1988).
2. (a) B. Salvato and M. Beltramini, *Life Chem. Rep.*, **8**, 1 (1990). (b) W.P.J. Gaykema, W.G.J. Hol, J.M. Vereijken, N.M. Soeter, H.J. Bak, J.J. Beintema, *Nature (London)*, **309**, 23 (1984). (c) W.P.J. Gaykema, A. Volbeda, W.G.J. Hol, *J. Mol. Biol.*, **187**, 255 (1985). (d) A. Volbeda and W.G.J. Hol, *J. Mol. Biol.*, **209**, 249 (1989). (e) M. Beltramini, L. Bubacco, B. Salvato, L. Casella, M. Gullotti, S. Garofani, *Biochim. Biophys. Acta*, **1120**, 24 (1992).
3. D.A. Robb, in *Copper Proteins and Copper Enzymes*, R. Lontie, Ed., (CRC Press., Boca Raton, FL, 1984), vol. 2, pp. 207-241. (b) K. Lerch, *Life Chem. Rep.*, **5**, 221 (1987). (c) V.J. Hearing, *Methods Enzymol.*, **142**, 155 (1987). (d) G. Prota, *Med. Res. Rev.*, **8**, 525 (1988).
4. (a) P.M.H. Kroneck, F.A. Armstrong, H. Merckle, A. Marchesini, in *Ascorbic Acid: Chemistry, Metabolism and Uses*, P.A. Seib and B.M. Tolbert, Eds.,(Advanced Chemistry Series, Vol. 200, 1982), pp. 223-248. (b) A. Messerschmidt A. Rossi, R. Ladenstein, R. Huber, M. Bolognesi, G. Gatti, A. Marchesini, R. Petruzzelli, A. Finazzi-Agrò, *J. Mol. Biol.*, **206**, 513 (1989). (c) J.L. Cole, L. Avigliano, L. Morpurgo, E.I. Solomon, *J. Am. Chem. Soc.*, **113**, 8544 (1991). (d) L. Casella, M. Gullotti, G. Pallanza, A. Pintar, A. Marchesini, *Biol. Metals*, **4**, 81 (1991).
5. (a) S.I. Chan and P.M. Li, *Biochemistry*, **29**, 1 (1990). (b) G.T. Babcock and M. Wikström, *Nature*, **356**, 301 (1992). (c) B.G. Malmström, *Chem. Rev.*, **90**, 1247 (1990). (d) M Wikström and G.T. Babcock, *Nature*, **348**, 16 (1990).
6. (a) R.S. Himmelwright, N.C. Eickman, C.D. LuBien, K. Lerch, E.I. Solomon, *J. Am. Chem. Soc*, **102**, 7339 (1980). (b) M.E. Winkler, K. Lerch, E.I. Solomon, *J. Am. Chem. Soc.*, **103**, 7001 (1981). (c) D.E. Wilcox, A.G. Porras, Y.T. Hwang, K. Lerch, M.E. Winkler, E.I. Solomon, *J. Am. Chem. Soc.*, **107**, 4015 (1985).
7. G.L. Woolery, L. Porras, M. Winkler, E.I. Solomon, T.G. Spiro, *J. Am. Chem. Soc.*, **106**, 86 (1984).
8. (a) L. Casella and L. Rigoni, *J. Chem. Soc. Chem. Commun.*, 1668 (1985). (b) L. Casella, M. Gullotti, G. Pallanza, L. Rigoni, *J. Am. Chem. Soc.*, **110**, 4221 (1988).
9. L. Casella and L. Rigoni, *Rev. Port. Quim.*, **27**, 301 (1985).
10. (a) K.D. Karlin, P.L. Dahlstrom, S.N. Cozzette, P.M. Scensny, J. Zubieta, *J. Chem. Soc. Chem. Commun.*, 881 (1981). (b) K.D. Karlin, Y. Gultneh, J.P. Hutchinson, J. Zubieta, *J. Am. Chem. Soc.*, **104**, 5240 (1982). (c) K.D. Karlin, J.C. Hayes, Y. Gultneh, R.W. Cruse, J.W. McKown,

J.P. Hutchinson, J. Zubieta, *J. Am. Chem. Soc.*, **106**, 2121 (1984). (d) R.W. Cruse, S. Kaderli, K.D. Karlin, A.D. Zuberbuhler, *J. Am. Chem. Soc.*, **119**, 6882 (1988).
11. (a) T.N. Sorrell, *Tetrahedron*, **45**, 3 (1989). (b) T.N. Sorrell, V.A. Vankai, M.L. Garrity, *Inorg. Chem.*, **30**, 207 (1991).
12. L. Casella, O. Carugo, M. Gullotti, S. Garofani, P. Zanello, *Inorg. Chem.*, submitted for publication.
13. L. Casella, M. Gullotti, M. Bartosek, G. Pallanza, E. Laurenti, *J. Chem. Soc. Chem. Commun.*, 1235 (1991).
14. O.J. Gelling, F. Van Bolhuis, A. Meetsma, B.L. Feringa, *J. Chem. Soc. Chem. Commun.*, 558 (1988).
15. (a) D.E. Nikles, M.J. Powers, F.L. Urbach, *Inorg. Chem.*, **22**, 3210 (1983). (b) A.W. Addison, T.N. Rao, E. Sinn, *Inorg. Chem.*, **23**, 1957 (1984). (c) P. Zanello, *Comments Inorg. Chem.*, **8**, 45 (1988). (d) L. Casella, M. Gullotti, E. Suardi, M. Sisti, R. Pagliarin, P. Zanello, *J. Chem. Soc. Dalton Trans.*, 558 (1990).
16. N.J. Blackburn, S.S. Hasnain, T.M. Pettingill, R.W. Strange, *J. Biol. Chem.*, **266**, 23120 (1991).
17. L. Casella, M. Gullotti, G. Pallanza, *Biochem. Soc. Trans.*, **16**, 821 (1988).
18. (a) R. Menif and A.E. Martell, *J. Chem. Soc. Chem. Commun.*, 1521 (1989). (b) R. Menif, A.E. Martell, P.J. Squattrito, A. Clearfield, *Inorg. Chem.*, **29**, 4366 (1990). (c) M.P. Ngwenya, D. Chen, A.E. Martell, J. Reibenspies, *Inorg. Chem.*, **30**, 2732 (1991).
19. L. Casella, M. Gulloti, R. Radaelli, P. Di Gennaro, *J. Chem. Soc. Chem. Commun.*, 1611 (1991).
20. L. Casella, M. Gullotti, R. Radaelli, D. Cavagnino, to be submitted for publication.
21. (a) L.L. Simàndi, *Catalytic Activation of Dioxygen by Metal Complexes*, (Kluwer Academic Publishers, Dordrecht, The Netherlands, 1992). (b) J. -P. Chyn and F.L. Urbach, *Inorg. Chim. Acta*, **189**, 157 (1991).
22. L. Casella, M. Gullotti, G. Pessina, A. Pintar, *Gazz. Chim. Ital.*, **116**, 41 (1986).
23. Z.K. Barnes, G.T. Babcock, J.L. Dye, *Biochemistry*, **30**, 7597 (1991).
24. L. Casella, M. Gullotti, F. Gliubich, L. De Gioia, to be submitted for publication.
25. (a) L. Casella, M. Gullotti, L. De Gioia, E. Monzani, F. Chillemi, *J. Chem. Soc. Dalton Trans.*, 2945 (1991). (b) L. Casella, M. Gullotti, E. Monzani, L. De Gioia, F. Chillemi, *Rend. Fis. Acc. Lincei, Ser. IX*, **2**, 201, (1991).
26. K. Shikama, *Coord. Chem. Rev.*, **83** 73 (1988), and references therein.

OXIDATION CATALYSIS; A DINUCLEAR APPROACH

Ben L. Feringa

Department of Organic and Inorganic Chemistry, Groningen Centre for Catalysis and Synthesis, University of Groningen, Nijenborgh 4, 9747 AG Groningen, The Netherlands

Catalytic oxidation mediated by bimetallic complexes is a rapidly expanding area of chemistry due to the potential of multimetal complexes for enhanced reactivity and selectivity compared to mononuclear analogs.[1] A wealth of papers on mononuclear transition metal catalyzed oxidation reactions, using oxygen donors such as alkylhydroperoxides, hydrogen peroxide, alkaline hypohalides or iodosylbenzene, has been reported in the last decades.[2] Important goals of both academic and industrial importance are;

i.   oxidations without the production of stoichiometric amounts of salts and high atom economy conversions.

ii.   direct oxygenation of unfunctionalized hydrocarbons i.e. hydroxylations, although selective epoxidations, dehydrogenations, phenol oxidations, oxidative coupling reactions etc. are also highly warranted.

Much effort is currently devoted to develop useful catalytic systems for mild and selective oxidations with the aid of molecular oxygen.[3] It is therefore of great interest to elucidate the factors that govern the (reversible) binding and activation of O_2 in various natural oxygen transport systems and mono- and di-oxygenases. It is not surprising that there is a major role for dinuclear copper complexes as model compounds for copper proteins.[4] Guided by nature intriguing model systems for the active site of copper-containing O_2-binding enzymes such as hemocyanin have been developed.[5-7] The synthesis and structural characterization of μ-1,2-(1) and $\eta^2:\eta^2$-(2) dioxygen adducts of dinuclear Cu(I) complexes, by Karlin[6] and Kitajima and co-workers[7] respectively, illustrate some of the most important achievements.

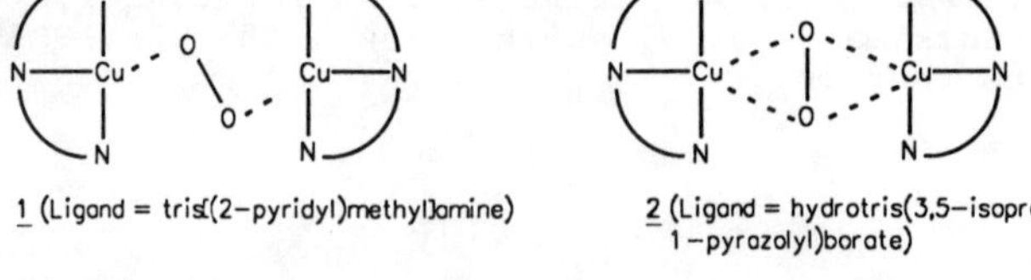

1 (Ligand = tris(2–pyridyl)methylamine)   2 (Ligand = hydrotris(3,5–isopropyl–1–pyrazolyl)borate)

Figure 1   Dioxygen adducts of dinuclear copper complexes.[6,7]

From K.D. Karlin and Z. Tyeklár, Eds., *Bioinorganic Chemistry of Copper* (Chapman & Hall, New York, 1993).

Following the pioneering work of Karlin and co-workers[8] on arene hydroxylation by molecular oxygen in a dinuclear Cu(I) complex several groups[9-14] have reported on related mimics for copper dependent monooxygenases (Scheme 1).

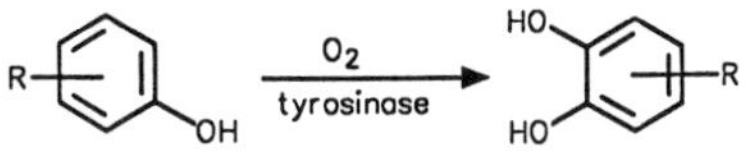

Scheme 1

Contrary to the successful development of oxidation catalysts based on metalloporphyrins,[15] synthetically useful catalysts, based on copper complexes, that act as mimics for copper containing monooxygenases[4] (e.g. tyrosinase, dopamine-β-hydroxylase) in the oxidation of __external__ substrates are scarce.[1c,16,17] This is even more surprising if one realizes that selective oxidations of organic substrates mediated by copper complexes have been known over a century.[18] For instance the Glaser oxidative coupling of acetylenes,[19] the oxidative dimerization of naphthols[20] and the formation of poly(phenylene ethers) from 2,6-disubstituted phenols[21] are all based on copper-amine complexes and molecular oxygen.

Considering bimetallic complexes that can act as catalysts for selective oxidations, the effect of the "second metal" in the catalyst complex is proposed to enhance the O_2 binding ability, to assist in oxygen-oxygen, or oxygen-R (R = alkoxide, halide, hydroxide), bond cleavage (as shown in **3**) or to interfere with substrate binding (as indicated in **4**).

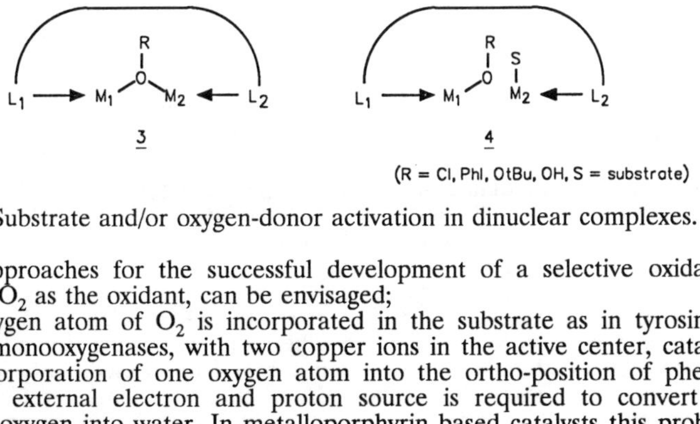

(R = Cl, PhI, OtBu, OH, S = substrate)

Figure 2   Substrate and/or oxygen-donor activation in dinuclear complexes.

Two approaches for the successful development of a selective oxidation catalyst, using O_2 as the oxidant, can be envisaged;

i.   one oxygen atom of O_2 is incorporated in the substrate as in tyrosinase. These monooxygenases, with two copper ions in the active center, catalyze the incorporation of one oxygen atom into the ortho-position of phenols and an external electron and proton source is required to convert the second oxygen into water. In metalloporphyrin based catalysts this problem is circumvented using "single oxygen" donors or external reducing agents.

ii.   both oxygen atoms of O_2 are used for the oxidation as is the case with dioxygenases.

Both approaches are used in our group employing either homo- or hetero-dinuclear copper complexes.

In this review recent results of oxygenations with dinuclear Cu complexes will be described with main emphasis on intramolecular oxidation studies of our laboratories.

OXIDATIVE HYDROXYLATION AND METHOXYLATION

As arene hydroxylation (Scheme 1) in dinuclear Cu(I) complexes seems to depend critically on the ligands employed, we studied dinucleating 2,6-bis[N-(2-pyridyl-ethyl)-formimidoyl]benzene (2,6-BPB) type ligands **5**, which have bidentate nitrogen-ligands available for each Cu(I) center in contrast to tridentate bis-(pyridylethyl)amine units present in Karlin's complexes.[8] Approaches using imidazolyl alkylimine chelating group have been reported by others.[9,11-14]

Copper(I) complex 6 (scheme 2) is extremely air sensitive in CH_2Cl_2, CH_3OH or CH_3CN solution and rapidly oxidizes to the hydroxy-phenoxy-bridged Cu(II) complex 8. X-ray analysis, independent synthesis and analysis of free ligand 9 confirmed the quantitative hydroxylation at the 1-position of the arene-moiety.[10]

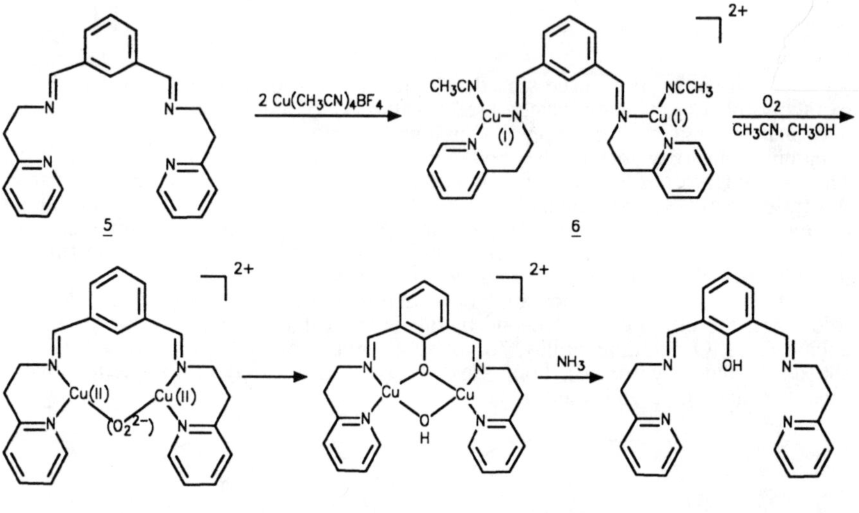

Scheme 2

A number of dinuclear Cu(II) complexes, structurally related to 8, have been reported in the literature.[22] No epoxidation of external substrates, like olefins, phenols or sulfides, was found using 6 in the presence of molecular oxygen. The mild and rapid arene-oxygenation observed here may be considered a mimic for copper-dependent monooxygenases. It is conceivable that the reaction of 6 with O_2 produces a dioxygen adduct 7 as reversible O_2 binding on dinuclear copper complexes is well precedented.[5,7,23] Furthermore we have observed reversible O_2 binding in the structurally closely related nitroaryl-substituted Cu(I) complex 10.[24]

The results show that arene-hydroxylation with dinuclear Cu(I) complexes and dioxygen is not specific to bidentate or tridentate ligand systems and that O_2 binding is possible with one chelating and one monodentate ligand attached to each Cu(I) center.

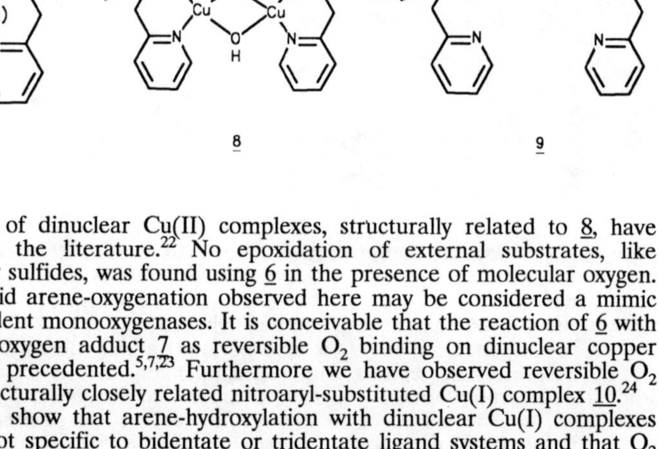

Figure 3 Structures of 10 and 11

308

It is remarkable that no monooxygenase activity is observed when pyridine is replaced for pyrazole (tridentate) or in some cases benzimidazoles (bidentate). Minor electronic effects but perhaps more important geometrical constraints seem to prevent hydroxylation to occur. This knowledge has been successfully applied by the groups of Waegell and Casella in the design of dinuclear copper complexes for the hydroxylation of exogenous phenols.[16,17]

In order to get more insight into the reactivity of the Cu(I) complexes and in particular in the dependency of the arene hydroxylation on the substituent pattern different X-substituents (alkoxy, halogen, alkyl) were placed at the arene-1-position in 11 which is most vulnerable to oxygenation (vide infra). Furthermore preliminary investigations show that both O_2-binding and hydroxylation are sensitive to the electronic effects of para-substituents (11, Y = OCH_3, CH_3, CO_2CH_3, NO_2).[24] When the ligand C-1 position was blocked with a methyl-substituent, as is present in complex 12, no arene hydroxylation was observed although rapid reaction with molecular oxygen takes place. As the major part of the ligand was recovered unchanged, presumably a bis(μ-hydroxy)dicopper(II) species 13 is formed, as was observed by Sorrell and co-workers[25] in the reaction of a related Cu(I) complex with O_2.

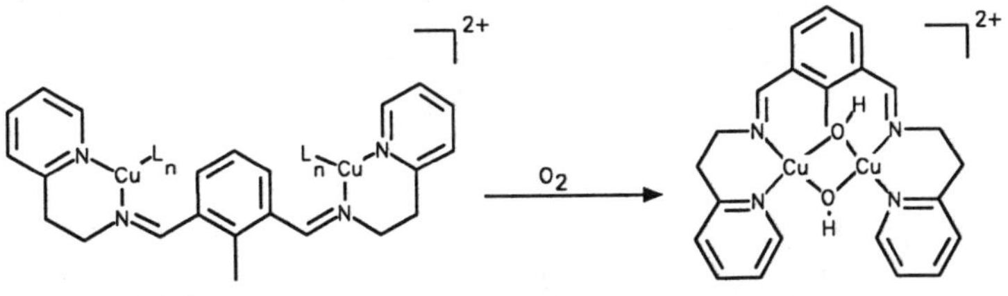

12 (L = CH_3CN, n = 0,1,2)   13

Scheme 3

Partial oxygenation of the pyridylethyl unit of the ligand cannot be excluded, however. This result is remarkable in view of the fact that Karlin and co-workers[26] found a methyl-migration upon exposure of complex 14 to oxygen (scheme 4) resulting in complex 15.

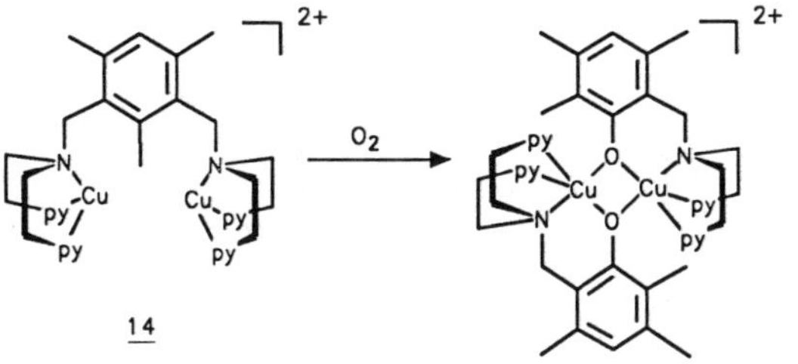

14

15

Scheme 4

In a different approach the bridging unit between the two Cu(I) centers was modified into a 3,6-disubstituted pyridazine leading to more rigid bis-Cu(I) complexes. Dinuclear Cu(I) complexes 17, with both 5,5- and 5,6-chelate ring sizes, were obtained from ligand 16. Although reversible oxygen binding to 17 was not observed, rapid oxidation and formation of H_2O_2 was found upon exposure to the air. The nature of the oxidation products has not been unrafled sofar. Using

309

dinuclear Cu(II) complex $\underline{18}$ a remarkable bis-α-methoxylation at the benzylic positions, with incorporation of two methoxy-substituents from the solvent, was found.

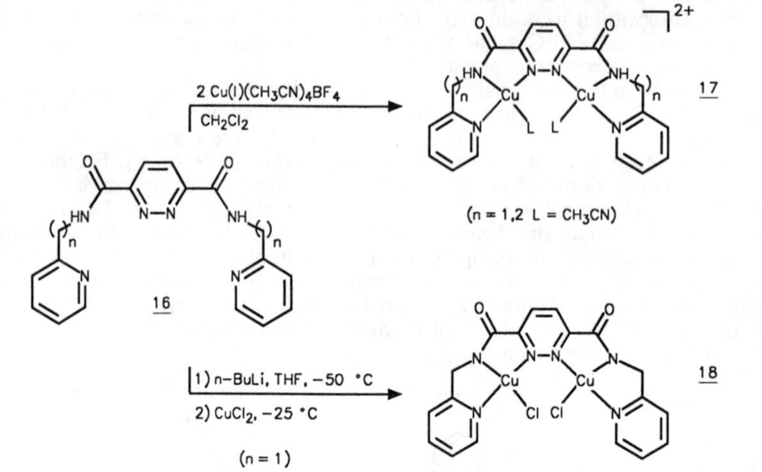

Scheme 5

Removal of the copper ions after the oxidation reaction provided bis-methoxylated product $\underline{23}$ in nearly quantitative yield[27] (scheme 6). Hydroxylation at the benzylic position followed by H_2O elimination to acylimine intermediate $\underline{19}$

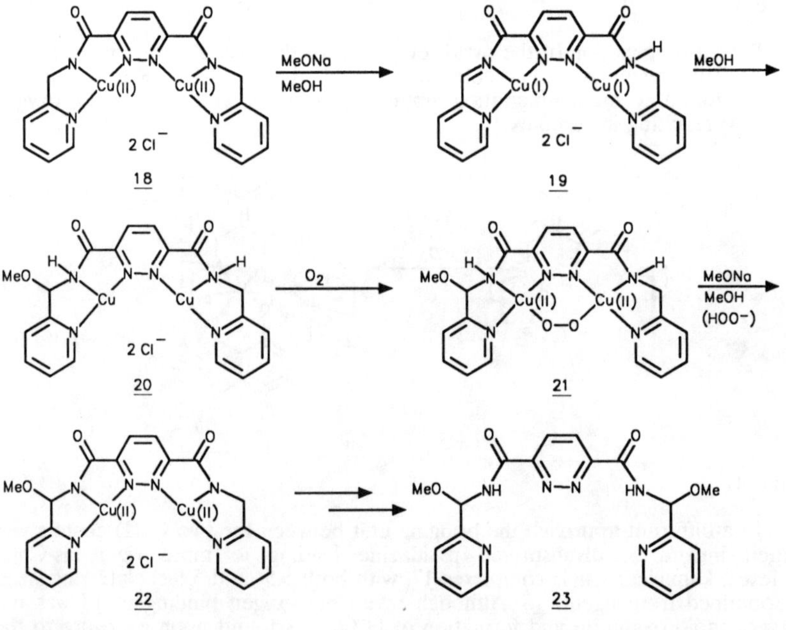

Scheme 6

310

could be envisaged but we prefer to rationalize the methoxylation via the mechanism outlined in scheme 6. The consecutive conversions involve electron transfer to Cu(II) and deprotonation to yield acylimine bound Cu(I) complex 19 which adds either methoxide or methanol. Reoxidation of the Cu(I) centers in 20 to Cu(II) by O_2 and repetition of the cycle provides bis-methoxylated product 23 (isolated after ammonia treatment). It is intriguing that selective dimethoxylation takes place under mild conditions in particular when compared to the more severe conditions in copper-mediated α-hydroxylation of N-salicyloyl-glycine.[28] Further mechanistic studies will be necessary, as well as assessment of the potential of this type of conversion for selective amine α-oxidation as an alternative to electrochemical methods[29] and the relevance for peptidylglycine-α-hydroxylating monooxygenase (PHM) mimics. Related oxidative pathways have been found in copper-mediated N-dealkylation of peptide ligands[30] in the presence of oxygen. Finally similar ligand oxidations might occur upon reaction of 17 with O_2.

An important observation was made when complex 18 was used in the oxidative coupling of 2,6-dimethylphenol 24. Several studies with in situ prepared Cu(II) catalysts for the PPO (25) formation have pointed to the involvement of dinuclear Cu(II) complexes in the crucial C-O coupling step whereas PPQ formation is proposed to be due to mononuclear complexes.[31] Under basic conditions dinuclear complex 18 is not only catalytically active for PPO formation but more important only 0.14% of quinone was found. The modest yields of polymer sofar are presumably due to competing ligand methoxylation.

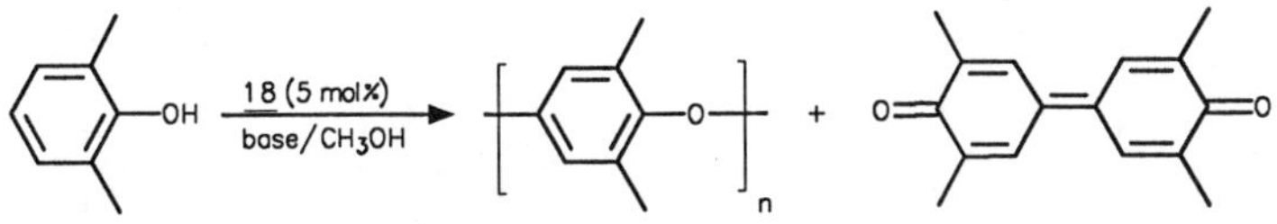

Scheme 7

These results experimentally confirm earlier proposals concerning activity and selectivity of dinuclear Cu(II) complexes in the formation of this important engineering polymer.

SELF-ASSEMBLY OF A HELICAL POLYNUCLEAR COPPER(I) COMPLEX

The design of self-assembling (supra)molecular structures is currently extensively studied and an important role for molecular helicity based on copper complexes has been revealed.

Using oligopyridines as ligands, spontaneous assembly of double helical polynuclear metal complexes has been demonstrated. Thus ligands with two to five 2,2'-bipyridine units separated by oxapropylene spacers form double helical Cu(I)-complexes[32] whereas similar molecular topology was observed with Ni(II), Pd(II), Cd(II) and Cu(I) complexes of various oligopyridines.[33] When the 1-methoxy-substituted ligand 2,6-bis[N-(2-pyridylethyl)formimidoyl]-1-methoxy-benzene(2,6-BPB-1-OCH$_3$, 26) was allowed to react with two equivalents of Cu(MeCN)$_4$BF$_4$ "chameleon-like" complexation behavior was observed (scheme 8). Dinuclear complexes 27a-27c are formed strongly dependent upon the amount of MeCN present. Acetonitrile complexation is reversible as heating of 27a at reflux in CH_2Cl_2 provided the acetonitrile free complex 27c whereas addition of 4 equivalents of MeCN reconverted it into 27a. Additions of excess (>10 equiv.) of MeCN to a solution of 27c in $CHCl_3$ provided, besides equimolar amounts of Cu(MeCN)$_4$BF$_4$, polynuclear Cu(I) complex 28.[34]

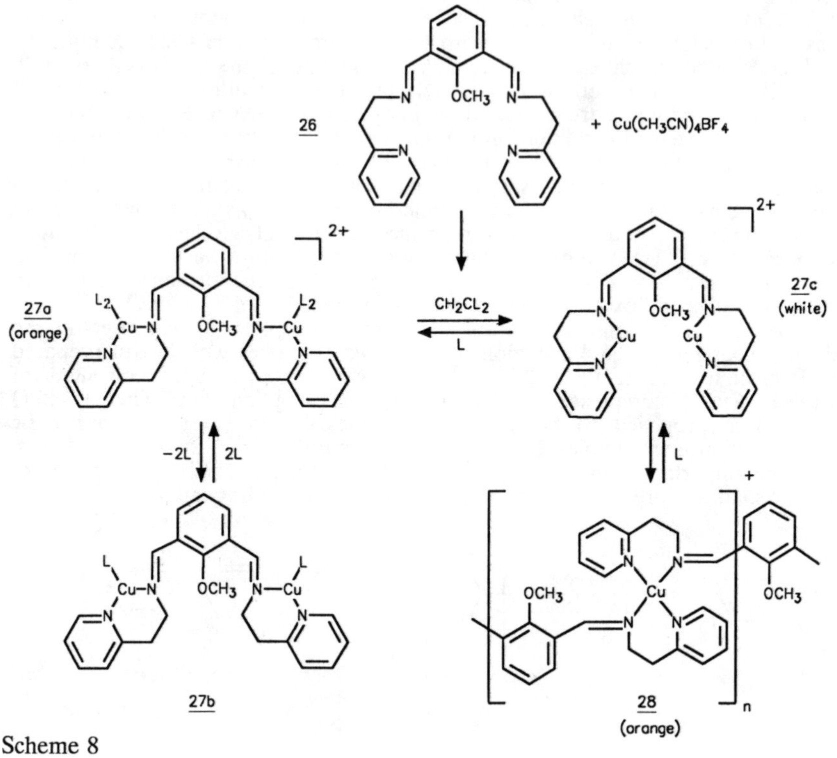

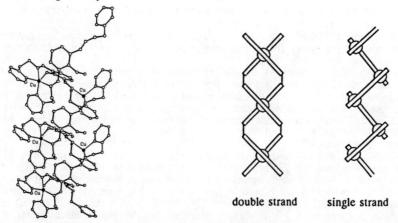

Scheme 8

The X-ray structure (figure 4) clearly shows that each Cu(I) atom is four-coordinated with two pyridylethylimine bidentate units from different ligands in a tetrahedral geometry.

double strand   single strand

Figure 4   Molecular structure of <u>28</u> and schematic representation of the helical structure of polynuclear Cu(I) complexes

312

In the linear Cu(I) coordination polymer two helices are found; single stranded one of which is left handed and the other right handed with the methoxy-substituents pointing outside the helix. A comparison with the Cu(I) coordinated double stranded helicates described by Lehn and co-workers[35] is schematically shown (figure 4).

The ability of ligands such as 26 to assemble in the presence of Cu(I) to single stranded helical complexes offers intriguing possibilities in the area of supramolecular copper-mediated catalysis. Furthermore these complexes are still very reactive towards molecular oxygen, although initial dissociation to mono- and oligo-nuclear complexes might be essential for their reactivity.

OXIDATIVE DEMETHYLATION AND DEHALOGENATION

Various enzymes including P450 dependent monooxygenases[36,37] ω-hydroxylases and ligninase[38] are known to catalyze the oxidative demethylation of arylethers. Model studies to elucidate mechanistic features as well as the development of catalysts for mild (oxidative) demethylation have been undertaken.

For this purpose 1-methoxy-substituted dinuclear Cu(I) complexes 29 and 30, which bear close structural analogies to complexes 6 and 12, were investigated.

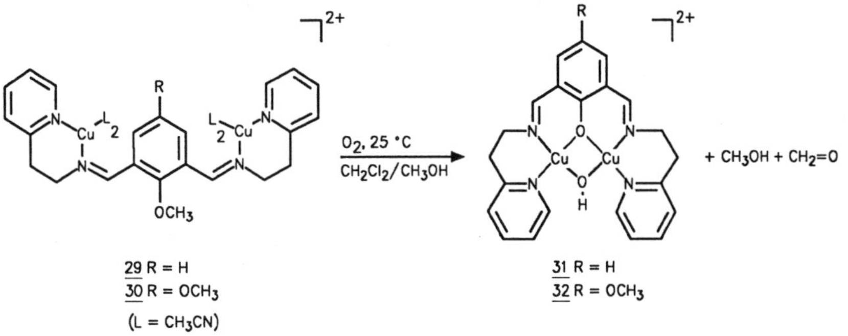

$$\xrightarrow[\text{CH}_2\text{Cl}_2/\text{CH}_3\text{OH}]{\text{O}_2,\ 25\ °\text{C}} \quad + \text{CH}_3\text{OH} + \text{CH}_2=\text{O}$$

29 R = H
30 R = OCH₃

(L = CH₃CN)

31 R = H
32 R = OCH₃

Scheme 9

In the presence of molecular oxygen the complex Cu₂(2,6-BPB-1-OMe) 29 rapidly oxidizes to 31, with a stoichiometry of O₂ uptake; Cu:O₂ = 2:1. An oxygen-induced demethylation of the anisole moiety takes place. ¹⁸O-labeling experiments support *dual pathways* with at least 60% aryl-oxygen and approximately 20% alkyl-oxygen bond cleavage; both leading to the same phenoxy-hydroxy-bridged dinuclear Cu(II) complex 31. In the *first route* an oxidative demethoxylation takes place (CH₃OH is liberated) whereas in the *second and minor route* oxidative demethylation occurs with the formation of formaldehyde. Although the nature of the intermediates in the oxidative demethylation (demethoxylation) is still unknown, the overall conversion is reminiscent of the arene-hydroxylations following O₂ binding, reported previously. A possible mechanistic pathway is given in scheme 10. Initial (reversible) oxygen binding to form 33 is followed by nucleophilic attack of the peroxodicopper(II) at the arene-1-position. Enhanced Lewis acidity in Cu(II) complex 33 and the presence of the 2,6-imine substituents make the arene moiety more vulnerable for nucleophilic attack. It has been proposed[23,26] that in the related arene-hydroxylations the peroxodicopper(II) entity is electrophilic in nature but it should be noted that the presence of electron withdrawing substituents and the ligating group might alter the mode of reaction in these types of complexes. The *major route* for the O₂-induced demethylation then proceeds via ipso-attack of the copper(II) peroxy species to form 35. Subsequent fragmentation and methoxide elimination from 35 results in phenoxy bridged complex 31. Consistent with the proposal is the slower ipso-oxidation of the more electron rich p-OCH$_3$ complex $\underline{30}$ as a result of the decreased rate of nucleophilic attack.

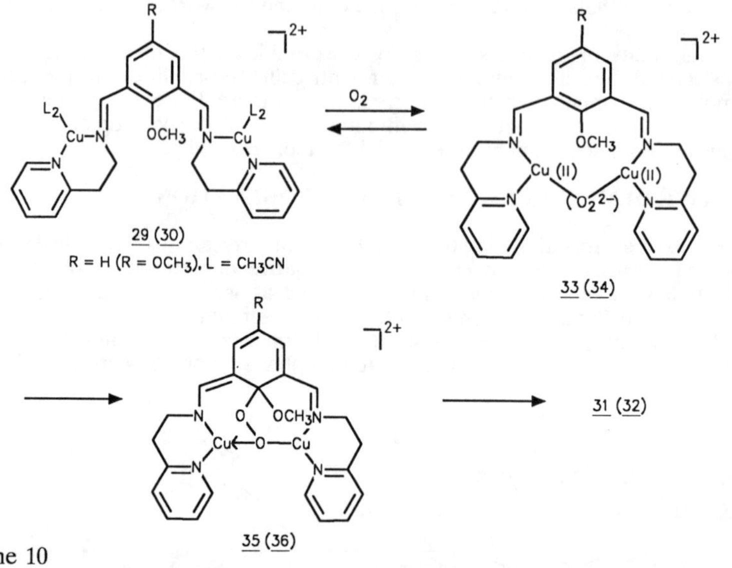

Scheme 10

The general accepted mechanism for demethylations based on cytochrome P450 dependent monooxygenases involves α-hydroxylation to a hemiacetal followed by fragmentation to phenol and carbonyl compounds, although other pathways have been suggested (scheme 11).[39]

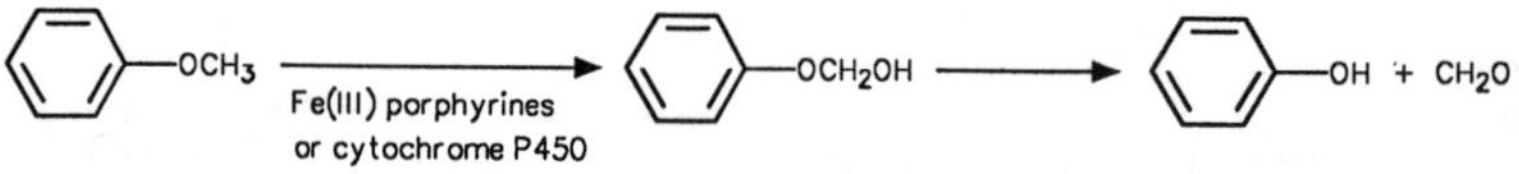

Scheme 11

The formation of $\underline{31}$ and formaldehyde (*minor route*) is reminiscent of such a demethylation mechanism. Furthermore an intriguing OCH$_3$, OCD$_3$ exchange was observed in complex $\underline{30}$ under ambient conditions in the presence of oxygen. Exclusive exchange at the 1-position, with the formation of 4-CH$_3$O-2,6-BPB-1-OCH$_3$ and 4-CH$_3$O-2,6-BPB-1-OCD$_3$ in a 1:1 ratio as well as the essential role of O$_2$ in the exchange process strongly point to the involvement of complex $\underline{37}$. Instead of attack of the peroxo-copper moiety (as in $\underline{35}$) ipso attack of deuteromethoxide has taken place and subsequent fragmentation provides $\underline{30}$ (1-OCH$_3$) or $\underline{38}$ (1-OCD$_3$). The significance of the experimental conditions (1h, $\underline{25}$ °C, CH$_2$Cl$_2$, CD$_3$OD 40:1 ratio) should be emphasized as generally Lewis acids or more severe conditions are essential to demethylate methoxy substituted arenes. The remarkable methoxide exchange further supports a nucleophilic mechanism. These experimental results indicate furthermore that electrophilic attack of peroxo-dicopper(II) might not be an exclusive pathway in arene hydroxylations.[40]

If the dinuclear copper complexes, described here, are capable of demethoxylating anisole-moieties in the presence of O$_2$, it is conceivable that oxidative dehalogenation is also possible. Oxidative dechlorination during a

314

intramolecular hydroxylation reaction mediated by a dinuclear copper complex was reported by Karlin.[41] A competing pathway to a dihydroxydicopper(II) complex was also observed, whereas highest yields of dechlorinated products were found in the presence of Zn as reductant.

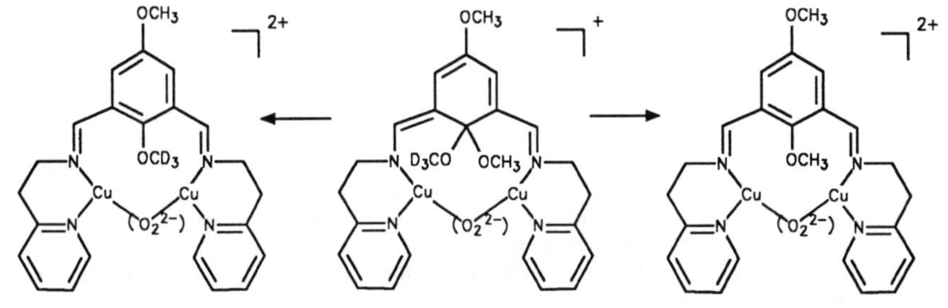

Scheme 12

Dehalogenation reactions are catalyzed by various enzymes using hydrolytic, reductive or oxidative pathways.[42] However, few enzymes are capable of oxidative dehalogenation of aromatic compounds. For instance phenylalanine hydroxylase (PAH, iron dependent) is known to defluorinate 4-fluoro-phenylalanine and converts a 4-chloro-phenylalanine to tyrosine (dechlorination) in the presence of O_2. A bacterial copper-dependent phenylalanine hydroxylase is also known. When dinuclear copper(I) complex 40, readily obtained from 1-bromo-substituted ligand 39, was allowed to react with O_2 at room temperature, high yields (80%) of dinuclear Cu(II) complex 31 were obtained.[43] Analysis for bromide showed 72% of the theoretical amount of Br⁻. In close analogy to Karlin's and our own observations with copper complexes showing monooxygenase activity is not too fetched to suggest formation of a $[Cu_2O_2]^{2+}$ complex 41 presumably followed by ipso-substitution to 42 and fragmentation with simultaneous Br⁻ elimination (scheme 13).

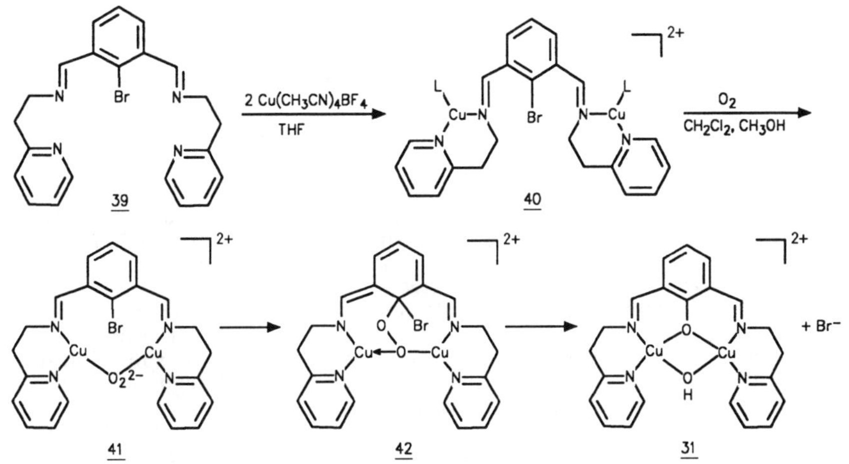

Scheme 13

315

Alternatively BrO⁻ might be formed initially with subsequent decomposition into Br⁻ and H_2O. It should be emphasized that an overall oxidative hydroxylation-debromination has taken place under mild conditions, i.e. room temperature and molecular oxygen. These copper containing systems can be considered interesting models to study the mechanism of metal-dependent oxidative dehalogenating enzymes.

CATALYTIC DEHYDROGENATION

During catalysis by monooxygenases, like tyrosinase, the incorporation of one oxygen atom into the ortho-position of phenols is accompanied by the formation of water involving an external electron- and proton-source. In an approach to design catalytically active complexes which could mimic tyrosinase, a copper(I)-copper(II) redox couple and a hydroquinone-quinone redox couple are incorporated in one system (scheme 14). The hydroquinone moiety should act as an electron shunt between an external reducing agent and the copper ions. As the aromatic nucleus is already hydroxylated, it might be expected that, after activation of O_2 either as a superoxo- or μ-peroxo-di-copper(II) intermediate, oxidation of an external substrate can occur.

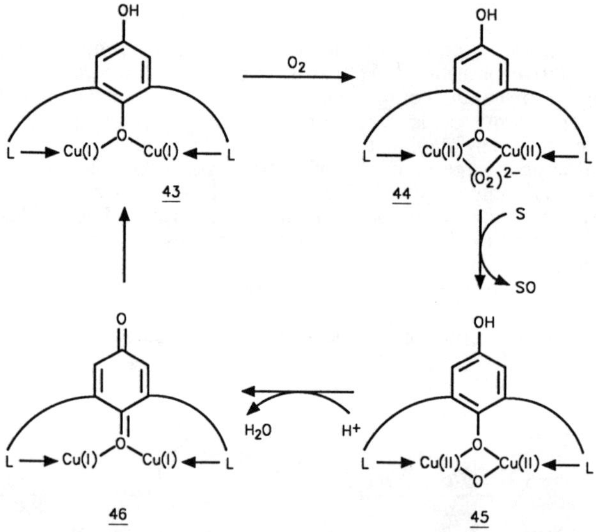

Scheme 14

This event then leads to a hydroquinone and hydroxy-bridged dinuclear Cu(II) complex 45 and an oxygenated substrate. It can be expected that the hydroquinone bridged complex 45 undergoes an internal electron transfer process to form a quinone-di-Cu(I) complex 46. Reduction of the quinone to the hydroquinone 43 via external reducing agents, i.e. ascorbic acid, zinc or by using electrochemical methods will complete the catalytic cycle. The electron-transfer system envisaged here is reminiscent of the quinone-based systems as found in the primary photochemical step in the bacterial photosynthesis[44] and in synthetic (metallo)porphyrin-quinone electron transfer systems.[45]

In order to form dinuclear copper complexes in which a hydroquinone (quinone) moiety is incorporated the new ligand 1,3-bis[N-2-(2'-pyridylethyl)-formimidoyl]-2,5-dihydroxybenzene 47 was prepared via a multistep route (scheme 15).[46] Dinuclear Cu(II) complex 48 was obtained by reaction of the bissodium salt of <u>47</u> with $Cu(ClO_4)_2 \cdot 6H_2O$.

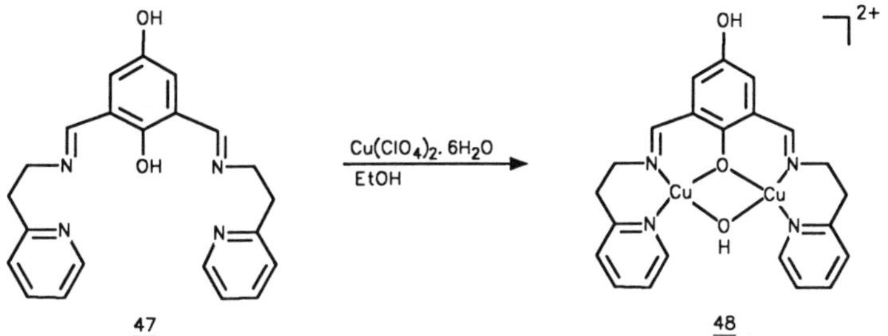

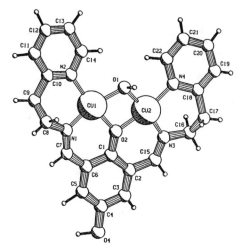

Scheme 15

Conclusive evidence for the presence of the hydroquinone and not a quinone moiety in <u>48</u> was obtained by X-ray structural analysis (figure 5).

Figure 5   Molecular structure of <u>48</u> (counter ions are omitted for clarity).

It is remarkable, and in contrast to expectation, that the hydroquinone unit in <u>48</u> is not oxidized to the corresponding quinone. Although the hydroquinone is ideally situated for electron transfer to the copper ions, this does not readily take place even in the presence of molecular oxygen. It has been suggested for related nickel complexes[47] that despite the fact that the metal ion acts as an electron sink, back-donation of electrons to the coordinated quinone might take place, strongly depending upon the redox potential of the quinone and metal and the quinone π-orbital and metal d-orbital levels. Cyclic voltametric studies under various conditions did not reveal reversible oxidation reduction patterns. As the hydroquinone oxidation in <u>48</u> does not take place a catalytic cycle as shown in scheme 14 is not feasible with the present system. Appropriate modifications in ligand <u>47</u> and complexes thereof that fulfill the requirements mentioned earlier might be obtained by reduction of the imine-functionality. It is expected that 2,6-dialkylamine-substituents will lower the oxidation potential sufficient for the quinone formation to occur. Such studies are currently in progress.

Considering the resistance towards "internal" oxidation it is even more remarkable that dinuclear copper complex 48 is an excellent catalyst for the oxidation of "external" hydroquinones to quinones and α-hydroxyketones to diketones using O_2 as the oxidant (illustrated in scheme 16).

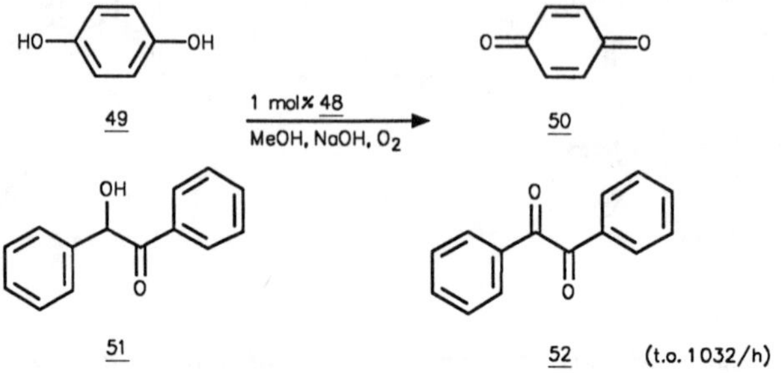

Scheme 16

Apparently the two-electron withdrawing imine substituents present in 47 sufficient increase the oxidation potential of the hydroquinone moiety to prevent intramolecular electron transfer to the Cu(II) ions. However, fast intermolecular electron transfer from 49 (or 51) apparently can take place. The stoichiometry with respect to substrate, base and O_2, and the proposed consecutive binding of phenolate and oxygen in binuclear Cu(II) complexes in the first step in catalytic phenol oxidations (vide supra) suggest the following mechanistic scheme (scheme 17). Substrate binding to 53 will be followed by two-electron transfer to yield a binuclear Cu(I) complex and subsequent O_2 binding will provide 54.

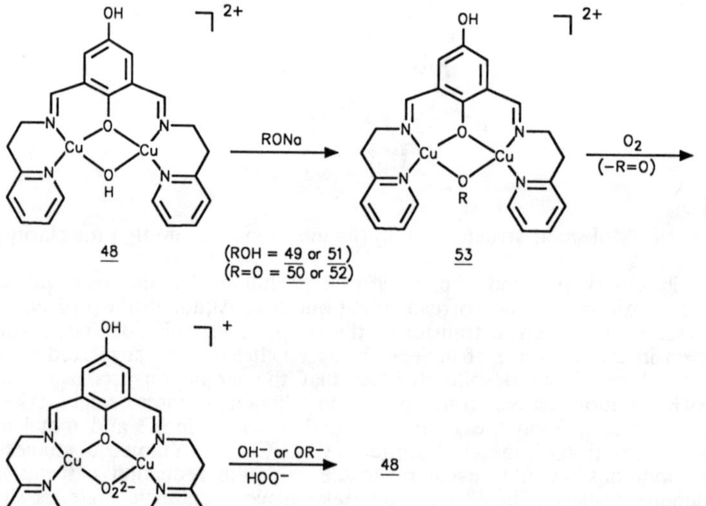

Scheme 17

318

As addition of base is essential for the oxidation to occur deprotonation of the substrate (ROH) either before or after binding to the copper ions presumably takes place. Alter-native mechanisms as shown in scheme 17 can be proposed; for instance a redox reaction of complex 54 with deprotonated substrate is an attractive possibility.

It should be emphasized that 1 mol of O_2/mol of substrate is consumed, indicating a two-electron transfer process contrary to four electron processes that lead to H_2O as was observed in oxidative phenol coupling[31] mediated by Cu(II) complexes. Furthermore, complex 48 catalyzes effectively the decomposition of H_2O_2 formed.

No oxygenation was achieved sofar with dinuclear hydroquinone complexes presented here as well as some structural analogs. The presence of a hydroquinone (or phenol) moiety in the ligands described here is only essential for bridging the Cu(II) ions but has as far as experimental evidence goes no role in the redox processes.

CHIRAL DINUCLEAR COPPER COMPLEXES

Besides mimicking monooxygenase activity and the important goal to achieve intermolecular oxidations of a variety of substrates with molecular oxygen a fascinating aspect is the use of chiral copper complexes for enantioselective oxidations. Enantioselective epoxidations using the Sharpless[1a] and Jacobsen[48] procedures as well as epoxidations with chiral metallo-porphyrin enzyme mimics[49] are the most prominent examples of oxidations mediated by chiral metal complexes. Chiral dinuclear copper complexes have sofar been unexplored for these purposes, although based on the oxygenation activity of several copper complexes interesting properties in stereoselective oxidations can be expected.

Starting from (S)-proline 55 the enantiomerically pure pentadentate ligand 56 was obtained in a multistep route. Reaction of 56 with Cu(ClO$_4$)$_2$ provided the phenoxy-hydroxy-bridged complex 57 with slightly distorted square planar geometry around each Cu(II) ion and a normal Cu-Cu separation of 2.971(3) Å. In accordance with a C_2-symmetric complex a trans-orientation of the N-benzyl-groups is found in 57.

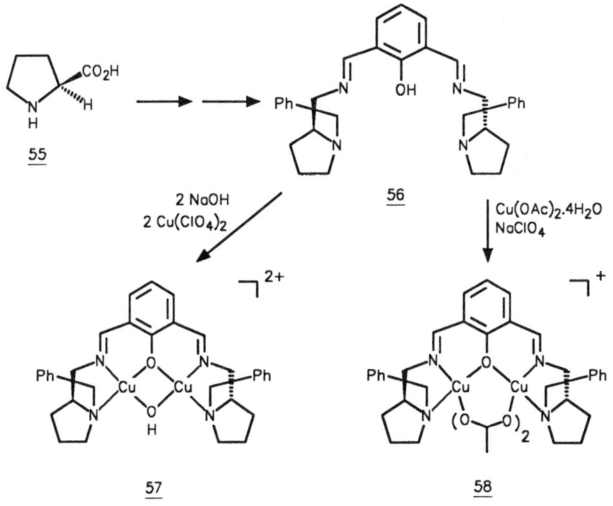

Scheme 18

319

Reaction of ligand $\underline{56}$ with $Cu(OAc)_2$ followed by treatment with $NaClO_4$ resulted in dinuclear copper complex $\underline{58}$. The molecular structure (figure 6) clearly shows the C_2- symmetry in the chiral complex with a less common bis(μ-acetato)- and phenoxo-bridges. A square-pyramidal arrangement around each Cu(II) ion and a Cu-Cu distance of 3.296(1) Å, typically for μ-phenoxo-bis(μ-acetato) bridged dinuclear metal complexes (i.e. Fe, Mn), is observed.[50]

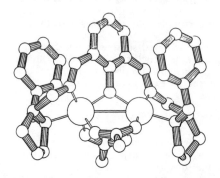

Figure 6   Molecular structure of $\underline{58}$ (counter ions are omitted for clarity).

In a further extension towards chiral dinuclear copper complexes with chiral clefts additional coordinating ligands were introduced (schematically shown in figure 7).

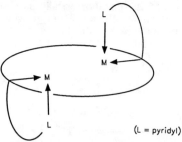

(L = pyridyl)

Figure 7   Additional coordinating ligand in a C_2-symmetric dinuclear complex.

As the N-benzyl moieties in complex $\underline{57}$ and $\underline{58}$ are rather flexible replacement of the benzyl groups by additional ligands will enforce binding in a C_2-symmetric arrangement. Higher selectivity during the approach of a prochiral substrate to the sterically more crowded and more rigid copper-centers can be expected.

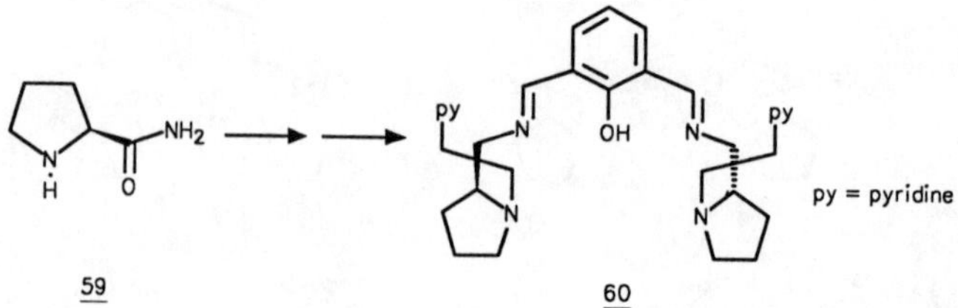

py = pyridine

$\underline{59}$ $\underline{60}$

Scheme 19

Starting with (S)-prolinamide <u>59</u> the multidentate C$_2$-symmetric ligand <u>60</u> was prepared in several steps. Both dinuclear Cu(II)- and Ni(II)-complexes were obtained with penta coordination around each metal ion. The molecular structure obtained for the dinuclear Ni(II) complex by X-ray analysis nicely illustrates the distorted square pyramidal arrangement (figure 8).

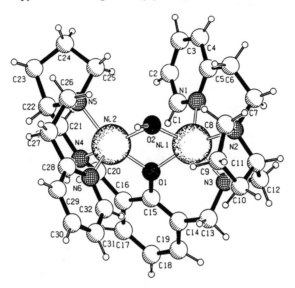

Figure 8   Molecular structure of dinuclear Ni(II) complex of <u>60</u> (counter ions are omitted for clarity).

Furthermore the C$_2$-symmetry with trans-orientation of the ligating pyridyl-ethyl moieties is clearly shown. The synthesis and structural analysis of these chiral dinuclear copper complexes clearly demonstrates that well defined geometries, with asymmetric environments around the metal centers, can be generated. In our view the approach via these new types of chiral ligands hold promise to develop future enantioselective oxidation catalysts.

CONCLUDING REMARKS

The design and synthesis of dinuclear copper complexes, with well defined geometries, offer exciting possibilities to elucidate structure-reactivity patterns in oxidations with molecular oxygen as well as to achieve effective mimics of oxygenating enzymes. It has been demonstrated that, strictly depending upon the nature of the ligands, arene-hydroxylation, demethoxylation and demethylation of anisole moieties, oxidative methoxylation, dehalogenation and phenol coupling reactions can be mediated by dinuclear copper complexes in the presence of O$_2$ as the oxidant. From the various studies from our group and others on the reactivity of dinuclear copper complexes in the presence of molecular oxygen it is clear that the rates and the type of oxygenation reaction are governed by i.e. electronic effects of the ligand substituents, geometrical constraints enforced by the ligand, redox-potentials of the dinuclear copper site and the substrates, medium effects and presumably also the nature of the dicopperperoxo-species.

Apparently minor differences in steric (or electronic) effects of the dinucleating ligands and "protection" of ligand positions prone to oxygenation with inert substituents can direct the oxygenation from an intramolecular to intermolecular mode. Furthermore oxygenations which usually require rather enforced conditions i.e. oxidative dehalogenation, hydroxylations and oxidative demethoxylation can be performed at room temperature with O_2. Although mainly intramolecular reactions were successful sofar with dinuclear copper complexes and O_2 the findings above clearly indicate that a wide range of selective intermolecular substrate oxygenations should be possible, provided we are able to achieve the proper ligand framework for the dinuclear copper center. As several key features for O_2 binding and activation are now known it will be essential to study effective (and cheap) procedures for the oxygenation and simultaneous reduction of the "second" oxygen to water in order to arrive at catalytic processes. We are even further challenged by the design of chiral dinuclear copper complexes for enantioselective and catalytic oxygenations with molecular oxygen.

ACKNOWLEDGEMENTS

I am grateful to my co-workers involved in this project: Dr. O.J. Gelling, Drs. R. Pieters, Drs. J. Alkema, Drs. M.T. Rispens, Drs. A. de Vries. The help of Dr. J.C. de Jong and the X-ray analyses of Drs. A. Meetsma, F. van Bolhuis and A. Spek are greatly appreciated.

REFERENCES

1. a. T. Katsuki and K.B. Sharpless, *J. Am. Chem. Soc.*, **102**, 5974 (1980). b. K.A. Jørgensen, R.A. Wheeler, R. Hoffmann, *J. Am. Chem. Soc.*, **109**, 3240 (1987). c. A.E. Tai, L.D. Margerum, J.S. Valentine, *J. Am. Chem. Soc.*, **108**, 5006 (1986). d. J.P. Collman, P. Denisevich, Y. Konai, M. Marrocco, C. Koval, F.C. Anson, *J. Am. Chem. Soc.*, **102**, 6027 (1980). e. R.H. Holm, *Chem. Rev.*, **87**, 1401 (1987). f. K.D. Karlin and Y. Gultneh, *Prog. Inorg. Chem.*, **35**, 219 (1987).

2. H. Mimoun, in *Comprehensive Coordination Chemistry*, G. Wilkinson, Ed., (Pergamon Press, 1987), vol 6, pp. 317-410.

3. J.T. Groves, in *Metal Ion Activation of Dioxygen*, T.G. Spiro, Ed., (Wiley, New York, 1980), p. 125.

4. a. *Copper Proteins and Copper Enzymes*, R. Lontie, Ed., (C.R.C. Boca Raton, Fl., 1984). b. E.I. Solomon, in *Metal Ions in Biology*, T. Spiro, Ed., (Wiley Interscience, New York, 1981), vol. 3. p. 41. c. R. Lontie and R. Witters, *Met. Ions Biol. Syst.*, **13**, 229 (1981). d. K. Lerch, *ibid* **13**, 143 (1981).

5. a. K.D. Karlin, R.W. Cruse, Y. Gultneh, J.C. Hayes, J. Zubieta, *J. Am. Chem. Soc.*, **106**, 3372 (1984). b. K.D. Karlin, R.W. Cruse, Y. Gultneh, A. Faroog, J.C. Hayes, J. Zubieta, *ibid*. **109**, 2668 (1987).

6. R.R. Jacobsen, Z. Tyeklar, A. Faroog, K.D. Karlin, S. Liu, J. Zubieta, *J. Am. Chem. Soc.* **110**, 3690 (1988).

7. N. Kitajima, K. Fujisawa, Y. Moro-oka, *J. Am. Chem. Soc.* **111**, 8975 (1989).

8. a. K.D. Karlin, P.L. Dahlstrom, S.N. Gozzette, P.M. Scensny, J. Ziebata, *J. Chem. Soc., Chem. Commun.*, 881 (1981). b. K.D. Karlin, J.C. Hayes, Y. Gultneh, R.W. Cruse, J.W. McKown, J.P. Hutchinson, J. Zubieta, *J. Am. Chem. Soc.*, **106**, 2121 (1984).

9.   a. L. Casella, L. Rigoni, *J. Chem. Soc., Chem. Commun.*, 1668 (1985). b. L. Casella, M. Gulotti, G. Palanza, L. Rigoni, *J. Am. Chem. Soc.*, **110**, 4221 (1988).
10.   O.J. Gelling, A. Meetsma, F. van Bolhuis, B.L. Feringa, *J. Chem. Soc., Chem. Commun.*, 552 (1988).
11.   T.N. Sorrell, *Tetrahedron*, **45**, 51 (1989).
12.   J.S. Thompson, *J. Am. Chem. Soc.*, **106**, 8308 (1984).
13.   a. R. Menif and A.E. Martell, *J. Chem. Soc., Chem. Commun.*, **15**, 21 (1989). b. R. Menif, A.E. Martell, R.J. Squattrito, A. Clearfield, *Inorg. Chem.*, **29**, 4366 (1990).
14.   M. Réglier, E. Amadéi, R. Tadayoni, B. Waegell, *J. Chem. Soc., Chem. Commun.*, 447 (1989).
15.   a. J.T. Groves, T.E. Nemo, R.S. Meyers, *J. Am. Chem. Soc.*, 6101, 1032 (1979). b. J.P. Collman, J.I. Brauman, B. Meunier, T. Hayashi, T. Kodadek, *J. Am. Chem. Soc.* **107**, 2000 (1985). c. I. Tabushi, *Coord. Chem. Rev.* **86**, 1 (1988).
16.   M. Réglier, C. Jorand, B. Waegell, *J. Chem. Soc., Chem. Commun.* 1752 (1990).
17.   L. Casella, M. Gulotti, R. Radaelli, P. Di Gennaro, *J. Chem. Soc., Chem. Commun.* 1611 (1991).
18.   C.R.H.I. de Jonge, in *Organic Synthesis by Oxidation with Metal Compounds*, W.J. Mijs and C.R.H.I. de Jonge, Eds., (Plenum Press, New York, 1986), pp. 423-444.
19.   C. Glaser, *Ber. Dtsch. Chem. Ges.*, **2**, 422 (1869).
20.   W. Brackman, E. Havinga, *Recl. Trav. Chim. Pays-Bas* **74**, 937 (1955).
21.   A.S. Hay, H.S. Blanchard, G.F. Endres, J.W. Eustance, *J. Am. Chem. Soc.* **81**, 6335 (1959).
22.   a. S.K. Mandal and K. Nag, *J. Chem. Soc., Dalton Trans.*, 2141 (1984). b. J. Lorösch and W. Haase, *Inorg. Chim. Acta*, **108**, 35 (1985). c. J.J. Grzybowsky, P.H. Merrell, F.L. Urbach, *Inorg. Chem.* **17**, 3078 (1979).
23.   Z. Tyeklar and K.D. Karlin, *Acc. Chem. Res.*, **22**, 241 (1989).
24.   J. Alkema, O.J. Gelling, B.L. Feringa, to be published.
25.   T.N. Sorrell, M.R. Malachowski, D.L. Jameson, *Inorg. Chem.*, **21**, 3250 (1982).
26.   K.D. Karlin, B.I. Cohen, R.R. Jacobson, J. Zubieta, *J. Am. Chem. Soc.*, **109**, 6194 (1987).
27.   R. Pieters, O.J. Gelling, B.L. Feringa, manuscript in preparation.
28.   P. Capdevielle and M. Maumy, *Tetrahedron Lett.* **32**, 3831 (1991).
29.   T. Shono in *Electroorganic Chemistry as a New Tool in Organic Synthesis*, (Springer Verlag, Berlin, 1984), ch. 2.
30.   K. Veera Reddy, S.-J. Jin, P.K. Arora, D.S. Sfeir, S.C.F. Maloney, F.L. Urbach, L.M. Sayre, *J. Am. Chem. Soc.*, **112**, 2332 (1990). See also: F. Wang and L.M. Sayre *J. Am. Chem. Soc.*, **114**, 248 (1992).
31.   a. F.J. Viersen, Ph.D. Thesis, University of Groningen, 1988. b. H.C. Meinders and G. Challa, *J. Mol. Catal.*, **7**, 321 (1980). c. S. Tsuruya, K. Nakamae, H. Yonezawa, *J. Catal.* **44**, 40 (1976).
32.   J.M. Lehn and A. Rigault, *Angew. Chem. Int. Ed. Engl.*, **27**, 1095 (1988).

33. a. E.C. Constable, M.D. Ward, M.G.B. Drew, G.A. Forsyth, *Polyhedron*, **8**, 2551 (1989). b. E.C. Constable, M.D. Ward, D.A. Tocher, *J. Am. Chem. Soc.*, **112**, 1256 (1990). c. E.C. Constable, S.M. Elder, J. Healy, M.D. Ward, D.A. Tocher, *ibid.* **112**, 4590 (1990).

34. O.J. Gelling, F. van Bolhuis, B.L. Feringa, *J. Chem. Soc., Chem. Commun.*, 917 (1991).

35. J.M. Lehn, A. Rigault, J. Siegel, B. Harrowfield, B. Chevrier, D. Moras, *Proc. Natl. Acad. Sci. USA* **84**, 2565 (1987).

36. G.A. Hamilton in *Molecular Mechanisms of Oxygen Activation*, O. Hayashi, Ed., (Academic Press, New York, 1974).

37. J.R. Lindsay Smith and P.R. Sleath, *J. Chem. Soc., Perkin Trans. II*, 621 (1983).

38. a. K. Miki, V. Renganathan, M.H. Gold, *FEBS Lett.*, **203**, 235 (1986). b. J.M. Palmer, P.J. Harvey, H.E. Schoemaker, *Phil. Trans. R. Soc. Lond. A.*, **321**, 495 (1987).

39. V. Ullrich, *Angew. Chem. Int. Ed. Engl.*, **11**, 701 (1972).

40. O.J. Gelling and B.L. Feringa *J. Am. Chem. Soc.* **112**, 7599 (1990).

41. M.S. Nasir, B.I. Cohen, K.D. Karlin, *Inorg. Chim. Acta*, **176**, 185 (1990).

42. S. Keuning, D.B. Janssen, B. Witholt, *J. of Bacteriology*, **163**, 635 (1985).

43. O.J. Gelling and B.L. Feringa *Recl. Trav. Chim. Pays-Bas* **110**, 89 (1991).

44. J. Deisenhofer, O. Epp, K. Miki, R. Huber, H. Michel, *J. Mol. Biol.*, **180**, 385 (1984).

45. M.P. Irvine, R.J. Harrison, G.S. Beddard, P. Leighton, J.K.M. Sanders, *Chem. Phys.*, **104**, 315 (1986).

46. O.J. Gelling, A. Meetsma, B.L. Feringa, *Inorg. Chem.*, **29**, 2816 (1990).

47. C. Benelli, A. Dei, D. Gatteschi, L. Pardi, *J. Am. Chem. Soc.*, **110**, 6897 (1988).

48. W. Zhang, J.L. Loebach, S.R. Wilson, E.N. Jacobsen, *J. Am. Chem. Soc.*, **112**, 2801 (1990).

49. J.T. Groves and R.S. Myers, *J. Am. Chem. Soc.*, **105**, 5791 (1983).

50. O.J. Gelling, Ph.D. Thesis, University of Groningen, 1990.

DIOXYGEN ACTIVATION AND TRANSPORT BY DINUCLEAR COPPER(I) MACROCYCLIC COMPLEXES

Arthur E. Martell, Rached Menif, Patrick M. Ngwenya and David A. Rockcliffe

Department of Chemistry, Texas A&M University, College Station, Texas 77843-3255, U.S.A.

Recent papers[1,2] reported the synthesis of a 24-membered macrocyclic tetra Schiff base, $MX_2(DIEN)_2$, 1, 3,6,9,17,20,23-hexatricyclo[23.3.1.1^{11,15}]triaconta-1(29),2,9,11(30),12(13),14,16,23,25,27-decane, $C_{24}H_{30}N_6$, by the 2+2 condensation of m-phthaldehyde and diethylenetriamine by the general procedure of Zagwinski et al.[3] This non-template one-step procedure for forming the macrocyclic ligand is straightforward and gives a high yield. The Schiff base product obtained is an equilibrium mixture of the parent tetra Schiff base and imidazole-containing lower Schiff bases in which the central nitrogen of the diethylenetriamine moiety has added to an imine (Schiff base) group. The two main constituents of this mixture are believed to be those illustrated in Scheme I (1, 1a). Incomplete formation of the macrocycle 1 because of its equilibrium with other species such as 1a caused no difficulty since the equilibrium was shifted on the addition of $Cu(CH_3CN)_4ClO_4$ to give a nearly quantitative yield of the binuclear Cu(I) complex, 2. When the solution of the orange complex 2, obtained in CH_2Cl_2 solution under argon, was exposed to dioxygen at ambient room temperature, the color gradually turned to green indicating the formation of the binuclear Cu(II) complex, 3. The green product was hydrolyzed in 6 M HCl and extracted with chloroform to give a 74% yield of the 2-hydroxy-m-phthaldehyde, 4. An $^{18}O_2$ tracer study showed that all of the phenolic oxygen came from the dioxygen that reacted with 2. Spectroscopic evidence indicated that the green complex formed in non-aqueous CH_2Cl_2 solution was the μ-hydroxy-μ-phenoxy dicopper(II) complex 3 (Scheme I). The insertion reaction, electron transfer and the formation of the bridging hydroxyl and phenoxy groups is visualized as involving the formation of an intermediate peroxo-bridged binuclear Cu(I)-dioxygen complex, illustrated in Scheme I by the formula in broken brackets. It is noted that all the elements needed for the bridging groups are present at the active site and water is not needed for the reaction. A possible way by which the peroxo-bridged binuclear Cu(I) dioxygen complex could rearrange with electron transfer to give the hydroxide and phenolate bridged binuclear Cu(II) complex is indicated by a and b.

It is noteworthy that a cobalt(II) dioxygen complex formed under the same conditions underwent a metal-centered oxidation to the corresponding Co(III) complex and no oxygen insertion took place.

Hydrogenation of 1 produced the macrocycle 5 which has more basic aliphatic amino donor groups than 1, as indicated by the potentiometric equilibrium

From K.D. Karlin and Z. Tyeklár, Eds., Bioinorganic Chemistry of Copper
(Chapman & Hall, New York, 1993).

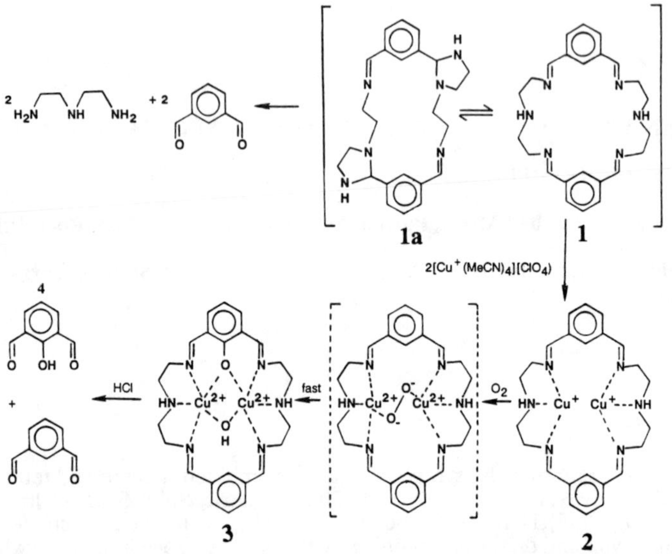

Scheme 1. Proposed Mechanism for the Cu(I)-Mediated Hydroxylation of a Coordinated Macrocyclic Ligand

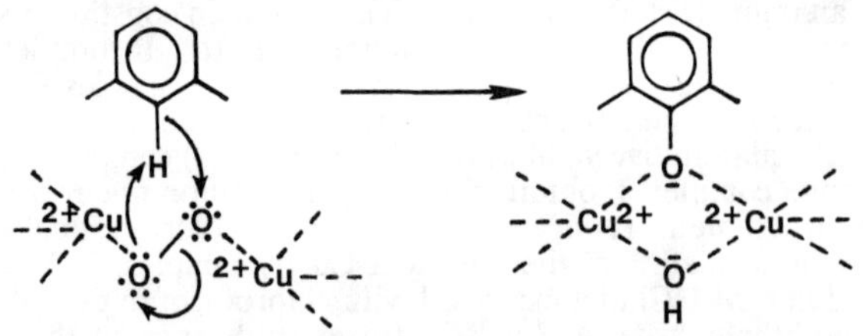

curves in Figure 1, and the protonation constants in Table 1. The Cu(II) curves in Figure 1 show that both mononuclear and binuclear complexes are formed. The facile protonation of the mononuclear complex indicates that the Cu(II) is coordinated to one diethylenetriamine moiety, while the protonation occurs on the other. A probable bonding arrangement for the diprotonated mononuclear Cu(II) chelate is indicated by formula **6**. The binuclear Cu(II) macrocyclic complex M_2L (Table 1) has strong tendencies to form mono and dihydroxo species. It is suggested that the hydroxo groups are bridging (are coordinated simultaneously to both metal centers) as indicated by formula **7** for the dihydroxo species. Similar behavior was suggested previously for the binuclear complex of a closely-related ligand, BISDIEN.[4]

326

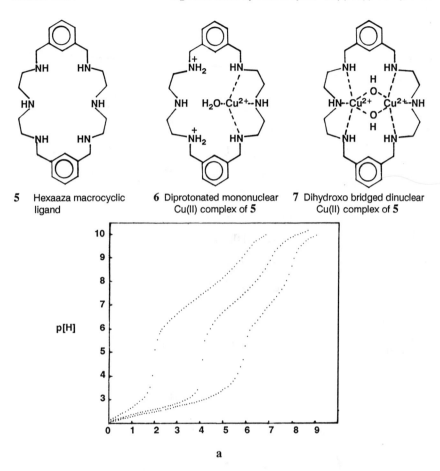

5 Hexaaza macrocyclic ligand

6 Diprotonated mononuclear Cu(II) complex of **5**

7 Dihydroxo bridged dinuclear Cu(II) complex of **5**

Figure 1. p[H] profiles of the hexahydrobromide of hexaazamacrocycle **5**, $C_{24}H_{38}N_6 \cdot 6HBr$, L, and for 1:1 and 2:1 molar rations of Cu(II) to ligand. Concentration of ligand $= 1.00 \times 10^{-3}M$, t $= 25.0$ °C, $\mu = 0.100$ M (KNO₃). a = moles of base (KOH) added per mole of ligand.

A very stable binuclear Cu(I) complex, **8**, was prepared by the (3+2) condensation of isophthalde-hyde and tris(aminopropyl)amine (TRTN) in very good yield (75%).[5] The ligand was combined with two equivalents of Cu(CH₃CN)₄ClO₄ to give the binuclear Cu(I) complex **8**, indicated by the crystal structure shown in Figure 2. This binuclear Cu(I) complex is unusual in that it does not react with dioxygen or with other donors that usually react with Cu(I), such as carbon monoxide, cyanide, azide, iodide, thiocyanate, or sulfide. This lack of reactivity cannot be due to the large spacing (4.44 Å) between the Cu centers because some of these donors fit nicely between the metal ions. Also, not even sol-

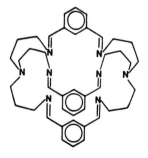

MX₃TRTN₂, **8**

Table 1. Protonation Constants and Cu(II) Binding Constants of the Hexaaza Macrocyclic Ligand L, **5**.[a]

Symbol	Equilibrium Quotient	Log K^X
K_1^H	[HL]/[H][L]	9.49
K_2^H	[H_2L]/[H][HL]	8.73
K_3^H	[H_3L]/[H][H_2L]	8.03
K_4^H	[H_4L]/[H][H_3L]	7.29
K_5^H	[H_5L]/[H][H_4L]	3.64
K_6^H	[H_6L]/[H][H_5L]	3.45
K_{ML}^M	[ML]/[M][L]	13.79
K_{MHL}^H	[MHL]/[H][ML]	8.69
$K_{MH_2L}^H$	[MH_2L]/[H][MHL]	7.32
$K_{M_2L}^M$	[M_2L]/[M][ML]	9.68
$K_{M_2(OH)L}^{OH}$	[M_2(OH)L][H]/[M_2L]	-7.26
$K_{M_2(OH)_2L}^{OH}$	[M_2(OH)$_2$L][H]/M_2(OH)L]	-8.40

[a] 25.0 °C, M = 0.100 M (KNO_3).

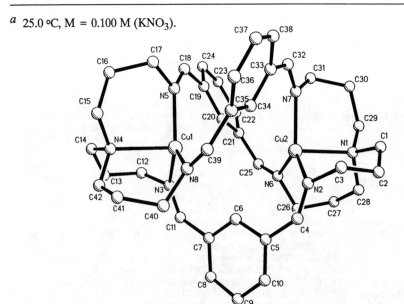

Figure 2. View of the dinuclear Cu(I) cryptate [$Cu_2(MX_3TRTN_2)$]$^{2+}$, **9**.

vent molecules are present in the cavity of the cryptate. On the basis of the evidence available, it was suggested that Cu(I) is very effectively coordinated by four nitrogen donors of very low basicity, three of which are "soft" imine nitrogens which form coordinate bonds to Cu(I) with considerable covalent character. It should also be pointed out that the Cu(I) centers in this inert binuclear complex are four-coordinate.

A binuclear copper cryptate was formed by combining two equivalents of Cu(II) with the cryptand **10** obtained by hydrogenation of the octa Schiff's base, **9**, synthesized by the $(3+2)$ condensation of isophthaldehyde with triaminotriethyl-amine (TREN) respectively, as indicated in Scheme 2.[6] The potentiometric equilibrium curves showing formation of the Cu(II) complex are shown in Figure 3

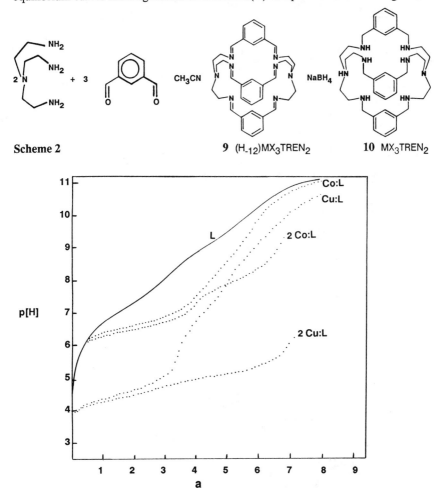

Scheme 2   **9** (H$_{-12}$)MX$_3$TREN$_2$ **10** MX$_3$TREN$_2$

Figure 3.   Potentiometric equilibrium profiles of 1.00 x 10^{-3} M ligand, L [(MX)$_3$(TREN)$_2$).6HBr], in the absence of metal ion and in the presence of 1:1 and 2:1 mole ratios of Cu(II) and Co(II) to L as indicated. p[H] = -log [H$^+$]; a = moles of base (0.100 M KOH) added per mole of ligand present; t = 25.0 °C; μ = 0.100 M, adjusted with KNO$_3$.

329

and the constants calculated from the potentiometric data are presented in Table 2. The data show considerable resemblance between the complexes formed from this new cryptand and those formed by the parent cryptand ligand, O-BISTREN. Formulas of the hydroxo-bridged binuclear macrobicyclic binuclear complexes represented by **11** and **12** show the similarities between the two complexes. The crystalline binuclear copper complex was found by X-ray analysis to be the carbonato-bridged structure **13** (Figure 4), wherein two of the oxygens of the carbonato group are coordinated to each of the copper(II) centers, while the third is bound to a hydronium ion.

Table 2. Logarithmic Formation Constants^a of Cu(II) Chelates of (MX)$_3$(TREN)$_2$^{b} and O-BISTREN^c

Equilibrium Quotient	Log K$_f$	
	(MX$_3$)(TREN)$_2$	O-BISTREN
[ML]/[M][L]	16.79	16.54
[M$_2$L]/[M]^2[L]	26.20	29.21
[MLOH][H]/[ML]	-9.70	-10.23
[M$_2$LOH][H]/[M$_2$L]	-4.58	-4.26
[MHL]/[ML][H]	8.65	8.78
[MH$_2$L]/[MHL][H]	7.29	7.70
[MH$_3$L]/[MH$_2$L][H]	6.23	6.87

^{a} t = 25.0 °C; μ = 0.100 M NO$_3$. ^{b} Reg.6. ^{c} Ref.7.

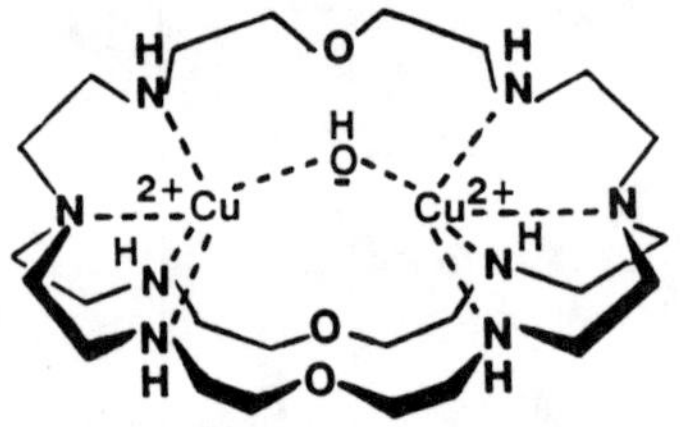

11 Cu$_2$(OH)O-BISTREN^{3+}

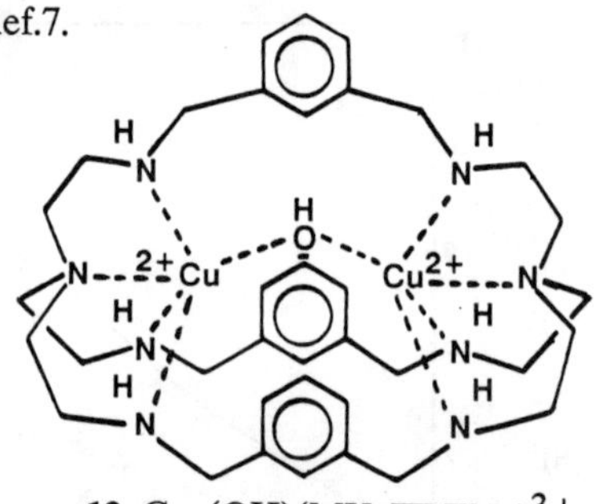

12 Cu$_2$(OH)(MX$_3$TREN$_2$)$^{3+}$

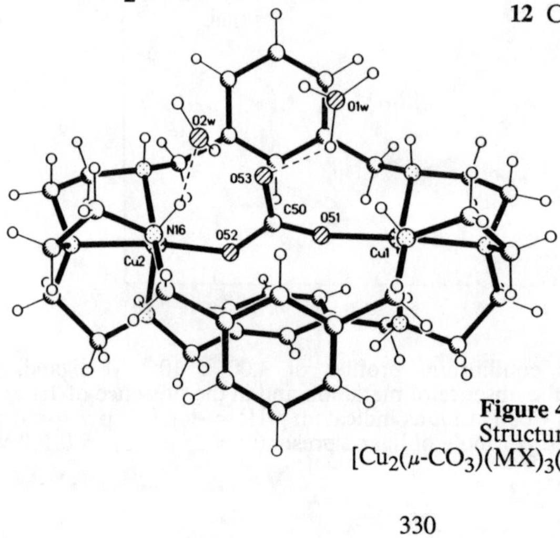

Figure 4.
Structure of the major component of
[Cu$_2$(μ-CO$_3$)(MX)$_3$(TREN)$_2$(H$_3$O)]Br$_3$.3H$_2$O, **13**

Because the instability of the dioxygen complex of the binuclear Cu(I) macrocyclic complex was considered as due at least in part to the proximity of the dioxygen to a site (aromatic C-H) susceptible to hydroxylation, it was decided to synthesize a binuclear copper(I) macrocyclic dioxygen complex with an unreactive atom near the binuclear center. Such a complex was prepared by the (2+2) condensation of furan-2,5-dialdehyde and diethylenetriamine, to give the macrocyclic tetra Schiff base ligand **14**, as indicated in Scheme 3.[8] This ligand was converted to the corresponding binuclear copper(I) complex, **15**, which was then treated with dioxygen to give the complex **16**. The deep red-brown complex was characterized spectrally with a moderately strong absorption band at around 500 nm (as well as a stronger one at ~360 nm). It is noted that the coordinated dioxygen (indicated in **16** in the form of a peroxo bridge between the metal centers) is not near any molecular group with which it might react. The nearest atoms are the ether oxygens of the furan rings. The formation of the dioxygen complex was followed by the 500 nm band, as indicted by the spectra illustrated in Figure 5. Figure 6 shows the rate of formation of the oxygen complex to be fairly rapid up to two hours when the curve levels off due to degradation of the dioxygen-containing species to the green Cu(II) complex, **17**. The binuclear Cu(I) complex thus formed is fairly long-lived. It was found that by following its spectral absorbtivity to have a half-life of about two hours at 5.0 °C and 45 minutes at 25.0 °C, indicating the time span of its existence in solution.

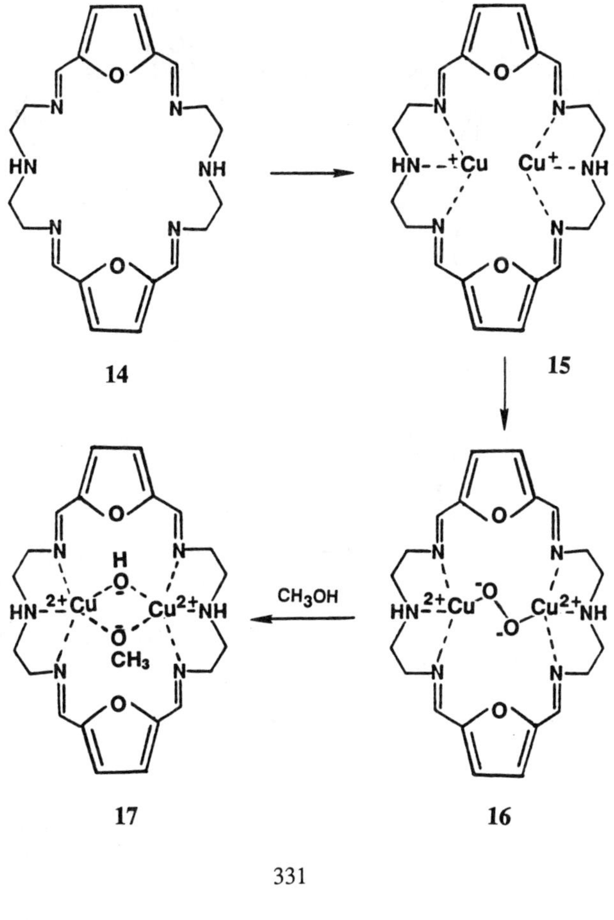

Scheme 3

14

15

17

16

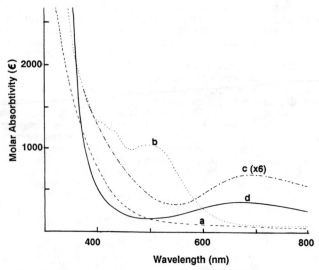

Figure 5. Absorption spectra of the copper-$(FD)_2(DIEN)_2$ complexes in 1:3 acetonitrile-methanol solution: (a, --) 1.84×10^{-4} M Cu(I) complex of $(FD)_2(DIEN)_2$ prior to oxygenation; (b, ...) system a during oxygenation; (c, ---), the irreversible degradation product of b, 1.05×10^{-3} M; and (d, —) the Cu(II) complex of $(FD)_2(DIEN)_2$, 1.84×10^{-4} M.

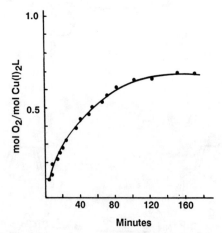

Figure 6. Oxygen absorption by Cu(I)-$(FD)_2(DIEN)_2$ complex (1.8×10^{-3} M) at 25.0 °C. $P_{O_2} = 153$ mm.

The stoichiometric and catalytic oxidation of various reducing substrates by the binuclear macrocyclic copper(I) dioxygen complex, **16**, was then investigated.[9] The substrates chosen for study are phenols, quinones, catechols, ascorbic acid, and 3,4-dimethylaniline. The substrates investigated and the oxidation products obtained are listed in Table 3 and the time course of their redox interactions with the

332

dioxygen complexes are given in Figures 7-10. It is noted that Cu(II) complexes may also oxidize these substrates and that there are two ways for copper(II) complexes to form: 1, by the degradation of the macrocyclic binuclear copper(I) dioxygen complex (16 → 17, Scheme 3) and 2, by the two-electron reduction of the dioxygen (the peroxo-bridge) which in the solvent employed would produce a copper(II) complex similar to or identical with 17. Therefore, to avoid this complication initial rates of the redox reactions illustrated in Figures 7-10 were estimated and are listed in Table 3.

The binuclear copper(II) complex of the macrocyclic ligand 14 (Scheme 3) was prepared from the macrocycle by treating it with two equivalents of CuCl$_2$ dissolved in methanol. The time course of the stoichiometric oxidations of the reducing substrates are also illustrated in Figures 7-10 and initial observed rates

Table 3.   Initial Rates in the Cu(I)-Dioxygen and Cu(II) Oxidations of Substrates

		Initial, Pseudo First Order Rate, s^{-1}	
Substrate	Product	Cu(I) + O$_2$	Cu(II)
Hydroquinone	Benzoquinone	1.1×10^{-4}	2.2×10^{-5}
t-Butylhydroquinone	*t*-butylbenzoquinone	1.2×10^{-4}	1.6×10^{-5}
2,6-Di-*t*-butylphenol	3,3′,5,5′-tetratertiarybutyldiphenoquinone	2.4×10^{-4}	5.8×10^{-5}
2,6-Dimethoxyphenol	3,3′,5,5′-dimethoxydiphenoquinone	7.4×10^{-3}	5.6×10^{-5}
3,5-Di-*t*-butylcatechol	3,5-ditertiarybutyl-1,2-benzoquinone	1.4×10^{-8}	~0
3,4-Dimethylaniline	3,4-dimethylnitrosobenzene	6.5×10^{-5}	~0
Ascorbic acid	Dehydroascorbic acid	5.93×10^{-4}	1.96×10^{-5}
4-*t*-butylcatechol	γ lactone of maconic acid methyl ester	1.05×10^{-8}	~0

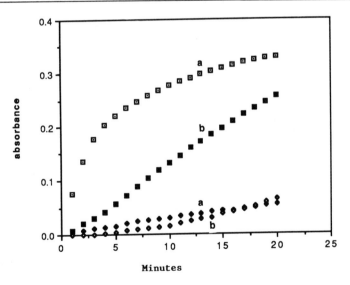

Figure 7.   Time course for the formation of 3,3′,5,5′-tetramethoxydiphenoquinone (a) and 3,3′,5,5′-tetra-*t*-butyldiphenoquinone (b) during the oxidation of 2,6-dimethoxyphenol and 2,6-di-*t*-butylphenol with [Cu$_2$(FD)$_2$(DIEN)$_2$] + O$_2$ (upper curves) and Cu$_2^{II}$-(FD)$_2$(DIEN)$_2$ (lower curves).

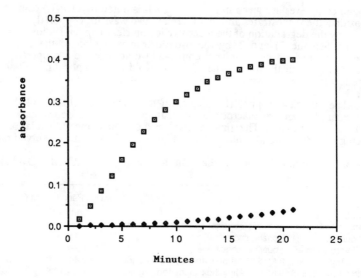

Figure 8. Time course for the formation of dehydroascorbic acid during the oxidation of ascorbic acid with $[Cu_2(FD)_2(DIEN)_2]$ + O_2 (upper curve) and Cu_2^{II}-$(FD)_2(DIEN)_2$ (lower curve).

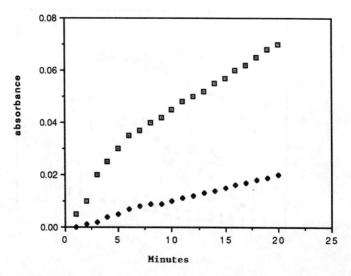

Figure 9. Time course for the formation of benzoquinone during the oxidation of hydroquinone with $[Cu_2(FD)_2(DIEN)_2]$ + O_2 (upper curve) and Cu_2^{II}-$(FD)_2(DIEN)_2$ (lower curve).

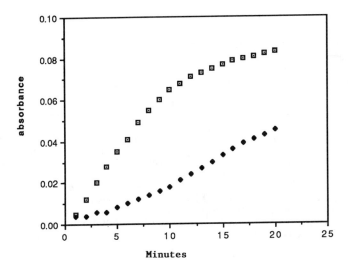

Figure 10. Time course for the formation of *t*-butylbenzoquinone during the oxidation of *t*-butylhydroquinone with [Cu$_2$(FD)$_2$(DIEN)$_2$] + O$_2$ (upper curve) and Cu$_2$^{II}-(FD)$_2$(DIEN)$_2$ (lower curve).

are listed in Table 3. Although the rates given are initial values, it is seen from the figures that they do not change much as the reaction proceeds. In the case of stoichiometric oxidations by the Cu(II) complex, the use of initial rates is probably not necessary, since the reduction product cannot be an oxidant for the reactions (the Cu(II) complex does not degrade and no free oxygen is present to combine with the Cu(I) complex formed). For every substrate listed the rate of Cu(II) oxidation is significantly lower than that of oxidation by the Cu(I) dioxygen complex. In the case of three reductants (the catechols and 3,4-dimethylaniline) oxidation by the copper(II) complex does not occur at all.

The reactions listed in Table 3 were also run in a catalytic mode, with the macrocyclic binuclear copper(I) complex as catalyst, in the presence of excess dioxygen, and substrate.[9,10] The reaction was followed by the amount of dioxygen taken up, as described previously.[11] The reaction was assumed to be over when no further oxygen was taken up. The results of the catalytic studies, presented in Table 4, show from 2-5 turnovers for five substrates. No catalytic activity was detected for the substrates which were not oxidized by the macrocyclic binuclear copper(II) complex represented by **17**. The proposed catalytic cycle shown in Scheme 4 is set up with ditertiarybutylphenol as the substrate. For the other substrates the cycles proposed are similar, differing only in the reducing substrates and the corresponding oxidation products. For the reductant shown in Scheme 4, ditertiarybutylphenol, the oxidation by the Cu(I) dioxygen complex is seen in Table 3 and Figure 7 to be about four times faster than the oxidation of the same substrate by Cu(II). Thus after the initial phase of the catalytic process one would expect that in the presence of excess substrate, the relative concentrations of the two oxidants in the cyclic catalytic process to adjust themselves so as to approach the relative concentrations determined by their relative rates as oxidants. In this case the concentration of the Cu(I) dioxygen complex as an oxidant need be only one fifth the concentration of all copper species present in order to keep up to the oxidation reaction of the copper(II) complex, which would be present at four times its concentration.

Table 4. Catalytic Activity of $[Cu_2(FD)_2(DIEN)_2]^{2+}$ in the Oxidation of Various Substrates

Substrate	Turnover/hr	Time of Reaction (min)
Hydroquinone	7	60
t-Butylhydroquinone	5	43
2,6-Di-t-butylphenol	25	5
2,6-Dimethoxyphenol	30	10
Ascorbic Acid	6	46

For extension of this research, it is planned to examine other reducing substrates that may react with this macrocyclic binuclear copper(I) dioxygen complex, **16**, such as benzene, toluene, N,N-dimethylaminobenzene, tyrosine and L-DOPA. Also, it is planned to synthesize binuclear copper(I) dioxygen complexes with analogous macrocyclic ligands containing electronegative bridging groups other than ether oxygen, such as would be obtained by replacing the furan ring by pyrrol, thiophene, and pyridine. It is also planned to determine the reactivity of **2** and dioxygen with ligands in which the H of the C-H group is replaced by a methyl, a fluoride, or an OH group in an effort to block the hydroxylation reaction.

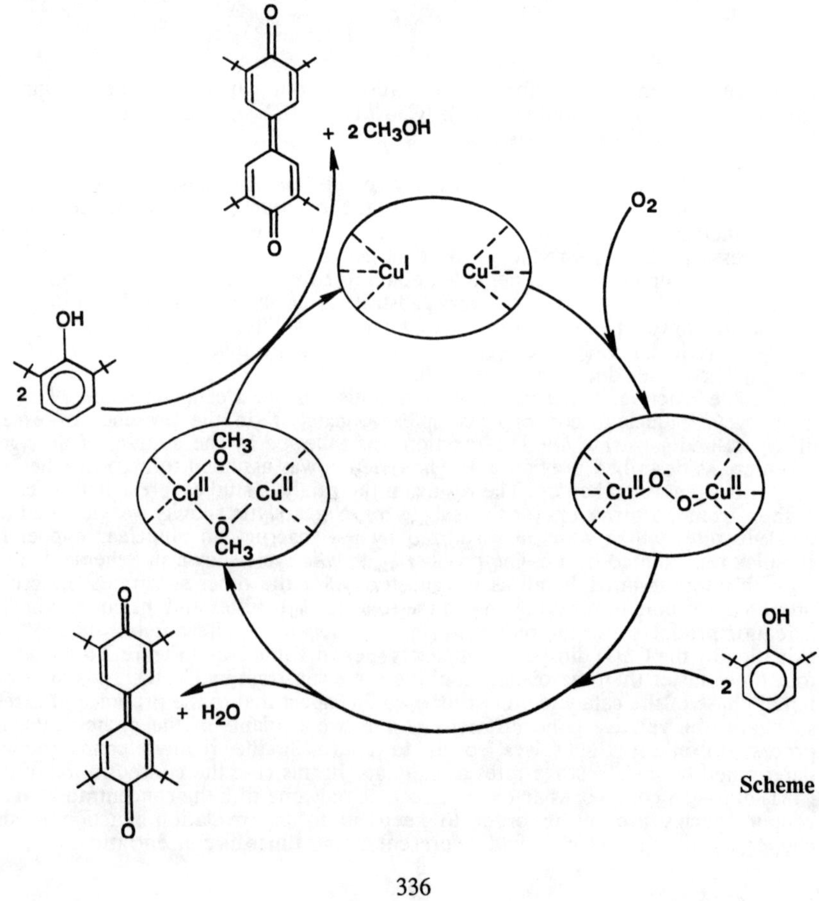

Scheme 4

Acknowledgement: This research was supported by the Office of Naval Research.

References
1. R. Menif and A.E. Martell, *J. Chem. Soc. Chem. Commun.*, 1552 (1989).
2. R. Menif, A.E. Martell, P.J. Squattrito, A. Clearfield, *Inorg. Chem.* **29**, 4723 (1990).
3. J. Zagwinski, J.M. Lehn, R. Meric, J.P. Vigneron, *Tetrahedron Lett.*, **28**, 3489 (1987).
4. R.J. Motekaitis, A.E. Martell, J.P. Lecomte, J.M. Lehn, *Inorg. Chem.* **22**, 609 (1983).
5. M.P. Ngwenya, A.E. Martell, J. Reibinspies, *J. Chem. Soc. Chem. Comm.* 1207 (1990).
6. R. Menif, J. Reibenspies and A.E. Martell, *Inorg. Chem.*, **30**, 3446 (1991).
7. R.J. Motekaitis, A.E. Martell, J.M. Lehn, E.I. Watanabe, *Inorg. Chem.*, **21**. 4253 (1982).
8. M.P. Ngwenya, D. Chen, A.E. Martell, J. Reibenspies, *Inorg. Chem.*, **30**, 2732 (1991).
9. D.A. Rockcliffe and A.E. Martell, *J. Chem. Soc. Chem. Commun.*, submitted.
10. D. A. Rockcliffe and A. E. Martell, to be published.
11. D. Chen and A.E. Martell, *Inorg. Chem.* **26**, 1026 (1987).

IMIDAZOLE-LIGATED COPPER COMPLEXES: SYNTHESIS, STRUCTURE, AND REACTIVITY

Thomas N. Sorrell, Martha L. Garrity, Joseph L. Richards, F. Christopher Pigge, and William E. Allen

Department of Chemistry, The University of North Carolina, Chapel Hill, NC 27599-3290, USA

A diverse group of copper-containing proteins carries out the transport and activation of the O_2 molecule in biological systems.[1-5] Extensive characterization of biomimetic analogs of two of those, namely hemocyanin and tyrosinase, has afforded a better understanding of the inorganic chemistry of peroxocopper complexes at the molecular level.[6-12]

Histidine is a ubiquitous ligand of copper ions in proteins, and Figure 1 illustrates the coordination sphere for each of the three types of copper-containing proteins. It is difficult to find cuproproteins that have no histidine residue coordinated to the metal ion, so it is not surprising that copper(I) complexes having imidazole ligands are becoming increasingly studied as model compounds for the active sites of dioxygen-binding and activating proteins.[6,7c]

The synthesis of chelating ligands having heterocyclic donors that *mimic* imidazole has had a richer history in the field of copper chemistry,[6a] and pyrazole, pyridine, and benzimidazole groups have found the most utility, because of their ready availability. 2-Vinylpyridine and 2-(chloromethyl)pyridine are each commerically available and require little manipulation to be incorporated into metal chelating structures.

Benzimidazoles are easily prepared from carboxylic acids and o-phenylenedi-amine,[13] and pyrazoles are the products of hydrazine and β-diketones.[14] Functionalized starting materials are therefore simply obtained, for the most part.

Except for the commercially available histidine and histamine, both of which can be used to form Schiff-base ligands,[9] imidazoles pose some problems for the synthetic chemist because they possess two nitrogen atoms attached to the same carbon atom. The standard technique used in the preparation of benzimidazoles cannot be employed for imidazole because the aliphatic analog of o-phenylenediamine is unstable. Construction of the imidazole ring therefore requires a condensation reaction between reactive nitrogen-containing intermediates and carbonyl compounds. Imidate esters, formamide, and α-aminoketones and aldehydes are the building blocks on which we have relied for the preparation of imidazole-containing chelates.

From K.D. Karlin and Z. Tyeklár, Eds., *Bioinorganic Chemistry of Copper* (Chapman & Hall, New York, 1993).

Figure 1   Examples of Cuproprotein Active Site Structures

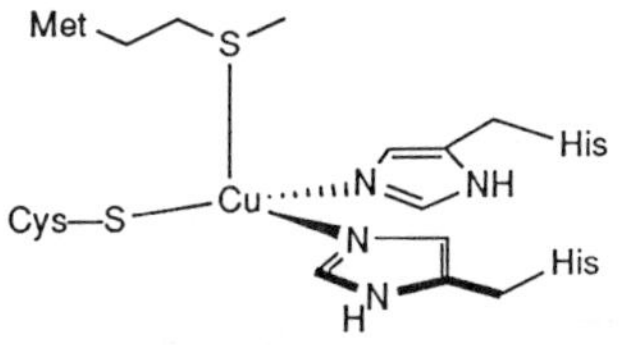

Type I: "Blue" Copper Proteins 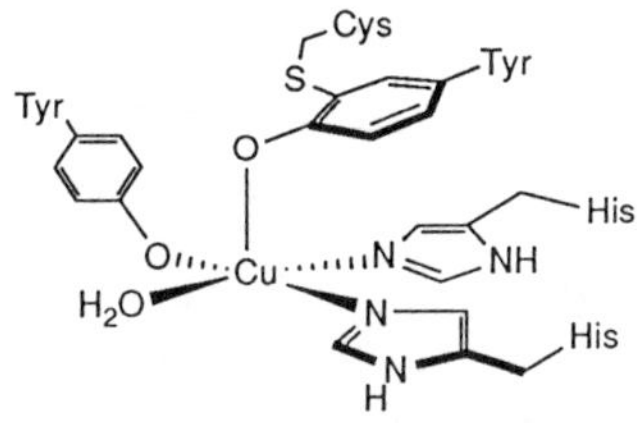   Type II: Galactose Oxidase

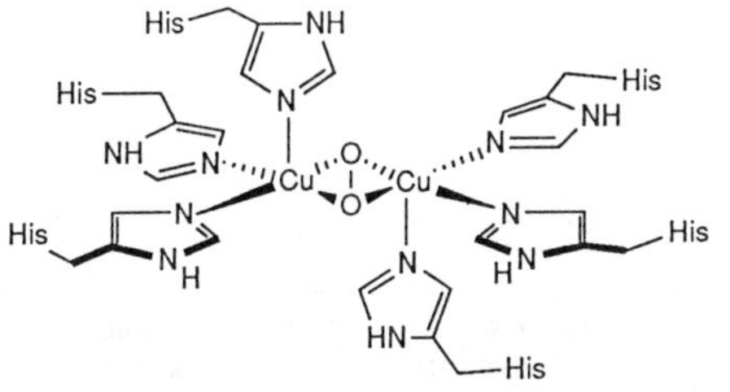

Type III: Oxyhemocyanin

LIGANDS BASED ON TRIS(IMIDAZOLYL)METHANE

Modeling the reversible binding of dioxygen by hemocyanin is a goal that has attracted much interest over the past two decades. Especially important to this field was the characterization in 1989 by Kitajima and co-workers of the first $\mu-\eta^2,\eta^2-$peroxo copper(II) complex,[8a] a discovery that dramatically changed the way we view the structure of the oxygenated form of hemocyanin. Theoretical[15] and experimental[16,17] work since that time has focused on understanding the detailed structural and reactivity properties of this unusual coordination mode, and subsequent crystallographic studies have confirmed this unit as the structure of the active site in oxyhemocyanin.[18]

Because the ligand employed by Kitajima's group is an anion with three nitrogen donors provided by pyrazole rings, we became interested in preparing an imidazole analog to see what changes, if any, could be expected from a structurally analogous ligand having no charge. In fact, we had prepared the prototype of this system several years ago, using a tris(imidazole) coordination unit to bind copper in the presence of carbon monoxide as a way to probe the luminescence properties of hemocyanin-carbonyl (HcCO).[19] Figure 2 shows a schematic view, based on an X-ray crystal structure, of the copper complex obtained from tris[2-(N-methylimidazolyl)]methoxymethane (timm). Figure 3 illustrates the observed luminescence spectra for the complex in both the presence and absence of carbon monoxide.

One of the more useful consequences of this earlier work on the model was the hypothesis that deoxyhemocyanin should also display a weak luminescence when irradiated with light at 280-300 nm. Because such a property had not been observed before, it proved worthwhile to examine hemocyanin from a variety of organisms. As expected, we were able to demonstrate that the deoxy form of the protein does luminesce,[20] and Beltramini and co-workers have subsequently examined the photophysics of the luminescence in detail.[21]

Figure 2 A schematic representation of the structure of $[Cu(timm)]_2{}^{2+}$ and its solution reactivity with MeOH and CO as determined by conductivity measurements.

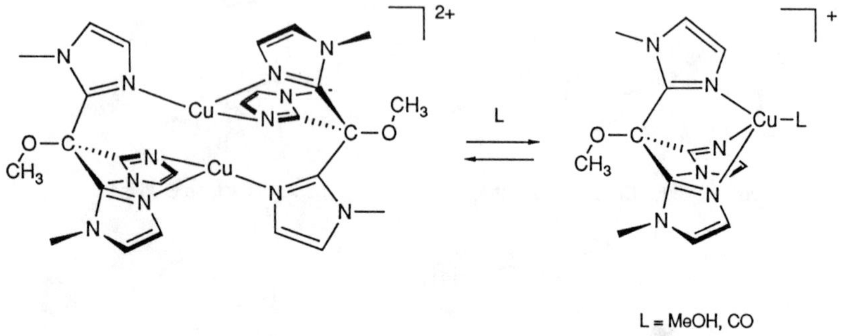

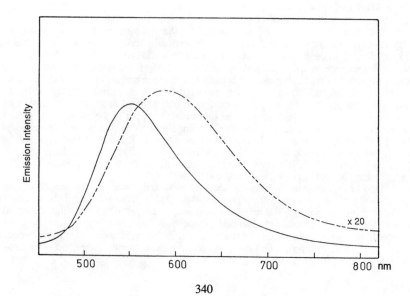

L = MeOH, CO

Tripod Ligands with Hindered Imidazoles

We have now begun an investigation into methods to synthesize hindered analogs of the original timm ligand. The general scheme to prepare the imidazole, outlined in Figure 4, makes use of α-aminoketones derived from pinacolone or 3-methyl-2-butanone. It works well only when the alkyl group on the nitrogen atom is isopropyl, *tert*-butyl, aryl, or benzyl.

Primary alkyl groups can be attached to nitrogen by alkylation of 4-isopropyl-imidazole, in which case two isomers are formed. A very tedious chromatographic separation is the only route to the desired 1-alkyl-4-substituted imidazole.

Figure 3 Luminescence spectra of $[Cu(timm)]_2{}^{2+}$ in the absence (— -- —) and presence (———) of carbon monoxide at 77K in a methanol—ethanol glass. The excitation wavelength is 280 nm.

340

Figure 4   Synthetic schemes for the synthesis of hindered 1,4-disubstituted imidazoles

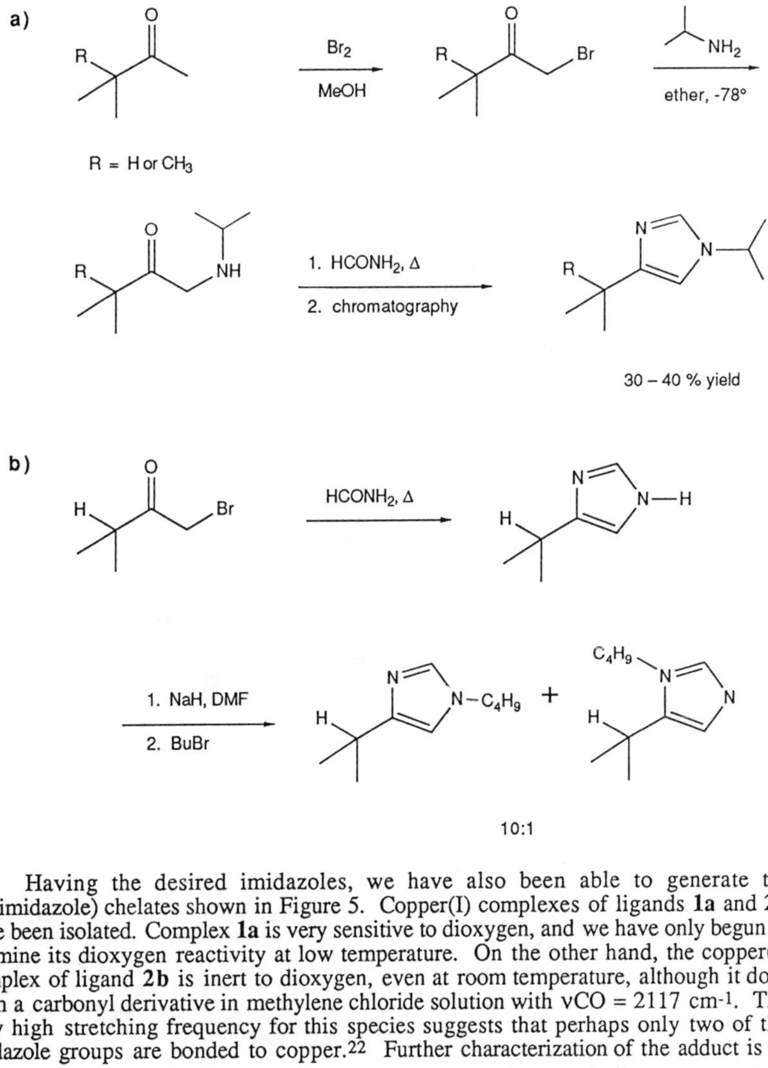

a)

R = H or CH$_3$

30 – 40 % yield

b)

10:1

Having the desired imidazoles, we have also been able to generate the tris(imidazole) chelates shown in Figure 5. Copper(I) complexes of ligands **1a** and **2b** have been isolated. Complex **1a** is very sensitive to dioxygen, and we have only begun to examine its dioxygen reactivity at low temperature. On the other hand, the copper(I) complex of ligand **2b** is inert to dioxygen, even at room temperature, although it does form a carbonyl derivative in methylene chloride solution with $vCO = 2117$ cm^{-1}. The very high stretching frequency for this species suggests that perhaps only two of the imidazole groups are bonded to copper.[22] Further characterization of the adduct is in progress.

The ligands shown at the bottom of Figure 5 (**1a, 1a'**, and **2a**) will provide a series to study the relationship between structure and reactivity of copper(I) complexes toward dioxygen. The methoxy and hydroxy ligands are neutral, but the alkoxide carries a negative charge with it, like tris(pyrazolyl)borate. With the isopropyl groups on the imidazole rings, the oxygen functionality of the ligand is shielded and is expected not to interact with the metal ion.

Figure 5 Synthesis of tris(imidazole)methanol derivatives of hindered imidazoles

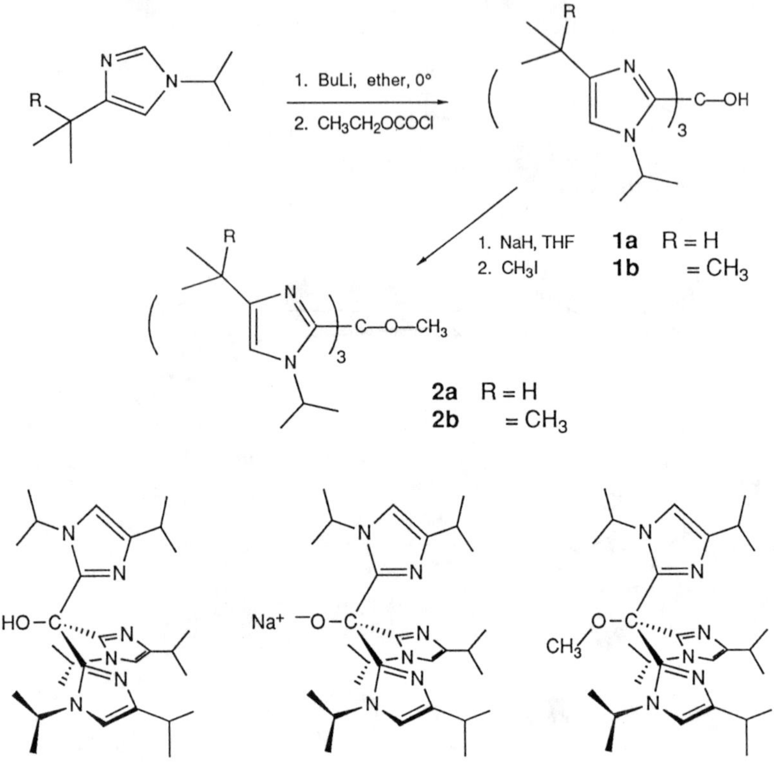

1a 1a' 2a

LIGANDS BASED ON IMIDAZOLYLETHYLAMINE

We have recently reported a synthesis of tridentate, imidazole-containing chelates.[6b] By that methodology, ligands **3**, **4a** and **4b** could be prepared by the routes shown in Figure 6. The key step is the conversion of the nitrile to the imidate ester which reacts with aminoacetaldehyde dimethylacetal to form the imidazole nucleus. *This is a reaction which we believe can be adapted to the synthesis of many diverse imidazole-containing chelating ligands.* The ligands were converted to the corresponding copper (I) complexes by reaction with $[Cu(CH_3CN)_4]BF_4$ in methanol or tetrahydrofuran under an inert atmosphere. The *n*-butyl derivatives of the ligands were used to circumvent solubility problems at low temperature.

To ensure that the imidazole-ligated Cu(I) complexes had structures analogous to previously characterized pyridine, pyrazole, and benzimidazole species,[6a] we carried out a crystal structure determination of {*N*,*N*-bis[2-(1'methyl-2'imidazolyl)ethyl]amine} copper(I) tetrafluoroborate (L = **4a**) and the structure is depicted in Figure 7.   The geometry is the expected T-shape which is common for complexes of this type of ligand, and the bond distances and angles vary little from those observed in related three-coordinate complexes.[6a]

Figure 6   The synthesis of imidazole chelates via imidate esters

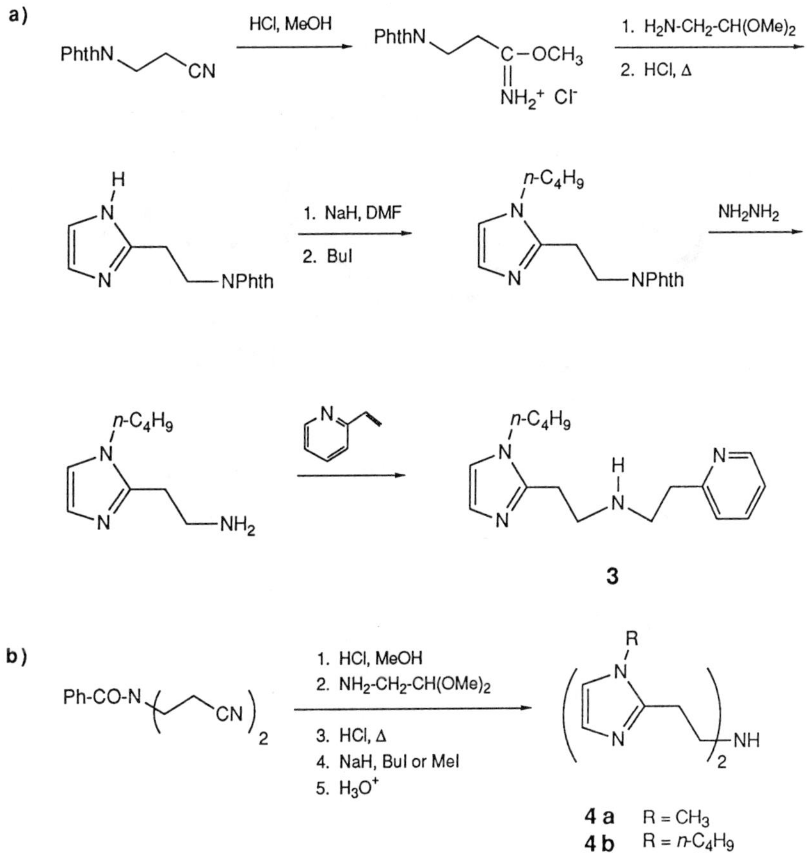

a)

b)

4 a   R = CH₃
4 b   R = n-C₄H₉

The reactivity of dioxygen with complexes CuL⁺ (L = **3** and **4b**) dissolved in CH$_2$Cl$_2$ at -78° differs from the derivatives studied before.[6b]   Thus, CuL⁺ (L = **5** and **6**) react with dioxygen to give *brown* species,[6,7] but CuL⁺ (L = **4b**) reacts instantaneously under the same conditions to give a *purple* species having a lifetime of less than a minute at -78°. Fortunately, the mixed imidazole-pyridine complex (L = **3**) gives an analogous purple species which is stable for up to two hours at low temperature (Figure 8). Manometry on the soluble complex at -78° in methylene chloride reveals a stoichiometry of 2 Cu per 1 O$_2$.[23]

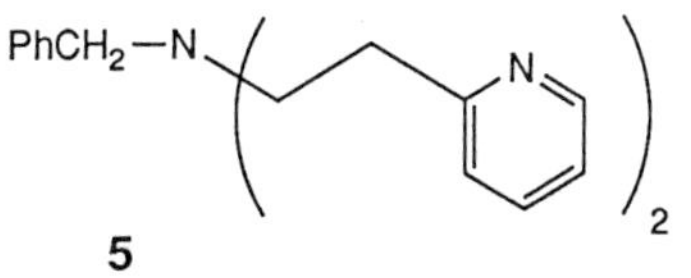

5 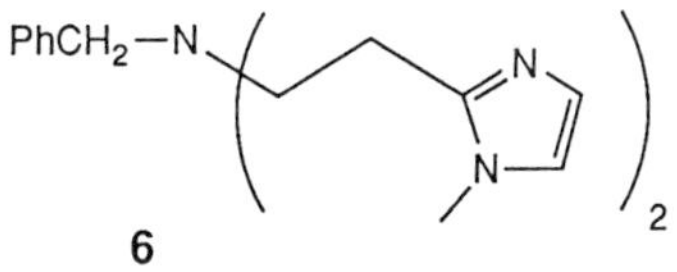   **6**

343

Figure 7   X-ray crystal structure of CuL+ (L = **4a**). Bond lengths (Å) and angles (deg): Cu-N11 = 1.898 (7), Cu-N21 = 1.899 (7), Cu-N3 = 2.213 (7), N11-Cu-N3 = 97.7 (3), N11-Cu-N21 = 164.9 (3), N21-Cu-N3 =97.7 (3).

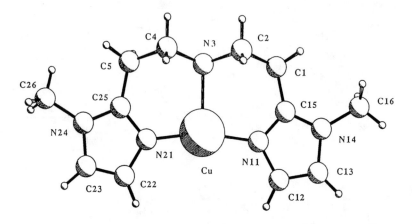

Figure 8   Visible spectra for the dioxygen reaction with complexes CuL+ (L =**3**) (———) and (L = **5**) (- - - - -) at -78 °C in CH$_2$Cl$_2$.

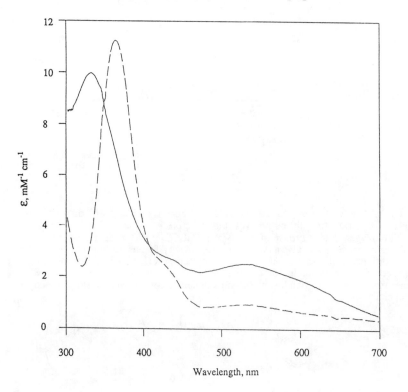

344

To characterize the putative peroxo complex further, we recorded the resonance Raman spectrum of the purple species $[LCu-O_2-CuL]^{2+}$ (L = 3).[24] We tentatively assign an enhanced band at 832 cm[-1] to the O-O stretch and bands around 450 cm[-1] to Cu-O (and Cu-N) stretching modes. Because the O-O stretching frequency is much higher than that observed for the known $\mu-\eta^2,\eta^2$–peroxocopper(II) complex (741 cm[-1]), [8a] but is the same as that for a structurally characterized trans-μ-1,2 peroxo complex (832 cm[-1]),[15,25] we believe that the structure of our dioxygen adduct has the form $[LCu-O-O-CuL]^{2+}$, in which each copper ion is square planar (A):

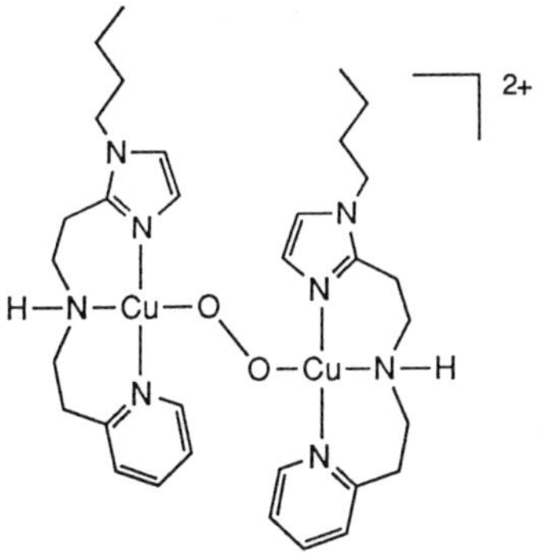

A 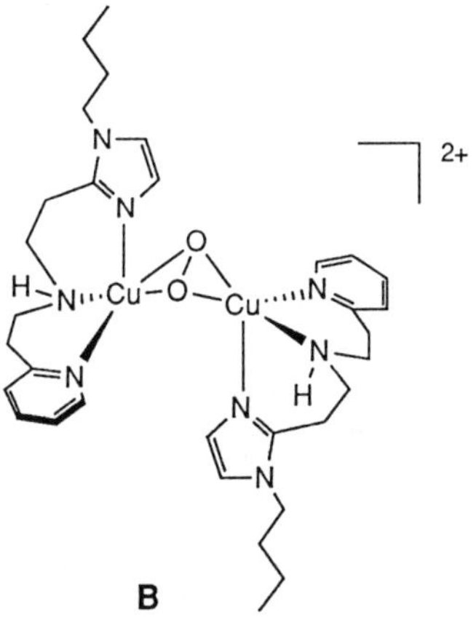   **B**

Karlin has proposed that the dioxygen adducts formed from copper(I) complexes of N-alkyl[bis(pyridylethyl)amine] ligands (e.g., 5) have a bent $\mu-\eta^2,\eta^2$ structure analogous to **B**. This hypothesis is based on EXAFS spectra and on their differences in reactivity, versus other types of dioxygen adducts, toward a variety of reagents including protons and triphenylphosphine.[17] However, the dioxygen adduct $[LCu-O_2-CuL]^{2+}$ (L = 3) has the same reactivity as that formed when L = 5 (Figure 9). It is also the same as the dinuclear copper complexes previously described by Karlin[7b] in that it is unreactive at -78° toward protons (up to 10 equivalents) and toward Ph3P.

In contrast, the only structurally characterized trans-μ-1,2 peroxo copper(II) complex[17,25] reacts immediately with 2 equivalents of acid, generating hydrogen peroxide, or with triphenylphosphine, liberating dioxygen and forming a copper(I) phosphine adduct.

Thus for the dioxygen adduct $[LCu-O_2-CuL]^{2+}$ (L = 3), we find it difficult to reconcile the observed reactivity patterns, which favors structure **B**, with νO-O, consistent only with structure **A**. It is intuitively unlikely for T-shaped Cu(I) complexes to react to form a structure like **B**, because the expected square planar geometry of the resulting copper(II) ion is readily accommodated by **A**. With a non-planar ligand like tris(pyrazolyl)borate, the copper(II) peroxo complex would have to be tetrahedral at copper if the perioxide group were to bind in a μ-1,2 fashion, so it is not surprising that the alternative side-bound geometry results.[26]

We also do not understand why removal of an alkyl group from the amine group of these tridentate ligands would change the structure of the subsequently generated dioxygen adduct so much. We are currently attempting to understand these enigmatic results by investigating other derivatives of these ligands. We believe that more data are needed to define the structure/reactivity relationships among the different types of copper-dioxygen complexes.

Figure 9 Comparison of the physical and reactivity properties of copper dioxygen complexes [LCu-O$_2$-CuL]$^{2+}$ at -78 °C in dichloromethane.

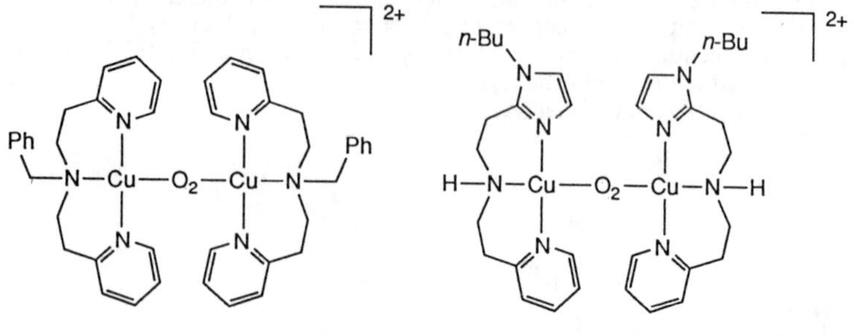

(L = 5)

brown

- stable for days
- inert to < 10 equiv H$^+$
- inert to Ph$_3$P

(L = 3)

violet

- stable for hours
- inert to < 10 equiv H$^+$
- inert to Ph$_3$P

SUMMARY AND CONCLUSIONS

We have explored two general methods for the synthesis of imidazole derivatives that should be useful for preparing chelating ligands. Sterically encumbered imidazoles having an isopropyl or *tert*-butyl group in the 4-position are readily obtained from the corresponding α-aminoketone and formamide.

Copper(I) complexes of imidazole ligands have properties that recommend their continued study as mimics for the active site of cuproproteins. The luminescence properties of hemocyanin are reproduced with a complex having three imidazole ligands to copper. Furthermore, copper(I) imidazole complexes react with dioxygen to give adducts with spectroscopic properties that are reminescent of those observed in natural systems, particularly oxyhemocyanin.

Acknowledgement is made to the University Research Council at UNC, the National Science Foundation (CHE-9100280), and the A. P. Sloan Foundation for financial support of this work.

References and Notes

1. A. Volbeda, W. G. J. Hol, *J. Mol. Biol.*, **209**, 249-279 (1989).
2. K. Lerch, *Life Chem. Rep.*, **5**, 221-234 (1987).
3. J. L. Cole, P. A. Clark, E. I. Solomon, *J. Am. Chem. Soc.*, **112**, 9534-9548 (1990).
4. N. Ito, S. E. V. Phillips, C. Stevens, Z. B. Ogel, M. J. McPherson, J. N. Keen, K. D. S. Yadav, P. F. Knowles, *Nature*, **350**, 87-90 (1991).
5. M. C. Brenner, J. P. Klinman, *Biochemistry*, **28**, 4664-4670 (1989).
6. a) T. N. Sorrell, *Tetrahedron*, **45**, 3-65 (1988). b) T. N. Sorrell, M. L. Garrity, *Inorg. Chem.*, **30**, 210-215 (1991).
7. a) K. D. Karlin, Y. Gultneh, *Prog. Inorg. Chem.*, **35**, 219-279 (1987). b) Z. Tyeklar, K. D. Karlin, *Acc. Chem. Res.*, **22**, 241-248 (1989). c) I. Sanyal, R. W. Strange, N. J. Blackburn, K. D. Karlin, *J. Am. Chem. Soc.*, **113**, 4692-4693 (1991).

8.   a) N. Kitajima, K. Fujisama, Y. Moro-oka, K. Toriumi, *J. Am. Chem. Soc.*, **111**, 8975-8976 (1989). b) N. Kitajima, T. Koda, S. Hashimoto, T. Kitagawa, Y. Moro-oka, *J. Am. Chem. Soc.*, **113**, 5664-5671 (1991).
9.   a) L. Casella, L. Rigoni, *J. Chem. Soc., Chem. Commun.*, 1668 (1985). b) L. Casella, M. Gullotti, G. Pallanza, L. Rigoni, *J. Am. Chem. Soc.*, **110**, 4221 (1988). c) L. Casella, M. Gullotti, M. Bartosek, G. Pallanza, E. Laurenti, *J. Chem. Soc., Chem. Commun.*, 1235 (1991).
10. a) O. J. Gelling, A. Meetsma, B. L. Feringa, *Inorg. Chem.*, **29**, 2816-2822 (1990). b) O. J. Gelling. B. L. Feringa, *J. Am. Chem. Soc.*, **112**, 7599-7604 (1990).
11. a) R. Menif, A. E. Martell, *J. Chem. Soc., Chem. Commun.*, 1521-1523 (1989). b) R. Menif, A. E. Martell, P. J. Squattrito, A. Clearfield, *Inorg. Chem.*, **29**, 4723-4729 (1990). c) M.. P. Ngwenya, D. Chen, A. E. Martell, J. Reibenspies, *Inorg. Chem.*, **30**, 2732-2736 (1991).
12. M. Reglier, C. Jorand, B. Waegell, *J. Chem. Soc., Chem. Commun.*, 1752 (1990).
13. K. Hofmann in *The Chemistry of Heterocyclic Compounds. Imidazole and its Derivatives*, Interscience, New York, 1953.
14. R. Fusco in *Pyrazoles, Pyrazolines, Pyrazolidines, Indazoles, and Condensed Rings*, R. H. Wiley, ed., Interscience, New York, 1967, p. 3.
15. P. K. Ross, E. I. Solomon, *J. Am. Chem. Soc.*, **113**, 3246-3259 (1991)
16. N. Kitajima, T. Koda, Y. Iwata, Y. Moro-oka, *J. Am. Chem. Soc.*, **112**, 8833 (1990).
17. P. P. Paul, Z. Tyeklar, R. R. Jacobson, K. D. Karlin, *J. Am. Chem. Soc.*, **113**, 5322-5332 (1991).
18. K. Magnus, H. Ton-That, *J. Inorg. Biochem.*, **47**, 20 (1992).
19. T. N. Sorrell, A. S. Borovik, *J. Am. Chem. Soc.*, **109**, 4255 (1987).
20. T. N. Sorrell, K. Lerch, M. Beltramini, *J. Biol. Chem.*, **263**, 9576 (1988).
21. M. Beltramini, A. Favero, P. di Muro, G. P. Rocco, B. Salvato, *J. Inorg. Biochem.*, **47**, 46 (1992).
22. T. N. Sorrell, D. L. Jameson, *J. Am. Chem. Soc.*, **105**, 6013 (1983).
23   Three measurements were performed, and the individual samples absorbed 0.85, 1.0, and 1.05 equivalents of dioxygen per 2 equivalents of copper.
24. Resonance Raman spectra were recorded by Dr. Robert Kessler in our departmental laser lab on a frozen sample of the adduct in CH_2Cl_2 at 77K. Excitation was carried out at 488 nm. Additional resonance Raman spectra of this adduct and its [18]O labeled analog are currently being done in collaboration with Professor Edward Solomon at Stanford University.
25. R. R. Jacobson, Z. Tyeklar, A. Farooq, K. D. Karlin, S. Liu, J. Zubieta, *J. Am. Chem. Soc.*, **110**, 3690 (1988).
26. Unfortunately, we do not know what vO-O is for the dioxygen adducts of copper complexes formed from the N-alkylated tridentate ligands described by Karlin[17] and by us[6b] because of their instability toward photodecomposition in the laser beam.

OXIDATION OF UNACTIVATED HYDROCARBONS: MODELS FOR TYROSINASE AND DOPAMINE β-HYDROXYLASE

Marius Réglier*, Edith Amadéi, El Houssine Alilou, Franck Eydoux, Marcel Pierrot and Bernard Waegell

Laboratoire d'Activation Sélective en Chimie Organique, URA CNRS 1409, case 532, Université d'Aix-Marseille III, Faculté des Sciences de Saint Jérôme, Avenue Escadrille Normandie-Niemen, 13397 Marseille Cedex 13, France.

INTRODUCTION

The reactivity of the so-called unactivated hydrocarbons is of current interest from the theoretical, experimental and industrial points of view, in the areas of chemistry and biochemistry. Two different categories of substrates are part of class of unactivated hydrocarbons: saturated alkanes and aromatic hydrocarbons. Quite interestingly, the C-H bond energy of the former (102 kcal/mole) is lower than the latter (110 kcal/mole). The activation process by transition metal-complexes requires some kind of association step and can eventually be visualized as an oxidative addition of a C-H bond to an electrophilic metallic species (Fig. 1). The most efficient strategy for C-H bond activation should therefore combine an increase of the electron density in the C-H σ* orbitals (which requires nucleophiles) and a decrease of the electron density in the C-H σ orbital (which requires electrophiles, for instance super acids or low-valent metal complexes of Ru, Ir, Rh, Re).[1] This combination can efficiently be achieved in radicals or high-valent metal complexes such as highly electrophilic metal oxo species (which can formally act as oxene sources). The occurrence and reactivity of such reactive intermediates has been clearly demonstrated in iron porphyrin complexes of the CP-450 family.[2]

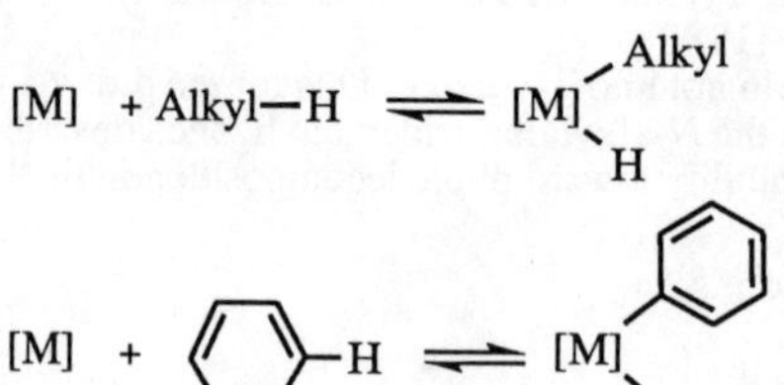

Figure 1. Activation of C-H bonds by transition-metal complexes.

Much less is known about the interaction of copper proteins with dioxygen (Tab. 1),[3] and the intimate nature of the reactive intermediates involved in the subsequent chemical oxidative process. The work we report here chemically models the enzymes

From K.D. Karlin and Z. Tyeklár, Eds., *Bioinorganic Chemistry of Copper*
(Chapman & Hall, New York, 1993).

tyrosinase (Tyr) [4] and dopamine β-hydroxylase (DBH) [5] which are both binuclear copper monooxygenases.

Table 1. The major binuclear copper-containing proteins which interact with dioxygen.

Enzymes [functions]	Reactions
Hemocyanin (Hc) [dioxygen carrier]	$Hc + O_2 \rightleftharpoons Hc(O_2)$
Tyrosinase (Tyr) [monooxygenase] [oxidase]	
Dopamine β-Hydroxylase (DBH) [monooxygenase]	

TYR, A COUPLED BINUCLEAR COPPER PROTEIN

Tyr is a type 3 (or binuclear coupled) copper-containing enzyme which catalyzes *ortho*-hydroxylation of phenols to catechols (phenolase activity) and oxidation of catechols to *ortho*-quinones (cathecolase activity). The structure of the Tyr active site is not exactly known. However, comparison of spectral properties (EPR, electronic absorption, CD, and resonance Raman) of Tyr with those of Hc [6] establishes a close similarity of the dicopper active site structure in these two proteins.[7] A X-ray structural study on the deoxy state of Hc indicates that 3 imidazole groups (histidine residues) are coordinated to each Cu(I) ion, and a copper-copper separation of 3.7 Å is observed. For oxy-Hc, several chemical and spectroscopic studies suggest a copper-copper distance of about 3.6 Å with an exogenous peroxo ligand bridged in a cis μ-1,2 fashion. To account for the diamagnetism of oxy-Hc, an additional hydroxy or aquo ligand has been suggested to bridge the two copper atoms. In oxy-Tyr the peroxo dicopper species **I** (Fig. 2) is accessible to phenol, which can coordinate and undergo hydroxylation to give catechol. Solomon et al. [8] have proposed an aromatic hydroxylation mechanism in which a rearrangement of square pyramid to trigonal bipyramid geometry at one copper atom takes place (**I** to **II**). This rearrangement results in a polarization of the peroxo ligand, which renders one of the oxygens electrophilic towards the aromatic ring of the phenol to provide the copper-catechol intermediate **III**. Recently, Kitajima et al. [9] have isolated a new type of dicopper peroxo complex where the peroxo ligand bridges the two copper atoms in a μ-η²:η² fashion. This new complex, which exhibits spectroscopic and structural features identical to those of oxy-Hc and oxy-Tyr, is a new model for dioxygen binding in Hc and Tyr.[10] Furthermore, Karlin et al. [11] have proposed that the copper dioxygen species **IV** has an electrophilic character and could be responsible for the hydroxylation of phenols.

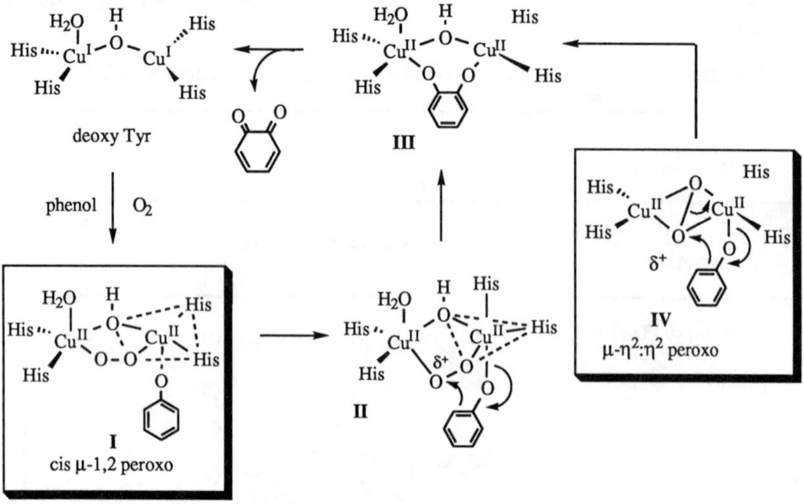

Figure 2. Intermediates proposed in phenol hydroxylation by Tyr.

DBH, A NON-COUPLED BINUCLEAR COPPER PROTEIN

DBH is a type 2 copper-ascorbate-dependent monooxygenase, which catalyzes the benzylic hydroxylation of dopamine to noradrenaline. Little information is available concerning the O_2 binding mode to the bimetallic Cu(I) sites. In contrast to the situation found in type 3 copper proteins (Hc and Tyr), all evidence coming from EPR,[12] ESEEM,[13] EXAFS [14] and biochemical studies [15] suggest that DBH contains two inequivalent copper atoms per active site. A Cu_A site is proposed to be at the core of the reductant site where ascorbate binds and delivers one electron at a time. A Cu_B center, at a distance greater than 5 Å, is involved in dioxygen fixation; it is responsible, by the intermediacy of a Cu(II) hydroperoxo species, for the hydroxylation of dopamine. During the hydroxylation, a homolytic cleavage of the peroxide bond, with concomitant benzylic hydrogen abstraction, has been proposed. These processes give rise to a Cu(II)-O• species and a benzylic radical, which recombine to give noradrenaline (Fig. 3).[16]

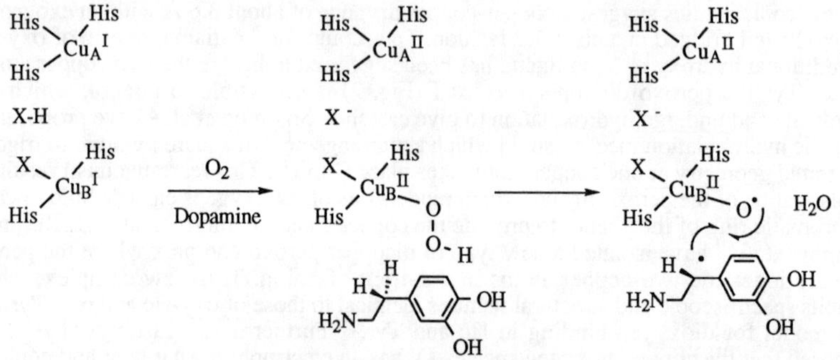

Figure 3. The mechanism proposed for DBH-catalyzed hydroxylation of dopamine.

CHEMICAL MODELING

There is presently a great deal of interest in chemically modeling the active site of the binuclear copper proteins which interact with dioxygen.[17] For the type 3 copper-containing proteins, these investigations are aimed at the synthesis of model complexes which exhibit structural and spectroscopic characteristics similar or identical to the original protein and/or which mimic the chemical reactivity. These features are related to the topology of the metallic center which proceeds from the structure and complexation sites of the synthetic ligand (Fig. 4). The results which are reported below essentially concern modeling, which aims to mimic the original enzyme's reactivity.

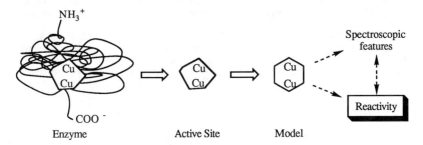

Figure 4. Chemical Modeling.

In modeling Tyr and DBH, the first question which arises is whether modeling should take into consideration mononuclear or binuclear complexes like in Tyr or DBH. For the Hc active site models, it was shown that it is not necessary to have a dinucleating ligand to obtain a binuclear peroxo-copper species.[9,18] However, if one wants to study a peroxo ligand bridging two copper atoms in cis μ-1,2 fashion (**I**, Fig. 2) it is optimal to have a binucleating ligand in which the two copper atoms will be at a distance of about 3.6 Å. For DBH, since only one copper atom forms the Cu(II)OOH species which hydroxylates dopamine, the use of mononuclear copper complexes appears reasonable. Similarly binuclear complexes are not necessary to study the reactivity of species such as Cu(II)-O•, which are involved in the hydroxylation process. The second question is how and what the different copper-oxygen species can be obtained (Fig. 5). As already reported, peroxo species can be generated by direct interaction of dioxygen with a Cu(I) complex.[18] Two ways are possible for the synthesis of a Cu(II)OOH species: 1) protonation of a peroxo species;[19-20] or 2) direct interaction of hydrogen peroxide with a Cu(II) complex.[20] Radical Cu(II)-O• species could be obtained either by homolytic cleavage of peroxo species,[21] or by analogy with the CP-450 peroxide shunt, by reaction of a Cu(I) complex with iodosylbenzene.[22]

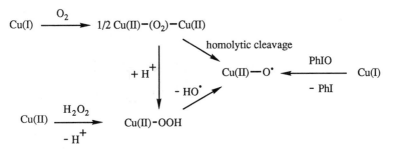

Figure 5. Possible ways to obtain a Copper-oxygen active species.

Binuclear Copper Models for the Active Site of Tyr [23]

As already mentioned, the distance between the two copper atoms should be around 3.6 Å to sterically accommodate the exogenous substrate to be hydroxylated, as well as the reactive oxygen metal species involved in this reaction, for exemple, the μ-peroxo dicopper species likely to be involved for Tyr (**I** and **IV**, Fig. 2). From the electronic point of view, such a species can only be formed if the metal has a filled π bonding d orbital of intermediate energy which can allow partial but not complete electron transfer from Cu(I) to O_2.[24] The oxido reductive process involves an electron transfer between metal and dioxygen and therefore requires a change in the coordination number, which should be easily accommodated by the nature and structure of the ligand. The desired complexes should be able to activate dioxygen without oxidative degradation of the ligand, so that exogenous substrates can be catalytically hydroxylated.

The ligand we propose features a biphenyl spacer on which two pyridine arms are fixed. We have chosen the biphenyl group as a spacer because of the flexibility due to the biphenyl atropisomerism. Molecular mechanics calculations on the dicopper complex performed with the BIOGROMOS program,[25] shows that the copper-copper distance of 3.6 Å could be obtained with a biphenyl dihedral angle of 60° (Fig. 6). The calculations show that a ten degree variation of the biphenyl angle gives rise to a variation of 0.73 Å for the Cu-Cu distance.

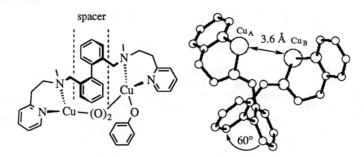

Figure 6. The representation of the desired binuclear copper(I) complex emphasizes the relation between the copper-copper distance and the biphenyl dihedral angle θ.

Two dicopper(I) complexes **1-2** (Fig. 7) based on this model have been synthesized by reaction of the appropriate ligand and 2 equiv. of [Cu(CH₃CN)₄]PF₆. The ligands were prepared in good yields by reduction with BH₃ of bis (N,N-methyl 2-(2-pyridyl) ethyl diphenamide (**1**), or by condensation of 2 equiv. of 2-(2-aminoethyl) pyridine and diphenaldehyde (**2**).

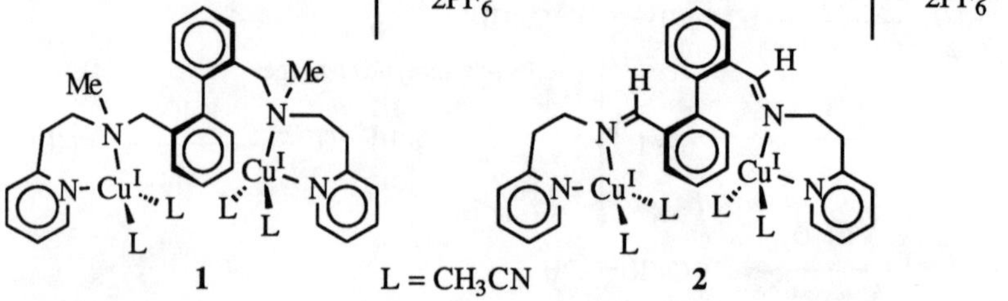

Figure 7. Binuclear copper(I) complexes **1-2** featuring the biphenyl spacer.

Under argon, dichloromethane (or acetonitrile) solutions of dicopper(I) complexes **1-2** are stable but in the presence of O_2, these complexes are rapidly oxidized to dicopper(II) complexes. During this reaction, the ligand of complex **1** undergoes a demethylation of the methyl amine part,[26] but the ligand of the bis imino dicopper(I) complex **2** is resistant to oxidation. Under these conditions, complex **2** gives complex **3** (Fig. 8). The formulation of **3** as a μ-oxo complex comes from: 1) the stoichiometry of oxygen uptake, which is consistent with a four-electron reduction of O_2; 2) the observation of oxygen atom transfer from **3** to triphenylphosphine, which affords triphenylphosphine oxide.

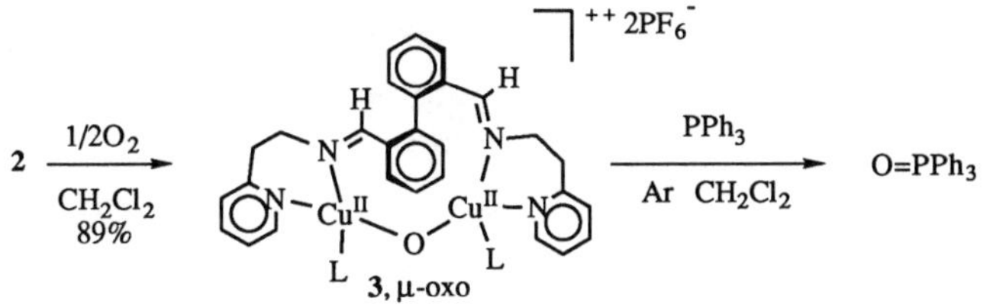

Figure 8. Reaction of the dicopper(I) complex **2** with dioxygen.

Complex **2** exhibits both catalytic phenolase and catecholase activities, with 2,4-di-*t*-butylphenol (DTBP) and 3,5-di-*t*-butylcatechol (DTBC), respectively. In the presence of triethylamine (2 equiv.) in dichloromethane, DTBP was oxidized to 3,5-di-*t*-butyl-*o*-quinone (DTBQ) with dioxygen by a catalytic amount of complex **2** (Fig. 9). A kinetic study showed a turnover of 16 h^{-1}, unfortunately the reaction stops after one hour, probably because the complex **2** is oxidized into the μ-oxo species **3** which is inactive in this oxidative process. In the absence of triethylamine, DTBP is transformed into its dimer **4** by an auto-oxidation process. The role of triethylamine is most probably to assist formation of a phenolate copper complex **IV** (Fig. 2) as postulated by Solomon.[8]

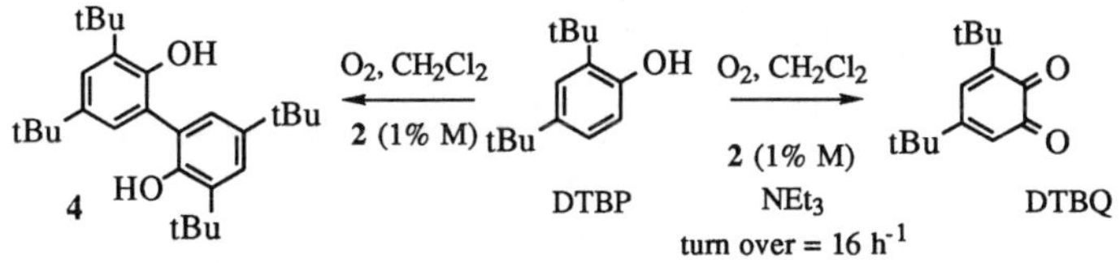

Figure 9. DTBP oxidation catalyzed by complex **2**.

In order to test for catecholase activity, 3,5-di-*t*-butylcatechol (DTBC) was added to the reaction of dioxygen and a catalytic amount of complex **2** (Fig. 10). Practically quantitative yields of DTBQ were observed after 1 day of reaction at room temperature. A kinetic study showed a turnover of 11 h^{-1}. Under the same conditions, the μ-oxo copper complex **3** led to the same catalytic transformation of DTBC into DTBQ with an identical rate.

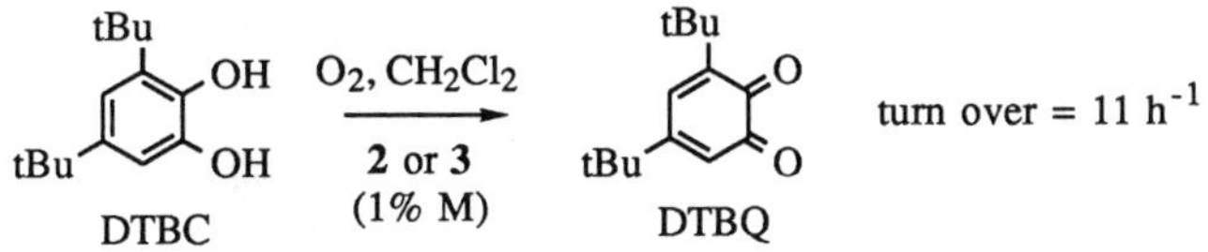

Figure 10. DTBC oxidation catalyzed by complexes **2** and **3**.

353

In order to characterize the catecholate dicopper species postulated in the Tyr mechanism (**III**, Fig. 2), we have prepared complex **5** by treatment of complex **2** with DTBQ in CH$_2$Cl$_2$ under argon. The same species (**5**) can be obtained by reaction of dicopper(II) complex **3** and DTBC (Fig. 11). The UV-vis spectrum of complex **5** exhibits two absorption assigned as catecholate-copper charge transfer transitions (400 nm, ε = 3000 and 558 nm, ε = 1200). Complex **5** is very sensitive to dioxygen and decomposes immediately to DTBQ. All attempts to crystallize this intermediate have failed, even using tetrachloro orthoquinone [27] which is known to give stable copper-catechol complexes.[28]

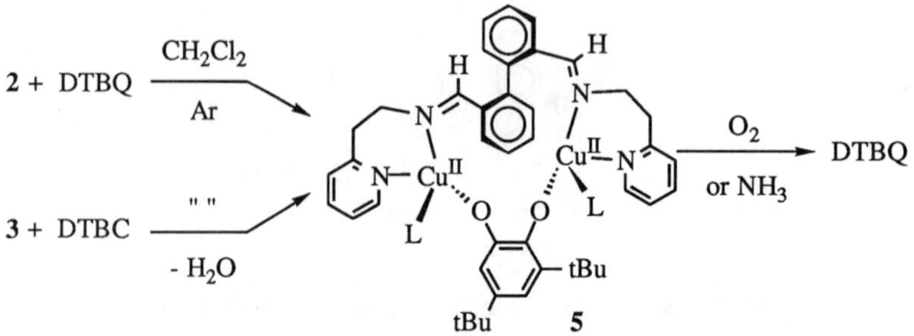

Figure 11. Formation of the dicopper-catecholate species **5**.

Our results can be rationalized as shown in Fig. 12. Species **V** and **5** are the common intermediates for the catecholase and phenolase processes. In the phenolase process, the peroxo species **V** reacts with phenol to form **VI**, which loses H$_2$O to give the catecholate copper complex **5**. This species reacts rapidly with dioxygen giving DTBQ and peroxo species **V**. In the catecholase process, the peroxo species **V** reacts with DTBC to give DTBQ and complex **3**, which further reacts with another DTBC molecule giving the catecholate copper complex **5**.

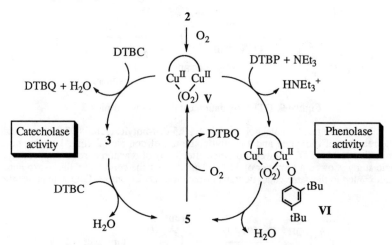

Figure 12. Intermediates involved in catecholase and phenolase processes.

As already mentioned, the structure of the peroxo species involved in the Tyr process is a matter of controversy (**I** and **IV**, Fig. 2). Molecular modeling shows that the biphenyl ligand is flexible, thus perfectly accommodating either a cis $\mu-1,2$ peroxo (Fig. 13) or a $\mu-\eta^2:\eta^2$ peroxo intermediate (Fig. 14). At this time it is difficult to provide evidence in favor of one or the other peroxo configuration for species **V**.

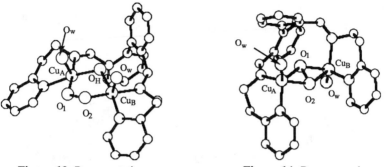

Figure 13. Representation of cis μ-peroxo species. $d_{CuCu} = 3.29$ Å for $\theta_{biphenyl} = 88°$

Figure 14. Representation of $\mu-\eta^2:\eta^2$ peroxo species. $d_{CuCu} = 3.59$ Å for $\theta_{biphenyl} = 53°$

O_w = water molecule, O_H = hydroxo bridge

This report of a chemical model featuring a catalytic phenolase and catecholase activity on an exogenous substrate in presence of dioxygen, had only been preceded by the one of Bulkowski [29] and was shortly followed by the model of Casella.[30]

Mononuclear Copper Complexes, a Possible Way to Model the DBH Active Site? [31]

In Tyr model binuclear copper(I) complexes reported to date, the initial association step between the enzyme model and the phenolic substrate can easily be achieved between the hydroxy group of the substrate and one of the copper atoms (**V** to **VI**, Fig. 12). For DBH models the initial association step is less obvious unless the phenethylamine (used as a simplified model of dopamine) is covalently bound to the ligand by amine, imine or amide links (Fig. 15). We have investigated the reactivity of such complexes, in order to probe for hydroxylation reactions on the bonded phenethylamine group.

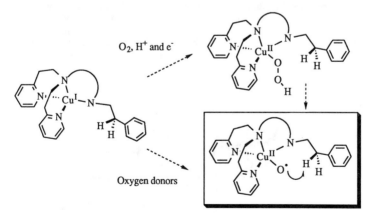

O_2, H$^+$ and e$^-$

Oxygen donors

Figure 15. Approach to the modeling of DBH Cu$_B$ center.

Amino copper(I) complexes **6a,b**.[32] As reported by Karlin et al.,[33] the copper(I) complex **6b** (n = 1) reacts with O_2 without hydroxylation of the ligand. However, we found that in the presence of iodosylbenzene in acetonitrile this complex affords, after decomplexation, 50% unreacted ligand **7b** and 50% of product **8b**, resulting from α-pyridine hydroxylation (Fig. 16). The same results are obtained with the copper(I) complex **6b** (n = 2). Under the same conditions, complex **6a** (n = 1) gives 50% unreacted ligand **7a** and 50% of alcohol **8a**, resulting from hydroxylation of the pyridylmethyl group.

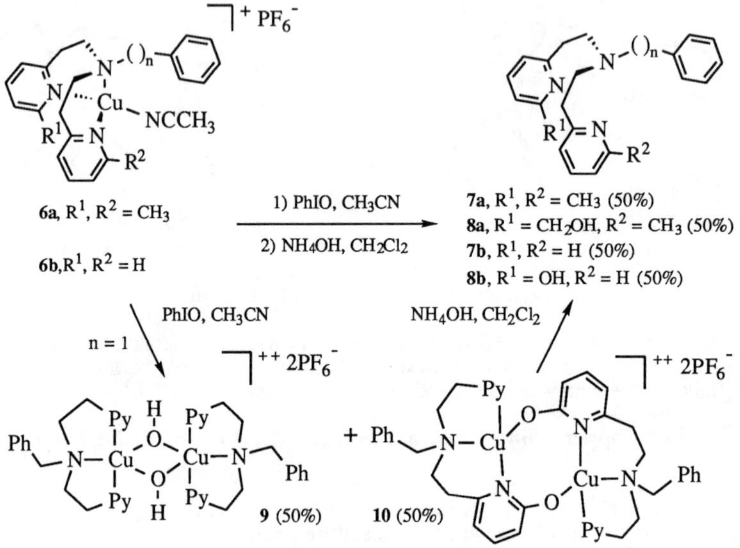

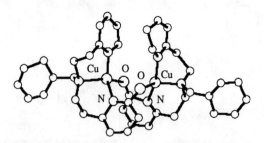

Figure 16. PhIO oxidation of amino copper(I) complexes **6a,b.**

During the PhIO oxidation of compound **6b** (n = 1), we have identified and isolated an intermediate copper(II) complex, which crystallizes into two forms **9** and **10**. The binuclear complex **9** is identical to the one described by Karlin,[32] isolated from the dioxygen oxidation of complex **6b** (n = 1). A crystal structure determination of complex **10** has shown that the two copper atoms are at a distance of 3.698 Å from each other; each is at the center of a distorted tetrahedron. The oxygen atoms have, surprisingly, ended up bonded to both a copper atom and to the pyridine carbon α to the nitrogen (Fig. 17).

Figure 17. Structural depiction of the dicopper(II) complex **10.**

According to what is established in iron and manganese porphyrin chemistry,[22] we propose the formation of a Cu(II)-O$^{\bullet}$ species **VII** formed by initial complexation of PhIO on the copper(I) complexes **6a-b**, followed by electron transfer and homolytic cleavage of the I-O bond (Fig. 18).

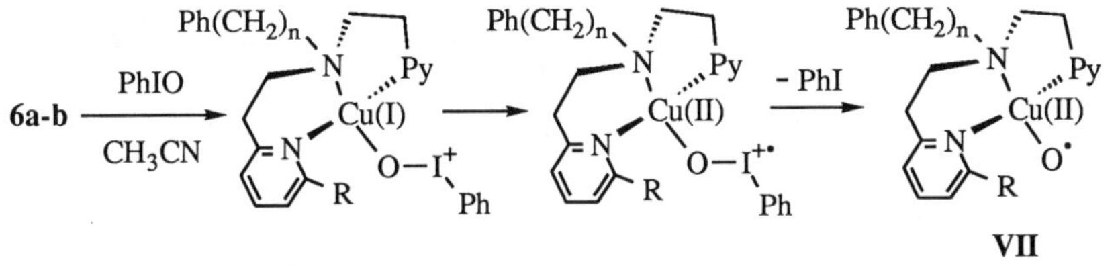

Figure 18. Formation of the Cu(II)-O$^{\bullet}$ species **VII** by interaction of copper(I)complexes **6a-b** and PhIO.

The X-ray structural analysis of complexes **9**[32] and **10** (Fig. 17) clearly indicates that the N-(CH$_2$)$_n$-Ph side chain is far apart from the dicopper center. Although it is not possible to draw definitive conclusions about the copper-oxygen species responsible for pyridine hydroxylation, it is reasonable to propose a structure of the Cu(II)-O$^{\bullet}$ species **VII** that is close to those of these complexes. This suggestion led us to consider a model of the reactive intermediate (Fig. 18) which incorporates sufficient proximity between the Cu(II)-O$^{\bullet}$ species and the carbon which will be hydroxylated. This model provides an explanation for the absence of hydroxylation on the N-(CH$_2$)$_n$-Ph side chain, which was initially expected. For complex **6b**, the key step is the addition of the Cu(II)-O$^{\bullet}$ species to the pyridine ring yielding intermediate **VIII**, which undergoes an oxido reductive process favored by proton abstraction. For complex **6a**, the hydroxylation can be explained by an abstraction of a methyl hydrogen atom, as it shown in Fig. 19.

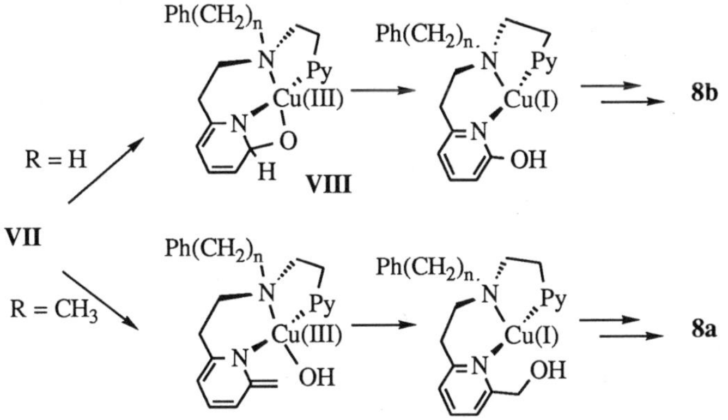

Figure 19. Mechanism proposed for the pyridine hydroxylation by PhIO.

Imino copper(I) complex 11. In order to test the proximity requirement, the pyridine ring was replaced by an imino group derived from benzylamine. In the reaction of **11** and PhIO in CH$_3$CN, the oxidation proceeds mainly as expected on position A (Fig. 20), and results in oxidative cleavage of the benzylic position. What is even more interesting is that the reaction proceeds not only with PhIO but also with dioxygen itself. In this case, we observe the oxidative cleavage of the two benzylic positions with a ratio A/B depending on the solvent (for CH$_3$CN A/B = 1/4 and for MeOH A/B = 4/1).

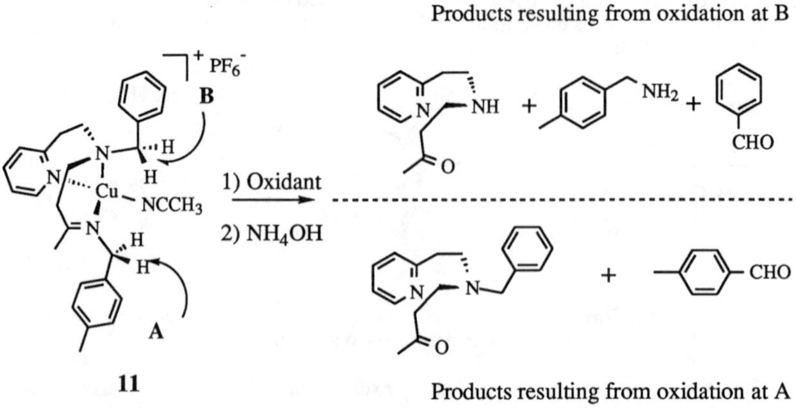

Products resulting from oxidation at B

Figure 20. Reaction products from imino copper(I) complex **11** and O_2 or PhIO.

The cleavage at position A observed upon oxidation with PhIO in CH_3CN could arise from direct hydroxylation by the mechanism already proposed for the pyridine hydroxylation (Fig. 19), involving the formation of a Cu(II)-O• species **IX** which abstracts an benzylic hydrogen α to the imino group (Fig. 21).

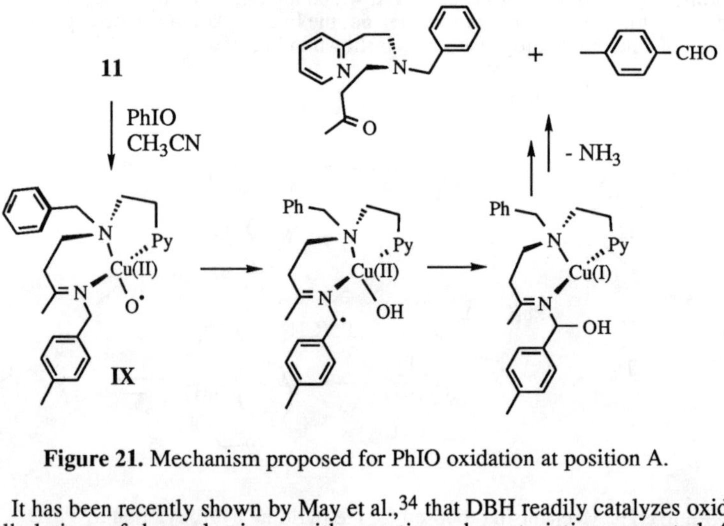

Figure 21. Mechanism proposed for PhIO oxidation at position A.

It has been recently shown by May et al.,[34] that DBH readily catalyzes oxidative N-dealkylation of benzylamines with reaction characteristics expected for a monooxygenase-catalyzed process. A mechanism involving the single electron transfer from nitrogen amine to the Cu(II)OOH species yielding, $N^{+•}$, HO^- and Cu(II)-O• species has been proposed. On the basis of these results, we propose a mechanism (Fig. 22), where heterolytic cleavage of the peroxide bond is assisted by a one electron transfer from the nitrogen amine (or imine) to the peroxo species **X**. This leads to two Cu(II)-O• species, **XI** and **XII**, which can abstract a benzylic hydrogen α to the $N^{+•}$.

358

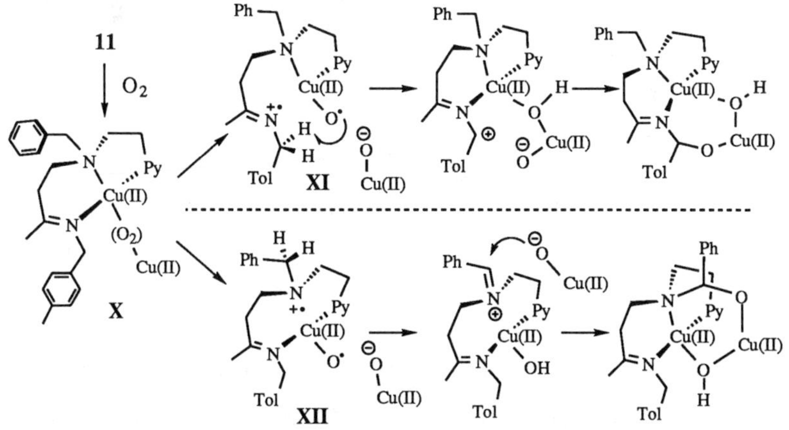

Figure 22. Mechanism proposed for the oxidation of complex **11** by dioxygen

<u>Amido copper(I) complexes **12** and **13**.</u> [35] A quite striking difference in chemical behavior between PhIO and O_2 was also observed with the copper complexes **12** and **13** where the ligand is built around a tripodal nitrogen and linked to the substrate by an amide group instead of an imine (Fig. 23).

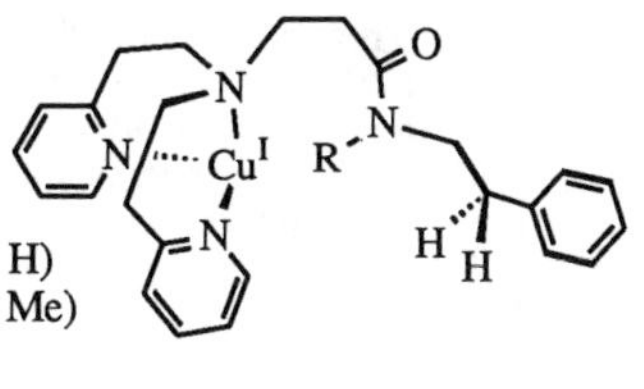

12 (R = H)
13 (R = Me)

Figure 23. Amido copper(I) complexes **12,13**.

The reaction of complex **12** with iodosylbenzene in CH_3CN affords, after decomplexation, 52% of the unchanged ligand **14** as well as the two monohydroxylated compounds **16** (26%) and **17** (22%) (Fig. 24).

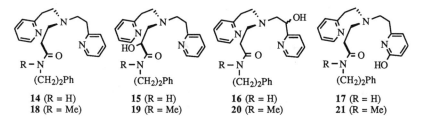

R—N	R—N	R—N	R—N
$(CH_2)_2Ph$	$(CH_2)_2Ph$	$(CH_2)_2Ph$	$(CH_2)_2Ph$
14 (R = H)	15 (R = H)	16 (R = H)	17 (R = H)
18 (R = Me)	19 (R = Me)	20 (R = Me)	21 (R = Me)

Figure 24. Products obtained by oxidation of amido copper(I) complexes **12** and **13**.

Compound **16** is a result of hydroxylation at the benzylic position of one pyridine. Compound **17** is from *ortho* hydroxylation of a pyridine ring, similar to that observed with the amine ligand. When the same reaction is performed with complex **13**, product **21**, resulting from pyridine ring hydroxylation, is the only one formed. This observation leads us to believe that compounds **17** and **21** come from the Cu(II)-O• species **XI** by a mechanism already described (Fig. 19). In the case of complex **12**, the Cu(II)-O• species **XI** (which is known to function as hydrogen atom acceptor[21,36]) can abstract the amide hydrogen to give a highly reactive Cu(III)OH species, **XII**, which hydroxylates the benzylic position.[37]

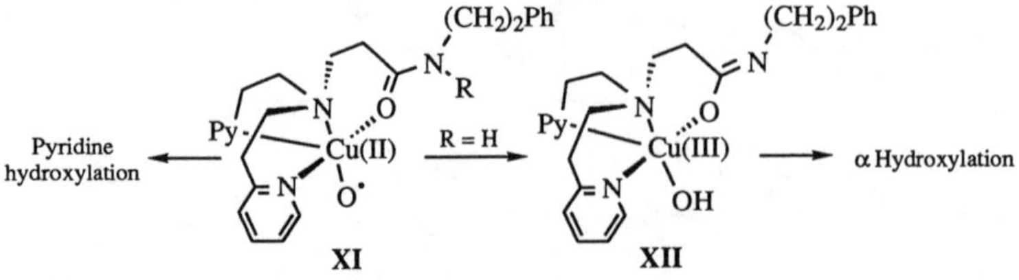

Figure 25. Mechanism proposed for PhIO oxidation of complexes **12,13**.

When a solution of the copper(I) complex **12** is exposed to O$_2$, a green solution is rapidly formed. Following demetallation using aqueous ammonia, analysis of the organic products indicated that 74% of unchanged ligand **14** and **16** (15%) were obtained in addition to the compound **15** (11%) which is the product hydroxylated α to the carbonyl group. At this time, it is difficult to provide evidence for the oxygenated copper species responsible for the hydroxylated products **15** and **16**. The copper(I) complex **12** is so reactive towards dioxygen that, our attempts to isolate and characterize the reactive Cu-O$_2$ species have been unsuccessful. Furthermore, the possibility involving formation of a Cu(II)OOH species (from proton abstraction of the amide group by a nucleophilic peroxo copper(II) species) can be excluded. When the N-methyl derivative **13** was used, the same hydroxylated products **19** (10%) and **20** (17%) were obtained. Another possibility is to consider that the hydroxylated products are due to reaction of a trans μ-peroxo or a μ-η^2:η^2-peroxide bridged copper dimer. These species could be able to abstract a reactive hydrogen in its vicinity, that is in the position β to the tripodal nitrogen atom.

X, Y = 2-pyridyl or phenethylamide

trans μ-peroxo   μ-η^2:η^2 peroxo

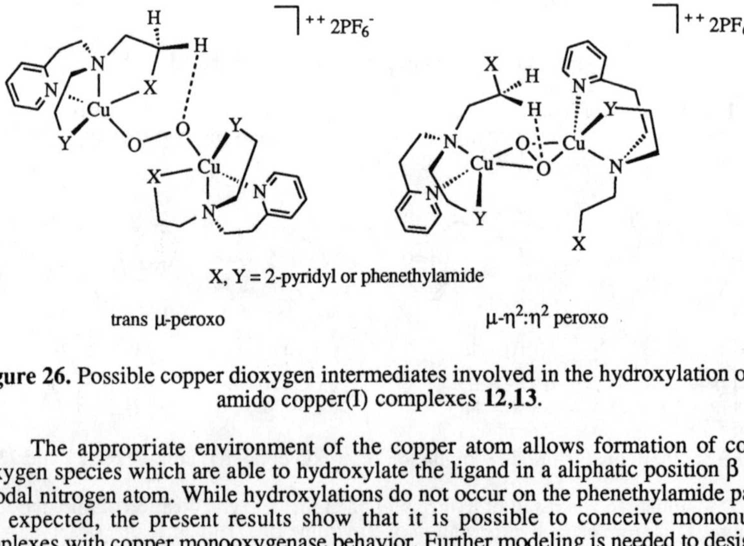

Figure 26. Possible copper dioxygen intermediates involved in the hydroxylation of the amido copper(I) complexes **12,13**.

The appropriate environment of the copper atom allows formation of copper-dioxygen species which are able to hydroxylate the ligand in a aliphatic position β to the tripodal nitrogen atom. While hydroxylations do not occur on the phenethylamide part, as was expected, the present results show that it is possible to conceive mononuclear complexes with copper monooxygenase behavior. Further modeling is needed to design the best copper ligand to achieve hydroxylation on the desired methylene group, as in DBH.

The approach used here needs to be improved to attain the goal of a copper complex with catalytic hydroxylation properties towards exogenous ligands. Thus, the problem to achieve association of dopamine into the copper coordination sphere still remains to be solved.

CONCLUSION

More examples, presently under investigation in our laboratory, will illustrate the importance of the topology about the copper atom. What is possible to say at this time is that more work needs to be done: numerous complexes still have to be synthesized with various combinations of the nature and length of the arms around the tripodal nitrogen, in order to understand how these factors influence the nature of the oxygenated intermediates and the related reactivity as DBH models. The reactivity of these systems is generally important as many of them are able to activate dioxygen even at low temperatures. The drawback of this high reactivity is that we have not been able to isolate any stable copper dioxygen complex and crystallize it to obtain an X-ray structure. Work along these lines is currently underway in our group, as well as the structure determination of DBH itself. This, as well as the synthesis of efficient suicide inhibitors should provide the information needed to further understand the mode of action of DBH.

ACKNOWLEDGMENTS

We are grateful to the Centre National de la Recherche Scientifique and the Ministère des Affaires Etrangères for their financial support.

REFERENCES

1. W. D. Jones and F. J. Feher, *Acc. Chem. Res.*, **22**, 91 (1989).
2. *(a) Cytochrome P-450: Structure, Mechanism and Biochemistry*, P. R. Ortiz de Montellano, Ed., (Plenum, New York, 1986). *(b)* D. Mansuy and P. Battioni, in *Activation and Functionalization of Alkanes*, C. L. Hill, Ed., (Wiley-Interscience, New York, 1989), pp. 195-218. V. W. Bowry, J. Lusztyk and K. U. Ingold, *J. Am. Chem. Soc.*, **113**, 5687 (1991). V. W. Bowry and K. U. Ingold, *ibid.*, **113**, 5699 (1991).
3. *(a) Metal Ions in Biology*, T. G. Spiro, Ed., (John Wiley and Sons, New York, 1981), vol. 3. *(b) Copper Proteins and Copper Enzymes*, R. Lontie, Ed. (CRC, Boca Raton, FL, 1984), vol. 2.
4. E. I. Solomon in ref. 3 *(a)*, pp. 91-97. D. A. Robb in ref 3 *(b)*, pp.207-241.
5. J. J. Villafranca, in ref 3 *(a)*, pp. 263-290. L. C. Stewart and J. P. Klinman, *Ann. Rev. Biochem.*, **57**, 551 (1988).
6. E. I. Solomon in ref. 3 *(a)*, pp. 81-90. W. P. J. Gaykema, W. G. J. Hol, J. M. Vereijken, N. M. Soeter, H. J. Bak and J. J. Beintema, *Nature*, **309**, 23 (1984). W. P. J. Gaykema, A. Volbeda and W. G. J. Hol, *J. Mol. Biol.*, **187**, 255 (1985). A. Volbeda and W. G. J. Hol, *J. Mol. Biol.*, **206**, 531 (1989). A. Volbeda and W. G. J. Hol, *J. Mol. Biol.*, **209**, 249 (1989).
7. E. I. Solomon, M. J. Baldwin and M. D. Lowery, *Chem. Rev.*, **92**, 521 (1992).
8. D. E. Wilcox, A. G. Porras, Y. T. Hwang, K. Lerch, M. E. Winckler and E. I. Solomon, *J. Am. Chem. Soc.*, **107**, 4015 (1985).
9. N. Kitajima, K. Fujisawa and Y. Moro-oka, *J. Am. Chem. Soc.*, **111**, 8975 (1989).
10. N. Kitajima, K. Fujisawa, C. Fujimoto, Y. Moro-oka, S. Hashimoto, T. Kitagawa, K. Toriumi, K. Tatsumi and A. Nakamura, *J. Am. Chem. Soc.*, **114**, 1277 (1992).
11. M. S. Nasir, B. I. Cohen and K. D. Karlin, *J. Am. Chem. Soc.*, **114**, 2482 (1992).
12. N. J. Blackburn, D. Collison, J. Sutton and F. E. Mabbs, *Biochem. J.*, **220**, 447 (1984). N. J. Blackburn, M. Concannon, S. Khosrow Shahiyan, F. E. Mabbs and D. Collison, *Biochemistry*, **27**, 6001 (1988).
13. J. Mc Cracken, P. R. Desai, N. J. Papadopoulous and J. J. Villafranca, *Biochemistry*, **27**, 4133 (1988).

14. S. S. Hasnain, G. P. Diakun, P. F. Knowles, N. Binsted, C. D. Garner and N. J. Blackburn, *Biochememistry*, **221**, 545 (1988). R. A. Scott, R. J. Sullivan, W. E. Jr. De Wolf, R. E. Dolle and L. I. Kruse, *Biochemistry*, **27**, 5411 (1988). W. E. Blumberg, P. R. Desai, L. Powers, J. H. Freedman and J. J. Villafranca, *J. Biol. Chem.*, **264**, 6029 (1989). T. M. Pettingill, R. W. Stange and N. J. Blackburn, *J. Biol. Chem.*, **266**, 16996 (1991). N. J. Blackburn, S. S. Hasnain, T. M. Pettingill and R. W. Stange, *J. Biol. Chem.*, **266**, 23120 (1991).
15. L. C. Stewart and J. P. Klinman, *Biochemistry*, **26**, 5302 (1987). M. C. Brenner, C. J. Murray and J. P. Klinman *Biochemistry*, **28**, 4656 (1989).
16. N. Anh and J. P. Klinman, *Biochemistry*, **22**, 3090 (1983). S. M. Miller and J. P. Klinman, *Biochemistry*, **24**, 2114 (1985).
17. J. M. Latour, *Bull. Soc. Chim. Fr.*, 508 (1988). T. N. Sorrel, *Tetrahedron*, **45**, 3 (1989). Z. Tyeklár and K. D. Karlin, *Acc. Chem. Res.*, **22**, 241 (1989). K. D. Karlin and J. Zubieta, in *Biological and Inorganic Copper Chemistry*, K. D. Karlin and J. Zubieta, Eds. (Adenine Press, New York, 1982 and 1984). D. E. Fenton, *Pure and App. Chem.*, **61**, 903 (1989). K. D. Karlin and Z. Tyeklár, in *Advances in Inorganic Biochemistry*, G. L. Eichhorn and L. G. Marzilli, Eds., (Elsevier, New York, 1992), vol. 9.
18. J. S. Thompson, in *Biological and Inorganic Copper Chemistry*, K. D. Karlin and J. Zubieta, Eds., (Adenine Press, Guilderland, New York, 1986), vol. 2, pp 1-10. J. S. Thompson, *J. Am. Chem. Soc.*, **106**, 8308 (1984). R. R. Jacobson, Z. Tyeklár, A. Farooq, K. D. Karlin, S. Liu and J. Zubieta, *J. Am. Chem. Soc.*, **110**, 3690 (1988). I. Sanyal, R. W. Strange, N. J. Blackburn and K. D. Karlin, *J. Am. Chem. Soc.*, **113**, 4692 (1991).
19. M. Mahroof-Tahir, N. N. Murthy, K. D. Karlin, N. J. Blackburn, S. N. Shaikh and J. Zubieta, *Inorg. Chem.*, **31**, 3001 (1992).
20. K. D. Karlin, P. Ghosh, R. W. Cruse, A. Farooq, Y. Gultneh, R. R. Jacobson, N. J. Blackburn, R. W. Strange and J. Zubieta, *J. Am. Chem. Soc.*, **110**, 6769 (1988).
21. N. Kitajima, T. Koda, Y. Iwata and Y. Moro-oka, *J. Am. Chem. Soc.*, **112**, 8833 (1990).
22. T. J. McMurry and J. T. Groves, in ref. 2(a), pp. 1-28. B. Meunier, *Bull. Soc. Chim. Fr.*, **4**, 578 (1986). D. Mansuy, *Pure and Applied Chem.*, **59**, 759 (1987).
23. M. Réglier, C. Jorand and B. Waegell, *J. Chem. Soc. Chem. Commun.*, 1752 (1990).
24. K. D. Karlin and Y. Gultneh, *J. Chem. Ed.*, **62**, 983 (1985).
25. Program supplied by BIOSTRUCTURE, Strasbourg, France.
26. M. Réglier, unpublished results.
27. E. H. Alilou, M. Giorgi, M. Pierrot and M. Réglier, *Acta Cryst.*, **C48**, 1612 (1992).
28. K. D. Karlin, Y. Gultneh, T. Nicholson and J. Zubieta, *Inorg. Chem.*, **24**, 3727 (1985).
29. J. E. Bulkowski, U.S. Pat. 4,545,937.
30. L. Casella, M. Guillotti, R. Radaelli and P. Di Gennaro, *J. Chem. Soc. Chem. Commun.*, 1611 (1991).
31. E. Amadéi, Thèse, Marseille 1991.
32. M. Réglier, E. Amadéi, R. Tadayoni and B. Waegell, *J. Chem. Soc. Chem. Commun.*, 447 (1989).
33. K. D. Karlin, Y. Gultneh, J. C. Hayes and J. Zubieta, *Inorg. Chem.*, **23**, 519 (1984).
34. K. Wimalasena and S. W. May, *J. Am. Chem. Soc.*, **109**, 4036 (1987).
35. E. Amadéi, E. L. Alilou, F. Eydoux, M. Pierrot, B. Waegell and M. Réglier, *J. Chem. Soc. Chem. Commun.*, in press. E. H. Alilou, E. Amadéi, M. Pierrot and M. Réglier, *J. Chem. Soc. Dalton Trans.*, in press.
36. O. Reinaud, P. Capdevielle and M. Maumy, *J. Chem. Soc. Chem. Commun.*, 566 (1990). P. Capdevielle, and M. Maumy, M., *Tetrahedron Lett.*, 3831 (1991).
37. C. M. Che, V. W-W. Yam and T. C. W. Mak, *J. Am. Chem. Soc.*, **112**, 2284 (1990).

COPPER-PTERIDINE CHEMISTRY. STRUCTURES, PROPERTIES, AND PHENYLALANINE HYDROXYLASE MODELS

Osamu Yamauchi,[a] Akira Odani,[a] Hideki Masuda,[b] and Yasuhiro Funahashi[a]

[a]Department of Chemistry, Faculty of Science, Nagoya University, Nagoya 464-01, Japan
[b]Department of Applied Chemistry, Nagoya Institute of Technology, Nagoya 466, Japan

INTRODUCTION

Pterins are derivatives of 2-amino-4-hydroxypteridine and are widely distributed in nature.[1] They were originally isolated from the pigments of butterfly wings,[1,2] and today a number of derivatives are known as cofactors for enzymes involved in the synthesis of neurotransmitters, amino acids, and nucleic bases (Figure 1). The tetrahydro form of folic acid, which is a pterin derivative having an aminobenzoylglutamate group attached to the pterin's 6-methyl group (**1**), is involved in C$_1$ unit transfer in the biosynthesis and metabolism of nucleic bases and some amino acids.[3] Biopterin (**2**) in its tetrahydro form is the cofactor for aromatic amino acid hydroxylases, phenylalanine hydroxylase, tyrosine hydroxylase, and tryptophan hydroxylase.[4] Phenylalanine hydroxylase (PAH), which is the most studied of the three, converts phenylalanine to tyrosine by incorporating the oxygen atom from dioxygen into the benzene ring, which is the first step toward the synthesis of dopamine, epinephrine, etc. and the metabolism of phenylalanine. Deficiency of PAH and also of biopterin results in hyperphenyl-alaninemia and phenylketonurea,[4-6] because the oxidative degradation of phenylalanine by PAH through tyrosine is blocked. Oxomolybdenum enzymes such as xanthine oxidase require a cofactor called molybdopterin,[7] whose structure has been concluded to be **3** having two vicinal thiol groups most probably as molybdenum binding sites.[8] Direct molybdenum-pterin interactions have been shown for a Mo(VI)-xanthopterin complex by X-ray structural analysis.[9]

PAH and other aromatic amino acid hydroxylases from mammals require nonheme iron in addition to biopterin, but the structures of the iron site and the reaction mechanisms are yet to be established.[4,10,11] On the other hand, PAH isolated from *Chromobacterium violaceum*[12] was found to contain 1 mole of Cu per mole of PAH in place of iron for mammalian PAH.[13] The copper site was reported to react with the tetrahydro form of 6,7-dimethylpterin (**4**),[14] and electron spin-echo envelope modulation spectral[15] and X-ray absorption spectral[16] studies indicated that the Cu site has two coordinated histidine imidazoles. Until recently, however, there have been almost no detailed studies on the basic modes of Cu(II)-pteridine ring interactions and the roles of metal ions in O$_2$ activation.

Pterins are usually sparingly soluble in water and organic solvents,[17] and this prevented detailed studies on metal complex formation. Folic acid, for example, is

From K.D. Karlin and Z. Tyeklár, Eds., *Bioinorganic Chemistry of Copper*
(Chapman & Hall, New York, 1993).

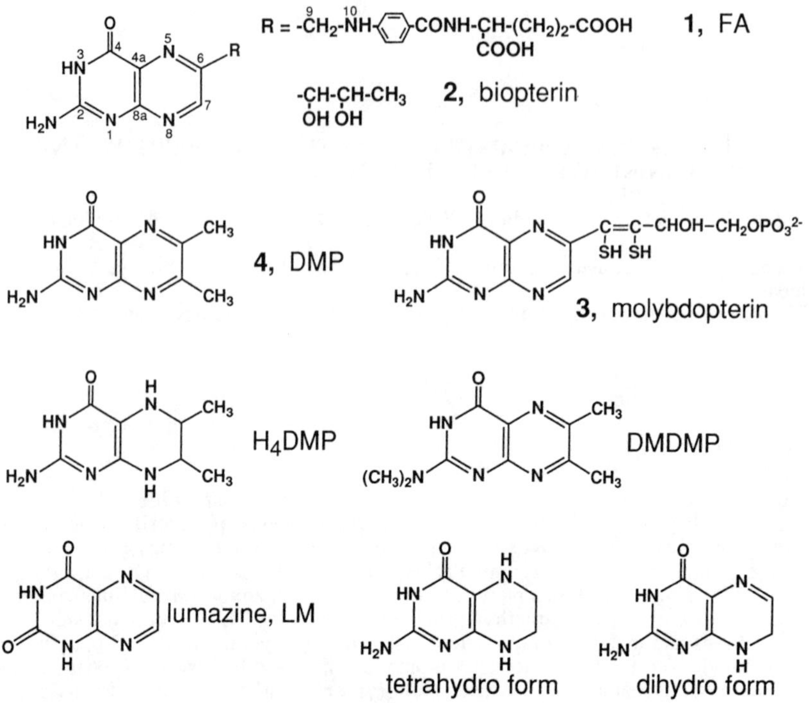

Figure 1. Structures and names (abbreviations) of pteridine derivatives.

easily soluble in alkaline solution but almost insoluble in acid-neutral solution. Since Cu(II) and other metal ions are hydrolyzed in alkaline solution, it is necessary to avoid hydrolysis in the reaction with folic acid dissolved at high pH. Use of an additional strong ligands may prevent hydrolysis and precipitate formation, and in our Cu(II)-pteridine chemistry we used Cu(II) as the complexes of 2,2'-bipyridine (bpy) and 1,10-phenanthroline (phen). Considering that the Cu site of PAH probably has two coordinated imidazoles, Cu(bpy)(NO$_3$)$_2$ etc. may serve as PAH Cu site models showing interesting reactivities toward pteridine and other ligands as cofactors or as substrates.

COPPER(II)-PTERIDINE COMPLEX FORMATION

The pterin ring offers the N(1), C(4)=O, and N(5) atoms as possible Cu(II) binding sites, and there are structural studies showing bindings through these groups. From stability constant measurements of folic acid and other pteridine derivatives, Albert concluded that pterins coordinate to Cu(II) through C(4)=O and N(5) atoms, giving the complexes similar to those of 8-hydroxyquinoline.[18] Lumazine (LM) in its deprotonated and neutral forms was reported to give Cu(LM)$_2$ and Cu(LM)$_2$X$_2$ (X = Cl, Br), respectively, with O(4) and N(5) coordinated.[19] A ternary complex of LM, Cu(bpy)(LM), was prepared by mixing LM dissolved in aqueous NaOH with Cu(bpy)(NO$_3$)$_2$ in CH$_3$OH.[20] Burgmayer et al. prepared a number of binary and ternary metal complexes including Cu(II) complexes containing pterin or 2-(ethylthio)-4-oxopteridine (ethp) and determined the structures of Cu(ethp)$_2$(phen) and Cu(tppb)(pterin) (tppb = tris(3-phenylpyrazolyl)-hydroborate) in the solid state.[21] Isolation of complexes was successful when metal acetates rather than metal chlorides were used, which indicates that the acetate ion is important as a base. On the other hand, tetrahydropterins are strong reducing agents, and trihydropterins are unstable radical species. Both undergo redox reactions with Cu(II), so that no stable complex has been isolated.

Low solubility of pterins is to a large extent due to extensive hydrogen bonding involving the amino and hydroxy (oxo) groups.[17] Acylation of the 2-amino group therefore greatly enhances the solubility. Karlin et al. synthesized some multidentate pterin ligands, one of which contains a bis[2-(2-pyridyl)ethyl]amine moiety at the C(6) position, and succeeded in solubilizing them in organic solvents by introducing a pivaloyl residue into the 2-amino group.[22] They obtained soluble Cu(I) and Cu(II) complexes exhibiting a reversible one-electron redox reaction and determined the structure of a Cu(II) complex (5).

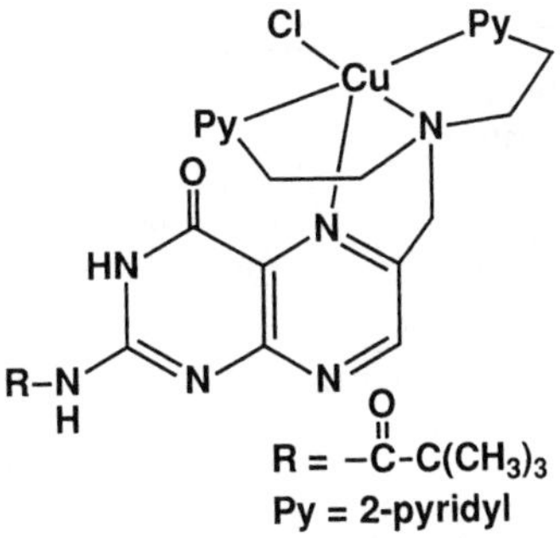

R = −C(=O)-C(CH$_3$)$_3$
Py = 2-pyridyl

5

Cu(bpy)$^{2+}$–Folic Acid Interactions

From a spectroscopic study Fridman and Levina[23] proposed that FA in Cu(II)–glycinate–FA interacts with Cu(II) at the glutamate moiety. A ternary system of 1:1:1 Cu(II)–bpy–FA gives a stable blue solution at pH 8-9, which undergoes a color change to red at pH > 10. The red solution turns green upon standing in the open air and gives green crystals involving pterin-6-carboxylate (PC), Cu(bpy)(PC)(H$_2$O) (6).[24] The oxidative cleavage of the FA side chain is observed for the Cu(bpy)(FA) and Cu(phen)(FA) systems but not for Cu(en)(FA) (en = ethylenediamine), indicating that the presence of π-accepting ligands affects the reactivity of the pterin moiety. The redox potentials which are higher for Cu(bpy)(FA) and Cu(phen)(FA) than for Cu(en)(FA) by ca. 80-100 mV may also favor the reaction. Because the pterin nucleus is electron deficient, the α-carbon of the C(6) substituent is susceptible to various oxidations: FA is oxidized at the C(6)-C(9) bond by H$_2$O$_2$[25] and alkaline permanganate[26] to give pterin-6-carboxaldehyde and PC, respectively. Since the above reaction occurs only at high pH, it is inferred to be initiated by deprotonation from the N(10)H group followed by rapid electron transfer from N$^-$ to Cu(II), giving Cu(I) and a nitrogen radical. The reaction may be tentatively described as follows (eq (1)):

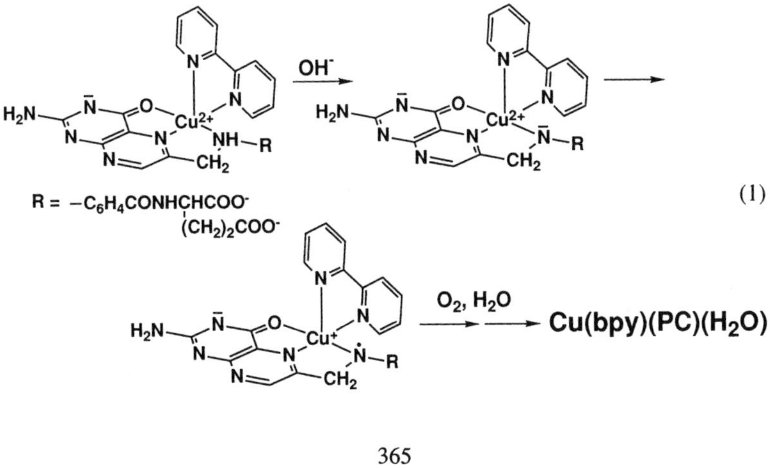

R = −C$_6$H$_4$CONHCHCOO$^-$
          (CH$_2$)$_2$COO$^-$

(1)

O$_2$, H$_2$O   ⟶   Cu(bpy)(PC)(H$_2$O)

STRUCTURES OF COPPER(II) COMPLEXES

The coordination modes of pterin derivatives have been established for several binary and ternary Cu(II) complexes. The complex of 2-(N-dimethyl)-6,7-dimethylpterin (DMDMP), [Cu(DMDMP)(NO$_3$)$_2$] (7), prepared in ethanol was found to have two DMDMP molecules equatorially bound to Cu(II) through O(4) and N(5) without deprotonation.[27] Complex 5 described above has the pterin ring coordinated through N(5) in the axial position with the Cu-N distance of 2.33 Å due to the effective side chain coordinating groups.[22] In the ternary complex of LM, [Cu(bpy)(LM)](NO$_3$)·H$_2$O (8), LM coordinates to Cu(II) through N(5) in the equatorial plane and O(4) in the axial position, the rest of the coordination plane being occupied by the two nitrogen atoms of bpy.[20] This structure shows that the N(1) atom of the coordinated LM of a neighboring complex can bind with Cu(II).

In view of the fact that FA is cleaved by Cu(bpy)$^{2+}$ and Cu(phen)$^{2+}$ but not by Cu(en)$^{2+}$ and that spectroscopic studies on Cu-containing PAH indicate that the Cu site has two coordinated imidazoles,[15,16] ternary complexes involving a pterin and bpy, phen, or a related ligand may serve as models for understanding the structures of the Cu center and Cu–pterin cofactor–substrate interactions. [Cu(bpy)(PC)(H$_2$O)] (6), which is the reaction product from the Cu(bpy)$^{2+}$–FA system, has a unique structure, where the Cu(II) ion is in a square-planar geometry with the two nitrogen atoms of bpy, the N(5) atom of PC, and a water oxygen atom in the coordination plane.[28] The O(4) and a 6-carboxylate oxygen atom coordinate to Cu(II) through long axial bonds (2.499 and 2.391 Å, respectively) to make the pterin ring perpendicular to the Cu(II) plane. The same bonding mode has been found for [Cu(en)(PC)(H$_2$O)]·H$_2$O.[29] The observed orientation of the pterin ring in the Cu(II) coordination sphere is probably due to the steric hindrance arising from the substituent at C(6) and stabilization by two axial bonds.

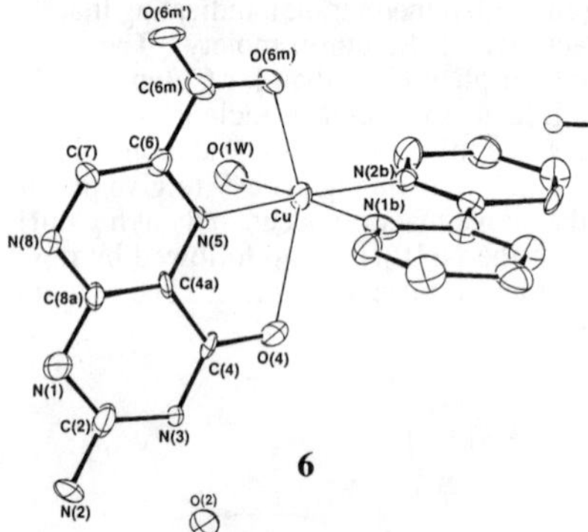

6

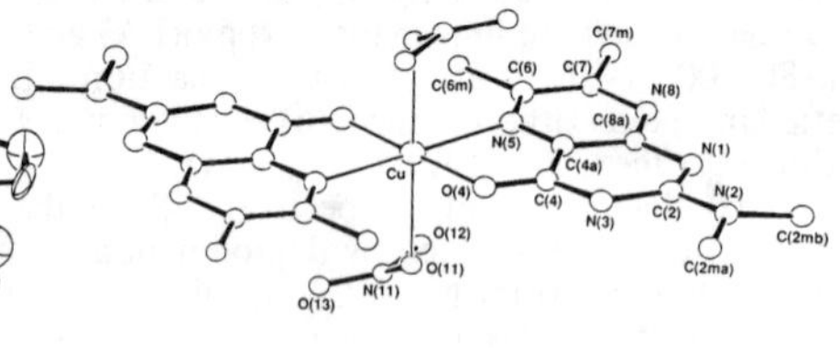

7

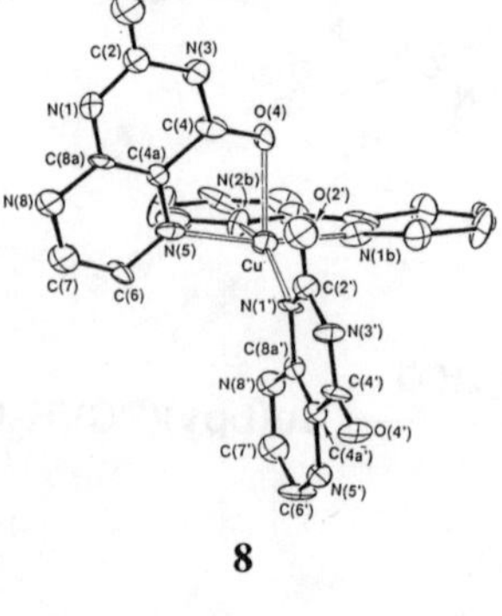

8

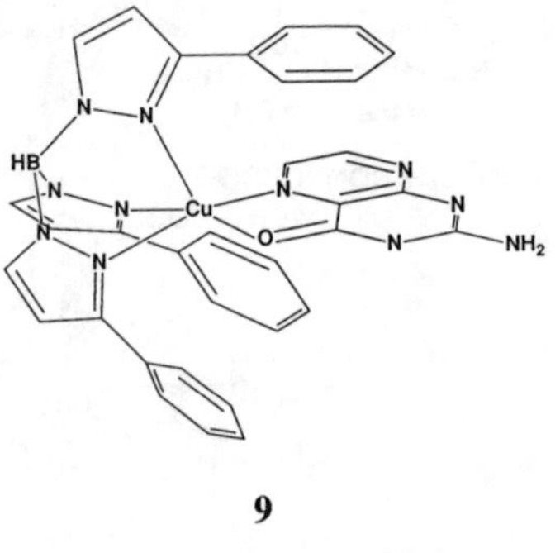

9

366

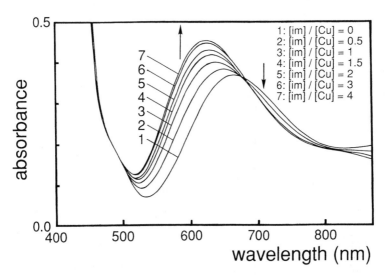

Figure 2. Absorption spectral changes of Cu(bpy)(PC)(H₂O) due to addition of im.

Preferred Formation of Quaternary Complex Cu(bpy)(PC)(im)
The solution equilibrium in the quaternary system is expressed by the stability constants defined by the following equation (charges are omitted for simplicity)

$$p\text{Cu} + q(\text{bpy}) + r(\text{PC}) + s(\text{im}) + t(\text{H}) \underset{\beta pqrst}{\rightleftharpoons} \text{Cu}_p(\text{bpy})_q(\text{PC})_r(\text{im})_s(\text{H})_t$$

$$\beta pqrst = \frac{[\text{Cu}_p(\text{bpy})_q(\text{PC})_r(\text{im})_s(\text{H})_t]}{[\text{Cu}]^p[\text{bpy}]^q[\text{PC}]^r[\text{im}]^s[\text{H}]^t} \qquad (3)$$

where p, q, r, s, and t refer to the numbers of moles of Cu(II), bpy, PC, im, and proton (H), respectively, in the complex. The log β_{11100} and log β_{11110} values for Cu(bpy)(PC) and Cu(bpy)(PC)(im) were determined to be 13.83 and 17.57, respectively, by pH titration at 25 °C and $I = 0.1$ M (KNO₃) (Table III).[20] Figure 3 illustrates the species distributions as a function of pH for the 1:1:1:1 Cu(II)-bpy-PC-im system (5 mM) calculated from the log β_{pqrst} values; Cu(bpy)(PC)(im) predominates at pH 6-10 in the quaternary system (ca. 75 % at pH 8.0). In the 1:1:1 Cu(II)-bpy-PC system (1 mM) Cu(bpy)(PC) is the major species at pH 6-8.5 (more than 70 % at pH 7.0).[20] These calculations support that combination of ligands around Cu(II) in Cu(bpy)(PC) and Cu(bpy)(PC)(im) is thermodynamically favored. A further insight into the stepwise stability constants[20] reveals that the value of 4.83 log units for the formation of Cu(bpy)(PC) from Cu(bpy) and PC is nearly as large as that (4.95) for the formation of Cu(PC) from free Cu(II) and PC. This shows that the bpy molecule in Cu(bpy) does not interfere with the bonding of PC, which is also confirmed for the PC-Cu(terpy) bonding where terpy (= 2,2',6',2"-terpyridine) occupies three coordination sites.[20]

Though statistically unfavored, Cu(bpy)(PC) with one water molecule equatorially coordinated exhibits a remarkable affinity for im as the third ligand to bind with Cu(II). Again, the stepwise constant (3.74) for formation of Cu(bpy)(PC)(im) from Cu(bpy)(PC) and im is the same as that (3.77) for formation of Cu(bpy)(im) from Cu-

Table III. Stability Constants for the Quaternary System $Cu_p(bpy)_q(PC)_r(im)_s(H)_t$ at 25 °C and $I = 0.1$ M (KNO_3)[1])

system	species	$pqrst$	$\log \beta_{pqrst}$
Cu-bpy-PC-im	Cu(bpy)(PC)(im)	11110	17.57(1)
Cu-bpy-PC	Cu(bpy)(PC)	11100	13.83(1)
Cu-bpy-im	Cu(bpy)(im)	11010	12.767(2)
	Cu(bpy)(im)$_2$	11020	15.686(2)
	Cu(bpy)(im)(OH)	1101-1	4.607(8)
Cu-PC-im	Cu(PC)(im)	10110	8.96(1)
	Cu(PC)(im)$_2$	10120	12.43(1)
	Cu(PC)(im)$_3$	10130	15.13(3)
Cu-bpy	Cu(bpy)	11000	9.0(2)
	Cu(bpy)$_2$	12000	14.724(6)
	Cu(bpy)(OH)	1100-1	2.04(2)
	Cu(bpy)(OH)$_2$	1100-2	-7.68(5)
Cu-PC	Cu(PC)	10100	4.947(6)
	Cu(PC)(OH)	1010-1	-2.00(1)
Cu-im	Cu(im)	10010	4.223(2)
	Cu(im)$_2$	10020	7.675(3)
	Cu(im)$_3$	10030	10.484(5)
	Cu(im)$_4$	10040	12.44(1)
	Cu(im)(OH)	1001-1	-3.27(2)
	Cu(im)(OH)$_2$	1001-2	-11.29(2)
	Cu(im)$_2$(OH)	1002-1	-0.23(1)
Cu-OH	Cu(OH)	1000-1	-7.223(5)
	Cu$_3$(OH)$_4$	3000-4	-21.05(3)
bpy	(bpy)(H)	01001	4.394(1)
PC	(PC)(H)	00101	7.272(4)
	(PC)(H)$_2$	00102	10.202(9)
im	(im)(H)	00011	6.994(1)

1) Taken from Ref. 20. Values in parentheses denote estimated standard deviations.

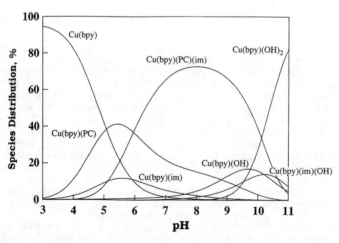

Figure 3. Calculated species distributions as a function of pH for the 1:1:1:1 Cu-bpy-PC-im system (5 mM).

(bpy) and im. Since bpy is a good π-accepting ligand and oxidized PC is an electron deficient molecule, coordination of these molecules does not lower the electronegativity of Cu(II). In line with this the above observations indicate that combination of bpy and PC around Cu(II) favors the third ligand both sterically and electronically.

REDOX REACTIONS BETWEEN Cu(II) AND TETRAHYDROPTERINS

The fact that all the aromatic amino acid hydroxylases require tetrahydrobiopterin and iron or copper indicates that a redox reaction between them is essential for activation of dioxygen. A number of studies point to the formation of an intermediate hydroxylating species involving the enzyme, dioxygen, and the cofactor.[4,10,11,32-34]   4a-Peroxotetrahydropterin has been implicated as such an intermediate, and the important roles played by iron in further activation in the hydroxylation step by mammalian PAH have been proposed by Benkovic et al.[11,32,33] Viscontini et al. studied formation of pterin radicals[35] and hydroxylation of phenylalanine by model systems with tetrahydropterin and iron complexes, and proposed a mechanism involving an iron-containing intermediate.[35,36]

Fe(III) and Cu(II) are normally one-electron oxidants and can oxidize tetrahydropterins to trihydropterins. Studies have been reported on PAH models, some of which use Cu, Fe, and other metal salts and complexes.[35,37-39] Because trihydropterins are unstable radicals, however, it has not been possible to detect one-electron redox reactions between the metal ions and tetrahydropterins both in natural and model systems, and the mechanisms of oxidation to quinonoid dihydropterins coupled with O_2 activation in the enzymatic reaction still remain to be established. Viscontini et al.[35] and Bobst[40] first reported the ESR spectra of protonated trihydropterin by oxidizing tetrahydropterin with H_2O_2 in CH_3OH-CF_3COOH where the trihydropterin radical is sufficiently stable. The radical species was concluded to have a spin density on N(5) by simulation of the spectrum. In order to find out if the one-electron redox reaction occurs with a metal ion, we measured the ESR spectrum of a solution of $Cu(NO_3)_2$ in CH_3CN added to an equimolar amount of H4DMP·2HCl in CH_3OH under N_2. In this solution H4DMP is protonated, and the H3DMP radical may be stabilized as H4DMP$^{•+}$. The Cu(II) ESR signals completely disappeared, and a signal due to a radical species was observed (Figure 4).[41] The hyperfine spectral pattern was simulated and fitted to that of H4DMP$^{•+}$, which is very similar to the spectrum reported by Bobst.[40] This shows that Cu(II) and H4DMP undergo a one-electron redox reaction to give Cu(I) and H4DMP$^{•+}$. Cu(bpy)$^{2+}$ reacted with neutralized H4DMP in CH_3OH-CH_3CN but exhibited the radical signals only after addition of CF_3COOH. An aqueous solution of 1:1 Cu(bpy)$^{2+}$–H4DMP·2HCl (pH 7.2) containing an excess of spin trapping agent DMPO (= 5,5-dimethyl-1-pyrroline N-oxide) was apparently ESR-silent but gave the signals due to the spin adduct of HO$^{•}$ upon bubbling O_2.[41] These observations indicate that O_2 is activated by a possible Cu(bpy)$^{+}$–H3DMP$^{•}$ complex which may be stable in the absence of O_2. The redox reactions between Cu(II) and H4DMP may be tentatively written as follows:

$$Cu(bpy)^{2+} + H4DMP \xrightarrow{-H^+} [Cu(bpy)^{+}\text{-}H3DMP^{•}] \xrightarrow{O_2}$$

$$[Cu(bpy)^{2+}\text{-}(^{-}OO\text{-})H3DMP] \qquad (4)$$

where $[Cu(bpy)^{2+}$–$(^{-}OO$–$)H3DMP]$ is regarded as a catalytic intermediate, and HO$^{•}$ trapped by DMPO possibly originates from this.

The 1:1 Cu(bpy)$^{2+}$–H4DMP systems (2mM) in buffered solution converted phenylalanine ethylester (20mM) to *o*-, *m*-, and *p*-tyrosines in small yields under the conditions employed (0.2-1.9 % of Cu(bpy)$^{2+}$–H4DMP depending on buffers used).[42] Since no appreciable reactions occurred in the absence of Cu(bpy)$^{2+}$, the observed hydroxylation mimics that of Cu-containing PAH, although no attempts were made to improve the yields. There was observed only weak site selectivity of hydroxylation at

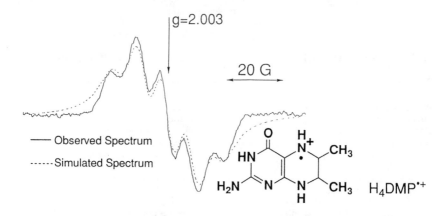

Figure 4. Observed and simulated ESR spectra of 1:1 Cu(NO3)2-H4DMP·2HCl under N2 at room temperature. Solvent: CH3OH-CH3CN.[41]

the meta position, and this may reflect various orientations of the side chain benzene ring of phenylalanine at the Cu center.

ACKNOWLEDGMENT

We are grateful to Professor Sadao Matsuura, Fujita Health University College, for helpful suggestions on the synthesis of pterins. We thank Dr. Masahiro Kohno, JEOL, for simulation of ESR spectra. Thanks are due also to Masayuki Hayashi, Mikinobu Ando, Rie Suzuki, Hideo Itagaki, and Miki Yamazaki for assistance with the experiments. This work was supported by Grants-in-Aid for Scientific Research on Priority Areas (Nos. 03241101 and 04225102) from the Ministry of Education, Science, and Culture of Japan, to which our thanks are due.

REFERENCES AND NOTES

1. J. C. Nixon, in *Folates and Pterins*, R. L. Blakley and S. J. Benkovic, Eds., (Wiley-Interscience, New York, 1985), Vol. 2, pp.1-42.
2. W. Pfleiderer, *J. Heterocycl. Chem.*, **29**, 583 (1992).
3. (a) R. L. Blakley, *The Biochemistry of Folic Acid and Related Pteridines*, A. Neuberger, E. L. Tatum, Eds. (North-Holland, Amsterdam, 1969). (b) D. Voet and J. G. Voet, *Biochemistry* (John Wiely & Sons, New York, 1990).
4. S. Kaufman and D. B. Fisher, in *Molecular Mechanisms of Oxygen Activation*, O. Hayaishi, Ed., (Academic, New York, 1974), pp. 285-369.
5. R. G. H. Cotton, in *Folates and Pterins*, R. L. Blakley and V. M. Whitehead, Eds., (Wiley-Interscience, New York, 1986), Vol. 3, pp. 359-412.
6. T. Nagatsu, S. Matuura, T. Sugimoto, *Med. Res. Rev.*, **9**, 25 (1989).
7. (a) S. P. Cramer and E. I. Stiefel, in *Molybdenum Enzymes*, T. G. Spiro, Ed., (Wiley-Interscience, New York, 1985), pp. 411-441. (b) S. J. N. Burgmayer and E. I. Stiefel, *J. Chem. Educ.*, **62**, 943 (1985).
8. (a) J. L. Johnson and K. V. Rajagopalan, *Proc. Natl. Acad. Sci. U. S. A.*, **79**, 6856 (1982). (b) S. P. Kramer, J. L. Johnson, A. A. Ribeiro, D. S. Millington, K. V. Rajagopalan, *J. Biol. Chem.*, **262**, 16357 (1987).
9. S. J. N. Burgmayer and E. I. Stiefel, *J. Am. Chem. Soc.*, **108**, 8310 (1986).

10. R. Shiman, in *Folates and Pterins*, R. L. Blakley and S. J. Benkovic, Eds., (Wiley-Interscience, New York, 1985), Vol. 2, pp. 179-249.

11. T. A. Dix and S. J. Benkovic, *Acc. Chem. Res.*, **21**, 101 (1988).

12. H. Nakata, T. Yamauchi, H. Fujisawa, *J. Biol. Chem.*, **254**, 1829 (1979).

13. S. O. Pember, J. J. Villafranca, S. J. Benkovic, *Biochemistry*, **25**, 6611 (1986).

14. S. O. Pember, S. J. Benkovic, J. J. Villafranca, M. Pasenkiewicz-Gierula, W. E. Antholine, *Biochemistry*, **26**, 4477 (1987).

15. J. McCracken, S. O. Pember, S. J. Benkovic, J. J. Villafranca, R. J. Miller, J. Peisach, *J. Am. Chem. Soc.*, **110**, 1069 (1988).

16. N. J. Blackburn, R. W. Strange, R. T. Carr. S. J. Benkovic, *Biochemistry*, **31**, 5298 (1992).

17. W. Pfleiderer, in *Folates and Pterins*, R. L. Blakley and S. J. Benkovic, Eds., (Wiley-Interscience, New York, 1985), Vol. 2, pp. 43-114.

18. A. Albert, *Biochem. J.*, **54**, 646 (1953).

19. M. Goodgame and M. Schmidt, *Inorg. Chim. Acta*, **36**, 151 (1979).

20. A. Odani, H. Masuda, K. Inukai, O. Yamauchi, *J. Am. Chem. Soc.*, **114**, 6294 (1992).

21. J. Perkinson, S. Brodie, K. Yoon, K. Mosny, P. J. Carroll, T. V. Morgan, S. J. N. Burgmayer, *Inorg. Chem.*, **30**, 719 (1991).

22. M. S. Nasir, K. D. Karlin, Q. Chen, J. Zubieta, *J. Am. Chem. Soc.*, **114**, 2264 (1992).

23. Y. D. Fridman and M. G. Levina, *Koord. Khim.*, **11**, 61 (1985).

24. T. Kohzuma, A. Odani, Y. Morita, M. Takani, O. Yamauchi, *Inorg. Chem.*, **27**, 3854 (1988).

25. D. Roberts, *Biochim. Biophys. Acta*, **54**, 572 (1961).

26. D. R. Seegar, D. B. Cosulich, J. M. Smith, Jr., M. E. Hultquist, *J. Chem. Soc.*, **1949**, 1753.

27. T. Sugimori, H. Masuda, H. Itagaki, Y. Funahashi, A. Odani, O. Yamauchi, unpulished results. Formula, $CuC_{20}H_{26}N_{12}O_8$; mol. wt., 626.06; monoclinic, space group $C2/c$; $a = 14.163(3)$, $b = 11.813(3)$, $c = 16.771(8)$ Å; $\beta = 102.46(2)°$; $Z = 4$; $R = 0.111$.

28. T. Kohzuma, H. Masuda, O. Yamauchi, *J. Am. Chem. Soc.*, **111**, 3431 (1989).

29. O. Yamauchi, Y. Funahashi, A. Odani, H. Masuda, and K. Inukai, *J. Inorg. Biochem.*, **47**, 40 (1992). Details will be published elsewhere.

30. B. Fischer, J. Strähle, M. Viscontini, *Helv. Chim. Acta*, **74**, 1544 (1991).

31. J. H. Bieri, *Helv. Chim. Acta*, **60**, 2303 (1977).

32. T. A. Dix and S. J. Benkovic, *Biochemisty*, **24**, 5839 (1985).

33. T. A. Dix, D. M. Kuhn, S. J. Benkovic, *Biochemistry*, **26**, 3354 (1987).

34. P. F. Fitzpatrick, *Biochemistry*, **30**, 6386 (1991).

35. A. Ehrenberg, P. Hemmerich, F. Müller, T. Okada, M. Viscontini, *Helv. Chim. Acta*, **50**, 411 (1967).

36. (a) M. Viscontini, H. Leidner, G. Mattern, T. Okada, *Helv. Chim. Acta*, **49**, 1911 (1966). (b) M. Viscontini and G. Mattern, *Helv. Chim. Acta*, **53**, 372 (1970).

37. M. Viscontini and G. Mattern, *Helv. Chim. Acta*, **53**, 377 (1970).

38. L. I. Woolf, A. Jakubovic, E. Chan-Henry. *Biochem. J.*, **125**, 569 (1971).

39. S. Ishimitsu, S. Fujimoto, A. Ohara, *Chem. Pharm. Bull.*, **32**, 752 (1984).

40. A. Bobst, *Helv. Chim. Acta*, **51**, 607 (1968).

41. Y. Funahashi, M. Hayashi, T. Kohzuma, A. Odani, H. Masuda, and O. Yamauchi, submitted for publication.

42. To be published.

DESIGN AND SYNTHESIS OF MODEL SYSTEMS FOR DIOXYGEN BINDING AND ACTIVATION IN DINUCLEAR COPPER PROTEINS

C. F. Martens,[a] R. J. M. Klein Gebbink,[a] P. J. A. Kenis,[a]
A. P. H. J. Schenning,[a] M. C. Feiters,[a] J. L. Ward,[b] K. D. Karlin,[b] and
R. J. M. Nolte[a]

[a]Department of Organic Chemistry, Nijmegen SON Research Center, University of Nijmegen, Toernooiveld, NL-6525 ED Nijmegen, The Netherlands

[b]Department of Chemistry, Johns Hopkins University, Baltimore, MD 21218, U.S.A.

INTRODUCTION

Following the spectroscopic[1,2] and crystallographic[3-6] characterization of the dinuclear copper site of hemocyanin, the coordination chemistry of models for such sites was developed[7,8]. Nature employs the dinuclear copper site for binding of dioxygen, as found in the hemocyanins, the oxygen transport proteins of molluscs and arthropods. It is also employed in enzymes, such as tyrosinase[9], and, in combination with copper ions representing other types of biological copper, laccase[10] and ascorbate oxidase[11], for the activation of dioxygen. Interestingly, protons and chloride ions affect the cooperativity of the oxygen binding by *Panulirus interruptus* (spiny lobster)[12] and *Limulus polyphemus* (horseshoe crab)[13] hemocyanin, respectively, whereas the aggregation of *Octopus dofleini* (Pacific octopus) hemocyanin subunits to form a functional unit is controlled by magnesium ions[14]. Recent crystal structures and spectroscopic studies[5,6,15] of oxy- and deoxy-hemocyanin under a variety of circumstances have given indications that these effects may operate at the level of control of the Cu-Cu distance.

Much of the chemistry of models for the dinuclear copper site has been directed towards the mimicking of the reversible dioxygen binding[7,16] or activation of dioxygen.

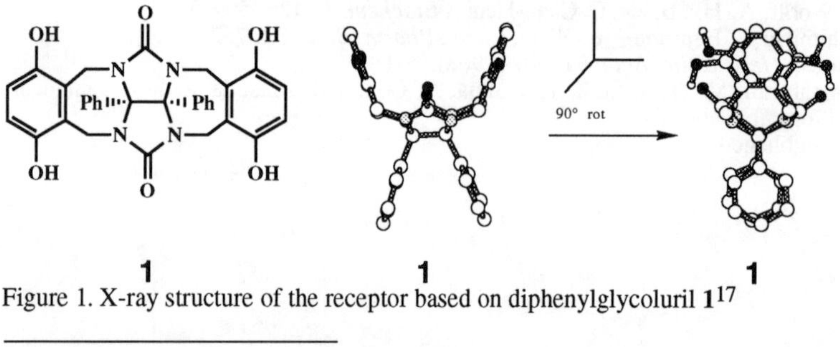

1   **1**   **1**

Figure 1. X-ray structure of the receptor based on diphenylglycoluril **1**[17]

From K.D. Karlin and Z. Tyeklár, Eds., *Bioinorganic Chemistry of Copper*
(Chapman & Hall, New York, 1993).

There are no examples of models that mimic the substrate binding site of enzymes. We describe here molecules that combine the dinuclear copper site with a receptor based on diphenylglycoluril, which has been shown to have different affinities for the isomers of dihydroxybenzene[18]. The receptor 1, as shown in Fig.1, has a higher affinity for resorcinol as compared to catechol and hydroquinone due to a combination of π-stacking and hydrogen-bonding properties.

Chart 1.

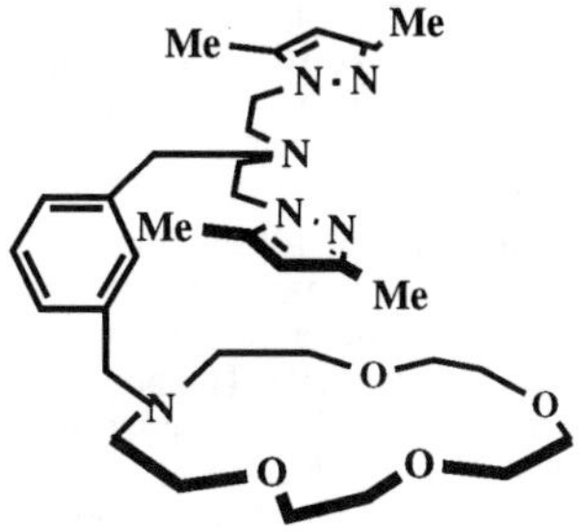

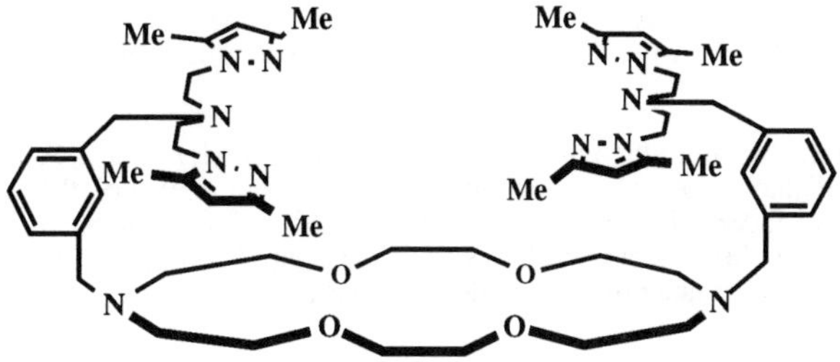

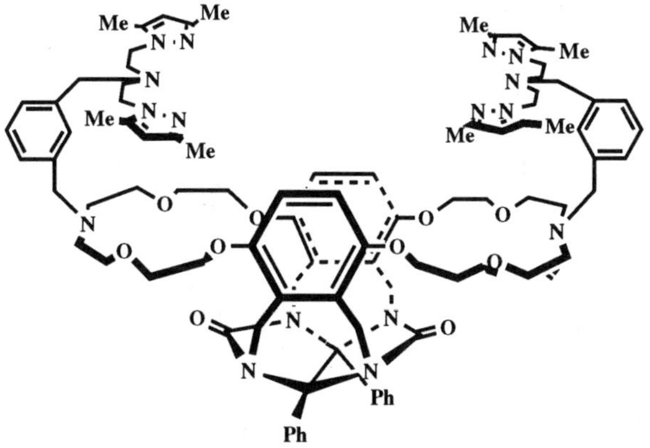

Because the link between the receptor moiety and the ligand sets was made through aza crown ethers with m-xylene spacers (Chart 1, 2), the complexes of these ligands, without the diphenylglycoluril receptor moiety, were also synthesized and studied.

PYRAZOLE LIGANDS, ALCOHOL OXIDATION

We have synthesized a number of ligands based on the bis{2-(3,5-dimethyl-1-pyrazolyl)ethyl}amine ligand, which is linked via a m-xylene spacer to a variety of crown ethers[19]. The compounds (Chart 1) include 2, where it is linked to mono-aza 15-crown-5, and 3, where two ligand sets are linked to the nitrogens of 1,10-diaza-18-crown-6. In 4, two ligand sets are linked to a diaza-crown-ether basket based on the diphenylglycoluril unit[20].

Interestingly, the introduction of a benzyl substituent on the amine nitrogen in the bis{2-(3,5-dimethyl-1-pyrazolyl)ethyl}amine ligand affected the redox potential of its copper complex, as judged from the electrochemistry. Upon benzylation, a positive shift of $E_{1/2}$ was observed, from 0 to 300 mV vs. the Fc/Fc+ couple in acetonitrile. The benzyl substituent makes the amine nitrogen a softer base, which is more suited for coordination to Cu(I). Likewise, the mononuclear copper complex of 2 was found to have an $E_{1/2}$ of 290 mV, with an equilibrium potential of 500 mV, showing that the metal was in the Cu(II) state. For the dinuclear copper complex 3, the situation is more complicated: the cyclic voltammetry displayed a broadened trace, which upon analysis indicated oxidation in two consecutive steps, with $E_{1/2}$ values of 290 and 310 mV, respectively. The equilibrium potential was 150 mV, showing that the dinuclear Cu(II) complex had been fully reduced to the Cu(I) state.

As a consequence of the higher reduction potential, the Cu(II) complexes of 2, 3, and 4 undergo reduction in alkanolic solvents. After complex formation in methanol, the green color of these complexes fades. The intensity of the d-d transition absorption band at 700 nm can be used to follow this process. With the binuclear copper complex of 3, the decay is particularly rapid. The production of formaldehyde during the reaction could

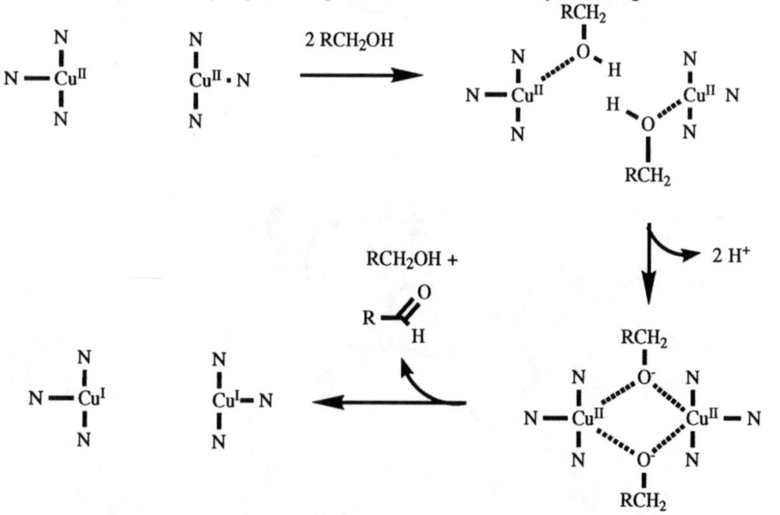

Figure 2. Proposed mechanism for the reduction of the Cu(II) complex of 3.

be detected by the fuchsine test. Clearly, a reduction mechanism requiring the proximity of two Cu(II) ions as acceptors for the 2 electrons from the oxidation of methanol to formaldehyde is operative, as shown in Fig.2. In this reaction, the 2-electron transfer 'gear', typical of organic redox reactions, is matched with the 1-electron 'gear' of transition metals. We propose that the redox reaction takes place in a μ-alkoxo-bridged complex. This is in agreement with the observation that the entropy of activation of the reaction is negative ($\Delta S^{\neq}$ = -120 J/Mol.K), pointing to the formation of an ordered complex in the rate-limiting step. The overall reaction is entropy driven (ΔS^{o} = 63 J/mol.K).

Having observed the rapid reduction of the dinuclear copper complex of **3**, we looked for ways to stimulate the reduction of the mononuclear complex of **2** by somehow forcing the copper sites of two molecules to come close together, allowing the formation of the μ-alkoxo-bridged complex. 15-Crown-5 forms sandwich complexes with K^{+} ions[21]. Upon titration of the diamagnetic Zn(II) complex of **2** with potassium picrate, an upfield shift of the ^{13}C resonances of the crown ether carbons was observed. It was concluded that 1:1 complexes were present at high K^{+}:**2** ratios, whereas sandwich complexes predominated at low ratios (Fig.3). The effect of K^{+} ions on the reduction of the dinuclear copper complex was followed by UV-vis at 700 nm. The rate of reduction was found to correlate with the extent of formation of the sandwich complex, with a maximum 5-fold acceleration in the range for the K^{+}:**2** ratio of 0.5-1 (Fig.3). This suggests that addition of K^{+} ions at low concentrations induces the formation of an active dinuclear complex from a relatively inactive monomer, which is reminiscent of the Mg^{2+}-induced aggregation of inactive hemocyanin subunits to form an oxygen-binding decamer[14].

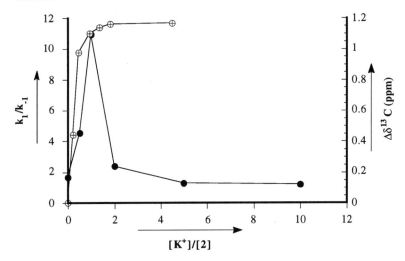

Figure 3. Reduction of complex **2** as a function of the concentration of K^{+} ions (●, ordinate on the left); ^{13}C NMR upfield shift displacement ($\Delta\delta$) for a ring CH_2 carbon atom of compound **2** as a function of the concentration of K^{+} ions (⊕, ordinate on the right).

Following our observation that the dinuclear copper complex of **3** is active in the oxidation of primary alcohols to aldehydes, it was to be expected that the dinuclear copper complex of **4** would be particularly active in the oxidation of primary alcohols that contain a dihydroxybenzene moiety fitting into the receptor part of the molecule. Indeed, 3,5-dihydroxybenzylalcohol was oxidized by **4** at a rate too fast to be followed without stopped-flow kinetics. The product, 3,5-dihydroxybenzaldehyde, was detected by HPLC and TLC. Alcohols without the dihydroxybenzene moiety only gave a very small reaction rate. We conclude that we have synthesized and characterized a receptor that is able to select the molecules which further react with the dinuclear copper site.

PYRIDINE LIGANDS, OXYGEN COMPLEXES

Pyridine ligands are more promising than pyrazole ligands as far as reversible oxygen binding to their copper complexes is concerned[7]. By analogy to the series presented for the pyrazole ligands (Chart 1), a series of crown ethers with pyridine ligands **5-7** (Chart 2) was synthesized[22]. The properties of the mononuclear copper complex of **5** were compared to those of the dinuclear complex of **6**, this time with respect to the formation of oxygen complexes. The EXAFS results are shown in Table 1. The EXAFS spectrum of the dinuclear Cu(I) complex of **6** is given in Fig.4.

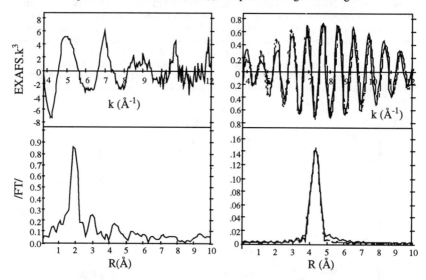

Figure 4. EXAFS (upper panels) and phase-corrected Fourier transforms (lower panels) of **6**. 2 Cu(I). Left panels: raw data; right panels: Fourier-filtered shell. Dashed line: simulation by the programme EXCURVE[23] with parameters of Table 1, bottom right.

Table 1. Parameters derived from single scattering EXAFS simulations of isolated shells

Compound	Main shell	Distance (Å)	Variance (Å)	Cu-Cu shell
5. Cu(I)	5 ± 1 N/O	2.00	0.12	none identified
5. Cu(II)(OH)$_x$	4.5 ± 1 N/O	2.00	0.07	3.3 Å
6. 2 Cu(I)	4 ± 1 N/O	1.97	0.12	4.3 Å

Chart 2.

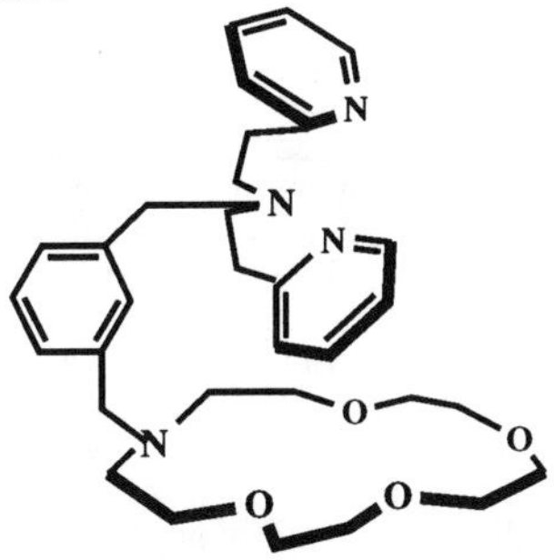

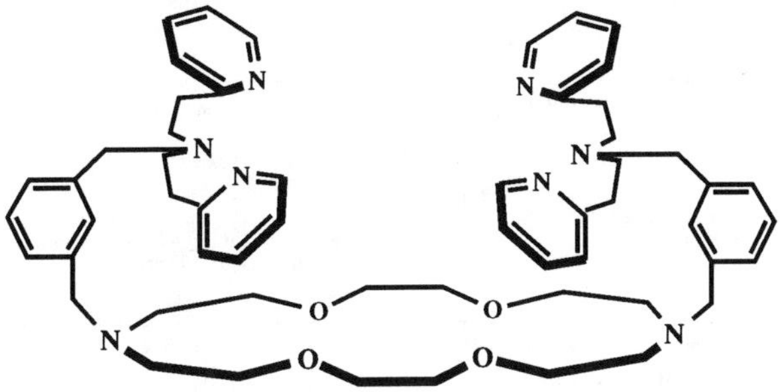

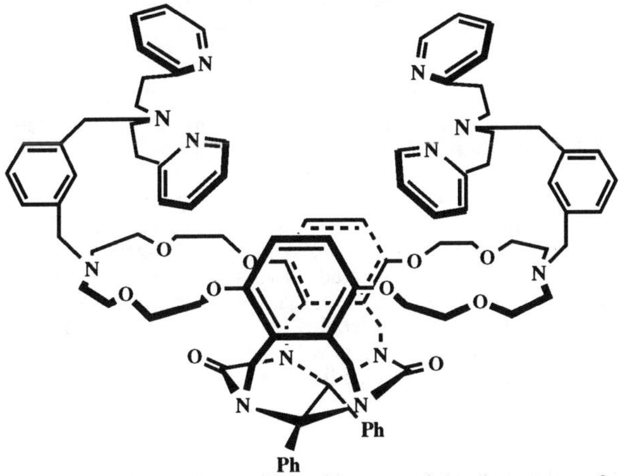

All complexes show a clear signature of the ring atoms of coordinating pyridine moieties which is noted at 3.0 and 4.3 Å in the Fourier transform (which represents a radial distribution function, Fig.4, bottom left) and can be analysed by multiple scattering simulations[24]. The results in Table 1 have been obtained by single scattering approximation as a means of identifying the shells. In all Cu(I) complexes studied, the copper ion appears to be coordinated by one or two solvent molecules (acetonitrile) in addition to the pyridine nitrogens and probably the amine nitrogen. A clear 4.3 Å Cu-Cu distance was detected in the dinuclear Cu(I) complex of 6 (Fig.4, right). This means that a significant proportion of the molecules has the copper ions preorganized at a distance suited for oxygen binding in a trans-μ-1,2 configuration[25].

The low temperature UV-vis experiments (Fig.5) show that the mononuclear Cu(I) complex of **5** absorbed at 360 nm. Upon exposure to dioxygen, a stable adduct was formed, absorbing at 360 and 452 nm. Warming to room temperature yielded a μ-hydroxo-bridged complex (indicated by IR-spectroscopy), in which the 452 nm band had completely disappeared (Fig.5, left panel). The EXAFS spectrum of this complex at 20 K shows a 3.3 Å Cu-Cu distance (Table 1), which is intermediate between that of a typical μ-hydroxo-bridged complex, viz. 2.9 Å[26], and a μ-η^2:η^2 peroxo complex, viz. 3.5 Å[27]. The dinuclear Cu(I) complex of **6** shows absorption at 350 nm, and only a transient oxygen-adduct formation (Fig.5, right panel). Possibly the bound oxygen is activated to such an extent that the solvent, dichloromethane, is attacked. Work is in progress to characterize the oxygen activation chemistry of the dinuclear Cu(I) complex of the receptor molecule **7** with suitable substrates.

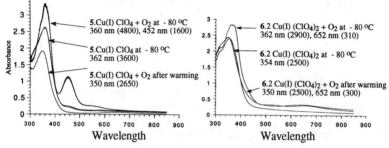

Fig.5. Low-temperature (- 80 °C) UV-vis spectra of **5**. Cu(I) (left) and **6**. 2 Cu(I) (right) in dichloromethane. ε-values, in L.Mol^{-1}.cm^{-1}, in parentheses.

CONCLUSIONS

We have designed copper coordination compounds that mimic some aspects of dinuclear copper proteins, i.e. cooperativity, oxygen binding and activation. In addition, a new reaction of dinuclear copper sites is described, viz. the oxidation of alcohols by the copper complexes of the pyrazole ligands **2**, **3**, and **4**, whose redox potential is modified by benzylation. This reaction is fast for the dinuclear copper complexes with pyrazole ligands **3** and **4**. It is slow for the mononuclear complex with pyrazole ligands and crown ether moieties **2**. However, the reaction is accelerated if two molecules of **2** are brought together by forming a sandwich complex with K$^+$-ions. The receptor moiety of the more complex molecule **4** is capable of selecting molecules for reaction with the dinuclear copper site. The copper complexes of crown ethers with pyridine ligands **5** and **6** bind dioxygen, but in the case of the dinuclear complex **6** the dioxygen adduct is unstable.

ACKNOWLEDGEMENTS

The authors acknowledge the Dutch Organization for Scientific Research (NWO) and the National Science Foundation (NSF, USA) for support. Special thanks are due to R.W.M. Aben and J.W. Scheeren, for assistance with high-pressure synthesis, H.-F. Nolting and C. Hermes (EMBL Outstation at DESY, Hamburg, Germany) for EXAFS beam time and assistance, and to J.G.M. van der Linden and J. Heck (Inorganic Chemistry, Nijmegen) for valuable discussions.

REFERENCES

1. E.I.Solomon, in *Metal Ions in Biology, Vol.2, Copper Proteins*, T.G.Spiro, Editor (Wiley Interscience, New York, 1981), pp. 41-108.
2. T.G. Spiro, G.L. Woolery, J.M. Brown, L. Powers, M.J. Winkler, and E.I. Solomon, in *Copper Coordination Chemistry: Biochemical and Inorganic Perspectives*, K.D. Karlin and J. Zubieta, Eds., (Adenine Press, Guilderland,1983), pp.23-41.
3. W.P.J. Gaykema, A. Volbeda and W.G.J. Hol, *J.Mol.Biol.*, **187**, 255 (1985).
4. A. Volbeda and W.G.J. Hol, *J.Mol.Biol.*, **209**, 249 (1989).
5. K. Magnus and H. Ton-That, *J.Inorg.Biochem.*, **47**, 20 (1992).
6. B. Hazes and W.G.J. Hol, manuscript in preparation.
7. Z. Tyeklár and K.D. Karlin, *Acc.Chem.Soc.*, **22**, 241 (1989).
8. T.N. Sorrell, *Tetrahedron*, **45**, 3 (1989).
9. K. Lerch, *Life Chem.Rep.*, **5**, 221 (1987).
10. B. Reinhammar and B.G. Malmström, in *Metal Ions in Biology, Vol.2, Copper Proteins*, T.G.Spiro, Editor (Wiley Interscience, New York, 1981), pp. 109-149.
11. A. Messerschmidt, A. Rossi, R. Ladenstein, R. Huber, M. Bolognesi, G. Gatti, A. Marchesini, R. Petruzzelli, A. Finazzi Agró, *J.Mol.Biol,*. **206**, 513 (1989).
12. H.A. Kuiper, M. Coletta, L. Zolla, E. Chiancone, M. Brunori, *Biochim.Biophys.Acta*, **626**, 412 (1980).
13. M. Brouwer, C. Bonaventura, J. Bonaventura, in *Physiology and Biology of Horseshoe Crabs: Studies on Normal and Environmentally Stressed Animals*, (A.R.Liss Inc., New York, 1982) pp. 257-267.
14. K.I. Miller and K.E. van Holde, *Comp.Biochem.Physiol.*, **B 73**, 1013 (1982).
15. M.C. Feiters, *Comm.Inorg.Chem.*, **11**, 131 (1990).
16. N. Kitajima, K. Fujisawa, C. Fujimoto, Y. Moro-oka, S. Hashimoto, T. Kitagawa, K. Toriumi, K. Tatsumi, A. Nakamura, *J.Am.Chem.Soc.*, **114**, 1277 (1992).
17. J.W.H. Smeets, R.P. Sybesma, F.G.M. Niele, A.L.Spek, W.J.J. Smeets, R.J.M. Nolte, *J.Am.Chem.Soc.*, **109**, 928 (1987).
18. R.P. Sybesma, A.P.M. Kentgens, R.J.M. Nolte, *J.Org.Chem.*, **56**, 3199 (1991).
19. C.F. Martens, A.P.H.J. Schenning, R.J.M. Klein Gebbink, M.C. Feiters, J.G.M. van der Linden, J. Heck, R.J.M. Nolte, *J.Chem.Soc., Chem.Commun.*, accepted for publication.
20. R.P. Sybesma and R.J.M. Nolte, *J.Org.Chem.*, **56**, 3122 (1991).
21. M.J. Calverley and J. Dale, *Acta Chem.Scand.*, **B 36**, 241 (1982).
22. C.F. Martens, P.J.A. Kenis, R.J.M. Nolte, K.D. Karlin, manuscript in preparation.
23. S.J. Gurman, N. Binsted, I. Ross, *J.Phys.C: Solid State Phys.*, **19**, 1845 (1986).
24. N.J. Blackburn, R.W. Strange, A. Farooq, M.S. Haka, K.D. Karlin, *J.Am.Chem.Soc.*, **110**, 4263 (1988).
25. R.R. Jacobsen, Z. Tyeklár, A. Farooq, K.D. Karlin, S. Liu, J.Zubieta, *J.Am.Chem.Soc.*, **110**, 3690 (1988).
26. C.F. Martens, A.P.H.J. Schenning, M.C. Feiters, G. Beurskens, P.T. Beurskens, R.J.M. Nolte, *Inorg.Chim.Acta*, **190**, 163 (1991).
27. N. Kitajima, K. Fujisawa, Y. Moro-oka, K. Toriumi, *J.Am.Chem.Soc.*, **111**, 8975 (1989).

COPPER DIOXYGENATION CHEMISTRY RELEVANT TO QUERCETIN DIOXYGENASE

Gábor Speier

Department of Organic Chemistry, University of Veszprém, 8201 Veszprém, Hungary

INTRODUCTION

Biological oxygenations catalyzed by oxygenases are very important processes in nature for the metabolism of various organic substances. The oxygenases are metal-containing proteins and a fair number of them utilizes copper at their active sites.[1] The dioxygenases as a subclass of these enzymes degrade cyclic organic substrates such as tryptophan, catechol, protocatechuic acid and quercetin and have the characteristics of incorporating both oxygen atoms of dioxygen into the substrate. Biomimetic chemistry is a useful tool

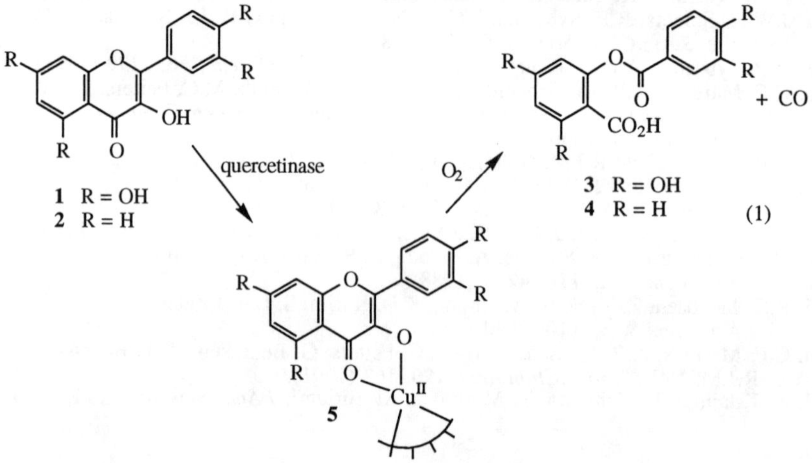

1 R = OH
2 R = H

3 R = OH
4 R = H

5

+ CO

(1)

for determination of the possible structure of the active site, binding mode of the substrate and the reaction pathways of the enzyme actions. Quercetin 2,3-dioxygenase is a copper-

From K.D. Karlin and Z. Tyeklár, Eds., *Bioinorganic Chemistry of Copper* (Chapman & Hall, New York, 1993).

containing dioxygenase[2-6] with copper(II) in the active site, which catalyzes the oxygenolysis of 3-hydroxyflavones (**1,2**) to the corresponding depsides (**3,4**) as a result of the oxidative cleavage of the heterocyclic ring (equation 1). Quercetin was assumed to coordinate to Cu^{II} as a chelating ligand at the 3-hydroxy and 4-carbonyl groups.[7-9] In a few cases copper(II) compounds have been used in model catalytic oxygenations of 3-hydroxyflavones and related compounds.[10-12] Since no copper flavonolate or quercetin complexes were known in the literature we set the ultimate goal to prepare and characterize such complexes which would serve as structural models and give possibilities to study relevant stoichiometric and catalytic oxygenation reactions in order to disclose the possible pathway of the enzymatic reaction.

PREPARATION OF Cu^I AND Cu^{II} FLAVONOLATE COMPLEXES

Although there were suggestions for the coordination of quercetin to copper(II) in quercetin dioxygenase in the literature[7,8] no well-characterized model compounds have been described. In order to obtain information of the possible binding of the substrate and to use them in oxygenation reactions to study the oxidative ring splitting of the heterocyclic ring in the substrate, first we attempted to prepare stable copper(I) and copper(II) flavonolate complexes with different auxiliary ligands.

Preparation of Copper(I) Flavonolate Complexes

Since it is well known that Cu^I is relatively soft and it favors soft ligands such as phosphines or aromatic amines,[13] we tried to prepare copper(I) flavonolate complexes with triphenylphosphine (PPh3), 1,2-bis(diphenylphosphino)ethane (diphos), pyridine (py), 2,2'-bipyridine (bpy) and N,N,N',N'-tetramethylethylenediamine (tmeda) as auxiliary ligands. Flavonol (flaH, **2**) was reacted with sodium in THF or methanol to give sodium flavonolate, which on further reaction with copper(I) chloride in the presence of PPh3 or diphos results in the formation of [Cu^I(fla)(PPh3)2] (**6**) and [Cu^I(fla)(diphos)] (**7**) in ~80%

yields (equation 2). Compounds **6** and **7** show characteristic absorption bands in the visible region at around 420 nm due to the flavonolate ligand.[14] Strong IR bands at 1560 and 1525 cm^{-1} assigned to ν(CO), showing a decrease of 40-70 cm^{-1} compared to flavonol [ν(CO)=1602 cm^{-1}], arise as a result of the formation of a five-membered rings.[15] Due to line broadening no ^{1}H and ^{31}P NMR spectra could be obtained at room temperature. The crystal structure of **6** is shown in Figure 1. The complex has a distorted tetrahedral geometry around the copper with the flavonol chelating through the 3-hydroxy and 4-carbonyl groups with Cu-O bond lengths of 2.051(4) and 2.167(5) Å.[16]

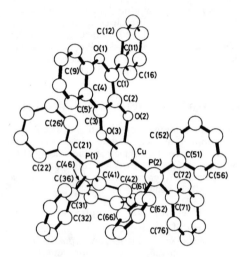

Figure 1. Molecular structure of [Cu^I(Fla)(PPh$_3$)$_2$] (**6**) (Reproduced with permission from ref. 16. Copyright 1990. Royal Society of Chemistry).

Earlier studies have shown that 1,2-diketones react with metallic copper (Cu^0) to give copper(I) semidione species due to electron transfer from Cu^0 to the ketones.[17] Because of the keto-enol tautomerism of 3-hydroxyflavones[18] (equation 3) the keto form **8** do exist in

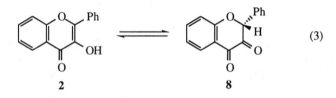

$$(3)$$

$$\text{2} \qquad\qquad\qquad \text{8}$$

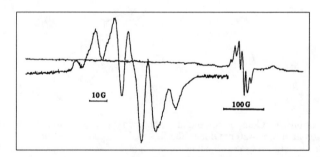

Figure 2. Solid state ESR spectrum of [Cu^I(fla$^\bullet$)(PPh$_3$)$_2$] (**9**).

384

a small concentration and its reaction with Cu^0 in the presence of PPh_3 led to the radical species **9**. It could not be isolated in pure form because of the not negligible consecutive conversion to **6** and hydrogen evolution (equation 4). The product contains approximately

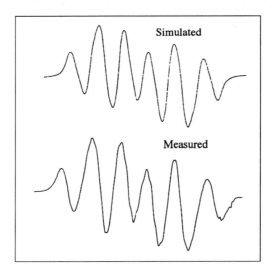

20% **9** depending on the reaction time. However, the presence of **9** in the reaction mixture could be proved by ESR spectroscopy (Figure 2 and 3). The solid state spectrum of **9** indicates the presence of an organic radical with a g value of 2.0074 showing a fine structure (Figure 2). The solution spectrum of **9** in acetonitrile (Figure 3) exhibits hyperfine structure and the coupling constants determined by simulation were found to be: $a_{Cu}=9.2$, $a_P(2)=12.0$, $a_H(1)=2.8$, $a_H(2)=1.4$ and $a_H(2)=0.6$ G. The formation of the one-

Figure 3. ESR spectrum of $[Cu^I(fla^\bullet)(PPh_3)_2]$ (**9**) in acetonitrile.

electron reduction product **9** from **2** may be important as a radical intermediate in the enzyme reaction since the formation of an endoperoxide by the reaction of a radical species with dioxygen, as proposed in some studies, seems to be a reasonable assumption for the enzyme reaction.[19]

The reactions of sodium flavonolate with copper(I) chloride in the presence of py, bpy or tmeda in THF or methanol as solvent did not give complexes similar to those with phosphines (equation 5) but $[Cu^{II}(fla)_2]$ (**10**) and unidentified products were formed in rather low yields.[20]

$$\text{(structure)} \quad + \text{ CuCl } + \begin{array}{l} \text{py} \\ \text{bpy} \\ \text{tmeda} \end{array} \xrightarrow{\text{THF or MeOH}} [Cu^{II}(fla)_2] + \begin{array}{l} \text{unidentified} \\ \text{products} \end{array} \quad (5)$$

$$\mathbf{10}$$

In an other attempt mesitylcopper(I)[21] was reacted with flavonol in the presence of py or tmeda (equation 6) under careful conditions. Again $[Cu^{II}(fla)_2]$ was formed in somewhat

$$0.2\left(\text{structure}\right)_5 + \text{structure} + \begin{array}{l} \text{py} \\ \text{tmeda} \end{array} \xrightarrow{\text{MeCN}} \left\{ \begin{array}{l} [Cu^{II}(fla)_2] \\ + \\ \text{unidentified products} \\ \text{and mesitylene} \end{array} \right. \quad (6)$$

better yield along with mesitylene. When bpy is used as auxiliary ligand, a different product, probably $[Cu^{I}(Fla)(bpy)]$ (**11**) was formed first, which converts to $[Cu^{I}(bpy)_2]_2[Cu^{I}(Fla)_3)]$ (**12**) in 38% yield (equation 7). **12** is very air-sensitive even in the solid state. The compound is diamagnetic and ESR-silent.. X-ray studies on disordered crystals gave only preliminary results (Figure 4). Two copper(I) ions are surrounded with two bpy in a tetrahedral geometry while another copper(I) has four fla ligands in a distorted octahedral arrangement with Cu-O distances between 2.166 and 2.066 Å.

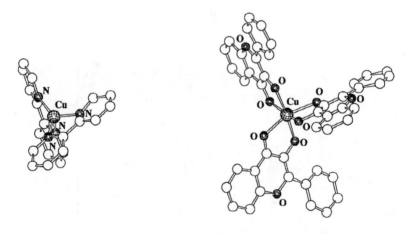

Figure 4. Preliminary crystal structure of $[Cu^{I}(bpy)_2]_2[Cu^{I}(fla)_3]$ (**12**).

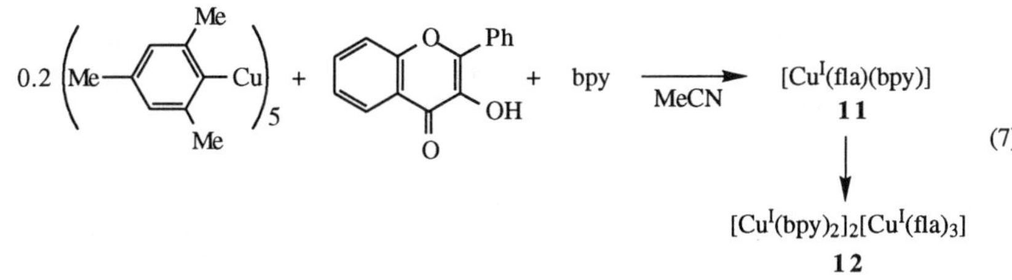

$$[Cu^I(fla)(bpy)]$$
$$\textbf{11}$$

$$(7)$$

$$[Cu^I(bpy)_2]_2[Cu^I(fla)_3]$$
$$\textbf{12}$$

Preparation of Copper(II) Flavonolate Complexes

Because of the oxidation state of copper in quercetinase is believed to be two,[6] the preparation, characterization and oxygenation studies of copper(II) flavonolate complexes seemed to be of great interest. Bis(methoxo)copper(II)[22] reacted with flavonol in dichloromethane at room temperature to give pure bis(flavonolato)copper(II) (**10**) in 83% yield (equation 8). The complex shows a characteristic absorption band in the visible

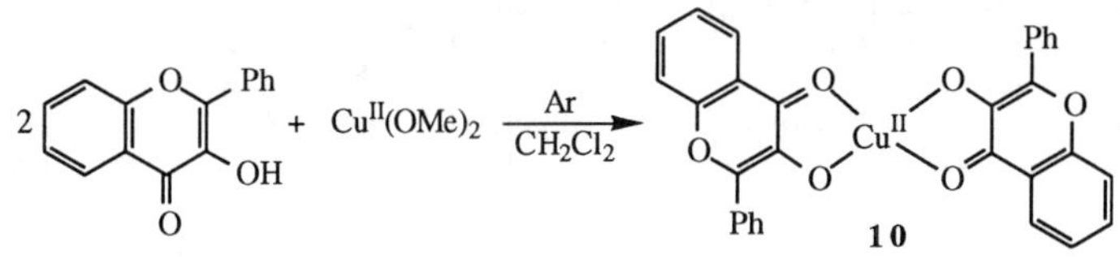

spectrum at 426 nm due to the flavonolate ligand and an IR absorption band at 1536 cm^{-1}

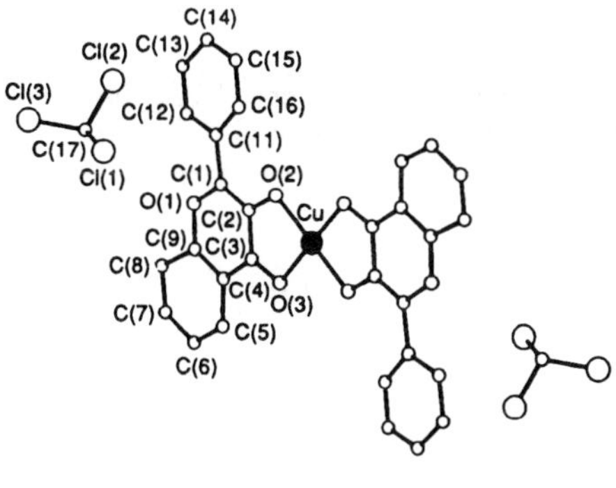

Figure 5. Crystal structure of $[Cu^{II}(fla)_2 \cdot 2CHCl_3]$ (**10**). (Reproduced with permission from ref. 20. Copyright 1991, Royal Society of Chemistry).

assigned to ν(CO). It is paramagnetic with μ_{eff}=2.10 BM and solid state ESR parameters of $g_{\parallel}$=2.2518 and $g_\perp$=2.0849. The crystal structure of $[Cu^{II}(fla)_2 \cdot 2CHCl_3]$ is shown in Figure 5. The complex possesses high symmetry with *trans* coordination of the flavonolate ligands in a square planar geometry. The two 3-hydroxychromone moieties and the central Cu^{II} show high planarity to within 0.074 Å. The phenyl rings on the heterocycle show a torsion angle of 6.4(1)°. The copper-oxygen bond distances are in the range of 1.901–1.944 Å, somewhat shorter than those in $[Cu^I(fla)(PPh_3)_2]$ (2.051–2.167 Å).[16] In the solid form and in solutions at ambient temperature, it is extremely stable toward molecular oxygen and was rapidly formed as a byproduct in the course of reactions where Cu^{II} and flavonolate ligands were present (see equations 5 and 6). The mixed ligand copper(II) flavonolate complex $[Cu^{II}Cl(fla)(py)]$ (**13**) could be prepared from $[Cu_2Cl_2(OMe)_2(py)_2]^{23}$ and flavonol in dichloromethane at room temperature (equation 9) in 88% yield. Complex **13** exhibits a absorption band at 425 nm assignable to the

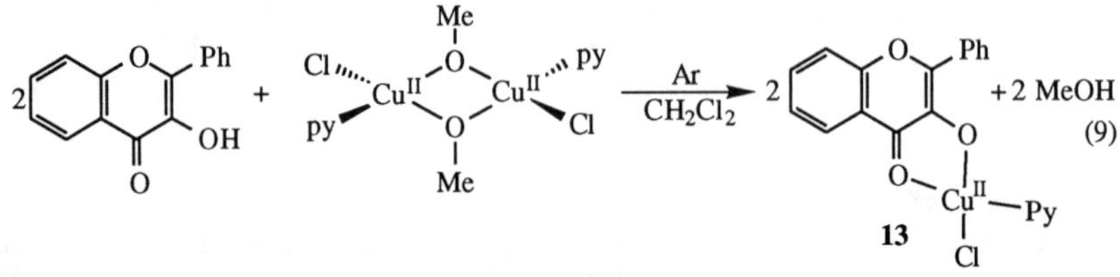

$$(9)$$

flavonolate ligand and the $\nu(CO)$ band at 1537 cm^{-1} in the infrared spectrum. It is paramagnetic (μ_{eff}=1.95 BM) with solid state ESR parameters of g_1=2.2231, g_2=2.0717 and g_3=2.0576. Recrystallization from chloroform resulted in decomposition with formation of compounds $CuCl_2(py)_2$ and $[Cu^{II}(fla)_2]\cdot 2CHCl_3$.

Attempts were made to prepare mixed-ligand copper(II) flavonolate complexes with chelating monoanionic ligands such as acetylacetonate and 8-hydroxyquinoline. The reaction of bis(acetylacetonato)bis(methoxo)dicopper(II)^{24} with flavonol did not give $[Cu^{II}(acac)(fla)]$ (equation 10) but $[Cu^{II}(fla)_2]$ in low yield. The displacement of 8-

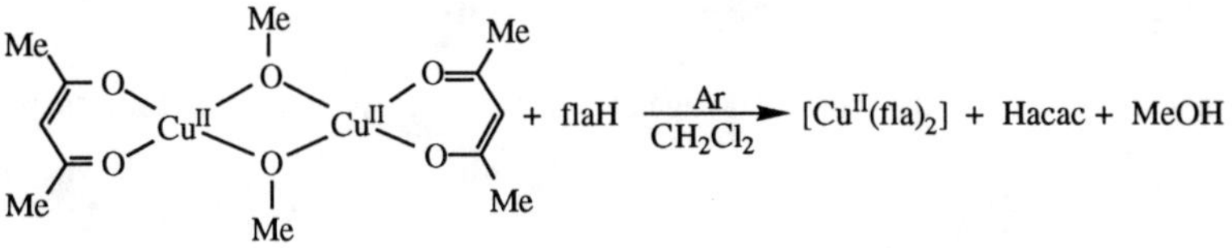

$$[Cu^{II}(fla)_2] + Hacac + MeOH \quad (10)$$

hydroxyquinoline in bis(8-hydroxyquinoline)copper(II)^{25} with flavonol failed even at higher temperature and long reaction times (equation 11). The reaction of bis(chloro)bis(8-

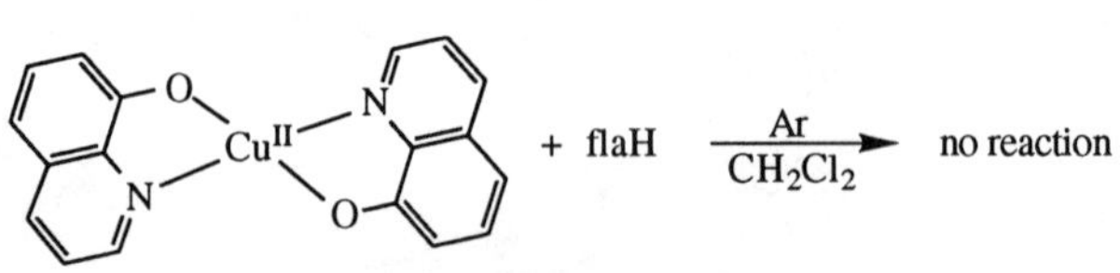

$$\text{no reaction} \quad (11)$$

hydroxyquinoline)dicopper(II)^{26} with flaNa in THF (equation 12) was also of no use for the preparation of mixed-ligand copper(II) flavonolate complexes.

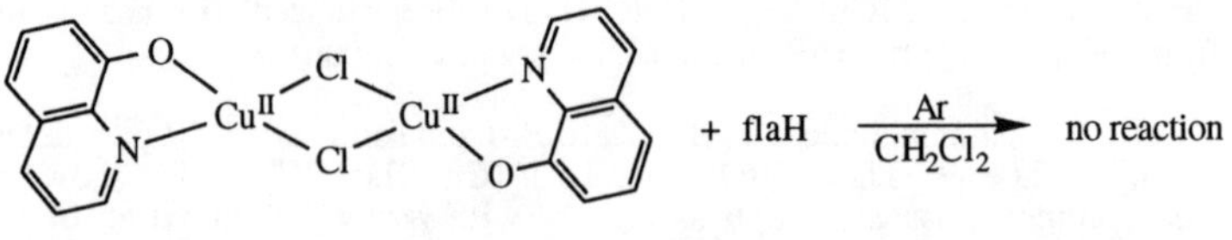

$$\text{no reaction} \quad (12)$$

Considerations for the preparation of cationic copper(II) flavonolate complexes $[Cu^{II}(fla)(N,N)]^+$ with chelating N,N-ligands such as bpy and tmeda turned out to be more successful. By reacting equimolar amounts of $[Cu^{I}(MeCN)_4]ClO_4^{27}$, flavonol and tmeda with dioxygen at room temperature in acetonitrile, green crystals of $[Cu^{II}(fla)(tmeda)]ClO_4$ (**14**) deposited slowly from the homogeneous solution in 60% yield (equation 13).

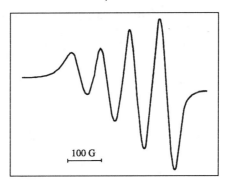

Compound **14** is a very stable mixed-ligand copper(II) flavonolate complex with probably square planar geometry. The paramagnetic compound with μ_{eff}=1.87 BM exhibits ESR parameters of g=2.062 (solid state) and g=2.105 (in toluene) (see Figure 5). In solution the fine structure of the spectrum showed coupling constants of a_{Cu}=85.0 G. The visible spectrum indicates the characteristic absorption band at 418 nm due to the flavonolate

Figure 6. ESR spectrum of [Cu^{II}(fla)(tmeda)]ClO$_4$ in toluene.

ligand. In the IR spectrum a broad band at ~1100 cm^{-1} could be assigned to v(Cl-O). In a similar manner, [Cu^{II}(fla)(bpy)]ClO$_4$ (**15**) could be obtained by the reaction of [Cu^I(MeCN)$_4$]ClO$_4$, flaH and bpy with dioxygen. Green microcrystals were formed in ~90% yield. Compound **15** is paramagnetic with μ_{eff}=1.96 BM which is in the range of 1.8–2.1 typically found for magnetically isolated, mononuclear copper(II) complexes. It showed absorption in the visible region at 418 nm assignable to the flavonolate ligand and v(Cl-O) absorption band at ~1100 cm^{-1} in the infrared spectrum.

OXYGENATION OF COPPER FLAVONOLATE COMPLEXES

For the elucidation of the possible mechanism of the ring cleavage reaction of the coordinated flavonolate ligand stoichiometric oxygenations of the copper flavonolate complexes **6, 9, 14** and **15** were carried out. Furthermore, catalytic oxygenations of flavonol were studied with some of the above mentioned complexes to check whether they are effective or selective catalyst for the enzyme-like process.

Stoichiometric Oxygenations of Copper Flavonolate Complexes

The oxygenation of [Cu^I(fla)(PPh$_3$)$_2$] (**6**) were carried out in dichloromethane or acetonitrile at room temperature. In a rather long reaction time a greenish compound and carbon monoxide with a small amount of carbon dioxide were formed according to equation 14 . On recrystallization, the greenish colored product gave a colorless complex

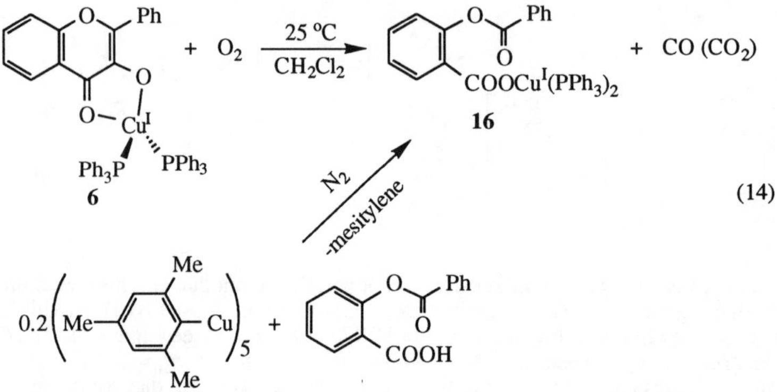

$$(14)$$

[Cu^I(depside)(PPh$_3$)$_2$] (**16**). The yields were around 90% and only a small portion of the compound was further oxidized to a copper(II) compound. **16** is diamagnetic with characteristic IR absorption bands of $v(CO)$ 1710 cm^{-1} and $v(CO_2)$ 1584 and 1390 cm^{-1}, respectively. The ^{31}P NMR spectrum showed a singlet at -3 ppm proving chemically

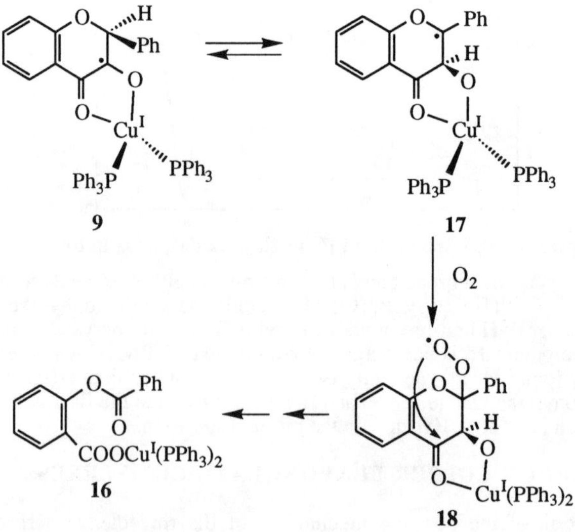

$$(15)$$

equivalent phosphines in solution. 16 could also be obtained by an independent procedure starting from mesitylcopper(I), PPh$_3$ and O-benzoylsalicylic acid[28] (equation 14). It is remarkable that during the very long oxygenation reaction time (~50 h) at room temperature practically only the coordinated flavonolate ligand was selectively oxygenated leaving Cu^I and PPh$_3$ unattacked.

Equation 4 shows the formation of [Cu^I(fla·)(PPh$_3$)$_2$] (9) by the reaction of flavonol with metallic copper in the presence of PPh$_3$. By carrying out the same reaction under dioxygen a much faster reaction took place (equation 15). After a short period of time the yellow colored solution turned to intense red which diminished later. We assume that the reason for that is the transient formation of a peroxy radical (18) as a result of the reaction of the radical species 9 or 17 with dioxygen. In 18 the peroxy group could attack the 4-carbonyl group of the coordinated flavonolate leading to an endoperoxide[19] which could assumedly break down to 16 and carbon monoxide. However, from the reaction mixture only 16 could be isolated in moderate yields. This oxygenation reaction of the coordinated flavonolate ligand seems to have a somewhat different pathway leading to the same final product.

Surprisingly, the flavonolate ligand coordinated to copper(II) in [Cu^{II}(fla)(tmeda)]ClO$_4$ (14) or in [Cu^{II}(fla)(bpy)]ClO$_4$ (15) showed enhanced inertness toward dioxygen. Higher temperatures were needed to facilitate the oxygenation and the formation of O-benzoyl-salicylato(tmeda)copper(II) perchlorate complex (19) and carbon monoxide (equation 16). Complex 19 could be also prepared in 85% yield by the oxygenation of a mixture of

$$\text{(16)}$$

[Cu^I(MeCN)$_4$]ClO$_4$, O-benzoylsalicylic acid and tmeda. 19 shows IR absorption bands at v(CO) 1720, v(CO$_2$) 1610, 1380 and v(Cl-O) 1085 cm^{-1}. It is paramagnetic with μ$_{eff}$=1.85 BM and ESR parameters in toluene solution of g=2.133 and a$_{Cu}$=64.6 G. In a similar way, the oxygenation of [Cu^{II}(fla)(bpy)]ClO$_4$ (15) resulted in the cleavage product O-benzoylsalicylato(2,2'-bipyridine)copper(II) perchlorate (20). It could also be prepared from O-benzoylsalicylic acid, [Cu^I(MeCN)$_4$]ClO$_4$, bpy and dioxygen in 70-80% yields. The IR data of v(CO)=1720, v(CO$_2$)=1600 and 1400 cm^{-1}, and v(Cl-O)=~1100 cm^{-1} and the magnetic susceptibility of μ$_{eff}$=1.84 BM support the structure.

Preliminary kinetic measurements on the oxygenation of [Cu^{II}(fla)(tmeda)]ClO$_4$ in DMF at 100-120°C were done by following the reaction rate with electron absorption spectroscopy. The course of the reaction monitored at 426 nm is presented in Figure 7, showing the decrease of the concentration of [Cu^{II}(fla)(tmeda)]ClO$_4$ (14) with time. Plots

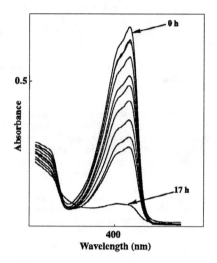

Figure 7. Oxygenation of [Cu^{II}(fla)(tmeda)]ClO$_4$ (**14**) in DMF followed spectrophotometrically.

of log concentration of [Cu^{II}(fla)(tmeda)]ClO$_4$ versus time give an approximately straight line indicating first order dependence on the complex **14** (Figure 8). Investigations on the

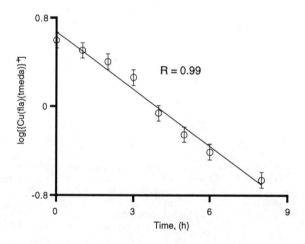

Figure 8. Dependence of the oxygenation rate on the [Cu^{II}(fla)(tmeda)]ClO$_4$ (**14**) concentration in DMF at 120°C and atmospheric dioxygen pressure.

influence of dioxygen pressure on the reaction rate in a relatively short range of dioxygen pressure gave a first order dependence on dioxygen concentration, as well (Figure 9).

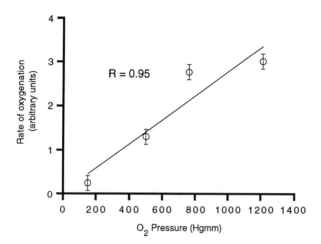

Figure 9. Dependence of the oxygenation rate of $[Cu^{II}(fla)(tmeda)]ClO_4$ (**14**) on the dioxygen pressure.

These very preliminary data seem to support the assumption that the rate law of the oxygenation of **14** obeys second order kinetics (equation 17).

$$\text{Oxygenation rate} = k_2 \, [\{Cu^{II}(fla)(tmeda)\}^+][O_2] \qquad (17)$$

The oxygenation reactions of **6**, **9** and **14** show that the flavonolate ligand (also in one-electron reduced form in **9**) coordinated either to copper(I) or copper(II) can be selectively oxygenated to the depside in stoichiometric reactions. However, the elucidation of the detailed, fine mechanism of the cleavage reaction needs further studies.

Catalytic Oxygenations by Copper Flavonolate Complexes

The copper(I) and copper(II) flavonolate complexes $[Cu^{I}(fla)(PPh_3)_2]$ (**6**), $[Cu^{I}(fla)(diphos)]$ (**7**), $[Cu^{II}Cl(fla)(py)]$ (**13**) and $[Cu^{II}(fla)_2]$ (**10**) were used as catalysts in the oxygenation of flavonol in acetonitrile and DMF at 80 °C. The products formed are shown in equation 18. The oxygenation reactions were very selective and in each case O-benzoylsalicylic acid (21) or its hydrolized derivatives, benzoic acid (23), salicylic acid (22) and dimethylbenzamide (24) along with carbon monoxide and carbon dioxide were formed in 20-95% yields depending on the solvent and reaction time. By the use of copper flavonolate complexes with phosphines (6,7) oxidation of the phosphines to phosphine oxides also occurred to a certain extent, while in the case of 13 probably 10 is the actual catalyst (it could be isolated after the catalytic experiments from the solutions).[29] These qualitative data show however that copper flavonolate complexes are selective catalysts in the oxidative ring cleavage reaction of flavonol but for better understanding the mechanistic features of the reaction different models and more detailed investigations are needed.

AKNOWLEDGEMENTS

The author thanks the collaborators in the cited papers, Mr. I. Lippai, Dr. A. Rockenbauer (Budapest), Prof. G. Huttner and dr. L. Zsolnai (Heidelberg) for the ESR and x-ray measurements and the Hungarian Research Fund (OTKA # 2326) for financial support.

REFERENCES

1. K. D. Karlin, Z. Tyeklár and A. D. Zuberbühler, in *Bioinorganic Catalysis*, J. Reedijk, Ed., (Marcel Dekker, New York, 1992), Ch. 9, p.261.
2. D. W. S. Westlake, G. Talbot, E.R. Blakely and F.J. Simpson, *Can. J. Microbiol.*, 5, 62 (1959).
3. F. J. Simpson, G. Talbot and D. W. S. Westlake, *Biochem. Biophys. Res. Commun.*, 2, 621 (1959).
4. S. Hattori and I. Noguaki, *Nature*, 184, 1145 (1959).
5. H. Sakamoto, *Seikagaku (J. Jpn. Biochem. Soc.)*, 35, 663 (1963).
6. T. Oka, F. J. Simpson and H. G. Krisknamurthy, *Can. J. Microbiol.*, 16, 439 (1977).
7. E. Makasheva and N. T. Golovhina, *Zh. Obschch. Khim.*, 43, 1640 (1973).
8. M. Thomson and C. R. Williams, *Anal. Chim. Acta*, 85, 375 (1976).
9. K. Takamura and M. Ito, *Chem. Pharm. Bull.*, 25, 3218 (1977).
10. M. Utaka and M. Hojo, Y. Fujii and A. Takeda, *Chem. Lett.*, 635 (1984)
11. M. Utaka and A. Takeda, *J. Chem. Soc., Chem Commun.*, 1824 (1985).
12. É. Balogh-Hergovich and G. Speier, *J. Mol. Catal.*, 71, 1 (1992).
13. B. J. Hathaway, in *Comprehensiv Coordination Chemistry*, G. Wilkinson, R. D. Gillard and J. A. MacCleverty, Eds, (Pergamon Press, Oxford, (1987), Vol.5, p.536.
14. L. Jurd and T. X. Geissman, *J. Org. Chem.*, 21, 1395 (1956).
15. L. M. Bellamy, *Ultrarot Spektrum und chemische Konstitution*, (Dr. Dietrich Steinkopf Verlag, Darmstadt, 1966), p.112.
16. G. Speier, V. Fülöp and L. Párkányi, *J. Chem. Soc., Chem. Commun.*, 512 (1990)
17. G. Speier and Z. Tyeklár, *Transition Met. Chem.*, 17, 348 (1992).
18. P. Sohár, L. Vargha and J. Kuszmann, *Acta Chim. Hung.*, 70, 79 (1971).
19. V. Rayananda and S. B. Brown, *Tetrahedron Lett.*, 22, 4331 (1981).
20. É. Balogh-Hergovich, G. Speier and G. Argay, *J. Chem. Soc., Chem.Commun.*, 551 (1991).
21. T. Tsuda, T. Yazawa, K. Watanabe, T. Fujii and T. Saegusa, *J. Org. Chem.*, 46, 192 (1981).
22. R. W. Adams, R. L. Martin and G. Winter, *Aust. J. Chem.*, 20, 773 (1967).
23. H. Finkbeiner, A. S.Hay, H. S. Blanchard and G. F. Endres, *J.Org.Chem.*, 31, 549(1966).
24. J. A. Bertrand and R. I. Kaplan, *Inorg. Chem.*, 4, 1657 (1965).
25. R. G. W. Hollingshead, in *Oxine and Its Derivatives*, (Butterworths, London, 1965), Vol. 1.
26. Prepared from CuCl, 8-hydroxyquinoline by oxygenation in acetonitrile.
27. B. J. Hathaway, D. G. Holah and J. D. Postlethwaite, *J. Chem. Soc.*, 3215 (1961).
28. A. Einhorn, L. Rothlauf and R. Seufert, *Chem. Ber.*, 44, 3309 (1911).
29. G. Speier in *Dioxygen Activation and Homogeneous Catalytic Oxidation*, L. I. Simándi, Ed., Elsevier, Amsterdam, 1991, p.269.

Nitrogen Oxide (NO$_x$) Chemistry and Biochemistry

TWO CRYSTAL FORMS OF *A. CYCLOCLASTES* NITRITE REDUCTASE

Elinor T. Adman and Stewart Turley

Department of Biological Structure SM-20,
University of Washington, Seattle WA 98195, USA

Nitrite reductase is one of a series of copper-containing proteins found in a denitrification pathway, in which nitrate is eventually converted to nitrogen gas[1]. The reaction catalyzed by NIR, converting nitrite to nitric oxide (NO_2^- to NO) is also carried out by a heme protein, cytochrome cd_1, which is detected in about twice as many organisms as is the copper protein. NIR has been biochemically characterized in at least eight organisms (see Table 1). Copper content and subunit composition appear to vary, but there is mounting evidence in addition to the antibody studies cited above that the first four of the NIRs listed in Table 1 are more similar to each other than previously believed (see notes in Table 1). NO is the primary product of NIR using ascorbate/phenazine methylsulfonate or a cupredoxin as the reducing agent[2,3]; N_2O can be produced when hydroxylamine is present. Jackson et al[4] have shown that NO acts as an inhibitor of its own production, and in a separate study, that ^{18}O is not incorporated into product for *A. cycloclastes* and *P. aureofaciens* NIR, while it may be for *R. sphaeroides* NIR[5].

Antibody serum, produced against NIR from *Achromobacter cycloclastes*, detected NIR from certain strains of *Pseudomonas*, *A. faecalis*, *A. xylosoxidans*, *P. aureofaciens*, *Bacillus azotoformans*, and *Corynebacterium nephridii*, while it did not detect dithiocarbamate inhibitable NIR from isolates of *Bacillus sp*, *Bradyrhizobium japonicum*, two Rhizobium strains, *Rhodopseudomonas sphaeroides forma sp denitrificans*, and two Pseudomonas types[6]. However, in another earlier study antibodies to NIR from *R. spheroides* detected *A. cycloclastes*, *A. denitrificans*, *Ps denitrificans* NIR so perhaps *R. spheroides* NIR is more closely related to *A. cycloclastes* NIR as well.[7]

For the first two of the NIRs in Table 1 a cupredoxin has been shown to be involved in reducing the nitrite reductase before NO_2^- can be reduced. Cupredoxins have been isolated for the third and fourth NIRs, again making it likely that these systems are related. Kashem et al[3] have shown that complex kinetics are involved in the reduction of cupredoxin in this process. We hope to show in future studies that the Type I copper is the site of initial reduction, *and* of specific interaction with the cupredoxin, followed by, or concomitant with, transfer of the reducing electron to the type II Cu which we believe is the site of substrate (NO_2^-) binding.

From K.D. Karlin and Z. Tyeklár, Eds., *Bioinorganic Chemistry of Copper*
(Chapman & Hall, New York, 1993).

TABLE 1. PHYSICALLY CHARACTERIZED CU-NITRITE REDUCTASES

SOURCE	MOL WT	COPPER/- monomer	UV-VIS (nm)	e-DONOR
A.cyclo- clastes[a]	3 x 36000	2 (I,II)	464 590 700	cpdxn
A. faecalis[b]	3 x 36000	2 (I,II)	457 587 700	cpdxn
P. aureo- faciens[c]	85000 : 2 x 40000	2 (I only)	474 595 780	cpdxn
A. xylos- oxidans[d]	70000 : 2 x 37000 ?	2 (I only)	470 594 780	cpdxn
Rps. sphaer oides[e]	80000 : 1 x 37000 1 x 39500	2 (I & II)	450 590	?
B. halo denitri- ficans[f]	82000 : 2 x 40000	2 (I & II)	454 595	?
Nitroso monas europa[g]	120000 : 4 x 35000	1		?
R. palus tris[h]	120000 : 2 x 68000?	3-5	418 522 552	?

(a) [8] , [9]; (b)[10] originally reported as 120000: 4 x 30000, but DNA sequence suggests 36000 (M. Nishiyama, personal communication); (c)[11] (type II signal also reported in some preparations); (d)[12]; low angle X-ray scattering measurements reported at this meeting suggest this is also a trimer (Abraham, Eady and Hasnain); (e)[13]; (f)[14]; (g)[15] (number of subunits from copper stoichiometry: 3.72/120000 MW; (h) [16]

In spite of probable similarities in tertiary structure, characteristic differences among members of the immunologically related class remain, namely that the Cu chromo-phores must differ. The *A. cycloclastes* NIR is green, while the *A. xylosoxidans* and *P aureofaciens* NIRs are blue. The strength of the extinction coefficients is characteristic of Type I coppers, due to a Cu-cysteinyl bond, making it unlikely that the differences are due to the second copper atom which is coordinated by only histidines. Hence it will be important to undertake the structural determination of a member of the 'blue class of NIRs in order to determine the likely cause of this difference

Figure 1 Two subunits of *A. cycloclastes* nitrite reductase, viewed from interior of trimer.

The structure of *Achromobacter cycloclastes* NIR, (crystallized in a cubic unit cell, space group $P2_13$, at pH 5.8 from phosphate buffer (20mM) and ammonium sulfate) determined at 2.3 A[8], contains three identical subunits, each having two β barrel folds, and two copper atoms. The N-terminal domain contains ligands appropriate for a type I copper, a Cys, His, Met and a second His N-terminal to the other three ligands. This domain also contains two of the three histidines which coordinate the second copper atom found on the surface of the subunit; the third coordinating histidine comes from domain two of the adjacent subunit in the trimer. Figure 1 is a ribbon drawing illustrating two of the three molecules in the trimer. In this view the monomer to the right has domain one to the front, while the monomer to the left (because it is rotated around the threefold axis of the trimer running nearly vertically in this view) has domain two at the front, thus exposing the two separate parts of the Cu-II binding site. Figure 2 is a view from outside the trimer, towards Cu-II, which is deep inside the channel formed by the interface between these two molecules. This view is rotated roughly 180° about the vertical axis of Figure 1 and tilted somewhat forward.

As pointed out previously[17], there are similarities of this structure to that of ascorbate oxidase[18], described elsewhere in this volume. Topologically the NIR domains are like those of ascorbate oxidase, although the latter is a dimer of subunits each containing three β barrel domains. In ascorbate oxidase the *third* domain contains the type I copper, as well as four histidine ligands to a novel trinuclear copper center, the remaining four histidine ligands being provided from the first domain of the protein. The

Figure 2 Two of the three subunits of NIR, viewed from outside the trimer, into one of three active site channels in the trimer.

three dimensional relationship between the two domains binding the trinuclear center is similar to that between domains of different subunits in NIR binding the type II Cu: in each case copper atom(s) are bound at the interface thus formed. The Type I Cu is about 12.5A from the type II Cu in NIR just as the Type I site in AOase is 12-13A from the nearest copper of the trinuclear center. In each case this distance is a consequence of adjacent amino acids binding the copper centers. In NIR His135/Cys136 bind the Type II/Type I sites respectively; in AO, His508/Cys509/His510 bind the trinuclear center, Type I center and trinuclear center.

Ascorbate oxidase is reduced by ascorbate, and in turn reduces O_2 to water, a 4 electron process. NIR is reduced by cupredoxin and reduces NO_2^- to NO, a one electron process (pH optimum at 6.2[2]). By analogy with ascorbate oxidase, and from preliminary studies previously described in which we soaked NO_2^- into crystals of NIR[8], we believe that Cu-II is the site at which NO_2^- is reduced. Also by analogy with ascorbate oxidase we note two residues which may also be important in the function of NIR: His 255 and Asp 98. Each of these is in the immediate environment of Cu-II, without being liganded to it, at least in the cubic form of the crystal for which the original structure was determined.

The original crystals of NIR contained only one subunit in the asymmetric unit, the molecule making use of the space group three-fold symmetry to crystallize. Varying the pH of the crystallization solution to pH 6.8 (20mM phosphate, 45% saturated ammonium sulfate) it has been possible to obtain crystals in a second form, an orthorhombic space group $P2_12_12_1$, which contain the entire trimer in the asymmetric unit. It has also been possible to obtain crystals at pH 6.2 (20mM ADA (N-[2-acetamide [-2-imino diacetic acid]), 46% saturated ammonium sulfate) and see if there are crystallographically detectable differences in the structure (hopefully in the region of His 255/Asp 98 and the type II Cu center) at the pH optimum of reduction of NO_2^- to NO, .

X-ray diffraction data were collected on an area detector to a resolution of 2.8 and 2.5Å for the pH 6.8 and pH 6.2 crystals respectively. Cell dimensions are a=99.47Å, b=115.3Å and c=115.9Å for the two forms. The structure was solved using the Patterson--correlation rotation function[19], using as a search model the trimer constructed from the 2.3A partially refined cubic structure (unpublished results). The rotated and translated model has been partially refined against each of the two data sets, using X-PLOR without any simulated annealing steps, resulting in an agreement factor of 0.187 for NIR68, and 0.216 for NIR62. The agreement of the structure factors between data sets is 0.13, which is the sort of agreement one sees between native and heavy atom derivative sets. Thus one might eventually expect some small but significant structural differences. No solvent has yet been modelled. Details of the structure determination will be published elsewhere.

The packing of the trimers in the new space group differs from that in the cubic form. The N and C termini of the subunit designated "A" in the trimer are involved in new packing interactions in the orthorhombic crystal form, and thus differ significantly in conformation from molecules B and C in the trimer. This difference ws detectable at the early stages of the structure solution as a direct superposition of the atoms in their original position. In particular, residues 334-340 have been displaced by the N terminal residues of a neighboring subunit "A", and at present have an uncertain arrangement. Rigid body refinement of the subunit positions relative to each other formed part of the structure solution; however, aside from the aforementioned differences in the A subunit, it is difficult to see any detectable differences elsewhere.

While the agreement factors are quite good for a model at this stage of refinement, the current bond lengths and angles at each copper center disagree too much to be reported at this time. It is entirely likely that they will agree when the refinement is complete, or if they do not agree, we will have a better understanding of why they do not. In view of the spectroscopic differences between this Type I site and the cupredoxins it is important to not obscure the literature with uncertain geometrical values. Moreover, at this stage of refinement, the errors are ~.5Å, making direct comparison of the coordinate sets not useful at this time.

One of the most fundamental *crystallographic* indications of differences in structure between different sets of conditions (when crystals are isomorphous) is a difference Fourier electron density map. It is calculated using coefficients $|F_A - F_B|exp$ $(i\alpha)$, where the phase, α, comes from MIR phases or from current model phases, and the structure factors, $F_{A,B}$ are those measured under conditions A and B, e.g. pH 6.8 and pH 6.2. If differences in coordinates are small they will be manifested by a pair of peaks, one positive and one negative on either side and close to the atomic site, indicating a slope in the electron density, or a shift in the coordinates. If the differences are large, there will be negative density where atoms are in structure B, and positive density where they are in structure A. If no change in position occurs, but there is a change in effective

Figure 3 Difference electron density between pH 6.8 and pH 6.2 data sets, contoured at 2σ. Positive contours are thicker than negative.

occupancy (e.g. diminished mobility, or actual change in average site occupancy, such as gain or loss of solvent) there may be a single peak, positive or negative.

Such a difference Fourier map was computed for the two data sets and is shown above in Figure 3. Both positive and negative contours greater than 2σ are shown: the positive (thick) contours represent density that is present in the pH 6.8 form and not in the pH 6.2, while negative (thin) density is present in the pH 6.2 form and not at pH 6.8. Three conclusions can be drawn from this map at the present time: 1) the changes are small 2) many of the interpretable significant changes happen near the Cu-II site and 3) the changes are not identical for each of the type II copper sites.

Figure 4, a close up of the difference density at the Type I Cu site (which incidently illustrates the similarity of this site to previously published cupredoxin geometry[20]) shows in subunit A, positive density more or less between His 145 and the Cu; in subunit B, positive density equidistant from Cu, Cys and His 145; and in subunit C, positive density reaching between Cu and Met 150. This density is ill-shaped, and just barely at a significant level. It is not presently clear just how it should be interpreted, although a small change in the occupancy at the copper site is one possibility.

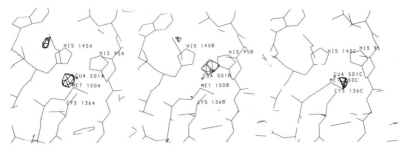

Figure 4 Difference density between pH 6.8 and pH 6.2 orthorhombic forms of NIR at the type I Cu site, subunits **A**, **B**, and **C**.

Figure 5 is a close-up view of the differences at the type II site. The only clear indication of a shift peak is at residue 255C, which shows a positive and negative peak on either side of the histidine ring. Taken at face value, this indicates that His 255 may move slightly towards the Cu in the pH 6.2 form. The large positive peak to one side of the Cu (meaning more density in the pH 6.8 form) is the largest in the difference map. It is not indicative of a shift in the copper position per se, although there may be a small shift contained therein, but is suggestive of a more tightly bound water in the higher pH form.

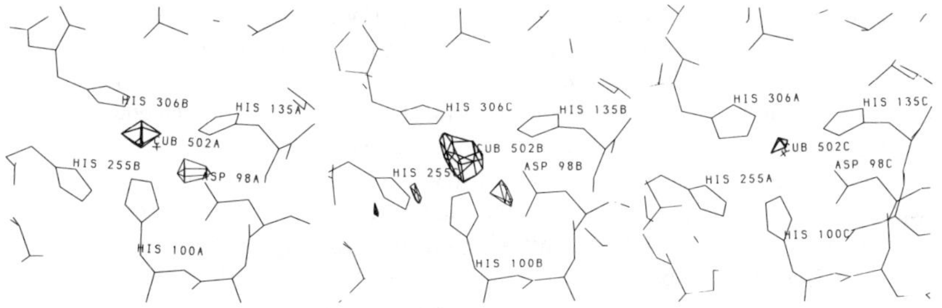

Figure 5 pH 6.8 minus pH 6.2 difference density at three Cu-II sites. (a) B/A interface; (b) C/B interface; (c) A/C interface. Thicker contours, positive (pH 6.8); thinner contours negative (pH 6.2).

The differences at the Cu-II site are quite similar for the A and B molecules (A/B and B/C interface) and least for the C site (C/A interface). A possible explanation for this suggests that the differences might be meaningful. The largest difference occur at the interface least restricted by packing interactions. The site which appears to have the least differences is at the interface made between subunit A and C, where the C-terminal tail of subunit A forms a new interaction, from residue 333 onward, with subunit C. The latter half of that interaction is displaced by direct contacts of the N-terminus of a symmetry related subunit A. Thus, one function of the C-terminal tail portion of the interactions between subunits might be to *maintain* a spacing between subunits that facilitates whatever requirements there are that determine the pH optimum for the reaction. The new packing interaction seen in this crystal form disallows this flexibility in some manner. At present this is rather speculative interpretation of the difference map, and the real details will have to await completion of the refinement of the structure.

The presence of a water bound to the type II Cu as a fourth ligand has been problematic. The evidence has largely been that the electron density for the type II Cu lies clearly out of the plane formed by the $N^{\varepsilon 2}$ atoms of the histidine ligands to the Cu leaving a site poised for a fourth ligand, and that in a difference electron density map calculated from the observed data and calculated model shows some density at about 2A from the copper in the fourth ligand site. It is however by no means as clearly defined as well ordered solvent further out in the solvent channel.

Why should there be a water involved at all? It may be that it is not, but that at pHs away from the optimum it may be bound to the Cu to a greater extent, making replacement by nitrite less favorable. Or, it (and chains of solvent in the solvent channel) might act as proton shuttles.

Protons are consumed in the overall stoichiometry of the reaction; it is not clear that proton donation in the active site is required. However, if it were, it is tempting to speculate that something with a pK like that of His could be involved. His 255 does not appear to move close enough to ligate Cu; however it is on the side of the copper which is exposed to the solvent channel and therefore likely to be near the nitrite (substrate). Hence if proton transfer to the nitrite facilitates reduction of the nitrite to NO, His 255 would appear to be poised to do this. Or, another way of looking at it might be that a more positively charged His facilitates binding of the negatively charged nitrite, and subsequent transfer of an electron to the nitrite, even without proton transfer.

Another residue which bears commenting on is Asp 98, visible in Figure 5. At the pHs of study, this residue should be negatively charged. Why should there be a negatively charged residue near this copper? Here one must speculate. A negatively charged residue could stabilize the $E-Cu^{+}-NO^{+}$ intermediate postulated by Jackson et al[6], or it may help orient a solvent molecule needed for proton transfer. Site directed mutagenesis studies in progress will help to understand the role of these two key residues.

In summary, diffraction studies on a new crystal form of nitrite reductase have shown that new packing interactions change the conformation of the C-terminus of one of the molecules in the asymmetric unit which in turn provides a plausible explanation for the small differences between the three Cu-II sites at two different pHs. Further studies are being carried out on binding NO_2^{-} and other anions in this crystal form.

ACKNOWLEDGMENTS

Thanks to Jeff Godden for stimulating discussions, to Jean LeGall and Ming Liu for providing purified *A. cycloclastes* nitrite reductase, to Teruhiku Beppu and Makoto Nishiyama for their continued interest and preliminary information on *A. faecalis* nitrite reductase, and to NIH support on GM31770.

REFERENCES

1. W.J. Payne, in *Denitrification in the Nitrogen Cycle*, H.L. Golterman, Ed., (Plenum, New York, 1985), 47-65.
2. H. Iwasaki and T. Matsubara, *J. Biochem.*, **71**, 645 (1972).
3. M.A. Kashem, H.B. Dunford, M.-Y. Liu, W.J. Payne, J. LeGall, *Biochem. Biophys. Res. Commun.*, **145**, 563 (1987).
4. M.A. Jackson, J.M. Tiedje, B.A. Averill, *FEBS Lett.*, **291**, 41 (1991).
5. R.W. Ye, I. Toro-Suarez, J.M. Tiedje, B.A.Averill , *J. Biol.Chem.*, **266**, 12848 (1991).
6. M.S. Coyne, A. Arunakumari, B.A. Averill, J.M. Tiedje, *Applied and Environmental Microbiology*, **55**, 2924 (1989).
7. W.P. Michalski, D.J.D. Nicholas, *Phytochemistry*, **27**, 2451 (1988).
8. J.W. Godden, S. Turley, D.C. Teller, E.T. Adman, M.-Y. Liu, W.J. Payne, J. LeGall, *Science*, **253**, 438 (1991).
9. M.-Y. Liu, M.-C. liu, W.J. Payne, J. LeGall, *J. Bacteriology*, **166**, 604 (1986).
10. T. Kakutani, H. Watanabe, K. Arima, T. Beppu, *J. Biochem.*, **89**, 463 (1981).
11. W.G. Zumft, D.J. Gotzmann, P.M.H. Kroneck, *Eur. J. Biochem.*,**168**, 308 (1987).
12. M. Masuko, H. Iwasaki, T. Sakurai, S. Susuki, A. Nakahara, *J. Biochem.*, **96**, 447 (1984).
13. W.P. Michalski, D.J.D. Nicholas, *Biochim. Biophys. Acta*, **828**, 130 (1985).
14. G. Denariaz, W.J. Payne, J. LeGall, *Biochim. Biophys. Acta*, **1056**, 255 (1991).
15. D.J. Miller, P.M. Wood, *J. Gen. Microbiol.*, **129**, 1645 (1983).
16. M. Preuss, J.H. Klemme, *A. Naturfosch.*, **38c**, 933 (1983).
17. E.T. Adman, *Advances in Protein Chemistry*, **42**, 145 (1991).
18. A. Messerschmidt, A. Rossi, R. Ladenstein, R. Huber, M. Bolognesi, G. Gatti, A. Marchesini, R. Petruzzelli, A. Finazzi-Agro, *J. Mol. Biol.*, **206**, 513 (1989).
19. A.T. Brunger, *J. Mol. Biol.* **203**, 803 (1988).
20. J. Han, E.T. Adman, T. Beppu, R. Codd, H.C. Freeman, L. Huq, T.M. Loehr, J. Sanders-Loehr, *Biochemistry*, **30**, 10904 (1991).

CHARACTERIZATION OF MONONUCLEAR COPPER-NITROGEN OXIDE COMPLEXES: MODELS OF NO_x BINDING TO ISOLATED ACTIVE SITES IN COPPER PROTEINS

William B. Tolman,[a] Susan M. Carrier,[a] Christy E. Ruggiero,[a] William E. Antholine,[b] and James W. Whittaker[c]

[a]Department of Chemistry, University of Minnesota, 207 Pleasant St. S.E., Minneapolis, MN 55455, [b]Medical College of Wisconsin, 8701 Watertown Plank Road, Milwaukee, WI 53226, [c]Department of Chemistry, Carnegie Mellon University, 4400 Fifth Avenue, Pittsburgh, PA 15213.

Detailed examination of the interactions of simple nitrogen oxides, $N_xO_y{}^{n-}$, with metal ions has become increasingly important for understanding the role of $N_xO_y{}^{n-}$ species in numerous biochemically and environmentally significant processes. Metal-mediated reactions of nitrogen oxides are ubiquitous within the global nitrogen cycle, in which the various forms of nitrogen are interconverted and distributed throughout the world.[1] For example, metalloenzymes containing molybdenum, iron, or copper are involved in every stage of denitrification, the dissimilatory process by which some bacteria use $NO_3{}^-$ and $NO_2{}^-$ as terminal electron acceptors to release gaseous NO, N_2O, and N_2 (Figure 1).[2] Environmental consequences of denitrification include the depletion of sources of nitrogen necessary for plant growth and the production of the greenhouse gas and ozone destroyer N_2O. Moreover, excess $NO_3{}^-$, a pollutant that contributes to eutrophication of lakes and rivers, can be removed from waste water by denitrification in a useful bioremedial application.[3] Dissimilatory nitrogen oxide reduction is particularly fascinating because it represents a metabolic respiration route analogous to the dioxygen consumption performed by most organisms. A key intermediate in

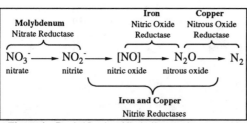

Figure 1 . Denitrification (dissimilatory nitrogen oxide reduction).

denitrification, nitric oxide (NO), has also been found to perform unusual functions in mammalian systems. Recent discoveries demonstrating that NO is a potent vasodilator, neurotransmitter, and cytotoxic agent has led to a resurgence of interest in this simple inorganic molecule.[4] The apparent binding of NO to the heme iron of guanylate cyclase in its activating role[5] and to other metalloproteins during destructive processes[6] provides added impetus for close study of the binding and activation of nitrogen oxides by metal ions in biology.

Copper proteins isolated from several types of bacteria have been shown to catalyze key reactions in denitrification. Conversion of $NO_2{}^-$ to NO and/or N_2O is

From K.D. Karlin and Z. Tyeklár, Eds., *Bioinorganic Chemistry of Copper*
(Chapman & Hall, New York, 1993).

performed by the nitrite reductases (NiR), copper-containing representatives of which have been identified in several bacterial strains, including *Alcaligenes* sp.[7] and *Achromobacter cycloclastes*.[8] Copper ion content, active site structural and spectroscopic features, and observed product ratios vary among the enzymes isolated from different sources. These, and other, aspects have been most extensively studied in the NiR from *Achromobacter cycloclastes*.[8-12] A recent 2.3-Å resolution X-ray crystal structure determination revealed that each monomer in the trimeric protein contains two well-separated copper ions (Cu···Cu = 12.5 Å), one with ligands typical of type 1 electron transfer centers, the other coordinated to three histidines and an apparent aquo moiety in an unusual tetrahedral geometry.[8] Preliminary Fourier difference maps obtained from crystals soaked in nitrite indicate that new electron density develops at the tetrahedral site, implying that nitrite binds and is subsequently reduced at that single copper ion. Mechanistic experiments have been interpreted to indicate that dehydration to form an electrophilic copper nitrosyl (E-Cu$^+$-NO$^+$) follows nitrite binding, with subsequent loss of NO and/or N-N bond formation and reduction affording the gaseous products.[12] Despite recent efforts, however, these proposed reaction steps, as well as details such as the mode of nitrite binding, the structure of the putative copper nitrosyl, and the role of copper, NO$_2^-$, and NO in N-N bond formation to yield N$_2$O, are not known with certainty.

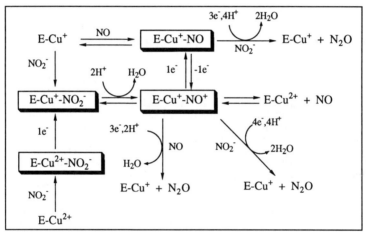

Figure 2. Possible nitrite reduction pathways for copper nitrite reductase (key putative intermediates enclosed in boxes).

Copper-N$_x$O$_y$ species have also been postulated to result from treatment of electron transfer or dioxygen-activating proteins with nitrogen oxides in studies performed in order to probe the enzymes' active site structures and functions.[13-16] For example, hemocyanin,[13] tyrosinase,[13b-h] laccase,[14] ascorbate oxidase,[13b] galactose oxidase,[15] and cytochrome c oxidase[14b,16] have been treated with NO and NO$_2^-$ to yield various spectroscopically characterized nitrogen oxide adducts. Postulates concerning the structures and mechanisms of dioxygen activation by these proteins have been put forth based on these results, despite difficulties in interpretation of the data acquired for the purported N$_x$O$_y$ complexes. Evidence supporting formation of simple NO adducts to mononuclear copper sites in a variety of other proteins, including NiR from *Achromobacter cycloclastes*,[12c,17] has also been reported.[13b] However, limited spectroscopic methods have been applied toward their characterization, preventing unambiguous structural assignments and restricting mechanistic studies. The importance of gaining a detailed understanding of these putative Cu-NO species is underscored by the recent discovery that CuZn-superoxide dismutase prolongs the effects of the endothelium-derived relaxing factor (NO) by mediating NO/NO⁻ redox interconversions, making it at least conceivable that other copper proteins also may be effected by the NO released by arginine-dependent NO-synthase.[18]

A firm chemical foundation on which to base structural and mechanistic hypotheses for the reactions of nitrogen oxides with copper proteins is notably absent, only scant precedent having been provided to date in the copper coordination chemistry literature for many of the proposed Cu-N_xO_y intermediates. In marked contrast to heme-iron systems, for which an array of well-characterized synthetic iron porphyrin[19] and relevant nonporphyrin iron, ruthenium, and osmium complexes[20] that bind and/or interconvert nitrogen oxides have been scrutinized for comparative purposes, synthetic analogs of copper protein-nitrogen oxide species are rare.[21] The goal of our research is to address this deficiency, with initial emphasis on the synthesis, structural and spectroscopic characterization, and study of the biomimetic reactivity of mononuclear models of the copper NiR active site. In particular, we have focused our attention on compounds that will provide precedent for the putative nitrite and nitrosyl dentrification intermediates enclosed in boxes in Figure 2. In addition to being important in nitrite reduction pathways,[12] these adducts resemble possible products of other copper protein-nitrogen oxide reactions.[13-17] We anticipate that insights gained from our modeling studies will therefore contribute to an understanding of both the environmentally important denitrification process in particular and nitrogen oxide interactions with biological copper centers in general.

A MODEL OF THE NITRITE REDUCTASE SUBSTRATE ADDUCT

Coordination of nitrite to the active site copper ion is the probable first step in the enzymatic nitrite reduction mechanism. Numerous possible structures may be envisioned for this initial enzyme-substrate adduct since nitrite is well-known to exhibit multiple bonding modes in its transition metal complexes. This phenomenon is exemplified in the relatively few known Cu(II)-nitrite complexes that have been structurally characterized.[22] These compounds are of limited use as stuctural, spectroscopic, or functional NiR models due to their nonbiomimetic ancillary ligand environments, however, so we have begun to investigate nitrite binding to copper ions complexed to tris(pyrazolyl)hydroborate ligands.[23] These ligands were chosen because their facially coordinating group of pyrazolyl donors closely approximates the array of histidine residues bonded to the catalytic center in NiR and because such ligands have been used with great success to model other copper protein active sites that contain analogous tris-imidazolyl coordination environments.[24] Strategic placement of large organic substituents at the 3-position prevents polynuclear complex formation,[25] a key prerequisite for successful modeling of the mononuclear NiR active site.

Mixing K[HB(t-Bupz)₃], CuCl₂, and NaNO₂ in MeOH cleanly afforded [HB(t-Bupz)₃]Cu(NO₂) (**1**, Figure 3),[26] a

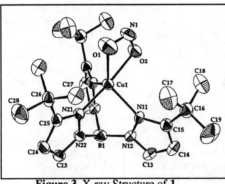

molecule which represents a speculative model[27] for the substrate adduct of copper NiR. The nitrite ion binds in a bidentate fashion via both oxygen atoms relatively symmetrically (the Cu-O distances differ by 0.19 Å, which is a small value compared to more asymmetric *O,O*-nitrite complexes of Cu^{22} and other transition metals). Most unusual is the coordination geometry of the Cu(II) ion in **1**, which is best described as distorted trigonal bipyramidal with the axial bonds being Cu1-O2 and Cu1-N21 (O2-Cu1-N21 angle = 170°). A nitrite bite angle of 59° represents the

Figure 3. X-ray Structure of **1**.

major distortion from an ideal trigonal
bipyramidal arrangement. The frozen solution
EPR spectrum of the complex (Figure 4) contains
a rhombic signal which is consistent with
retention of the unusual geometry in solution and
which sharply contrasts with the axial signal
normally observed for 5-coordinate square
pyramidal Cu(II).

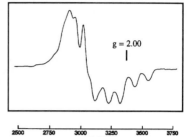

Figure 4. EPR spectrum of **1**.

The nitrite binding geometry in **1** differs
significantly from the N-bonded arrangement
observed in iron-porphyrin adducts proposed as
models for heme iron NiRs.[19] It is therefore
tempting to speculate that this is only one of many
important differences among the copper and iron
NiR reaction mechanisms and intermediates. However, we have yet to observe
conversion of the O,O-nitrite to a nitrosyl complex, a reaction that would necessarily
involve extensive molecular rearrangements. The suitability of **1** as a possible functional
model of NiR is thus unclear, and it would appear that in order to evaluate the
significance of the nitrite binding mode in the mechanism of denitrification, complexes
containing nitrite bound to copper (I or II) via nitrogen must be prepared.

A MONONUCLEAR COPPER NITROSYL COMPLEX

The importance of copper nitrosyl intermediates in denitrification (Figure 2) and in
structure/function probes of copper proteins, the possible role of copper proteins in the
reactions of biologically produced NO, and the rarity of known copper nitrosyl
complexes[28] together provide great impetus for examining biologically relevant Cu-NO
compounds. At the outset of our work, only one example of a structurally characterized
copper nitrosyl complex had been reported, the dicopper complex **2** (Figure 5).[21b]
Compound **2** was prepared by treating a dicopper(I) complex with $(NO^+)(PF_6^-)$; thus, **2**
can be formulated as having a Cu(I)-(NO⁺)-Cu(I), Cu(II)-(NO·)-Cu(I), or Cu(II)-(NO⁻)-
Cu(II) unit. The complex has ν(NO) at 1536 cm⁻¹, a symmetrical structure containing
copper ions with square pyramidal geometries, and an electronic absorption feature with
$\lambda_{max} = 730$ ($\varepsilon = \sim500$), all consistent with the
latter dicopper(II) formulation. The metal ions
appear to be antiferromagnetically coupled since
the compound exhibits no EPR spectrum and
has $\mu_{RT} = 0.59\ \mu_B$/Cu.

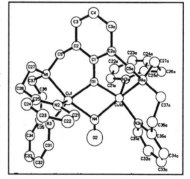

The isolation and characterization of **2** is
a significant discovery, yet because of its
dinuclear structure it is an insufficient model of
adducts to mononuclear copper proteins[13b] or of
possible dentrification intermediates (the
catalytic center of NiR is 12.5 Å distant from its
probable type I copper electron donor).[8] We
thus set as our goal the preparation of a
mononuclear copper nitrosyl complex which we
postulated would be accessible by treating a
suitably sterically hindered Cu(I) precursor with
NO(g). By analogy to the results of Paul et
al.,[21b] the product was anticipated to be
comprised of a Cu(II)-(NO⁻) unit.

Figure 5. Structure of **2** (reprinted
with permission from ref. 21b).

Synthesis and Molecular Structure

A cuprous complex of the bulky [HB(t-Bupz)₃]⁻ anion was chosen as starting
material for the synthesis. We found that treatment of Tl[HB(t-Bupz)₃] with CuCl in THF

afforded a novel dimer, {[HB(t-Bupz)$_3$]Cu}$_2$ (**3**) (Figure 6).[29] Each Cu(I) ion has a linear two-coordinate geometry with the expected[30] short Cu-N$_{pz}$ distances (1.86-1.88Å; Cu· · · Cu = 3.284 (8) Å). The bridging [HB(t-Bupz)$_3$]⁻ groups donate one pyrazole ring to each copper and the third pyrazole dangles free. The absence of an inversion center in the molecule is evident in the view approximately down the Cu· · · Cu line (right hand drawing of Figure 6), both uncoordinated pyrazole rings being located on the same (left) side of the vertical plane along the dimetal vector.

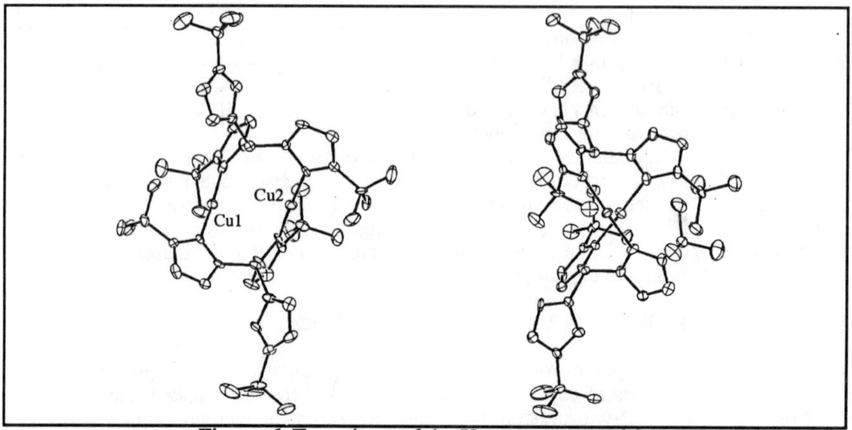

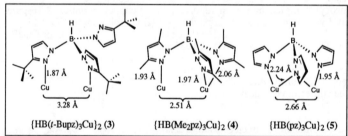

Figure 6. Two views of the X-ray structure of **3**.

Figure 7. Comparison of structurally characterized [tris(pyrazolyl)hydro-borate]Cu(I) dimers (only half of each dimer shown, see ref. 31).

It is interesting to compare the structure of **3** with those of the previously reported analogous complexes {[HB(3,5-Me$_2$pz)$_3$]Cu}$_2$ (**4**) and {[HB(pz)$_3$]Cu}$_2$ (**5**) (Figure 7).[31] In the latter, least hindered case, each Cu(I) ion is four-coordinate and one pyrazole ring of each [HB(pz)$_3$]⁻ ligand bridges the metals in a symmetric fashion. Addition of methyl groups to the pyrazole rings (compound **4**) causes the bridging pyrazole to shift to one of the Cu(I) ions and reduces the metals' coordination number to three. Finally, in **3** the steric bulk has forced one pyrazole ring to dissociate, giving two-coordinate Cu(I) ions with an increased Cu· · · Cu distance. These profound effects of the differing steric bulk of the ligands in **3-5** are also reflected in the solution properties of the complexes. Although **4** apparently dissociates in solution, variable temperature ^{1}H NMR spectra of both **3** and **5** suggest that they undergo facile pyrazole exchange processes. In **5**, intramolecular bridge-terminal pyrazole interchange is extremely rapid (limiting NMR spectrum not reached at -120°), whereas exchange of bound and unbound pyrazoles in **3** is slow enough to allow observation of the low-temperature limiting spectrum at -40°C

($\Delta G^{\ddagger} > 50$ kJ mol^{-1}). It is intriguing to view the thermodynamically stable structures of **3-5** (ignoring the differences in pyrazole substituents) as snapshots taken during such kinetically facile intramolecular fluxional processes.

Colorless solutions of **3** in CH$_2$Cl$_2$ or toluene turned deep red upon exposure to NO (1 atm).[29] An intense band at 1712 cm^{-1} in IR spectra of the red solutions, identified as ν(NO) based on its energy decrease upon isotopic substitution [ν(^{15}NO) = 1679 cm^{-1}], and an X-ray structure of crystals isolated upon cooling indicated that the sought-after

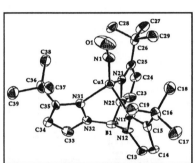

Figure 8. X-ray Structure of **6**.

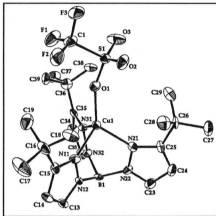

Figure 9. X-ray structure of **9**.

nitrosyl complex (**6**) had formed (Figure 8). Notable structural features of the molecule include an approximately tetrahedral geometry about copper, a short Cu1-N1 distance [1.759 (6) Å] typical for transition metal nitrosyl complexes,[32] and an intermediate Cu1-N1-O1 bend angle of 163.4 (6)° (between 180° linear and 120° "bent"). Because it is the only structurally characterized mononuclear copper nitrosyl compound, the structural parameters for **6** provide an important benchmark for comparison with the yet to be determined geometries of NO adducts to isolated copper sites in proteins.

Electronic Structure

Can **6** be adequately described as having a Cu(II)-(NO$^-$) unit? The vibrational spectroscopic and X-ray structural evidence is equivocal, making more detailed spectroscopic studies essential for answering this question (keeping in mind that any extreme resonance form such as Cu(II)-(NO$^-$) can only be an approximate description because of the relatively high degree of covalency in transition metal nitrosyl bonds).[32] To best address the issue, the spectroscopic signatures of tetrahedral [HB(Rpz)$_3$]Cu(II)X complexes, where X$^-$ is a non-redox active ligand (vs. a nitrosyl, which can readily form NO$^+$, NO$^·$, or NO$^-$), must be understood. We therefore synthesized and characterized a series of such compounds, focusing on the cases where R = *t*-butyl and X = Cl$^-$ (**7**), Br$^-$ (**8**), and CF$_3$SO$_3^-$ (**9**). All three compounds have tetrahedral geometries similar to **6**, as revealed by

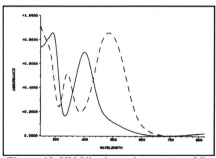

Figure 10. UV-Vis absorption spectra of **7** (—) and **8** (---).

similarities of the spectroscopic properties of **7** and **8** to crystallographically characterized analogs[33] and by an X-ray structure of **9** (Figure 9).[26] Solutions of the halide complexes **7** and **8** are orange-red as a result of the intense absorption bands shown in Figure 10. The large shifts of the λ_{max} values to lower energy upon substitution of Cl⁻ by Br⁻ (a stronger

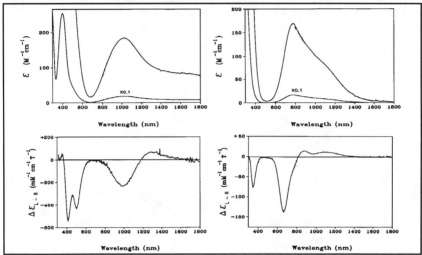

Figure 11. Room temperature absorbance (top) and 4K MCD (bottom) spectra of **7** (left) and **10** (right).

to a weaker field ligand) is consistent with assignment of these bands to LMCT transitions. Although the exact nature of these transitions [e.g., $X(\pi^*) \rightarrow Cu^{2+}(d)$ vs. $pz(\pi^*) \rightarrow Cu^{2+}(d)$], is currently unclear, they corroborate the presence of Cu(II) in the complexes.

Further support for tetrahedral Cu(II) centers in **7-9** is apparent from their visible and near-IR absorbance and low-temperature magnetic circular dichroism (MCD) spectra, shown for **7** and, for purposes of comparison, for the 5-coordinate square pyramidal [HB(t-Bupz)$_3$]Cu(OAc) (**10**) in Figure 11.[26,34] Ligand field transitions with corresponding intense MCD features are observed for both compounds, but at significantly longer wavelength (>900 nm) for **7**, consistent with its lower coordination number. Compounds **7-9** also exhibit diagnostic EPR spectroscopic properties, shown for **7** and **9** along with simulated spectra and derived parameters in Figure 12. Rhombic spectra with distinct Cu hyperfine features in the high field component are indicative of d_{z^2} ground states for the complexes, in distinct contrast to the axial spectra associated with the $d_{x^2-y^2}$ ground states of tetragonal Cu(II) compounds.[35] Similar rhombic EPR spectra have been observed for some type 1 copper proteins and model compounds.[33,36]

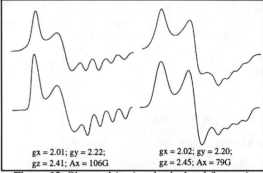

$gx = 2.01; gy = 2.22;$
$gz = 2.41; Ax = 106G$

$gx = 2.02; gy = 2.20;$
$gz = 2.45; Ax = 79G$

Figure 12. Observed (top) and calculated (bottom) EPR spectra of **7** (left) and **9**(right).

        The spectroscopic properties of **7-9** contrast with those of **6**, however, suggesting that the Cu(II)-(NO⁻) formulation for the latter is *not* appropriate. Like the halide compounds **7** and **8**, nitrosyl **6** has a deep red color due to a a relatively intense ($\varepsilon \sim 1200$) absorption band at 498 nm. To shed light on the nature of this absorbance feature, we prepared an analog of **6** containing the [HB(3,5-Ph₂pz)₃]⁻ ligand.[34] The IR spectrum for this analog exhibits a $\nu(NO)$ 8 cm⁻¹ higher in energy than the corresponding band for **6**, demonstrating that the phenyl-substituted ligand is electron withdrawing compared to the *t*-butyl-substituted case. Similar results for analogous complexes of CO corroborate this idea.[24] Importantly, the λ_{max} for the electron-poor [HB(3,5-Ph₂pz)₃]⁻ compound is shifted to 30 nm (1300 cm⁻¹) *higher* energy than **6**, in contrast to what would be expected if the visible absorption band in the nitrosyl complexes was a LMCT transition. A more appropriate assignment of the band which is consistent with its upward energy shift upon removal of electron density from the metal center is as a MLCT transition. Such transitions are more likely to occur in complexes of Cu(I) than Cu(II).[37] Additional significant differences between the nitrosyl complexes and **7-9** are apparent in the near-IR absorbance and MCD spectra (Figure 13). The absence of ligand field features in the spectra of **6** clearly argue against a Cu(II) formulation for the complex.

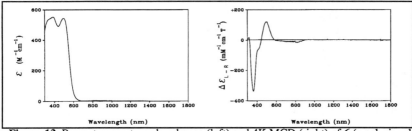

Figure 13. Room temperature absorbance (left) and 4K MCD (right) of **6** (weak signal in the near-IR due to trace amount of paramagnetic impurity).

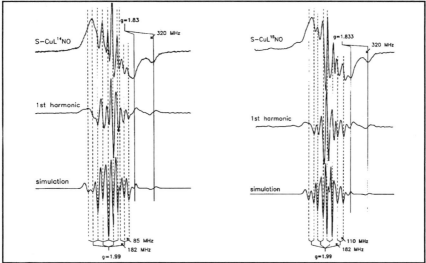

Figure 14. S-band (3.46 GHz) EPR spectra, first harmonics, and simulations of **6**-¹⁴NO (left) and **6**-¹⁵NO (right). Derived parameters agree with those obtained for X-band data.[34]

Finally, the nitrosyl compounds exhibit highly unusual EPR spectra that corroborate our conclusion that their electronic structure differs significantly from the Cu(II)-(X⁻) compounds. X (9.1 GHz)- and S (3.46 GHz)-band EPR spectra for ^{14}NO and ^{15}NO derivatives of **6** were recorded and simulated using a common set of parameters; shown in Figure 14 are the S-band data along with first harmonic spectra, our current simulations, and the g and A parameters derived therefrom.[34] The most important features to note are (i) $g_{//} < g_{\perp}$, consistent with the tetrahedral geometry of the complex, but with $g_{//} < 2.0$, which is unexpected for Cu(II), (ii) unusually large hyperfine coupling to the nitrosyl nitrogen (30 G in the ^{14}NO compound), suggesting extensive spin delocalization onto the NO group, and (iii) observation of the spectrum only at temperatures below 40 K, indicating onset of a relaxation mechanism at higher temperatures which is unusual for Cu(II) complexes.

In sum, comparison of the combined spectroscopic data for the nitrosyl complexes and **7-9** leads to the conclusion that the Cu(II)-(NO⁻) formulation for the former is not correct. Instead, we favor a Cu(I)-(NO·) valence bond description, again remembering that such an extreme structure is a formalism for a bond which probably has a relatively high degree of covalency. This is perhaps made more explicit in a molecular orbital description for the complex, which may be derived by using the scheme proposed previously for C_{3v} complexes with {MNO}[10] electron counts (**6** has {CuNO}[11]).[32a,b] In such a scheme (Figure 15), the unpaired electron in **6** is localized in a set of nearly degenerate molecular orbitals derived from antibonding combinations of metal d_{xz}, d_{yz} and nitrosyl π^* orbitals, above a fully occupied set of essentially metal d levels. Much of the experimentally determined spectroscopic data can be rationalized qualitatively by reference to this MO picture for the complex. The full d orbital set (2e, 3e, and 4a$_1$ in Figure 15) rules out ligand field transitions, thus explaining the featureless near-IR absorbance and MCD spectral data. LMCT transitions are also prevented by the filled d orbitals, supporting assignment of the visible absorption band at 498 nm as a MLCT transition from the metal d to the primarily NO-centered 4e orbital(s). Finally, the large hyperfine coupling to the nitrosyl nitrogen in the EPR spectra of **6** is consistent with the unpaired electron in the complex being substantially delocalized onto the nitrosyl ligand as implied by the MO scheme. This delocalization into a closely spaced set of orbitals might also explain the low $g_{//}$ value, the positive spin-orbit coupling constant (λ) resulting from the less than half filled 4e set in the ground state of the complex causing a negative g shift (versus positive shifts normally seen for the more than half filled shell of Cu(II) where $\lambda = -830$ cm^{-1}).[38] More quantitative explanations of the spectroscopic data await the results of additional experiments and more detailed calculations.

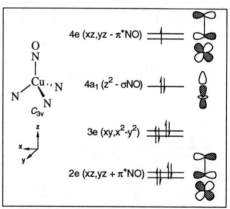

Figure 15. Qualitative molecular orbital diagram for **6** (see ref. 32a).

Reactivity

The results of preliminary reactivity studies are summarized in Figure 16. The NO ligand in **6** is quite labile and can be removed simply by allowing solutions of the complex to stand under Ar or N$_2$ (or, more rapidly, by purging). Readdition of NO to purged solutions reproduced the original UV-Vis absorption spectrum, implying that nitrosyl formation is a reversible reaction and that **3**, NO, and **6** are in dynamic equilibrium.

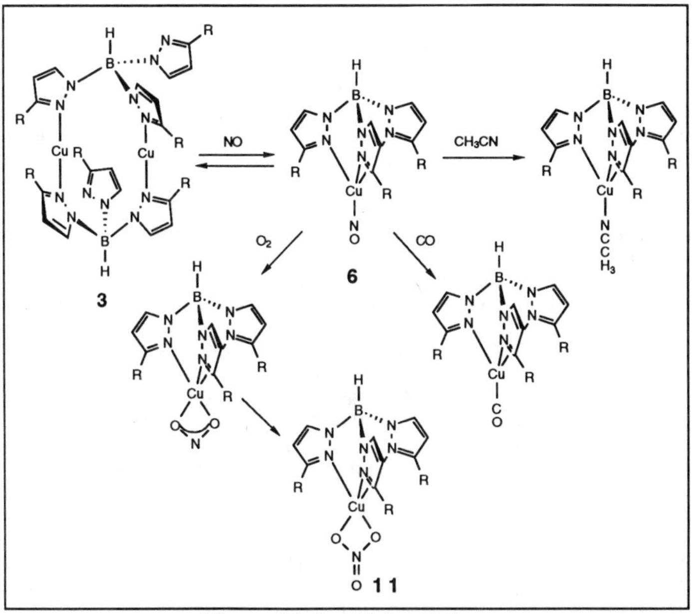

Figure 16. Reactivity of **6**.

Manometric experiments showed that stoichiometric binding is not achieved under 1 atm of NO and that substantially increased percentages of NO coordination occur at lower temperatures (77% at -40 °C vs. 42% at 22 °C). This result is consistent with the expected positive $\Delta S°$ for the equilibrium. The NO group can also be ejected by addition of ligands known to coordinate to Cu(I), such as CH_3CN and CO (Figure 16). Complex **6** is exceedingly air sensitive, affording the known complex [HB(t-Bupz)$_3$]Cu(NO$_3$) (**11**)[39] upon exposure to excess O_2. More careful addition of ~ stoichiometric quantities of O_2 yielded the air stable nitrite complex **1** (identified by EPR and UV-Vis spectroscopy), suggesting that **11** is formed via a mechanism involving prior formation of **1** followed by further oxidation by NO_2 (formed from O_2 + NO). Further mechanistic studies designed to test this hypothesis and identify possible intermediates in this reaction are planned.

CONCLUSIONS

With the synthesis and characterization of the Cu(II)-(NO$_2^-$) (**1**) and Cu(I)-(NO·) (**6**) complexes, significant progress has been made toward providing chemical precedence for biological species proposed to be important in denitrification and in structural and mechanistic probes of copper proteins. The isolation and definition of the molecular and electronic structure of the novel mononuclear nitrosyl complex **6** is particularly significant, both from a fundamental chemical perspective because of its intrinsically interesting properties and from a biological view because of its role as a model of putative copper protein-NO adducts. With regard to its properties, spectroscopic and physical data for **6** are significantly different from those observed for the only other structurally characterized copper nitrosyl complex **2** (Figure 5).[21] Whereas relatively strong binding accompanied by redox changes occurs in the preparation of **2**, we observe weak and reversible coordination of nitric oxide to a single Cu(I) ion to afford a molecule with properties suggesting little electron transfer between the metal and NO. How the differences between the properties of the mono- and dinuclear nitrosyl complexes, as well as the differences in their underlying molecular and electronic structural bases, relate to the biologically relevant reaction pathways traversed by these compounds is an important issue that needs to be addressed in future research.

Finally, **6** provides the first chemical precedent for a possible key intermediate in denitrification reactions catalyzed by the copper NiR from *Achromobacter cycloclastes*. Although most mechanistic proposals include as a key first step the dehydration of a $Cu(I)-(NO_2^-)$ adduct to form a reactive electrophilic nitrosyl labeled as $Cu(I)-(NO^+)$,[12] a reduced species analogous to **6** may be a key intermediate in the reaction manifold (Figure 2). Such a species may form upon reaction of NO with reduced NiR or by one electron reduction of the $Cu(I)-(NO^+)$ adduct. Either possibility is consistent with the observation of the involvement of NO in N_2O generation by the enzyme.[12c] Future work will focus on such functional modeling issues and will address the competence of **6** and other model compounds to produce denitrification products.

ACKNOWLEDGEMENTS

Funding for this research was provided by the Exxon Education Foundation (W.B.T.), the National Institutes of Health (GM47365-01 to W.B.T., DMB-9105519 to W.E.A., and GM46749 to J.W.W.), the Searle Scholars Program (W.B.T.), the National Biomedical ESR Center in Milwaukee (RR01008), and the University of Minnesota.

REFERENCES

1. P. M. H. Kroneck, J. Beuerle, W. Schumacher, in *Degradation of Environmental Pollutants by Microorganisms and their Metalloenzymes*, H. Sigel, A. Sigel, Eds. (Marcel Dekker, New York, 1992), vol. 28, pp. 455-505.

2. (a) C. C. Delwiche, Eds., *Denitrification, Nitrification, and Atmospheric Nitrous Oxide* (John Wiley & Sons, New York, 1981). (b) W. J. Payne, *Denitrification* (John Wiley & Sons, New York, 1981). (c) L. I. Hochstein, G. A. Tomlinson, *Ann. Rev. Microbiol.* **42**, 231-261 (1988). (d) W. G. Zumft, A. Viebrock, H. Korner, in *The Nitrogen and Sulphur Cycles*, J. A. Cole, S. J. Ferguson, Eds. (Cambridge University Press, Cambridge, MA, 1988), pp. 245-280. (e) P. M. H. Kroneck, W. G. Zumft, *Denitrification in Soil and Sediment* , 1-20 (1990). (f) A. H. Stouthamer, in *Biology of Anaerobic Microorganisms,* A. J. B. Zehnder, Eds. (Wiley, New York, 1988), pp. 245-301.

3. R. B. Mellor, J. Ronnenberg, W. H. Campbell, S. Diekmann, *Nature* **355**, 717-719 (1992).

4. (a) J. R. J. Lancaster, *Amer. Sci.* **80**, 248-259 (1992). (b) T. G. Traylor, V. S. Sharma, *Biochemistry* **31**, 2847-2849 (1992). (c) S. H. Snyder, *Science* **257**, 494-496 (1992). (d) D. S. Bredt, S. H. Snyder, *Neuron* **8**, 3-11 (1992). (e) L. J. Ignarro, *Biochem. Pharm.* **41**, 485-490 (1991).

5. S. A. Waldman, F. Murad, *Pharmacol. Rev.* **39**, 163-196 (1987).

6. J.-C. Drapier, C. Pellat, Y. Henry, *J. Biol. Chem.* **266**, 10162-10167 (1991).

7. (a) M. Sano, T. Matsubara, *Inorg. Chim. Acta* **152**, 53-54 (1988). (b) M. Masuko, H. Iwasaki, T. Sakurai, S. Suzuki, A. Nakahara, *J. Biochem.* **96**, 447-454 (1984). (c) T. Kakutani, H. Watanabe, K. Arima, T. Beppu, *J. Biochem.* **89**, 453-461 (1981). (d) T. Kakutani, H. Watanabe, K. Arima, T. Beppu, *J. Biochem.* **89**, 463-472 (1981). (e) S. Suzuki, T. Sakurai, A. Nakahara, M. Masuko, H. Iwasaki, *Biochim. Biophys. Acta* **827**, 190-192 (1985).

8. J. W. Godden et al., *Science* **253**, 438-442 (1991).

9. D. M. Dooley, R. S. Moog, M.-Y. Liu, W. J. Payne, J. LeGall, *J. Biol. Chem.* **263**, 14625-14628 (1988).

10. H. Iwasaki, S. Noji, S. Shidara, *J. Biochem.* **78**, 355-361 (1975).

11. M.-Y. Liu, M.-C. Liu, W. J. Payne, J. LeGall, *J. Bacteriol.* **166**, 604-608 (1986).

12. (a) C. L. Hulse, B. A. Averill, J. M. Tiedje, *J. Am. Chem. Soc.* **111**, 2322-2323 (1989). (b) R. W. Ye, I. Toro-Suarez, J. M. Tiedje, B. A. Averill, *J. Biol. Chem.*

266, 12848-12851 (1991). (c) M. A. Jackson, J. M. Tiedje, B. A. Averill, *FEBS Lett.* **291**, 41-44 (1991).

13. (a) B. Salvato et al., *Biochemistry* **28**, 680-84 (1989). (b) A. C. F. Gorren, E. de Boer, R. Wever, *Biochim. Biophys. Acta* **916**, 38-47 (1987). (c) A. J. M. Schoot Uiterkamp, *FEBS Lett.* **20**, 93-96 (1972). (d) A. J. M. Schoot Uiterkamp, H. van der Deen, H. C. J. Berendsen, J. F. Boas, *Biochim. Biophys. Acta* **372**, 407-425 (1974). (e) R. S. Himmelwright, N. C. Eickman, E. I. Solomon, *Biochem. Biophys. Res. Commun.* **81**, 237-242 (1978). (f) R. S. Himmelwright, N. C. Eickman, E. I. Solomon, *Biochem. Biophys. Res. Commun.* **86**, 628-634 (1979). (g) H. van der Deen, H. Hoving, *Biochemistry* **16**, 3519-3525 (1977). (h) J. Verplaetse, P. V. Tornout, G. Defreyn, R. Witters, R. Lontie, *Eur. J. Biochem.* **95**, 327-331 (1979).

14. (a) C. T. Martin et al., *Biochemistry* **20**, 5147-5155 (1981). (b) G. Rotilio, L. Mosrpurgo, M. T. Graziani, M. Brunori, *FEBS Lett.* **54**, 163-166 (1975). (c) D. J. Spira, E. I. Solomon, *Biochem. Biophys. Res. Commun.* **112**, 729-736 (1983).

15. J. W. Whittaker, unpublished observations.

16. G. W. Brudvig, T. H. Stevens, S. I. Chan, *Biochemistry* 5275-5285 (1980).

17. S. Suzuki et al., *Biochem. Biophys. Res. Commun.* **164**, 1366-1372 (1989).

18. M. E. Murphy, H. Sies, *Proc. Natl. Acad. Sci. USA* **88**, 10860-10864 (1991).

19. (a) H. Nasri, Y. Wang, B. H. Huynh, W. R. Scheidt, *J. Am. Chem. Soc.* **113**, 719-721 (1991). (b) H. Nasri, J. A. Goodwin, W. R. Scheidt, *Inorg. Chem.* **30**, 185-191 (1990). (c) M. G. Finnegan, A. G. Lappin, W. R. Scheidt, *Inorg. Chem.* **29**, 181-185 (1990). (d) H. Nasri, Y. Wang, B. H. Huynh, F. A. Walker, W. R. Scheidt, *Inorg. Chem.* **30**, 1483-1489 (1991). (e) M. H. Barley, M. R. Rhodes, T. J. Meyer, *Inorg. Chem.* **26**, 1746-1750 (1987).

20. (a) D. Sellmann, I. Barth, F. Knoch, M. Moll, *Inorg. Chem.* **29**, 1822-1826 (1990). (b) D. W. Pipes, M. Bakir, S. E. Vitols, D. J. Hodgson, T. J. Meyer, *J. Am. Chem. Soc.* **112**, 5507-5514 (1990). (c) F. Bottomley, in *Reactions of Coordinated Ligands,* P. S. Braterman, Ed. (Plenum Press, New York, 1989), vol. 2, pp. 115-222.

21. (a) P. P. Paul, K. D. Karlin, *J. Am. Chem. Soc.* **113**, 6331-6332 (1991). (b) P. P. Paul et al., *J. Am. Chem. Soc.* **112**, 2430-2432 (1990).

22. (a) N. W. Isaacs, C. H. L. Kennard, *J. Chem. Soc. (A)* 386-389 (1969). (b) K. A. Klanderman, W. C. Hamilton, I. Bernal, *Inorg. Chim. Acta* **23**, 117-129 (1977). (c) D. L. Cullen, E. C. Lingafelter, *Inorg. Chem.* **10**, 1264-1268 (1971). (d) S. Takagi, M. D. Joesten, P. G. Lenhert, *J. Am. Chem. Soc.* **97**, 444-445 (1975). (e) S. Takagi, M. D. Joesten, P. G. Lenhert, *Acta Crystallogr.* **B31**, 596-598 (1975). (f) S. Klein, D. Reinen, *J. Sol. State Chem.* **32**, 311-319 (1980). (g) D. Mullen, G. Heger, D. Reinen, *Sol. State Commun.* **17**, 1249-1252 (1975). (h) I. M. Procter, F. S. Stephens, *J. Chem. Soc. (A)* 1248-1255 (1969). (i) A. Walsh, B. Walsh, B. Murphy, B. J. Hathaway, *Acta Crystallogr.* **B37**, 1512-1520 (1981). (j) F. S. Stephens, *J. Chem. Soc. (A)* 2081-2087. (1969).

23. (a) S. Trofimenko *Prog. Inorg. Chem.* **34**, 115-210 (1986). (b) A. Shaver, in *Comprehnsive Coordination Chemistry,* G. Wilkinson, R. D. Gillard, J. A. McCleverty, Eds. (Pergamon Press, Oxford, 1987), vol. 2, pp. 245-259. (c) K. Niedenzu, S. Trofimenko, *Top. Curr. Chem.* **131**, 1-37 (1986).

24. N. Kitajima et al., *J. Am. Chem. Soc.* **114**, 1277-1291 (1992).

25. (a) S. Trofimenko, J. C. Calabrese, J. S. Thompson, *Inorg. Chem.* **26**, 1507-1514 (1987). (b) S. Trofimenko, J. C. Calabrese, P. J. Domaille, J. S. Thompson, *Inorg. Chem.* **28**, 1091-1101 (1989).

26. W. B. Tolman, *Inorg. Chem.* **30**, 4878-4880 (1991).

27. J. A. Ibers, R. H. Holm, *Science* **209**, 223-235 (1980).

28. (a) N. D. Yordanov, V. Terziev, B. G. Zhelyazkowa, *Inorg. Chim. Acta* **58**, 213-216 (1982). (b) M. Mercer, R. T. M. Fraser, *J. Inorg. Nucl. Chem.* **25**, 525-534 (1963). (c) M. P. Doyle, B. Siegfried, J. J. Hammond, *J. Am. Chem. Soc.* **98**, 1627-1629 (1976).

29. S. M. Carrier, C. E. Ruggiero, W. B. Tolman, G. B. Jameson, *J. Am. Chem. Soc.* **114**, 4407-4408 (1992).

30. For example, see: (a) Hendriks, H. M. J.; Birker, P. J. M. W. L.; Rijn, J. v.; Verschoor, G. C.; Reedijk, J. *J. Am. Chem. Soc.* **1982**, *104*, 3607-3617. (b) Engelhardt, L. M.; Pakawatchai, C.; White, A. H. *J. Chem. Soc., Dalton Trans.* 117-123 (1985).

31. C. Mealli, C. S. Arcus, J. L. Wilkinson, T. J. Marks, J. A. Ibers, *J. Am. Chem. Soc.* **98**, 711-718 (1976).

32. (a) J. H. Enemark, R. D. Feltham, *Coor. Chem. Rev.* **13**, 339-406 (1974). (b) R. D. Feltham, J. H. Enemark, in *Topics in Stereochemistry*, R. D. Feltham, J. H. Enemark, Eds. (Wiley, New York, 1981), vol. 12, pp. 155-215. (c) D. M. P. Mingos, D. J. Sherman, in *Advances in Inorganic Chemistry*, A. G. Sykes, Eds. (Academic Press, Inc., 1989), vol. 34, pp. 293-377.

33. (a) N. Kitajima, K. Fujisawa, Y. Moro-oka, *J. Am. Chem. Soc.* **112**, 3210-3212 (1990). (b) N. Kitajima, personal communication.

34. S. M. Carrier, C. E. Ruggiero, W. B. Tolman, W. E. Antholine, and J. W. Whittaker, to be submitted for publication.

35. E. I. Solomon, M. J. Baldwin, M. D. Lowery, *Chem. Rev.* **92**, 521-542 (1992).

36. See M. L. Brader, D. Borchardt, M. F. Dunn, *J. Am. Chem. Soc.* **114**, 4480-4486 (1992) and references therein.

37. T. N. Sorrell, A. S. Borovik, *Inorg. Chem.* **26**, 1957-1964 (1987).

38. G. Palmer, *Biochem. Soc. Trans.* **13**, 548-560 (1985).

39. R. Han, G. Parkin, *J. Am. Chem. Soc.* **113**, 9707-08 (1991).

THE EPR-DETECTABLE COPPER OF NITROUS OXIDE REDUCTASE AS A MODEL FOR Cu$_A$ IN CYTOCHROME c OXIDASE: A MULTIFREQUENCY ELECTRON PARAMAGNETIC RESONANCE INVESTIGATION

P.M.H. Kroneck[a], W.E. Antholine[b], H. Koteich[b], D.H.W. Kastrau[a], F. Neese[a], and W.G. Zumft[c]

[a] Universität Konstanz, Fakultät für Biologie, W-7750 Konstanz, FRG, [b] Biophysics Research Institute, Medical College of Wisconsin, Milwaukee, WI 53226, USA, [c] Lehrstuhl für Mikrobiologie, Universität Karlsruhe, W-7500 Karlsruhe, FRG

INTRODUCTION

Nitrous oxide reductase (N$_2$OR) is the terminal reductase in a respiratory chain converting N$_2$O to N$_2$ in the denitrifying bacteria:

$$N_2O + 2H^+ + 2e^- \rightarrow N_2 + H_2O \qquad [1]$$

Principal aspects of the subject have been covered recently, and these reviews may be consulted for primary reference.[1,2] The high activity form of the enzyme from *Pseudomonas stutzeri* (N$_2$OR I) has two identical subunits, each carrying a mixed-valence [Cu(1.5)...Cu(1.5)], $S = 1/2$ complex.[3] A catalytically inactive derivative of the enzyme (N$_2$OR V) has also the mixed-valence EPR-detectable site.[4,5]

Cytochrome c oxidase (COX) is the terminal oxidase of the aerobic respiratory chain of certain bacteria and of eukaryotic organisms. The enzyme catalyzes the conversion of O$_2$ to H$_2$O:[6,7]

$$4Cytc^{2+} + 8H_i^+ + O_2 \rightarrow 4Cytc^{3+} + 4H_0^+ + 2H_2O. \qquad [2]$$

The minimum catalytically active unit consists of heme a and heme a_3, Cu$_A$ and Cu$_B$, and a third Cu atom designated Cu$_X$,[8] or Cu$_C$.[9] According to Steffens et al.[10] on the basis of ICP-AES measurements, COX from bovine heart contains 3 Cu, 2 Fe, 1 Zn, and 1 Mg, and COX from *Paracoccus denitrificans*, 3 Cu and 2 Fe. For Cu$_B$, it is generally accepted that it is antiferromagnetically coupled to heme a_3.

The EPR-detectable Cu sites in N$_2$OR I from *P. stutzeri* and bovine heart COX have similar spectroscopic properties.[2,11] This conclusion is derived from investigating both enzymes by resonance Raman,[12] MCD and EXAFS,[5,13,14] and ESEEM.[15] Furthermore, comparison of the N$_2$OR primary structure with that of the COX subunit II from several sources indicates that a potential Cu-binding motif with two histidines, two cysteines, and one methionine is highly conserved.[16-18] A binuclear mixed valence

From K.D. Karlin and Z. Tyeklár, Eds., *Bioinorganic Chemistry of Copper*
(Chapman & Hall, New York, 1993).

site with nitrogen from histidine, sulfur from cysteine and methionine coordinated to copper was proposed:[3]

$$N_{His}, S_{cys} - [Cu(1.5) - S_{cys} - Cu(1.5)] - N_{His}, S_{Met} \qquad [3]$$

What is not clear at this stage is the disposition of the third copper, Cu_X, with respect to the protein structure. However, this cannot be expected to be solved by the approach described here and will have to await the completion of studies specifically aimed at this point, which are under way in several laboratories.

Here we compare the copper EPR signals of COX and N_2OR recorded over the frequency range 2.4 - 9.1 GHz. N_2OR from *Pseudomonas stutzeri* was prepared by Zumft and co-workers[4,19] and COX was prepared by Buse and co-workers.[3,11,20] Since the resonances appear better resolved in the second derivative mode, this display was particularly useful for the resolution of several shoulders not easily identified in the first derivative display. The similarity between COX and N_2OR is more evident in this display mode.

RESULTS

^{63}Cu and ^{65}Cu Enriched Nitrous Oxide Reductase.

Seven lines in the $g_\parallel$ region are resolved for the naturally abundant mixed valent $[Cu(1.5)...Cu(1.5)]$ $S = 1/2$ copper site in nitrous oxide reductase[19] and for both isotopically substituted sites, $[^{63}Cu(1.5)...^{63}Cu(1.5)]$ and $[^{65}Cu(1.5)...^{65}Cu(1.5)]$ at X-band (Fig.

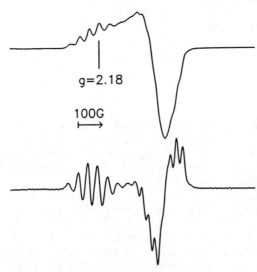

g=2.18

100G

1). Sharp features in the $g_\perp$ region as well as the $g_\parallel$ region are emphasized in the first harmonic display (Fig. 1). The couplings from the seven-line pattern in the $g_\parallel$ region and the high field couplings in the $g_\perp$ region are copper hyperfine lines because the couplings are sensitive to the particular isotope for copper, i.e. the hyperfine coupling for ^{65}Cu is 1.07 times the value for ^{63}Cu.[21] While a seven-line pattern is resolved, the relative intensities of the lines are not a 1, 2, 3, 4, 3, 2, 1 pattern expected for a binuclear copper site with equivalent Cu centers (Figs. 2,3, first derivative and the first harmonic of the first derivative of $g_\parallel$ region).

Fig. 1 X-band EPR spectra for N_2OR I substituted with ^{65}Cu at 15 K. (top) first derivative; (bottom) first harmonic of the first derivative.

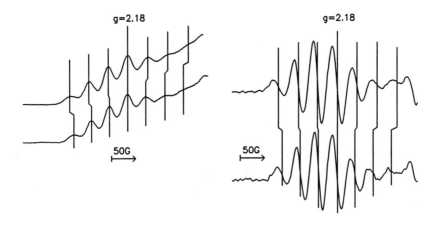

Fig. 2. X-band $g_\parallel$ region for N$_2$OR I at 15 K. (top) ^{65}Cu; (bottom) ^{63}Cu.

Fig. 3. X-band $g_\parallel$ region of first harmonic for N$_2$OR I at 15 K. (top) ^{65}Cu; (bottom) ^{63}Cu.

Simulation of the $g_\parallel$ Region of Nitrous Oxide Reductase.

Previous simulations of the mixed valent site in N$_2$OR and COX assumed two equivalent Cu centers and colinear axes for the g- and A-tensors.[3] If it is assumed that the angle (τ) formed from two Cu atoms and a bridging sulfur ligand is not 180° (Fig. 4), then the A-tensors for both Cu centers are not colinear with the g-tensor. The rotation of the Euler angle α about the z-axis was not varied because of small differences in g- and A-values, $|g_x - g_y| < |g_z - g_y|$, $|g_z - g_x|$, and $|A_x - A_y| < |A_z - A_y|$, $|A_z - A_x|$. Simulations of the $g_\parallel$ region obtained from varying the Euler angle β are shown in Figs. 5-8. Rotation of the Euler angle β, rotation about the x-axis, influences the spectral shape in the $g_\parallel$ region. If the angle τ is 60°, the peak-to-peak heights of high field lines in the $g_\parallel$ region are about 10% greater than the low field lines (Fig. 6). If the hyperfine tensors are not rotated equally, i.e., one copper hyperfine tensor is coincident with the g-tensor while the second copper hyperfine tensor is rotated 60°, the intensity of the lines in the $g_\parallel$

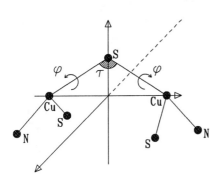

Fig. 4. Schematic for [Cu(1.5)...Cu(1.5)], $S = 1/2$ binuclear center with a sulfur bridging atom. N is nitrogen donor atom, S is sulfur donor atom, and Cu is copper.

region are distorted from the expected seven-line pattern seen in Fig. 2. The peak-to-peak heights of the seven-line pattern approach the same intensity for each line as the total Euler angle β increases (Figs. 6-8).

421

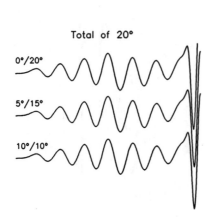

Total of 20°

0°/20°

5°/15°

10°/10°

Total of 60°

0°/60°

10°/50°

20°/40°

30°/30°

Fig. 5. Simulation of the $g_\parallel$ region for which the Cu-S-Cu angle τ is 20°. The Euler angle β, i.e., rotation about the x-axis for each copper, Cu(1)/Cu(2), is 0°/20°; 5°/15°, 10°/10°. $A_\parallel = 38$ G and $g_\parallel = 2.18$.

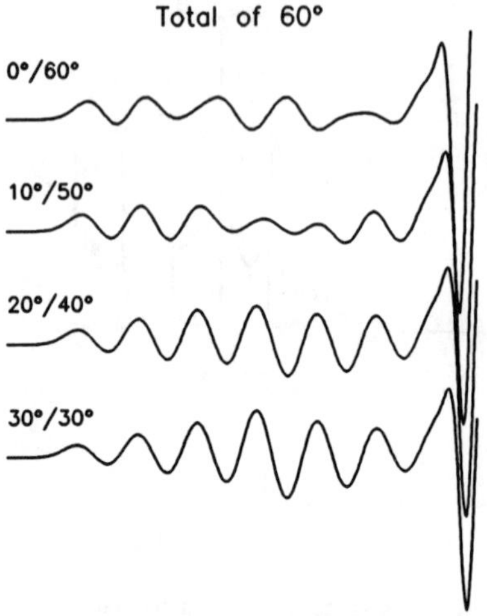

Fig. 6. Simulation of the $g_\parallel$ region for which the Cu-S-Cu angle τ is 60°. The Euler angle β for each copper, Cu(1)/Cu(2), is 0°/60°, 10°/50°, 20°/40°, and 30°/30°. g and A as in Fig. 5.

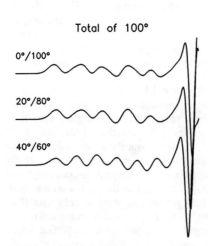

Total of 100°

0°/100°

20°/80°

40°/60°

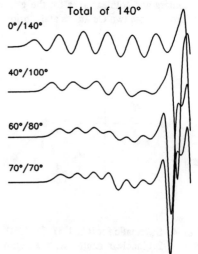

Total of 140°

0°/140°

40°/100°

60°/80°

70°/70°

Fig. 7. Simulation of the $g_\parallel$ region for which the Cu-S-Cu angle τ is 100°. The Euler angle β is 0°/100°, 20°/80°, and 40°/60°. g and A as in Fig. 5.

Fig. 8. Simulation of the $g_\parallel$ region for which the Cu-S-Cu angle τ is 140°. The Euler angle β is 0°/140°, 40°/100°, and 60°/80°, and 70°/70°. g and A as in Fig. 5.

The peak-to-peak heights for the first five lines of the seven-line pattern in the $g_{\parallel}$ region for the simulation in which the Euler angle β is 40° for one copper and -60° for the second copper (Fig. 7) fit the experimental data (Fig. 3) better than previous simulations.[3] While the best simulations to date still do not completely model the experimental spectrum and other Euler angles and tensors need to be adjusted, the Euler angle β appears to greatly affect the lineshape in the $g_{\parallel}$ region. These data are consistent with a total β angle of about 120° and a difference of about 20° in the angle for the first copper compared to the second copper, i.e., inequivalent coppers.

Analyses of the $g_{\parallel}$ region of cytochrome c oxidase.

If the [Cu(1.5)...Cu(1.5)] mixed valence center in N_2OR represents a model for the Cu_A site in COX, then a seven-line pattern in the $g_{\parallel}$ region is also expected for COX. As the temperature is increased from 30 to 100 K, the $heme_a$ EPR signal broadens, but the [Cu(1.5)...Cu(1.5)] signal, although broadened, is still detectable[3] (Fig. 9). Three hyperfine lines on the high field side of the S-band spectrum at 2.7 GHz are still resolved. The program SUMSPC92[22] was used to obtain the overmodulated first harmonic of the EPR spectrum. The over modulated spectrum is subtracted from the experimental data (middle spectrum, Fig. 9). The three high field lines are better resolved; the resolution is comparable to the resolution at 30 K. Moreover, additional lines on the low field side are well resolved (Fig. 9). Seven-line stick diagrams are drawn for a g_z and g_x. The solid lines in the patterns mark well resolved lines; dashed lines mark poorly resolved or overlapping lines. A seven-line pattern for the g_y has not been drawn, but this pattern overlaps both the low and high field patterns. A coupling of about 28 G at low field and 24 G at high field fit the seven line patterns. The spectrum for N_2OR at 30 K is shown for comparison with the spectrum of COX after subtraction of the over modulated spectrum (Fig. 9). The hyperfine couplings for N_2OR are about 38 and 27 G at low and high magnetic field, respectively. The remarkable similarity of the S-band lines enforces the hypothesis that the EPR detectable Cu sites in N_2OR and COX are similar.

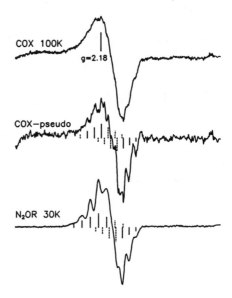

Fig. 9. S-band (2.8 GHz) EPR spectrum of bovine heart cytochrome c oxidase at 100 K (top) and of nitrous oxide reductase at 30 K (bottom). The overmodulated second harmonic spectrum is subtracted from the top spectrum to give the middle spectrum.

423

The low field lines with a hyperfine coupling of 30 G from the S-band spectra should be well separated from g_x and g_y at higher frequency, i.e. X-band (9.1 GHz). The EPR spectrum of cytochrome ba_3 was compared with the spectrum of bovine heart cytochrome aa_3 (Fig. 10). An overmodulated spectrum was subtracted from the experimental spectrum to enhance the resolution. Solid lines indicate clearly resolved lines and dashed lines complete the seven line pattern. Note, a heme Fe is superimposed on the low temperature copper spectrum. A coupling of 30 G is observed and the spectral width is consistent with a seven-line pattern.

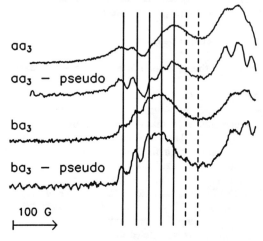

CONCLUSIONS

The pattern that identifies the EPR site in N_2OR as a binuclear center is the seven lines in the $g_{\parallel}$ region (Figs. 1-4). The hyperfine coupling ($A_{\parallel}$ = 38 G) is unusually small representing approximately 50% of the value expected for the blue type-1 Cu.[23] The spectra with a single isotope of either ^{63}Cu or ^{65}Cu rule out the possibility that the lines with natural abundant copper arise from a ^{63}Cu

Fig. 10. $g_{\parallel}$ region of X-band spectrum of bovine heart cytochrome c oxidase aa_3 and of *Thermus thermophilus* cytochrome ba_3. Spectra were overmodulated using a pseudomodulation method and subtracted from the original spectrum to give aa_3-pseudo and ba_3-pseudo, respectively. The spectrum for aa_3 was obtained at about 15 K and ba_3 at 50 K.

monomer (69% natural abundance) and a ^{65}Cu monomer (31 %) instead of the superimposition of the $[^{63}Cu(1.5)...^{63}Cu(1.5)]$, 47% $[^{63}Cu(1.5)...^{65}Cu(1.5)]$, 42.6%, and the $[^{65}Cu(1.5)...^{65}Cu(1.5)]$, 9.5%, configurations.[21] Since the copper hyperfine couplings for COX are about the same or less, fine structure in the COX spectra is also not attributed to resolution of the different isotopic forms. Moreover, the hyperfine lines throughout the spectrum are displaced upon substitution of ^{65}Cu and ^{63}Cu in N_2OR. Thus, the structure is unequivocally attributed to hyperfine values for copper.

The peak heights of seven lines in the $g_{\parallel}$ region attributed to the binuclear center are not 1, 2, 3, 4, 3, 2, 1 expected for two equivalent coppers. The central peaks are almost equal in intensity and the high field lines are less intense than the low field lines. Initial attempts to explain the lineshape consist of simulating the $g_{\parallel}$ region by varying the Euler angles for the copper hyperfine tensor. A value of 120° for β, i.e., rotation about the x-axis, altered the peak heights to give about equal intensities (Figs. 5-8).

It is expected that more information concerning the relative geometries of the g- and A-tensors can be provided by simulations for which all tensors are properly rotated (work in progress). Preliminary results are consistent with (i) a non-linear geometry for Cu-X-Cu with X being a bridging atom (Fig. 4), and (ii) inequivalent copper centers in the binuclear site.

One reason why it has been difficult to interpret the spectrum of the EPR-detectable Cu in COX is that the middle g value for the heme$_a$ signal is superimposed on the $g_\parallel$ region of the spectrum. The heme$_a$ signal is broadened at higher temperature, but the [Cu(1.5)...Cu(1.5)] signal is still detected even though the hyperfine lines are broadened. The resolution in the [Cu(1.5)...Cu(1.5)] spectrum is enhanced by subtraction of the second harmonic or by subtraction of an overmodulated spectrum.[22] Four or five lines in the X-band spectrum for COX, especially for cytochrome ba_3, are resolved on the low field side in the $g_\parallel$ region. The hyperfine coupling of 28-30 G is less than the value of 38 G for N_2OR. This value, $A_\parallel{}^{Cu}$, of 30 G for COX and 38 G for N_2OR as well as different values for g_x, 2.00 for COX and 2.02 for N_2OR, suggest that there are differences in the configurations for the two sites. Nevertheless, the EPR parameters for the EPR detectable sites in N_2OR and COX are attributed to binuclear mixed valence [Cu(1.5)...Cu(1.5)] with $S = 1/2$ configuration.

ACKNOWLEDGEMENTS

We thank Dr. H. Beinert for his interest in the work and for numerous helpful discussions. We also thank Drs. G. Buse and G. C. M. Steffens for the sample of bovine heart cytochrome c oxidase, and Drs. K. Surerus and J. A. Fee for the EPR spectrum of cytochrome ba_3. The work was supported by the National Science Foundation, USA (WA, DMB-9105519; INT-8822596), Boehringer Mannheim Stiftung (DK) and the Deutsche Forschungsgemeinschaft (PK, WZ).

REFERENCES

1. P.M.H. Kroneck, J. Riester, W.G. Zumft, and W.E. Antholine, *Biol. Metals* **3**, 103 (1990).
2. W.G. Zumft, and P.M.H. Kroneck, *FEMS Symp. Ser.* **56**, 37 (1990).
3. W.E. Antholine, D.H.W. Kastrau, G.C.M. Steffens, G. Buse, W.G. Zumft and P.M.H. Kroneck, *Eur. J. Biochem.*, in press (1992).
4. J. Riester, W.G. Zumft, and P.M.H. Kroneck, *Eur. J. Biochem.* **178**, 751 (1989).
5. D.M. Dooley, M.A. McGuirl, A.C. Rosenzweig, J.A. Landin, R.A. Scott, W.G. Zumft, F. Devlin, and P.J. Stephens, *Inorg. Chem.* **30**, 3006 (1991).
6. M. Wikström, K. Krab, and M. Saraste, *Cytochrome Oxidase. A Synthesis.* (Academic Press, New York, 1981).
7. G.T. Babcock and M. Wikström, *Nature* **356**, 301 (1992).
8. G.L. Yewey, and W.S. Caughey, *Ann. NY Acad. Sci.* **550**, 22 (1988).
9. A. Müller, and A. Azzi, *J. Bioenerg. Biomembr.* **23**, 291 (1991).
10. G.C.M. Steffens, K. Biewald, and G. Buse, *Eur. J. Biochem.* **164**, 295 (1987).
11. P.M.H. Kroneck, W.E. Antholine, D.H.W. Kastrau, G. Buse, G.C.M. Steffens, and W.G. Zumft, *FEBS Lett.* **268**, 274 (1990).
12. D.M. Dooley, R.S Moog, and W.G. Zumft *J. Am. Chem. Soc.* **109**, 6730 (1987).
13. R.A. Scott, W.G. Zumft, C.L. Coyle, and D.M. Dooley, *Proc. Natl. Acad. Sci. USA* **86**, 4082 (1989).
14. J.A. Farrar, A.J. Thomson, M.R. Cheesman, D.M. Dooley, and W.G. Zumft, *FEBS Lett.* **294**, 11 (1991).
15. H. Jin, H. Thomann, C.L. Coyle, and W.G. Zumft, *J. Am. Chem. Soc.* **111**, 4262 (1989).
16. A. Viebrock, and W.G. Zumft, *J. Bacteriol.* **170**, 4658 (1988).

17. G. Buse, and G.C.M. Steffens, *J. Bioenerg. Biomembr.* **23**, 269 (1991).
18. W.G. Zumft, A. Dreusch, S. Löchelt, H. Cuypers, B. Friedrich, and B. Schneider, *Eur. J. Biochem.*, **208**, 31 (1992).
19. C.L. Coyle, W.G. Zumft, P.M.H. Kroneck, H. Körner, and W. Jakob, *Eur. J. Biochem.* **153**, 459 (1985).
20. W.G. Zumft, A. Viebrock-Sambale, and C. Braun, *Eur. J. Biochem.* **192**, 591 (1990).
21. W. Kaim, and M. Moscherosch, *J. Chem. Soc., Faraday Trans.* **87**, 3185 (1991).
22. J.S. Hyde, A. Jesmanowicz, J.J. Ratke, and W.E. Antholine, *J. Magn. Reson.* **96**, 1 (1992).
23. R. Malkin, and B.G. Malmström, *Adv. Enzymol.* **33**, 177 (1970).

MIXED-LIGAND, NON-NITROSYL Cu(II) COMPLEXES AS POTENTIAL PHARMACOLOGICAL AGENTS VIA NO RELEASE.

Danae Christodoulou,*[a] Chris M. Maragos,[a] Clifford George,[b] Deborah Morley,[c] Tambra M. Dunams,[a] David A. Wink[a] and Larry K. Keefer[a]

[a]Chemistry Section, Laboratory of Comparative Carcinogenesis, National Cancer Institute, Frederick Cancer Research and Development Center, Frederick, Maryland 21702, [b]Laboratory for the Structure of Matter, Naval Research Laboratory, Washington D.C. 20375, and [c]Cardiology Department, Temple University, Philadelphia, Pennsylvania 19140.

INTRODUCTION

Nitric oxide (NO) has recently been established as a key bioregulatory agent[1]. Its multifaceted nature has been demonstrated in vascular relaxation, antiplatelet action, macrophage induced cytostasis and cytotoxicity, and neurotransmission. NO is produced endogenously from the amino acid L-arginine[2-3] by the enzyme NO synthase (Fig. 1). Several forms of this enzyme have been isolated. The NO produced can activate soluble guanylate cyclase, and thereby influence cellular concentration of another secondary messenger, cyclic GMP, resulting in different physiological responses. Another type of NO synthase, which is induced by activated macrophages and other cells, synthesizes NO which acts as a cytotoxic agent against tumor cells and bacteria. A number of cofactors are involved in the action of the enzyme, such as Ca^{2+}/calmodulin, tetrahydrobiopterin, FMN, FAD and NADPH. NO synthase has high homology to cytochrome P_{450} reductase.[4-5] Inhibitors of NO synthase activity include hemoglobin, the superoxide radical, and glucocorticoids. Analogues of L-arginine, where the guanidino NH_2 group has been replaced with groups such as MeNH (L-NMMA), Me (L-NIO), O_2N-NH (L-NA), or H_2N-NH (L-NAA) act as inhibitors[6] of the endothelial NO synthase in vivo.

The role of nitric oxide in vasodilation has been studied extensively. Nitric oxide has been identified as the "Endothelium Derived Relaxing Factor" (EDRF),[7-8] although this conclusion has been challenged.[9] EDRF has also been shown to inhibit platelet aggregation and to cause disaggregation of aggregated platelets.[10] Evidence for the chemical identity of EDRF has been obtained by a chemiluminescence method. In this method[11], the chemiluminescent product of NO with ozone is measured directly. The same response has been detected from EDRF released from mammalian endothelial cells.[7] The levels of NO released by endothelial cells causing relaxation of vascular strips or inhibition of platelet aggregation and adhesion have also been correlated with the amount of NO detected by chemiluminescence. The action of both EDRF and NO on vascular smooth muscle and platelets occurs via stimulation of soluble guanylate cyclase and elevation of cyclic GMP levels.[12]

*Author to whom correspondence should be sent at current address: Johnson Matthey Inc., Biomedical Division, 1401 King Road, West Chester, Pennsylvania 19380.

From K.D. Karlin and Z. Tyeklár, Eds., *Bioinorganic Chemistry of Copper*
(Chapman & Hall, New York, 1993).

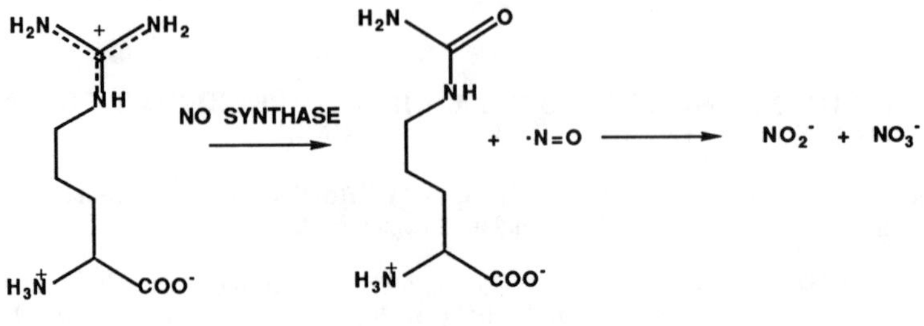

Fig. 1. Biosynthesis of nitric oxide.

In order to directly probe the role of nitric oxide in vascular relaxation and other physiological functions, we seek to develop novel medicinal agents that can carry and deliver NO in biological systems. We have explored the chemistry, structural properties and reactivity of a novel class of N-nitroso compounds that contain the gaseous molecule and can deliver it spontaneously, in a controlled manner, when dissolved in aqueous solutions.[13] These N-nitroso compounds display versatile pharmacological activity[14] which is attributed to their nitric oxide release.

Drago and coworkers have synthesized stable adducts of nitric oxide with nucleophiles[15] in the early sixties from the reaction of nitric oxide gas with various amines:

$$X^- \quad + \quad 2NO \quad \longrightarrow \quad X\text{-}N_2O_2^- \qquad (1)$$

These anions regenerate NO[16] upon hydrolysis; their rate of decomposition in aqueous solution obeys first order kinetics and their half-life depends on the residue X .[13] When X is a secondary amine, the extent of nitric oxide release upon decomposition, as determined by the chemiluminescence method,[13] approaches the theoretical amount of two (number of NO molecules per $X\text{-}N_2O_2^-$ ion). A good correlation between the extents of NO release from such $X\text{-}N_2O_2^-$ ions in aqueous media and their vasorelaxant potencies in isolated rabbit aorta has been observed.[13] The controlled, nonenzymatic generation of nitric oxide for biological purposes from the above class of compounds can be an advantageous research tool for the development of potential therapeutic agents.

We have been investigating the coordination properties of $X\text{-}N_2O_2^-$ ions, where X = a primary or secondary amine residue, or a polyamine, as a means of exploring the structural characteristics and reactivity of the $N_2O_2^-$ functional group present in the anions. Furthermore, we wished to determine whether coordination of nitric oxide/nucleophile adducts to metal centers might improve their pharmacological potency and/or stability. We have thus been investigating whether the unique properties of various metal ions can be employed to facilitate NO release to a target tissue, either via redox action controlled by the metal center or via interaction of biopolymer residues with possible vacant coordination sites on the metal center. Metal complexes of various ligand content could also provide a way of increasing the number of NO molecules potentially released in a single compound.

The $X-N_2O_2^-$ ions should exhibit ligand properties if the oxygen atoms of the $N_2O_2^-$ functional group are in the Z-configuration. This type of structure would resemble the reported $[O_3S-N_2O_2]^{2-}$ and $[O-N_2O_2]^{2-}$ structures, which contain cis oxygens but release primarily nitrous oxide, N_2O.[17-18] Evidence for the ligand properties of the $Et_2N-N_2O_2^-$ ion has been reported earlier by Longhi and Drago.[19] We have further investigated the interactions of this and other $X-N_2O_2^-$ ions. In the case of Cu^{2+} we have isolated and crystallized mixed-ligand complexes of various nuclearity, according to the general scheme of eq 2.

$$2Et_2N-N_2O_2^- \ + \ xCu^{2+} \ + \ yL \ \longrightarrow \ Cu_x(L)_y(Et_2N-N_2O_2)_2^{z-} \quad (2)$$

$$L = MeOH, \ OMe^-, \ OEt^-, \ OAc^-$$

As we will describe in this paper, these compounds are novel examples of copper complexes having nitric oxide as a ligand, oxygen bound rather than in the form of a classical metal nitrosyl, L_nM-NO. The structural properties and reactivity of these complexes might provide an insight into $Cu-N_2O_2$ species speculated to be involved in the mechanistic pathway of the denitrification process. A labile $Cu-NO$ intermediate, where the enzyme nitrite reductase forms NO prior to N_2O formation, has been implied.[20] So far, there are only very few examples of $LCu-NO$ complexes;[21-22] however, the structural and chemical properties of such species are of great importance. A plausible intermediate in the reduction pathway of NO_2^- to N_2O may involve a precursor that contains a N-N bond. The $N_2O_3^{2-}$ and $X-N_2O_2^-$ anions (where X = sulfite ion, amine) are examples of stable chemical entities that release N_2O and/or NO.

MATERIALS AND METHODS

The ligands, $Et_2NH_2^+ \ Et_2N-N_2O_2^-$ and $Na^+ \ Et_2N-N_2O_2^-$, were prepared according to modified literature procedures. The diethylammonium salt was produced in a Parr reactor at 60 psi NO pressure and the sodium salt was obtained by subsequent neutralization of this compound with sodium methoxide.[23] The copper complexes were synthesized anaerobically, in nonaqueous solutions, at low temperature (-30° C). Under these conditions the half-lives of starting materials and products can be prolonged. A series of the complexes has been generated by using one such compound as a synthon. Ligand exchange and rearrangement reactions suggest lability of the complexes in solution. However, in some cases thermodynamically stable products have been isolated by self assembly. All solid products were stored at -30° C and used fresh for NO determinations or in vitro experiments. Visible spectra and kinetic measurements were performed at 37° C and pH = 7.4.

The amounts of nitric oxide released upon dissolution of the compounds in aqeuous solution were determined by the chemiluminescence method, as described earlier.[13] Schematics of the apparatus used are shown in Fig. 2. The analyte was introduced into a thermostated reactor purged with argon. Deoxygenated phosphate buffer was then introduced via a septum to start the reaction. The nitric oxide evolved was quantified by sweeping purged gases continuously into a chemiluminescence detector. The area under the recorder trace was integrated and compared with nitric oxide gas standards.

For testing the vasoactivity of the compounds a standard vascular ring preparation was employed, as also was described earlier.[13] Segments of rabbit thoracic aorta were placed in a pH 7.4 modified Krebs buffer at 37 ° C and constricted with norepinephrine. Dose-response curves for the complexes were obtained in parallel runs with the free ligand in the range 10^{-11} M - 10^{-5} M. Schematics of this experiment are shown in Fig. 3.

Crystallographic studies were performed on single crystals of the complexes at low temperature. Detailed information on the structural data, as well as ORTEP drawings of the structures, will be published elsewhere.

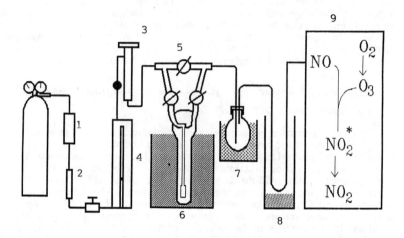

Fig. 2. Apparatus used for NO analysis. 1-2, oxygen traps. 3, gas injection port. 4, flowmeter. 5, reaction chamber. 6, thermostated bath. 7, first trap in dry ice-MeOH. 8, second trap above liquid nitrogen. 9, chemiluminescence detector.

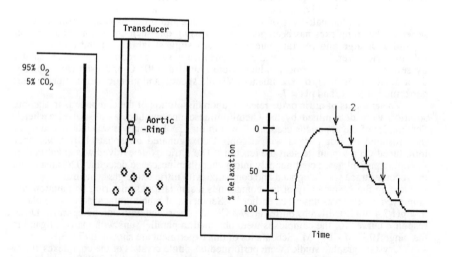

Fig. 3. Aortic ring test. 1, norepinephrine. 2, increasing doses of vasodilator.

RESULTS AND DISCUSSION

The synthesis of the complexes that contain bridging alkoxides as well as the bidentate Et_2N-$N_2O_2^-$ ligand is shown in Fig. 4. Among these, we have isolated the previously reported mononuclear $Cu(MeOH)(Et_2N$-$N_2O_2)_2$ [19] via a different route, and we have determined its crystal structure by X-ray diffraction analysis. Synthesis of a mixed-ligand complex, that contains bridging acetates along with the Et_2N-$N_2O_2^-$ ligand, was effected from the diethylammonium salt of the ligand. The complex was isolated as a monoanion, as shown in Fig. 5.

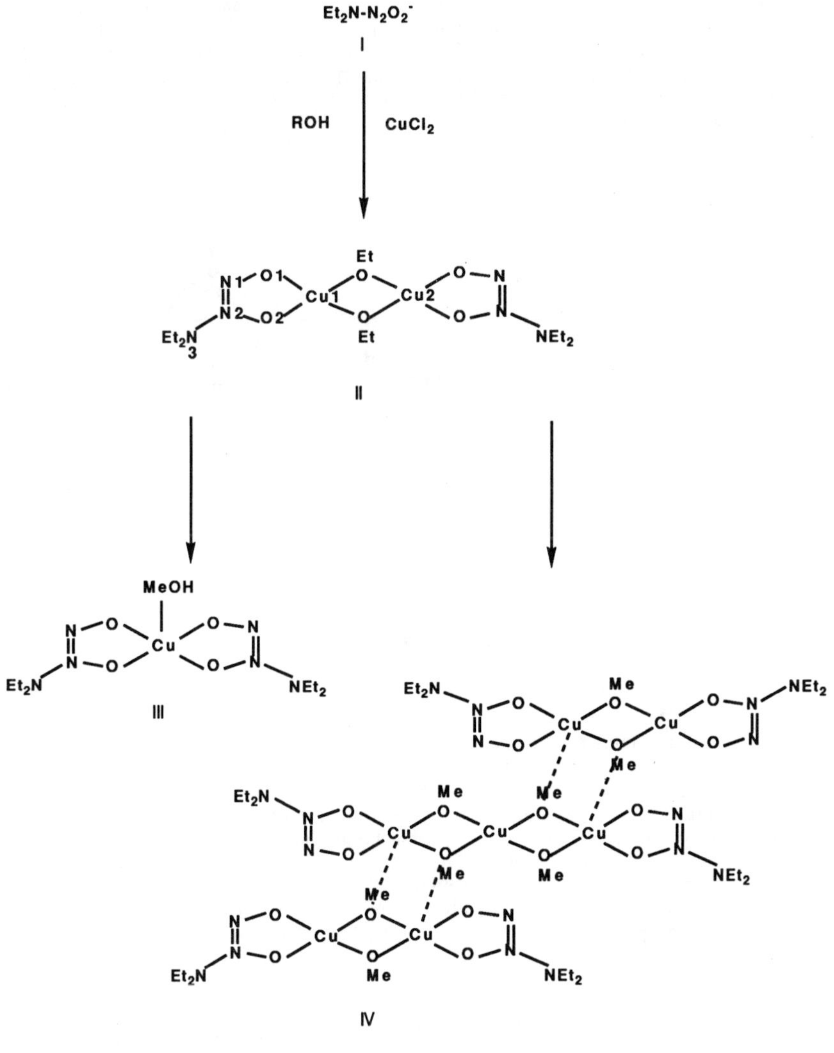

Fig. 4. Synthesis of alkoxo-bridged complexes.

Complexes II-V have been crystallized as blue solids and they have been characterized by elemental analysis, FT-IR, ^{1}H NMR and ESR spectroscopy. The crystal structures of the complexes as well as the one of the free ligand have been determined by X-ray diffraction analysis. Their magnetic properties have also been investigated. Complexes with bridging ligands exhibit aniferromagnetic coupling, although the behavior is rather complicated, due to additional intermolecular interactions between copper atoms and bridging alkoxides, at an average distance of 2.4Å. Intramolecular Cu-Cu distances are short and vary between 2.9-3.1Å, but they do not represent bonding distances. Complex V, although weakly coupled compared to the alkoxo-bridged compounds, exhibits pure antiferromagnetic behavior with $J = 14$ cm^{-1}. The ESR spectrum of V in MeOH is a "type II" Cu(II) signal with axial geometry, similar to the one of the mononuclear complex Cu(MeOH)(Et$_2$N-N$_2$O$_2$)$_2$, III, and suggests that the molecule dissociates in solution . This is consistent with the crystallographic results, that show one long interaction at 2.4Å between the η^1 acetates and each of the copper atoms. The cyclic voltammetry of the complexes has also been examined. In general, irreversible behavior has been observed resulting from ligand-based oxidation.

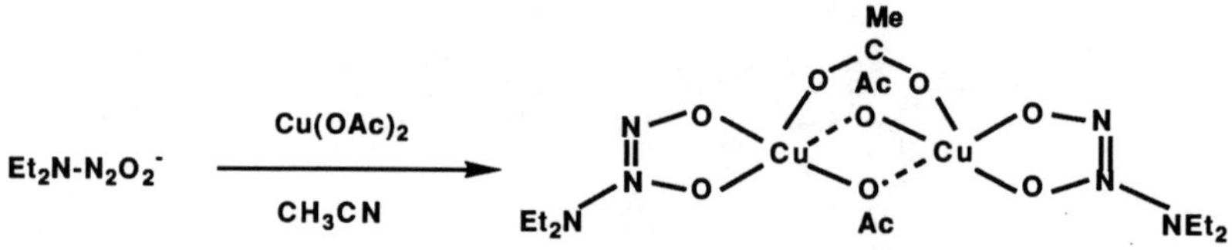

Fig. 5. Synthesis of Et$_2$NH$_2$$^+$ [Cu$_2$(OAc)$_3$(Et$_2$N-N$_2$O$_2$)$_2$]$^-$.

Structure determinations on these metal complexes demonstrate the planar, bidentate chelate nature of the N$_2$O$_2$$^-$ functional group contained in nitric oxide/secondary amine adducts, X-N$_2$O$_2$$^-$. These copper compounds are the first crystallographically characterized metal complexes of the ligand Et$_2$N-N$_2$O$_2$$^-$. In the N$_2O_2$$^-$ group, the N-N linkage has double bond character, in contrast to the (Et$_2$)N-N linkage, which is a single bond. The N-O distances are equivalent, with single bond character; both oxygens are functionalized and interact with the metal center. The structural parameters (averaged) within the N$_2$O$_2$$^-$ functional group in the free ligand and the copper complexes are summarized in Table 1. The numbering scheme used is the one shown in Fig. 4, structure II. In all the complexes, the Et$_2$N-N$_2$O$_2$$^-$ ligands are oxygen bound to the copper atoms, and arranged with their diethylamino moieties cis to each other. The trinuclear component of molecule IV is the only exception, having trans Et$_2$N-N$_2$O$_2$$^-$ ligands, as a result of crystallographically imposed symmetry. The molecule is situated on a center of symmetry and forms a layered trimer with two binuclear units via intermolecular copper-alkoxide interactions, as shown in Fig. 4.

432

(S,O) N3	$O_3S\text{-}N_2O_2{}^{2-}$ [17]	$O\text{-}N_2O_2{}^{2-}$ [18]	$Et_2N\text{-}N_2O_2{}^-$	$Cu_x(L)_y(Et_2N\text{-}N_2O_2)_2{}^{z-}$
dist., Å				
O1-N1	1.284(17)	1.347(4)	1.316(2)	1.311(6)
O2-N2	1.286(17)	1.322(4)	1.278(2)	1.324(6)
N1-N2	1.327(17)	1.264(5)	1.280(3)	1.274(6)
N2-N3(S,O)	1.791(12)	1.310(4)	1.428(3)	1.411(6)
angle, deg				
O1-N1-N2	112.2	112.9(3)	111.8(2)	112.3(4)
O2-N2-N1	125.1	122.5(3)	126.3(2)	123.6(4)
N1-N2-N3(S,O)	116.5	118.4(3)	113.5(2)	116.2(4)
O2-N2-N3(S,O)	118.4	119.1(3)	120.2(2)	120.0(4)

Table 1. Structural parameters in the $N_2O_2{}^-$ functional group.

The structural parameters of the $N_2O_2{}^-$ functional group are not significantly affected upon coordination. In all the complexes, the N-O distances are equivalent and similar to the ones found in $Na_2N_2O_3$.[18] However, in the free ligand, the N-O distance is slightly shorter than the terminal one. Alkylation of the $Et_2N\text{-}N_2O_2{}^-$ ion takes place on the terminal oxygen,[24] suggesting that O1 is the most electronegative oxygen in this ion.

The electronic spectra of the copper complexes are nearly identical, with a broad absorption centered at about 580 nm in the visible region, and the characteristic absorption of the $N_2O_2{}^-$ chromophore at 244 nm. Similarly, the UV absorption of this group is not significantly affected upon coordination. The behavior of the metal complexes in aqueous solution was similar to that of the free ligand, suggesting that nitric oxide release is controlled primarily by hydrolysis of the $Et_2N\text{-}N_2O_2{}^-$ ligand. Coordination of the ligand to copper in general prolonged the life of the derivative in nonaqueous medium, whereas the free ligand was more stable in aqueous solution. The amount of nitric oxide produced by decomposition of the complexes in aqueous medium is generally proportional to the number of $Et_2N\text{-}N_2O_2{}^-$ ligands present. Presumably this accounts for the increased vasorelaxant effects observed for these compounds in isolated rabbit aorta.

The extents of nitric oxide released by the complexes, as determined by the chemiluminescence method, are listed in Table 2. E_{NO} is the number of moles of nitric oxide generated per mole of compound decomposed upon dissolution at 37 ° C and pH = 7.4 for I, II, and V and pH = 5.0 for III, and IV. λ_{max} , and ε represent the absorption characteristics for the $N_2O_2{}^-$ chromophore, recorded in MeOH for the complexes and 0.01M NaOH for the free ligand.

COMPOUND	E_{NO}	λ_{max} (nm)	ε (mM^{-1}cm^{-1})
I	1.5±0.11	250	6.5
II	4.01±0.25	246	12.8
III	3.8±0.4	244	12.1
IV	12.6±0.7	248	37.7
V	3.95±0.10	242	12.5

Table 2. Nitric oxide content of the compounds and UV absorption characteristics of the $N_2O_2^-$ group.

The vasorelaxant effects of the complexes have been demonstrated in vitro with isolated rabbit aorta. In parallel studies with the free ligand, a significant increase in the vasorelaxant potencies of the complexes has been observed. Figure 6 illustrates the results of one such experiment. The vasorelaxant activities of complex IV and the ligand I are given as a function of concentration (dose) of the compound. Values for % relaxation are given as means ± SD for three (in I) to five (in IV) replicates. EC$_{50}$ (M), the concentration of compound that induces 50% relaxation in the aortic rings, was determined to be $9.9 \cdot 10^{-7}$ M for IV and $8.3 \cdot 10^{-6}$ M for I. The previously reported[13] value for I was $1.9 \cdot 10^{-7}$ M and in the same study sodium nitroprusside, a clinically used vasodilator, had EC$_{50}$ = $2.0 \cdot 10^{-6}$ M. Similar results, in which the potency has improved by an order of magnitude or better, have been observed with the other complexes, with the exception of V, which is the least stable in aqueous medium.

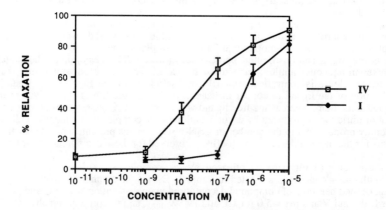

Fig. 6. Dose-response curve for % relaxation of compounds IV and I in isolated rabbit aorta.

CONCLUSION

Investigations of the ligand properties of the $Et_2N-N_2O_2^-$ ion led to the isolation of a variety of complexes of the general formula $Cu_x(L)_y(Et_2N-N_2O_2)_2^{z-}$. All contain the $N_2O_2^-$ functional group, best described as a bidentate chelating ligand since the oxygens are found in the Z-configuration and both interact with copper (II) ion as has been demonstrated by crystallographic studies. The ligand is bound via the oxygens to the metal centers rather than forming classical metal nitrosyls. These molecules release amounts of nitric oxide proportional to the ligand content and illustrate how biologically active ligands of this type can coexist with other biologically relevant ligands in the form of mixed-ligand complexes. The complexes suggest possibilities of modulating the reactivity as well as pharmacology of the $X-N_2O_2^-$ anions upon coordination and they present examples of oxygen coordination in the fundamental chemistry of copper (II) $-N_xO_y$ species.

ACKNOWLEDGMENTS

The authors are indebted to their colleagues who have contributed to the work presented here: Joseph E. Saavedra, (Chemistry Section, NCI-FCRDC) and William R. Dunham (Biophysics, University of Michigan).

REFERENCES

1.   S. Moncada, R. M. J. Palmer and E. A. Higgs, *Pharmacol. Rev.* , **43,** 109 (1991) and references therein.
2.   R. Iyengar, D. J. Stuehr and M. A. Marletta, *Proc. Natl. Acad. Sci. USA* , **84,** 6369 (1987).
3.   J. B. Hibbs, JR., R. R. Taintor, Z. Vavrin, D. L. Granger, J. C. Drapier, I. J. Amber and J. R. Lancaster, in *Nitric Oxide from L-Arginine: A Bioregulatory System*, S. Moncada and E. A. Higgs, Eds., (Elsevier, Amsterdam, 1990), pp. 189-223.
4.   K. A. White and M. A. Marletta, *Biochemistry*, **31,** 6627 (1992).
5.   D. J. Stuehr and M. Ikeda-Saito, J. Biol. Chem., **267,** 20547 (1992).
6. S. M. Gardiner, A. M. Compton, T. Bennett, R. M. J. Palmer and S. Moncada in *Nitric Oxide from L-Arginine: A Bioregulatory System*, S. Moncada and E. A. Higgs, Eds., (Elsevier, Amsterdam, 1990), pp. 489-491.
7.   L. J. Ignarro, G. M. Buga, K. S. Wood, R. E. Byrns and G. Chaudhuri, *Proc. Natl. Acad. Sci. USA* , **84,** 9265 (1987).
8.   R. F. Furchgott, in *Nitric Oxide from L-Arginine: A Bioregulatory System*, S. Moncada and E. A. Higgs, Eds., (Elsevier, Amsterdam, 1990), pp. 5-17.
9.   P. R. Myers, R. L. Minor, JR., R. Guerra, JR., J. N. Bates and D. G. Harrison, *Nature*, **345,** 161 (1990).
10. M. W. Radomski, R. M. J. Palmer and S. Moncada, *Proc. Natl. Acad. Sci. USA* , **87,** 5193 (1990).
11. M. J. Downes, M. W. Edwards, T. S. Elsey and C. L. Walters, *Analyst*, **101,** 742 (1976).
12. F. Murad, C. K. Mittal, W. P. Arnold, S. Katsuki and H. Kimura, *Adv. Cyclic Nucleotide Res.*, **9,** 145 (1978).
13. C. M. Maragos, D. Morley, D. A. Wink, T. M. Dunams, J. E.Saavedra, A. Hoffman, A. A. Bove, A. A.; L. Isaac, J. A. Hrabie, L. K. Keefer, *J. Med. Chem.*, **34,** 324 (1991).
14. L. K. Keefer, D. A. Wink, C. M. Maragos, D. Morley and J. G. Diodati, in *The Biology of Nitric Oxide*, S. Moncada, M. A.Marletta, J. B. Hibbs, JR., E. A. Higgs, Eds, (Portland Press, Colchester, UK), in press.
15. R. S. Drago, in *Free Radicals in Inorganic Chemistry*, American Chemical Society, (Advances in Chemistry Series, Number 36, Washington, D. C., 1962) pp. 143-149.

16. T. J. Hansen, A. F. Croisy and L. K. Keefer, in *N-Nitroso Compounds: Occurrence and Biological Effects*, H. Bartsch, I. K. O' Neill, M. Castegnaro and M. Okada, Eds., (International Agency for Research on Cancer, IARC Scientific Publications No. 41, Lyon, 1982) pp.21-29.

17. (a) E. G. Cox, G. A. Jeffrey, H. P. Stadler, *Nature*, **162**, 770 (1948). (b) E. G. Cox, G. A. Jeffrey and H. P. Stadler, *J. Chem. Soc.*, 1783 (1949). (c) G. A. Jeffrey and H. P. Stadler, *J. Chem. Soc.*, 1467 (1951).

18. H. Hope and M. R. Sequeira, *Inorg. Chem.*, **12**, 286 (1973).

19. R. Longhi and R. S. Drago, *Inorg. Chem.*, **2**, 85-88 (1963).

20. M. A. Jackson, J. M. Tiedje and B. M. Averill, *FEBS Lett.*, **291**, 41 (1991).

21. (a) P. P. Paul, Z. Tyeklar, A. Farooq, K. D. Karlin, S. Liu and J. Zubieta, *J. Am. Chem. Soc.*, **112**, 2430 (1990). (b) P. P. Paul and K. D. Karlin, *J. Am. Chem. Soc.*, **113**, 6331 (1991).

22. W. B. Tolman, *Inorg. Chem.* **30**, 4877 (1991).

23. R. S. Drago and B. R. Karstetter, *J. Am. Chem. Soc.*, **83**, 1819 (1961).

24. J. E. Saavedra, T. M. Dunams, J. L. Flippen-Anderson and L. K. Keefer, *J. Org. Chem.*, **57**, 6134 (1992).

Copper Oxidases

COPPER-CONTAINING ENZYMES: STRUCTURE AND MECHANISM

Joseph J. Villafranca, John C. Freeman, and Anne Kotchevar

Department of Chemistry, The Pennsylvania State University, University Park, PA 16802, USA

INTRODUCTION

Enzymes that require copper for their activity catalyze reactions of many diverse types. In recent years, we have been studying Cu-containing enzymes whose reaction mechanisms are complex and incompletely understood. This report covers studies on two enzymes with different Cu stoichiometry at their active sites as well as different spectroscopic properties. Kinetic data are presented to delineate the reaction mechanism of galactose oxidase and spectroscopic data reveal the different classes of copper in phenoxazinone synthase.

PHENOXAZINONE SYNTHASE

Phenoxazinone synthase catalyzes the oxidative coupling of two molecules of a substituted o-amino phenol to the phenoxazinone chromophore in the final step in the biosynthesis of actinomycin D (**Scheme I**). The reaction represents a six-electron oxidative coupling and appears to take place stoichiometrically in a series of three two-electron oxidations.[1] Cyclization occurs with the concomitant reduction of molecular

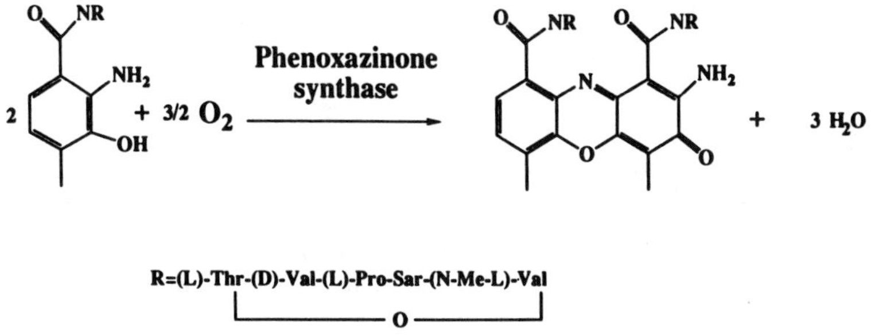

R=(L)-Thr-(D)-Val-(L)-Pro-Sar-(N-Me-L)-Val
└──────── O ────────┘

Scheme I: The biochemical reaction catalyzed by phenoxazinone synthase.

From K.D. Karlin and Z. Tyeklár, Eds., *Bioinorganic Chemistry of Copper*
(Chapman & Hall, New York, 1993).

oxygen to water. No evidence is available to establish whether electron transfer occurs in single electron steps throughout the enzymatic process, however.

The enzyme phenoxazinone synthase is naturally found in the bacterium *Streptomyces antibioticus* and has been cloned and overexpressed in *S. lividans* .[2] A recent analysis of the gene by our laboratory has led to a determination of the molecular mass of 67,500 da.

Experiments were conducted to determine the stoichiometric amount of copper necessary to fully activate phenoxazinone synthase. $CuCl_2$ was added to enzyme samples and the activity measured. The steady state specific activities are plotted against the copper content per subunit of the enzyme as shown in Figure 1. Maximum activity occurs at approximately 4-5 Cu atoms per subunit. Additional copper shows an inhibitory effect for reasons that remain unclear; this observation was also reported earlier.[3] Optical spectra of the enzyme show that the value of the extinction coefficient roughly follows the enzymatic activity indicating that the rise in the absorption signal at 598 nm (due to a blue copper center) is correlated with specific activity. The visible absorption envelope of phenoxazinone synthase is characteristic of a number of blue copper containing proteins and the assignment of the electronic transitions should be similar.

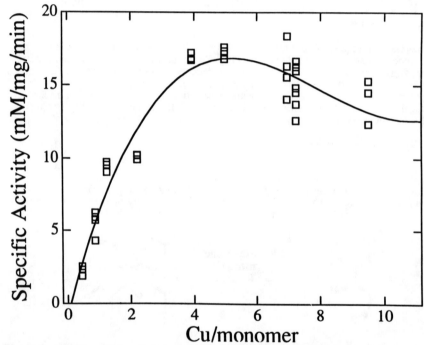

Figure 1: Specific activity of phenoxazinone synthase plotted against copper present per subunit.

The EPR spectra shown in Figure 2A indicate that the most tightly bound copper has characteristics of a type 1 copper center with typical $A_{||}$ values. When Cu^{2+} is successively added to apoenzyme the type 2 Cu EPR signal rapidly obscures the type 1 copper signal (Figure 2B); the type 1 copper signal is difficult to detect when the enzyme contains more than 4 copper atoms per monomer. There are two possible explanations for both the EPR and optical titration behavior. The first is that additional copper above 1 atom per monomer is bound sequentially in type 2 copper sites in an individual enzyme molecule. The second possibility is that there is a second type 1 copper site which when filled in a particular monomer causes the type 2 sites in that monomer to bind copper in a cooperative manner; this situation would be similar to that found with ceruloplasmin,[4] rather than the sequential binding pattern seen in ascorbate oxidase.[5] In the second case optical spectra and EPR spectra at differing copper contents would represent fully active monomers with 4-5 copper atoms and monomers with a single copper. Previous characterization of phenoxazinone synthase did not identify this protein as containing a blue copper center[1]: these observations can be explained by the present data showing that the type 2 center signal in high copper content samples obscures the EPR signal due to the type 1 center. It was noted however that the EPR spectrum in the high copper/monomer sample contained more than one signal. The ultraviolet region of the optical spectrum does not show a band at 330 nm that has in the past been attributed to type 3 centers. From our spin counting experiments, all the copper centers appear to be EPR active. Also, there is no bleaching in the 330 nm region under anaerobic conditions with the addition of substrate as seen in the other blue oxidases.[6] These results would preclude the presence of a type 3 binuclear copper center in phenoxazinone synthase.

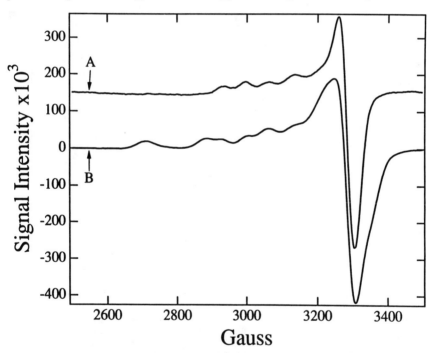

Figure 2: EPR spectra of phenoxazinone synthase. Spectrum A is the EPR signal of phenoxazinone synthase as isolated containing 0.8 coppers per monomer. The spectrum shows $g_{\parallel}$ of 2.24 and $A_{\parallel}$ of 70 G (0.0067 cm^{-1}). The $g_{\perp}$ value is 2.07. Spectrum B is the EPR signal of phenoxazinone synthase containing 5.2 copper per monomer. Two sets of $g_{\parallel}$ values can be assigned to the spectrum 2; one with $g_{\parallel} = 2.24$, A = 70 G (0.0067 cm^{-1}) typically associated with type 1 copper centers and one with $g_{\parallel} = 2.34$, A = 165 G (0.015 cm^{-1}) normally found associated with type 2 copper centers. Spin counting indicates 1.1 to 1.3 spins per copper.

The type 1 copper in phenoxazinone synthase is reduced by substrate (Figure 3) and this process is likely to be the path by which electrons enter the copper centers.

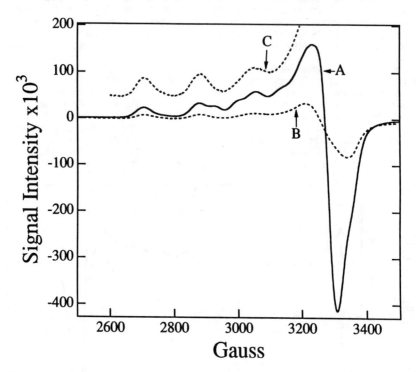

Figure 3: Anaerobic incubation of phenoxazinone synthase as followed by EPR spectroscopy. The solid line (spectrum A) is before the addition of substrate and the dashed line (spectrum B) is after substrate addition. Spectrum C is a five-fold expansion of the $g_{\parallel}$ region after substrate addition. After addition of substrate, the spectroscopic values are $g_{\parallel} = 2.28$, $A = 175$ G (0.016 cm^{-1}) and $g_{\perp} = 2.07$.

The data presented in this paper suggest two possible copper stoichiometries for the enzyme. The first consists of one type 1 center and 3-4 type 2 centers per monomer. This situation would mean that the one blue copper center would have an extinction coefficient of 4500 - 5000 M^{-1} cm^{-1} which is somewhat high but not unprecedented.[7] Establishing the exact stoichiometry of metal ion sites involved in copper-containing enzymes is typically a difficult process as exemplified by the cases of ascorbate oxidase[8] and dopamine β-hydroxylase[9] and a final determination for phenoxazinone synthase may not be realized until the crystal structure is solved.

A direct reduction of enzyme by substrate is assumed for ascorbate oxidase and for other blue oxidases.[10] **Scheme II** shows the oxidative cascade proposed by Begley.[1] Spin counting data from the anaerobic EPR experiment shown in Figure 3 indicates that four electrons from the ring system are transferred to the copper centers in phenoxazinone synthase under anaerobic conditions. Begley's group[1] demonstrated that the predicted product of the 4-electron oxidation (structure **5**) is oxidized readily in air to the final product 2-aminophenoxazinone (structure **6**). Thus, the enzyme most likely does not catalyze the final two electron transfer in **Scheme II**.

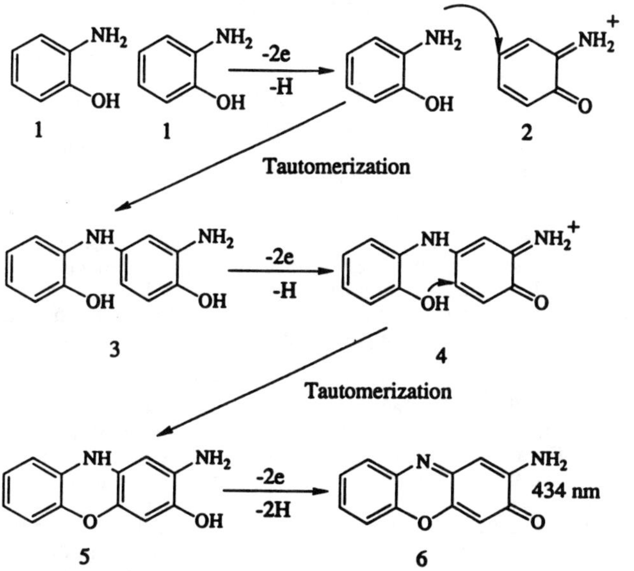

Scheme II: The reaction pathway for phenoxazinone formation by an oxidative cascade. The step showing formation of compound **6** (the six electron oxidation final product) from compound **5** (the four electron oxidation product) takes place readily in the presence of atmospheric oxygen and does not require enzyme.

Blue copper oxidases constitute a small group of enzymes responsible for substrate specific oxidation of quinols, phenols and amino phenols. These enzymes contain all three types of copper centers designated type 1 or blue, EPR detectable $\varepsilon_{600} = 10^3$, type 2, EPR detectable $\varepsilon_{600} = 10^1$, and type 3, EPR non-detectable, depending on their electron paramagnetic and visible spectral properties.[11] Additionally nitrous oxide reductase contains copper centers similar to type 1 (blue) and type 3.[12] Recent studies of sequences and crystal structures show significant homologies between small blue copper proteins (azurins, plastocyanins) and the blue copper oxidases and nitrous oxide reductase.[13-15] All of the blue oxidases will catalyze the oxidation of amino phenols to polymers and contain 4 to 8 coppers per subunit.[16,17] The copper content of phenoxazinone synthase makes it highly likely that an evolutionary relationship exists between the blue oxidases and phenoxazinone synthase.

The lack of type 3 copper centers in phenoxazinone synthase is unusual because the site of O_2 reduction is normally attributed to that type of center in the blue oxidases.[6,18] The requirement of type 3 copper centers for the reduction of O_2 is not strict as there are a number of copper containing proteins that use a type 2 site for the reduction of O_2 to water. Most notably, dopamine ß-hydroxylase has recently been shown to oxidize quinols to quinones while reducing O_2 to water.[19]

GALACTOSE OXIDASE

Galactose oxidase is an extracellular enzyme excreted by the fungus *Cladobotryum dendroides*,[20] It is a mononuclear Cu(II)-containing enzyme which catalyzes the oxidation of the C-6 alcohol in galactose to an aldehyde, removing the pro-S hydrogen, with the reduction of oxygen to hydrogen peroxide.[21-23] The enzyme will oxidize a variety of other primary alcohols[24] although it has high specificity for galactose. Galactose oxidase is composed of a single polypeptide chain of 68 kda and contains a tyrosyl radical which is a modified amino acid cofactor constructed by the cross linking of a cysteine and tyrosine residue in the copper binding site.[25-27]

The kinetic mechanism of galactose oxidase has been studied by us. Previous initial velocity data gave inconclusive results.[28,29] Due to the irreversibility of the chemical step and the fact that oxygen, one of the substrates, is never saturating, the application of other classical methods to determine the kinetic mechanism, such as product inhibition studies or equilibrium exchange, has been limited. At least 2 or 3 minimal mechanisms can be considered based on the initial velocity studies. Kinetic isotope effect studies were used in the current studies since they are ideally suited for distinguishing among them.[30-33]

Primary deuterium isotope effects are reported here at varying oxygen and galactose concentrations in order to address the details of the kinetic mechanism. The isotope effect studies were conducted using 1-*O*-methyl α-**D**-galactopyranoside and 1-*O*-methyl 6,6'-[^{2}H$_2$] α-**D**-galactopyranoside as substrates to prevent the mutarotation of the sugars from interfering in the kinetic analysis. The data were analyzed by fitting to standard hyperbolic curves with the exception of the measurements of V_{max}/K_{O_2}. The K_m for oxygen has been estimated to be greater than 3 mM which is beyond the range of saturation in this buffer and temperature. The slope of the tangent to the linear initial portion of the rate vs oxygen concentration plot is equal to V_{max}/K_m.[30] The values for V_{max}/K_{O_2} were obtained from the low oxygen data points. This analysis will distinguish between steady state kinetic mechanisms and a rapid equilibrium mechanism.

The reaction catalyzed by galactose oxidase is not reversible so a minimal mechanism can be written:

V_{max}/K_m measurements include all rate constants between the addition of the labeled substrate up to the first irreversible step. The isotope effect on V_{max}/K_{gal} and V_{max}/K_{O_2} will depend on the kinetic mechanism. Predictions can be made for the variation in the V_{max}/K_m isotope effects ($^D(V/K)$) as the other substrate is taken to zero and infinity. According to the predictions from equations derived by Klinman et. al.,[30] Cook and Cleland,[31] and Northrup,[32] if the mechanism is steady state ordered, the $^D(V/K)$ for the first substrate approaches one as the second substrate is taken to infinity, while the $^D(V/K)$ for the second substrate is invariant to the concentration of the first. In a rapid equilibrium ordered mechanism, the value of $^D(V/K)$ for either substrate will be equal and independent of the concentration of the other substrate. For a ping pong mechanism, the $^D(V/K)$ for the first substrate will be independent of the concentration of the second substrate and there will be no $^D(V/K)$ seen for the second substrate.

Figure 4 shows the isotope effect on V_{max}/K_m as a function of the oxygen concentration and 1-*O*-methyl α-**D**-galactopyranoside concentration. The isotope effects on V_{max}/K_{gal} and V_{max}/K_{O_2} are equal, within error, and have an average value of 8.7 ± 0.4. They are independent of the varied substrate concentration. Because the substrate used was dideuterated, there will be a small secondary isotope effect contribution on the overall isotope effect. This should only be, however, less than 10% of the total effect considering the magnitude of the isotope effect seen. Dividing through by the secondary isotope effect gives a kinetic isotope effect of about 7. This is close to an anticipated full isotope effect and suggests that the C-H bond cleavage fully limits catalysis. The magnitude of the kinetic isotope effect further suggests that the transition state of the reaction is located half way between the reactants and products.[33]

Galactose oxidase appears to have a rapid equilibrium kinetic mechanism where all of the steps, except the catalytic step, would be in equilibrium. For this nonsteady-state situation, all binding and dissociation steps are very rapid compared to the catalytic step, and the concentration of any of the intermediate enzyme forms never becomes constant. Also, previous data reported by others do not show a distinctive initial velocity pattern that intersects on the ordinate when the second substrate is varied which would be consistent with a rapid equilibrium ordered mechanism. Therefore from the data given above, the kinetic mechanism is best described as rapid equilibrium random.

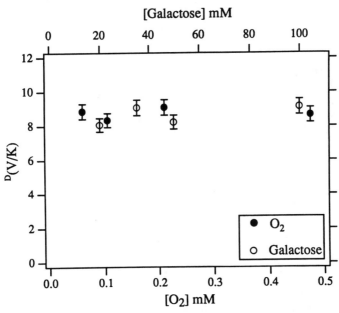

Figure 4. Dependence of the deuterium isotope effect for the galactose oxidase-catalyzed oxidation of 1-O-methyl α-**D** galactopyranoside on the concentration of oxygen and 1-O-methyl α-**D**-galactopyranoside.

The results are consistent with a chemical mechanism in which the C-6 pro-S hydrogen of galactose is abstracted as a hydrogen atom by the tyrosyl radical in the active form of the enzyme. A subsequent step involving electron transfer from the carbon-centered radical to the copper ion would complete the nominal two electron oxidation of C-6 to an aldehyde. The final step in the catalytic mechanism is the reduction of oxygen to produce hydrogen peroxide.

Bioinorganic Chemistry of Copper

REFERENCES

1.   C.E. Barry, P.G. Nayar, and T.P. Begley, *Biochemistry* **28**, 6323 (1989).
2.   G.H. Jones, and D.A. Hopwood, *J. Biol. Chem.* **259**, 14151 (1984).
3.   E. Katz, and H. Weissbach, *J. Biol. Chem.* **237**, 882 (1962).
4.   P. Aisen, and A.G. Morell, *J. Biol. Chem.* **240**, 1974 (1965).
5.   I. Savini, L. Morpurgo, and L. Avigliano, *Biochem. Biophys. Res. Commun.* **131**, 1251 (1985).
6.   L. Avigliano, G. Rotilio, S. Urbanelli, B. Mondovi, and A. Finazzi-Agro, *Arch. Biochem. Biophys.* **185**, 419 (1978).
7.   J.H. Freedman, and J. Peisach, *Anal. Biochem.* **141**, 301 (1984).
8.   M.H. Lee, and C.R. Dawson, *Arch. Biochem. Biophys.* **191**, 119 (1978).
9.   D.E. Ash, N.J. Papadopoulos, G. Colombo, and J.J. Villafranca, *J. Biol. Chem.* **295**, 3395 (1984).
10.   T.E. Meyer, A. Marchesini, M.A. Cusanovich, and G. Tollen, *Biochemistry* **30**, 4619 (1991).
11.   B. Reinhammar, and B.G. Malmstrom, in *Copper Proteins* , Spiro, T., Ed., (Wiley and Sons, New York,1982), pp 107-149.
12.   D.M. Dooley, R.S. Moog, and W.G. Zumft, *J. Am. Chem. Soc.* **109**, 67300 (1987).
13.   F.F. Fenderson, S. Kumat, E.T. Adman, M.Y. Liu, W.J. Payne and J. LeGall , *Biochemistry* **30**, 7180 (1991).
14.   A. Messerschmidt, and R. Huber, *Eur. J. Biochem.* **187**, 341 (1990).
15.   I. Ohkawa, N. Okada, A. Shinmyo, and T. Mitsuo,*Proc. Nat. Acad. Sci. U.S.A.* **86**, 1239 (1989).
16.   A. Marchesini, and P.M.H. Kroneck, *Eur. J. Biochem.* **101**, 65 (1979).
17.   J. Deinum, and T. Vanngard, *Biochim. Biophys. Acta.* **310**, 321 (1973).
18.   T. Sakurai, S. Sawada,S. Suzuki, and A. Nakahara, *Biochem. Biophys. Res. Commun.* **135**, 644 (1986).
19.   S.C. Kim, and J.P. Klinman, *Biochemistry* **30**, 8138 (1991).
20.   M.J. McPherson and P.F. Knowles, personal communication.
21.   J.A.D. Cooper, W. Smith, M. Bacila, and H. Medina, *J. Biol. Chem..*, **234**, 445 (1959).
22.   D. Amaral, L. Bernstein, D. Morse, and B.L. Horecker, *J. Biol. Chem.*, **238**, 2281 (1963).
23.   A. Maradufu, G.M. Cree, and A.S. Perlin, *Can. J. Chem.* , **49**, 3429 (1971).
24.   D.J. Kosman, M.J. Ettinger, R.E. Weiner, and E.J. Massaro, *Arch. Biochem. Biophys.*, **165**, 456 (1974).
25.   N. Ito, S.E.V. Phillips, C. Stevens, Z.B. Ogel, M.J. McPherson, J.N. Keen, K.D.S. Yadav, P.F. Knowles, *Nature* , **350**, 87 (1991).
26.   M.M. Whittaker and J.W. Whittaker, *J. Biol. Chem..*, **263**, 6074 (1988).
27.   M.M. Whittaker and J.W. Whittaker, *J. Biol. Chem.* , **265**, 9610 (1990).
28.   G.A. Hamilton, J. De Jersey, And P.K. Adolf, in *Oxidases and Related Redox Systems,* T.E. King, H.S. Mason, and M. Morrison, Eds., (University Park Press, Baltimore, Md. 1973), p. 103.
29.   L.D. Kwiatkowski, M. Adelman, R. Pennelly, and D.J. Kosman, *J. Inorg. Biochem.* , **14**, 209 (1981).
30.   J.P. Klinman, H. Humphries, and J.G. Voet, *J. Biol. Chem.* , **255**, 11648 (1980).
31.   P.F. Cook and W.W. Cleland, *Biochemistry*, **20**, 1790 (1981).
32.   D.B. Northrup, *Biochemistry*, **14**, 2644 (1975).
33.   D.B. Northrup, *Anal. Biochem.*, **132**, 457 (1983).

ACTIVE SITE LIGAND INTERACTIONS IN GALACTOSE OXIDASE

James W. Whittaker

Department of Chemistry, Carnegie Mellon University, 4400 Fifth Avenue, Pittsburgh, Pennsylvania 15213.

Redox catalysis by metalloenzymes displays a remarkable diversity in chemistry and in catalytic structures. The highly evolved catalytic active sites revealed by recent spectroscopic[1-4] and crystallographic[5-8] studies serve as an inspiration to synthetic chemists turning from design of molecular architecture to the engineering of functional molecules. Key catalytic principles on which this type of biomimetic complex can be built are presently emerging from active site studies on metalloenzymes, using approaches that probe the reaction mechanisms for the biological metal complexes. X-ray crystallography now routinely reveals structures of metalloenzyme active sites at near atomic resolution, establishing the geometric features with unparalleled accuracy. However, crystallographic data is still low resolution compared to the electronic level of detail that relates directly to the origins of chemical reactivity. Spectroscopy provides the essential link between the geometric factors revealed by crystallography and the electronic structures of metalloenzyme active sites. This role for spectroscopic studies, bridging between geometric and electronic structural descriptions, is particularly well represented by active site spectral studies on the fungal enzyme galactose oxidase.

Galactose oxidase catalyzes the oxidation of virtually any primary alcohol to the corresponding aldehyde, coupling that reaction to the reduction of O_2 to hydrogen peroxide[9]:

$$RCH_2OH \ + \ O_2 \ \longrightarrow \ RCHO \ + \ H_2O_2$$

The complete regioselectivity and the broad substrate specificity[9] have made this enzyme interesting for potential applications in organic synthesis[10] and as a model for synthetic alcohol oxidation catalysts. Early characterization of the protein identifying a single Cu ion and no additional cofactors[9] did little to explain the redox chemistry of galactose oxidase. The metal stoichiometry is anomalous in terms of a general principle of biological redox catalysis, that the number of metal ions equal the number of electrons involved in the chemistry. This principle is expressed, for example, in the structure of blue Cu proteins, in which a single Cu ion is associated with one-electron redox reactivity, and the oxygen carrier protein hemocyanin, which reversibly binds dioxygen as peroxide, a reversible two-electron reaction occurring at a binuclear copper active site. For galactose oxidase, both alcohol oxidation and O_2 reduction steps are two-electron redox processes, while single electron reactivity would be expected for the isolated cupric ion. This perplexing problem has been resolved recently by the identification of a free radical coupled complex in the

From K.D. Karlin and Z. Tyeklár, Eds., *Bioinorganic Chemistry of Copper*
(Chapman & Hall, New York, 1993).

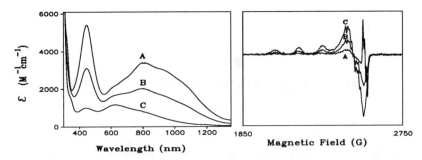

Figure 1. Optical absorption (Left) and EPR (Right) spectra for galactose oxidase species. (A) Redox-activated, (B) Native, (C) Reductively inactivated enzyme.

active site of galactose oxidase, composing a two-electron redox unit for alcohol oxidation chemistry[1]. The evidence for this new and interesting catalytic structure correlates a wide range of spectroscopic and chemical data defining a radical new principle for redox catalysis.

REDOX TRANSFORMATIONS OF GALACTOSE OXIDASE

The preparation of homogeneous redox modifications of galactose oxidase is essential for detailed spectroscopic studies. Early studies on the Native enzyme freshly isolated from *Dactylium dendroides* culture filtrates described complicated and variable results in basic quantitative characterization[11]. Spin quantitation of the EPR signal for the cupric center in the protein, optical absorption intensities, and specific activity all appeared to vary widely in an uncontrollable fashion from one preparation to the next[11]. The correlation of activity and optical spectra with the intensity of the cupric EPR signal provided an early clue suggesting a higher oxidation state copper complex might be involved in catalysis[12,13]. More recently the redox interconversions of galactose oxidase have been unravelled through a combination of systematic biochemical preparations and careful analytical characterization[1]. Brief treatment of galactose oxidase with a mild oxidant, ferricyanide, converts the enzyme to an intensely absorbing green form, exhibiting absorption bands across the entire visible region of the spectrum, where the molar extinction coefficient never falls below 3000 $M^{-1}cm^{-1}$, and the absorption extends into the near IR (Figure 1, Left, A). This spectrum is quite unlike the typical spectrum of a Cu^{2+} complex. Since this form exhibits the maximum specific activity, it has been identified as the Active form of galactose oxidase. Reduction of galactose oxidase under mild conditions with ferrocyanide converts the enzyme to a blue form characterized by relatively weak absorption across the visible region (Figure 1, Left, C). This form is catalytically devoid of activity, and so is described as the Inactive form of the enzyme. The inactivation is reversible and Active and Inactive enzyme may be readily interconverted by the appropriate redox agent[1].

These three species (Native, Active and Inactive enzyme) have been examined by quantitative EPR spectroscopy[1] to correlate with enzyme forms that have been described in the past (Figure 1, Right, A-C). The low temperature EPR spectrum of the Native enzyme is identical to that previously reported, with ground state parameters reflecting a predominantly $d_{x^2-y^2}$ ground state. Spin quantitation of this signal, obtained by double integration of the derivative EPR spectrum and comparison with a $Cu(ClO_4)_2$ spin standard, is variable, typically ranging from as low as 10 per cent to higher than 90 per cent of the value expected from the independent copper quantitation obtained by atomic absorption metal analyses. Oxidation virtually eliminates this EPR signal, with a minority fraction of about 10 percent of the copper contributing an impurity signal that is not eliminated by treatment with oxidants. Reduction paradoxically *increases* the intensity of the cupric EPR signal observed for the enzyme, with the full complement of enzyme-bound copper expressed in the quantitation of the EPR signal of the Inactive enzyme. The EPR signal intensity variation is exactly the opposite of what would be expected for a normal copper coordination complex which is paramagnetic in the oxidized cupric Cu^{2+} form and on reduction is converted to an EPR silent diamagnetic cuprous (Cu^+) complex.

Comparison of optical absorption and EPR spectra for the three complexes is not enough to resolve this puzzle, since there are no isosbestic points between the spectra that can be used to characterize the conversions. However, in CD spectra, at least five isodichroic points occur over the near UV-visibile spectral range. A redox interconversion of Active and Inactive enzyme may be followed through equilibration of the protein with ferri/ferrocyanide redox buffers with the potential of the solutions being poised by the ratio of redox components. This generates a series of CD spectra in which isodichroic points are cleanly maintained, indicating that no intermediate enzyme form occurs in the process. Further, the green and blue Active and Inactive enzyme forms isolated from the redox agents show exactly the same spectra and are intersected at their isodichroic points by the CD spectrum of the Native enzyme. This conclusively demonstrates that the Native form is simply a heterogeneous mixture of Active and Inactive enzyme. Thus, all the variability of the Native enzyme can be accounted for as the result of partial inactivation occurring during purification[1].

SPECTROSCOPIC ANALYSIS OF THE ACTIVE SITE COMPLEX

The availability of stable, homogenous and well defined redox modifications provided by controlled oxidation and reduction of galactose oxidase permit a spectroscopic characterization of each of these forms. The optical spectra of the blue, Inactive enzyme, which is associated with an isolated cupric center in the protein, may be assigned in terms of d→d and charge transfer transitions of a cupric coordination complex. The unusually strong CD observed within the broad, low energy absorption band (near 620 nm) is characteristic of a ligand field transition, where intrinsic magnetic dipole character contributes to a relatively large anisotropy $(\Delta\varepsilon/\varepsilon)$ between circular polarization and optical absorption intensities. The average energy of the d→d spectra is typical of a pentacoordinate pyramidal Cu^{2+} complex in model studies, and the strong CD can be assigned within an effective C_{4v} site symmetry as arising from the $d_{x^2-y^2} \rightarrow d_{xy}$ electronic transition. Spectral features near 330 nm may be assigned as arising from imidazole-to-copper[14] ligand-to-metal charge transfer (LMCT) from histidine ligands identified in EPR[11] and crystallographic[8] studies. The absorption band near 450 nm in the optical spectrum is also likely of LMCT origin, but is too low in energy to arise from imidazole coordination and is in the range for pyramidal cupric complexes of phenolates and thiolates. The intensity of this absorption band is, however, unusually low for a LMCT transition, suggesting that special geometric considerations may apply (see below). Spectroscopy thus provides an important electronic structural probe of metal interactions in the inactivated enzyme complex.

The optical spectrum of the green, active enzyme is clearly of very different nature than that of the blue inactive form. The active enzyme spectrum is associated with the unique, EPR silent oxidized copper complex of this enzyme form and is distinct from spectra for any cupric coordination complex previously reported. The intensities of the absorption bands are dramatically higher than those observed for d→d spectra of transition ions even in low symmetry complexes, indicating a different assignment is required for these dominating transitions for the redox activated enzyme. Earlier studies which provided the initial characterization the activation process led to an early proposal of a trivalent copper higher oxidation state complex in the redox activated protein[12,13]. This is an entirely plausible intrepretation of the experimental results available at the time, but must be reexamined in the light of additional spectroscopic data. For example, CD spectra for the active enzyme exhibit a band in the ligand field region of the spectrum near 560 nm which is associated with relatively low absorption intensity. In fact, the CD band occurs at a minimum in the absorption spectrum, indicating a large $\Delta\varepsilon/\varepsilon$ anisotropy for this spectral feature. The strong CD band near 620 nm is lost on redox activation, and the possibility that the 560 nm CD band arises from d→d transition within a cupric complex in the active

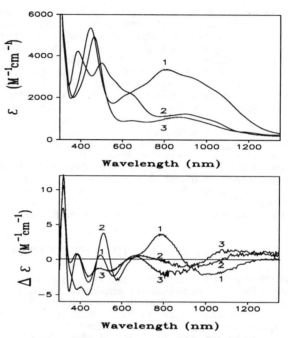

Figure 2. Optical absorption (Top) and circular dichroism (Bottom) spectra for ligand complexes of galactose oxidase. Active enzyme (1) complexes with sodium azide (2) and sodium cyanate (3).

enzyme provided the first spectroscopic clue to the assignment of the metal ion oxidation state in this form of the protein. However, the oxidation state of the metal ion in the activated enzyme is a key aspect in the chemistry of this active site requiring a more detailed analysis.

Copper Oxidation State Assignment

Exogenous ligand probes have been used to define the metal ion oxidation state. Azide binds tightly to both inactive and active enzyme forms, converting them to complexes that exhibit an intense near UV absorption band near 390 nm. This band allows the progress of a titration to be conveniently followed through the increase in absorption intensity. Azide binds to the active enzyme nearly an order of magnitude tighter than to the inactive form, suggesting that some active site structural change is associated with redox activation to account for this distinct chemistry. CD spectra of the azide complexes of Active and Inactive enzyme (Figures 2 and 3, Spectrum 2) appear quite similar in the Cu^{2+} ligand field region, both complexes exhibiting a single strong CD band in the range 550 - 600 nm. The metal ion oxidation state is reflected in the energy of the near UV transition, which has been assigned in a range of small molecule Cu-azide complexes and biological Cu metalloprotein azide adducts as arising from π LMCT for an equatorially coordinated anion. The LMCT transition energy is determined by the relative energies of ligand valence and metal d levels in the complex. Since metal d-orbital energies are stabilized in higher oxidation states, the LMCT transition energy is sensitive to metal ion oxidation state. The observation that the azide-to-copper LMCT occurs at virtually the same energy in the two complexes indicates that the metal ion oxidation state is the same.

The metal ion oxidation state in the active enzyme has been conclusively established by X-ray absorption studies of galactose oxidase complexes[15]. The X-ray absorption K-

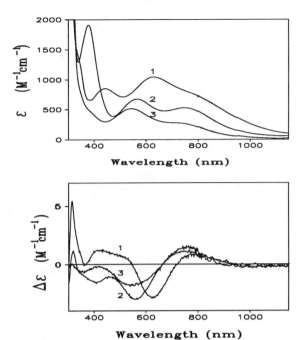

Figure 3. Optical absorption (Top) and circular dichroism (Bottom) spectra for ligand complexes of galactose oxidase. Inactive enzyme (1) complexes with sodium azide (2) and sodium cyanate (3).

edge transition, which occurs near 8900 eV for Cu, is sensitive to metal ion oxidation state, shifting approximately 10 eV to higher energy on successive oxidations of the metal ion. Comparing the X-ray absorption near edge structure (XANES) for galactose oxidase in its various redox modifications with inorganic models provides the basis for assigning the metal ion oxidation state in each complex. The XANES for Inactive galactose oxidase is typical of a simple cupric coordination complex in terms of transition energy and near edge structure. The Cu XANES for the redox activated enzyme is nearly superimposable with that of the Inactive form, again indicating that the oxidation state of the metal ion is unchanged on forming the Active enzyme complex. This leaves a puzzle in that the chemistry of redox activation involves removing one electron per active site of the Inactive enzyme, yet spectroscopically the metal ion appears not to be the site of oxidation. Clearly, some other group in the protein must be redox active. The oxidation of any group other than the metal ion in a single electron step will form what is correctly described as a free radical, establishing the basis for identifying a free radical and a cupric ion in the active site of galactose oxidase. The intense absorption features of the active enzyme can thus be identified as reflecting the presence of a free radical in the protein structure.

EPR spectra for the active enzyme (Figure 1, Right, A) show no indication of the presence of either a free radical or a cupric ion in this complex. A free radical EPR signal that occurs in active enzyme samples quantitates to less than 1 per cent of the active sites and so is clearly a minority species in these samples (however, see below). The absence of EPR signals from the cupric and free radical sites implies that interactions between these two centers must result in their mutual EPR inaccessibility. Since even weak dipolar coupling could broaden the spectra making them difficult to observe, EPR spectroscopy is not particularly sensitive to the strengths of these interactions. Magnetic susceptibility is sensitive to a much wider range of ground state splittings. Low temperature SQUID susceptibility data for galactose oxidase requires a diamagnetic ground state for the activated enzyme arising from strong antiferromagnetic exchange coupling of the cupric and free radical spins[15]. Higher temperature data appear to reflect thermal population of a low lying paramagnetic excited state at the higher temperature range, permitting an estimate of the strength of the antiferromagnetic coupling. The temperature dependence of the magnetic susceptibility leads to an estimate of 200 cm^{-1} as a lower limit for the magnitude of the singlet-triplet splitting in the complex. This implies direct coordination of the free radical to the cupric ion and leads to a description of the active enzyme as containing a *free radical coupled copper complex.*

This radical complex has been shown by stoichiometric substrate reactions to serve as a two-electron redox unit in galactose oxidase. An anaerobic sample of the green, Active enzyme titrated with slightly more than one equivalent of β-OCH$_3$-galactoside substrate becomes colorless. Both the intense absorption features of the radical and the relatively weak absorption of the cupric center are eliminated in this reaction. Since each of these components is a one-electron redox unit, this is consistent with two-electron reduction of the active site complex by substrate in a step independent of the presence of O$_2$. This reduction is clearly in keeping with the conversion of alcohol to aldehyde in this step, and implies a cuprous metal complex forms as an intermediate in the enzymatic mechanism[1]. X-ray absorption (XANES) studies on the substrate-reduced complex show the characteristic K-edge energy and lineshape of a cuprous coordination complex, confirming the Cu$^+$ oxidation state assignment[14]. This preformed cuprous complex is competent for O$_2$ reduction, and exposing the complex to air restores the intense absorption features of the Active enzyme complex.

Identification of the Radical Site in Galactose Oxidase

What is the nature of the unusual free radical-forming site in the protein? Intense optical absorption features are associated with the presence of the active site radical, suggesting that resonance Raman spectroscopy will be effective as a probe of its vibrational structure[17]. Excitation within the red absorption band for the active enzyme, at 649 nm, leads to strong resonance enhancement of a pattern of vibrational features in the Raman spectrum[2] at 1140, 1250, 1485, 1595 cm^{-1}. These vibrational features are the signature of a ligated tyrosine residue and have been characterized in a range of metal-phenolate models and biological metal complexes[18], indicating that a tyrosine residue is a component of the radical site. However, the spectra for galactose oxidase are unusual comparison with these other spectra in that the 1485 and 1595 cm^{-1} Raman bands, assigned as ring breathing modes for the coordinated phenolate, are dramatically lower by 15 and 5 cm^{-1} respectively than previously reported for any tyrosine complex. This indicates that the ligated tyrosine residue is significantly perturbed, lowering the frequencies of the ring breathing modes. The resonance Raman experiments identify tyrosine interactions in the active site, a perturbed tyrosine, with a second relatively unperturbed tyrosine ligand contributing to the intense radical absorption spectra of the Active enzyme.

The radical site that is defined by the spectroscopic studies described so far is masked in each case by interactions with the metal ion. Unmasking the radical by removing the copper and oxidizing the metal free apoenzyme permits the identification and characterization of the redox active

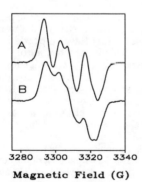

3280 3300 3320 3340

Magnetic Field (G)

Figure 4. EPR spectra for the free radical site in oxidized apo galactose oxidase containing β-[^{1}H,^{1}H]-tyrosine (A) or β-[^{2}H,^{2}H]-tyrosine (B).

protein residue in the active site[3]. The free radical has been revealed for spectroscopic characterization by oxidation of the metal free apoenzyme, which generates a sharp, free radical EPR signal near the free electron g-value, $g_e = 2.0023$. This radical is formed under the same mild oxidation conditions that generate the radical in the copper-containing protein, although the presence of the radical does not confer catalytic activity to the apoenzyme, since both radical and metal ion are required to form a catalytically active free radical coupled copper complex in the enzyme active site. These experiments clearly demonstrate that galactose oxidase is a free radical metalloenzyme, and indicate that the protein itself can participate in redox catalysis. The unmasking of the radical in galactose oxidase provides the direct evidence for free radical interactions which we had previously shown by indirect approaches.

Isotopic labelling experiments have allowed the identity of the radical to be established[3]. The average g-value of this radical EPR signal being characteristic of a phenolic radical[19] motivated the labelling of galactose oxidase with β-methylene deuterated tyrosine. The EPR signal of the radical in the oxidized apoenzyme of this selectively deuterated protein is found to be significantly perturbed (Figure 4), demonstrating that this form of galactose oxidase contains a tyrosine-derived radical. However, the radical is clearly distinct from tyrosine phenoxyl radicals in ribonucleotide reductase and phenolic model systems[19], consistent with assignment to a perturbed tyrosine residue. Careful simulation of the hyperfine structure of this EPR spectrum suggests this perturbation involves a modification at a position on the aromatic ring ortho to the phenolic hydroxyl group, eliminating one of the α-ring proton hyperfine interactions[4].

Comparison of Spectroscopic and Crystallographic Structural Information

A crystal structure has recently been reported for galactose oxidase at 1.7 Å resolution[8]. The oxidation state of the crystalline enzyme has not yet been conclusively established, but since the crystals are stabilized by hydroxylic compounds that normally quench the radical in the active enzyme it appears likely that the structure is of the Inactive form of galactose oxidase[4]. The active site structure revealed from the X-ray crystallographic studies shows an unusual protein modification, a cysteine-tyrosine covalently crosslinked at an aromatic ring carbon *ortho* to the phenolic hydroxyl group. (Figure 5) This unusual amino acid side chain forms one of the ligands to the metal center and appears to be a likely candidate for the radical forming site in the enzyme. Several lines of evidence support this identification. First, the thioether substitution would be expected to lower the redox potential of the phenol, stabilizing a free radical. Second, the EPR[3] and resonance Raman[2] spectral data on the radical in galactose oxidase is consistent with a perturbation of the tyrosine ring system, with elimination of an α-ring proton and a lowering of ring breathing mode vibrational frequencies. The covalent *ortho* substitution of the tyrosine side chain by a heavy atom would be expected to have exactly these effects. Thus, the complementary information on the existence and characteristics of a free radical in the active site of galactose oxidase from spectroscopic and chemical data, and the recognition

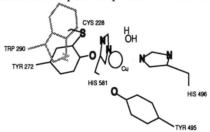

Figure 5. Crystallographic structure of the active site metal complex of galactose oxidase (after Ref. 8).

of a covalent crosslink in the active site from crystallography, combine to provide insight into the structure of the redox unit. The crystal structure also reveals additional features, not identified by spectroscopy. For example, the modified tyrosine residue that likely forms the radical site in galactose oxidase is overlain by a tryptophan indole ring that may participate in stabilizing the active site radical. A second tyrosine is coordinated axially in the approximately square pyramidal copper complex with a relatively long metal-oxygen bond distance, indicating relatively weak phenolate-metal interactions. This suggests an assignment for the low intensity optical transition near 440 nm as arising from LMCT for this axially coordinated tyrosine phenolate. A water ligand occupying a fifth ligation site in the neutral pH complex is replaced by acetate anion at low pH in ammonium acetate buffer, with the anion bound closer to the metal than the water it displaces.

EXOGENOUS LIGAND INTERACTIONS

Ligand perturbations allow the metal-free radical and metal-exogenous ligand interactions to be explored spectroscopically. A systematic investigation of ligand interactions in the active site of galactose oxidase reveals important new features of the catalytic complex. Optical absorption and CD spectra for both Active and Inactive forms of galactose oxidase recorded in the presence of N_3^- and OCN^- are shown in Figures 2 and 3. The trend that emerges in spectra for the Inactive enzyme anion complexes is that anion binding results in a decrease in intensity and a shift of the ligand field bands to higher energy. These data are consistent with conversion of the pyramidal complex of the unliganded site to a more tetragonal geometry on anion binding, implying displacement of a coordinated endogenous ligand. The ligand displaced in this reaction is most likely the weakly coordinated axial tyrosine phenolate and the observation that the 440 nm absorption band is eliminated on forming the anion complexes is consistent with this idea. The Active enzyme complexes also reflect a change in active site structure through a perturbation of the absorption spectra, but the results are more difficult to interpret in this case. The dominating feature near 450 nm is retained in the anion complexes, but the lower energy band is dramatically reduced in intensity. On binding azide a new absorption feature, the azide-to-Cu^{2+} LMCT band, grows in near 390 nm as the active site is titrated, indicating direct coordination of the anion to the cupric center. Metal-exogenous ligand interactions clearly can affect the radical site interactions in galactose oxidase, perhaps even modulating the metal-radical coupling and related redox characteristics.

Galactose oxidase is affected by temperature as well as exogenous anions. Enzyme frozen at cryogenic temperatures is thermochromic, reversibly changing color on warming. Inactive enzyme, which is blue in solution at room temperature, becomes pink at 77 K.

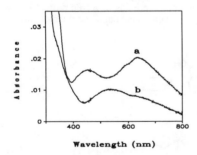
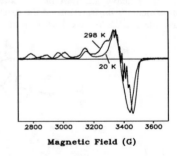

Figure 6. Optical absorption (Left) and EPR (Right) spectra for Inactive galactose oxidase in solution (a) and frozen at low temperature (b).

Active enzyme changes from green at room temperature to yellow at liquid N_2 temperature. The thermochromism of the Inactive enzyme can be observed in glassing solvents (50 per cent glycerol) as shown in Figure 6 (Left). The low temperature (200 K) spectrum closely resembles the anion bound spectrum observed in ligand titrations. Temperature-dependent changes in the EPR spectrum of the Inactive enzyme form can also be observed, as shown in Figure 6 (Right). The observed changes are consistent with the increased tetragonal character in the low temperature form. The room temperature spectrum clearly does not correspond to the spectrum of a tetrahedral cupric complex, ruling out that description which has recently been suggested[8] on the basis of the crystallographic data. The conversion to a yellow form of the Active enzyme also corresponds to a spectral change similar to that observed for the anion complexes. The most likely explanation for these observations is that the water that is coordinated in the active site at room temperature is deprotonated at low temperature to form a hydroxide complex which will be expected to exhibit spectral features similar to those observed for CN^- and other anions. The loss of the 440 nm band in these anion adducts indicates that the axially coordinated tyrosine is likely displaced in these complexes. This displacement, and the deprotonation of the coordinated

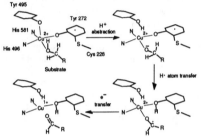

water, are expected to relate to mechanistically important proton transfer steps in the active site chemistry. These steps are outlined in Figure 7.

The binding of alcohol substrates to the enzyme is expected to follow the coordination chemistry established for anion binding, with the hydroxyl group directly interacting with the cupric ion. This proposed substrate complex has an analog in the crystallographically defined aquo complex of the inactive enzyme. Coordination of a hydroxyl group will increase its acidity, permitting an ionization with the proton

Figure 7. Substrate oxidation scheme.

transferring to an endogenous base in the active site. The loss of the optical absorption feature that we have tentatively assigned as the axial tyrosine (Tyr 495) to Cu^{2+} LMCT band on anion binding to the Inactive enzyme suggests that this relatively weakly coordinated phenolate may play a role as a proton acceptor in the catalytic mechanism. Deprotonation of the coordinated hydroxyl group would serve to activate the alcohol towards hydrogen atom transfer. The modified tyrosine residue on which the active site radical is localized is identified as the acceptor for β-methylene hydrogen atom abstraction from the alcohol in this picture, recognizing that significant unpaired electron density resides on the phenoxyl oxygen of the radical. The product of this hydrogen atom transfer, a ketyl radical, will reduce the cupric center and thus complete the substrate oxidation half-cycle of the turnover reaction. The anion interactions revealed in the interpretation of spectra for the anion complexes supports the essential features of this alcohol activation mechanism.

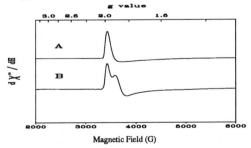

Figure 8. EPR spectra for bovine serum albumin (A) and galactose oxidase (B) nitric oxide complexes.

The oxygen reaction that reoxidizes the active site complex is less well understood. We have been interested in probing interactions between galactose oxidase and small molecules that can provide insight into these key aspects of the catalytic mechanism relating to dioxygen reactivity of the active site. Nitric oxide (NO) is a diatomic molecule which serves as an O_2 analog in probing the dioxygen binding sites of metalloenzymes, substituting for dioxygen in adducts with the heme iron in hemoglobin[20], for example. Galactose oxidase exposed to NO forms a complex exhibiting an unusual EPR spectrum (Figure 8). The EPR signal of this complex has a strikingly non-Curie temperature dependence, virtually disappearing by 50 K. In fact, an earlier study involving nitrogen temperature EPR experiments did not detect this NO complex and concluded that galactose oxidase does not form an NO adduct[21]. The EPR spectrum of the enzyme-NO complex is distinct from the spectra observed for NO frozen in ice or protein-containing buffers, and appears to be a specific modification. The signal of the specifically bound NO can be isolated from the other spectral components by selective saturation of the other signals at high power and low temperature. The treatment of the overlapping spectra obtained for the Inactive galactose oxidase NO complex is illustrated in Figure 9. Quantitation of individual spectral components can be achieved by deconvolution of the complex EPR spectrum of an enzyme sample exposed to NO. Successive subtraction of the spectra of nonspecifically bound NO (Figure 9a), active site cupric ion (Figure 9b), and the specifically bound NO adduct (Figure 9c) leaves insignificant residuals (Figure 9d), supporting this analysis. The amount of cupric EPR signal which must be subtracted in Figure 9b is virtually the same as for untreated Inactive galactose oxidase, indicating that nearly all the metal ion remains EPR active in this complex and is not reduced or otherwise masked on reaction with NO. Nonspecifically bound NO contributes a variable amount to these spectra. The signal remaining after removing these two components, Figure 9c, is the spectrum of the specifically bound NO galactose oxidase complex. Spin quantitation of the signal from the specifically bound NO complex consistenly yields approximately one S=1/2 per molecule

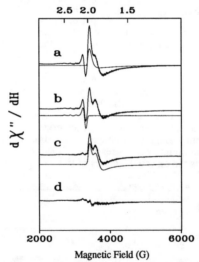

Figure 9. EPR spectral decomposition for a sample of galactose oxidase exposed to nitric oxide. (a-d) Successive steps in subtraction of spectral components. At each level the spectrum to be subtracted is displayed below the result of previous peeling step. (a) Starting spectrum, from which is subtracted in sequence (a) nonspecific NO spectrum; (b) active site cupric ion spectrum; (c) spectrum of specifically bound NO complex, isolated at high power; (d) records the small residuals following final subtraction.

of galactose oxidase, indicating a unique binding site for NO. Normally the metal ion would be considered the probable site for NO interactions, however the observation that the cupric EPR signal is unchanged in this complex indicates that this is not the case, since coupling between the NO radical and the cupric ion would eliminate the EPR signal for the metal center. Analysis of the EPR spectrum of the NO complex contributes insight into the nature of the NO complex. The EPR spectrum of the galactose oxidase NO complex is unusual in having all g-values below 2.00, consistent with localization of unpaired electron density on the NO. The basis for this assignment is that g-shifts below 2.00 result when the spin orbit coupling constant (λ) has a value less than zero. This is the situation expected for spin orbit coupling within the less-than-half-filled antibonding valence shell for the nitric oxide molecule. The localization of spin on NO and the absence of hyperfine splittings from Cu further support a nonmetal NO interaction site. While this information clearly does not restrict the NO binding site to the enzyme active site, the complex appears to be unique and an active site interaction is an attractive possibility. NO thus may be probing a non-metal O_2 interaction site in the enzyme. Interestingly, a similar NO complex has been reported for laccase[22], another enzyme with dioxygen interactions. The existence of a distinct O_2 interaction site off of the metal ion could account for a surprising aspect of the enzyme turnover kinetics of galactose oxidase. A recent detailed study of steady state kinetics for galactose oxidase[23] shows that oxygen and hydroxylic substrate are noncompetitive in their interactions with the enzyme, inconsistent with their binding to a single site in the protein. If, as we have outlined above, the hydroxylic substrate coordinates to the cupric ion during turnover, the dioxygen must have a distinct site for interactions. The specific NO binding reaction might probe this non-metal dioxygen interaction site, and therefore provide information on key aspects of oxygen redox steps in the catalytic mechanism.

CONCLUSIONS

Ligand interactions with galactose oxidase have already yielded significatnt insight into many features of the chemistry of the active site, contributing to a deeper understanding of active site reactivity and the elementary steps of substrate activation. In the work described here, anion binding experiments probe the coordination chemistry of the metal complex, while interactions with the paramagnetic gas NO may probe dioxygen interactions in the enzyme. Spectroscopic analysis of these complexes can be expected to continue to yield information on the electronic structural origins of the catalytic activity of galactose oxidase.

ACKNOWLEDGEMENTS

The author would like to gratefully acknowledge funding from the National Science Foundation (DMB-8717058) and the National Institutes of Health (GM 46749) in support of this research.

REFERENCES

1.   M.M. Whittaker and J.W. Whittaker *J. Biol. Chem.* **263**, 6074-6080 (1988).

2.   M.M. Whittaker, V.L. DeVito, S.A. Asher, and J.W. Whittaker *J. Biol. Chem.* **264,** 7104-7106 (1989).

3.   M.M. Whittaker and J.W. Whittaker *J. Biol. Chem.* **263,** 6074-6080 (1988).

4.   G.T. Babcock, M.K. El-Deeb, P.O. Sandusky, M.M. Whittaker, and J.W. Whittaker *J. Am. Chem. Soc.* 1992. 114: 3727-3734.

5.   J.W. Godden, S. Turley, D.C. Teller, E.T. Adman, M.-Y. Liu, W.J. Payne, and J. LeGall *Science* **253**, 438-442 (1991).

6.   A. Volbeda and W.G.J. Hol *J. Mol. Biol.* **209,** 249-279.

7.   A. Messerschmidt, R. Ladenstein, R. Huber, M. Bolognesi, L. Avigliano, R. Petruzelli, A. Rossi and A. Finazzi-Agró *J. Mol. Biol.* **224,** 179 (1992)

8.   N. Ito, S.E.V. Phillips, C. Stevens, Z.B. Ogel, M.J. McPherson, J.N. Keen, K.D.S. Yadev, and P.F. Knowles *Nature* **350**, 87-90 (1991).

9.   (a) G. Avigad, C. Asensio, D. Amaral, and B.L. Horecker *Biochem. Biophys. Res. Commun.* **4**, 474-477 (1961). (b)G. Avigad, D. Amaral, C. Asensio, and B.L. Horecker *J. Biol. Chem.* **237**, 2736-2743 (1962). (c) D. Amaral, L. Bernstein, D. Morse, and B.L. Horecker *J. Biol. Chem.* **238**, 2281-2284 (1963).

10.   A.M. Klibanov, B.N. Alberti, and M.A. Marletta (1982) *Biochem. Biophys. Res. Commun.* **108**, 804-808.

11.   D.J. Kosman, in *Copper Proteins and Copper Enzymes, Volume II*, R. Lontie, Ed., (CRC Press, Boca Raton, 1985), pp. 1-26.

12.   G.A. Hamilton, P.K. Adolf, J. de Jersey, G.C. DuBois, G.R. Dyrkacz, and R.D. Libby *J. Am. Chem. Soc.* **100**, 1899-1912 (1978).

13.   G.R. Dyrkacz, R.D. Libby, and G.A. Hamilton *J. Am. Chem. Soc.* **98**, 626-628 (1976).

14.   J.L. Hughey, IV, T.G., Fawcett, S.M. Rudich, R.A. Lalancette, J.A. Potenza, and H.J. Schugar *J. Am. Chem. Soc.* **101**, 2617-2623 (1979).

15.   K. Clark, J. E. Penner-Hahn, M. M. Whittaker, and J. W. Whittaker *J. Am. Chem. Soc.* **112**, 6433-6434 (1990).

16.   M. Sendova, J. Peterson, E.P. Day, M.M. Whittaker, and J.W. Whittaker (unpublished results)

17.   T.G. Spiro in *Chemical and Biochemical Applications of Lasers Volume I* C.B. Moore, Ed., (Academic Press, New York,1974) pp. 29-70.

18.   L. Que, Jr., R.H. Heistand, II, R. Mayer, and A.L. Roe *Biochemistry* **19**, 2588-2593 (1980).

19.   (a) B.A. Barry, and G.T. Babcock *Proc. Nat. Acad. Sci. USA* **84**, 7099-7103 (1987). (b) C. Bender, M. Sahlin, G.T. Babcock, B.A. Barry, T.K. Chandrashekar, S.P. Salowe, J. Stubbe, B. Lindström, L. Petersson, A. Ehrenberg, and B.-M. Sjöberg *J. Am. Chem. Soc.* **111**, 8079-8083 (1989) .

20.   T. Yonetoni, H. Yamamoto, J.E. Erman, J.S. Leigh, Jr., and G.H. Reed *J. Biol. Chem.* **247**, 2447-2455 (1972).

21.   A.C.F. Gorren, E. de Boer, and R. Wever *Biochim. Biophys. Acta* **916**, 38-47 (1987).

22.   C.T. Martin, R.H. Morse, R.M. Kanne, H.B. Gray, B.G. Malmström, and S.I. Chan *Biochemistry* **20**, 5147-5155 (1981).

23.   J.J. Villafranca, J. Freeman, and A. Kotchevar *J. Inorg. Biochem.* **47**, 35 (1992).

STRUCTURE AND REACTIVITY OF COPPER-CONTAINING AMINE OXIDASES

David M. Dooley,[a] Doreen E. Brown,[a] Alexander W. Clague,[a], Jyllian N. Kemsley,[a] Cynthia D. McCahon,[a] Michele A. McGuirl,[a] Petra N. Turowski,[a] W.S. McIntire,[b] Jaqui A. Farrar,[c] and Andrew J. Thomson[c]

[a]Department of Chemistry, Amherst College, Amherst, MA 01002, USA, ˑ
[b]Molecular Biology Division, Veterans Administration Medical Center, San Fransisco, CA 94121, USA,
[c]School of Chemical Sciences, University of East Anglia, Norwich NR4 7TJ, UK

INTRODUCTION

In recent years, considerable progress has been made toward understanding the active site structures and mechanisms of copper-containing amine oxidases. Copper-containing amine oxidases are one of the most widely distributed classes of "Type-2" copper enzyme. Copper amine oxidases have been highly purified from bacteria, yeasts, plants, and mammals.[1] Recently an inducible phenethylamine oxidase was purified from the K-12 strain of *E. coli* and shown to contain copper, the first example of a copper-containing amine oxidase in gram-negative bacteria.[2] Amine oxidases (hereafter this phrase will refer exclusively to the copper-containing enzymes, unless otherwise indicated) catalyze the oxidative deamination of primary amines:

$$RCH_2NH_2 + O_2 + H_2O -----> RCHO + H_2O_2 + NH_3$$

Substrate specificities depend on the enzyme source but nearly all biogenic primary amines are included, i.e. histamine, tyramine, tryptamine, dopamine, serotonin, norepinephrine, mescaline, 1,5-diaminopentane, 1,4-diaminobutane, spermine, and spermidine. Since the physiological function of many amine oxidases is the breakdown or transformation of biologically active amines, these enzymes may act as regulators of physiological amine concentrations and, therefore, participate in numerous biological processes.[1,3] For example, histamine is a potent effector of the cardiovascular and gastrointestinal systems, and is directly involved in inflammatory, allergic, and ischemic phenomena. Polyamines are implicated in the stimulation and control of cell proliferation, growth, protein and nucleic acid synthesis, and development. Polyamines may also be involved in enzyme regulation; their oxidation products are implicated as inhibitors of cell proliferation and may play a direct role in the immune response. Plant amine oxidases seem to be involved in the synthesis of alkaloids and other plant hormones, or in the controlled production of H_2O_2, which is needed for cell wall biosynthesis. Amine oxidases are key enzymes in carbon and nitrogen metabolism in microorganisms,

From K.D. Karlin and Z. Tyeklár, Eds., *Bioinorganic Chemistry of Copper*
(Chapman & Hall, New York, 1993).

permitting the organism to use a primary amine as a carbon or nitrogen source for biosynthesis. Another type of amine oxidase, known as lysyl oxidase, has been discovered in connective tissue.[4] Lysyl oxidases are responsible for the crosslinking of connective tissue (collagen and elastin) in aorta, lung, and cartilage. The severe pathology of dietary copper deficiency and lathyrism is associated with extremely low levels or inhibition of lysyl oxidase. Amine oxidases are therefore a very important class of Type-2 copper enzyme. A current structural model for the copper site in amine oxidases is illustrated in Fig. 1A.[5]

Amine oxidases have recently emerged as the first examples of what may prove to be a wholly new class of enzyme, that is, where a post-translationally modified amino acid side chain is present in the active site and has a redox role in catalysis. As discussed below, in 1990 Klinman and co-workers published convincing evidence that the carbonyl cofactor in bovine plasma amine oxidase was the quinone of tri-hydroxyphenylalanine (topa quinone or 6-hydroxydopa quinone), Fig. 1B.[6] Subsequent work has established that topa is derived from tyrosine.[7] Both copper and topa are required for amine oxidase catalysis. Several of the most interesting questions concerning the structure and function of amine oxidases relate to the different mechanistic roles of copper and topa, and the nature of the interactions between these two cofactors. Of particular interest is the possibility that an enzyme-bound radical, identified as a semiquinone, is an obligatory intermediate in the catalytic cycle.[8]

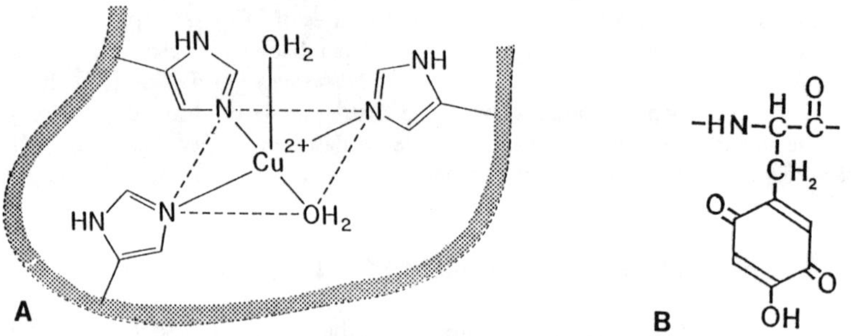

Figure 1. (A) Model for the Cu(II) sites in amine oxidases. (B) Topa quinone.

COPPER SITE STRUCTURE

The model for the Cu(II) sites in amine oxidases (Fig. 1A) is supported by a variety of spectroscopic results. Initial EXAFS experiments established that the first coordination shell of Cu(II) in bovine plasma amine oxidase (BPAO) was composed of N,O scatterers;[9] the overall EXAFS was very similar to that of the model $[Cu(imid)_4]^{2+}$. A subsequent study of porcine plasma amine oxidase (PPAO) was largely consistent with this picture, but indicated that the distances in the first coordination shell were split.[10] A heavy scatterer, originally attributed to an axial sulfur, is now thought to be Cl⁻ derived from the buffer.[11] The proposed model has received strong support from ENDOR experiments on porcine plasma amine oxidase,[12] and pulsed EPR studies (utilizing the electron spin echo envelope modulation, ESEEM, technique) of the bovine plasma and porcine kidney amine oxidases.[13] First, two populations of magnetically distinct histidyl imidazoles were identified as Cu(II) ligands for both native proteins from the ESEEM experiments.

Second, water was confirmed as a Cu(II) ligand. N_3^- and CN^- were shown to substitute for an equatorial H_2O ligand, displacing it from the first coordination shell of amine oxidase Cu(II). However, another H_2O molecule, with ESEEM parameters that are very similar to those of axially coordinated H_2O in Cu(II) model complexes, probably remains bound to the Cu(II) in the anion complex. Anion binding abolishes the distinction in magnetic coupling between the two populations of imidazoles. Thus the ESEEM results indicate that amine oxidase Cu(II) has both equatorial and axial water ligands.

In order to further define copper binding and the coordination structures of the Cu(II) sites, the paramagnetic effects of enzyme-bound Cu(II) on solvent water relaxation were investigated. The approach was to measure the nuclear magnetic relaxation dispersion (NMRD) profiles of amine oxidases as a function of temperature, substrate and inhibitor binding, and copper content.[5] For both porcine plasma and bovine plasma amine oxidase we found that the paramagnetic contribution to the relaxivity, T_{1P}^{-1}, the specific activity, and the intensity of the principal absorption band are linear functions of the copper content. These results are inconsistent with previous (and conflicting) claims that a one-Cu derivative of the bovine enzyme could be prepared that had either no, or full, activity. It was concluded that copper binding to the apo protein is either highly cooperative (i.e. the samples are mixtures of apo and fully-reconstituted proteins), or that the copper sites have equal affinity for Cu(II) and that population of the two sites is always equal, in both bovine and porcine plasma amine oxidase.

2 A

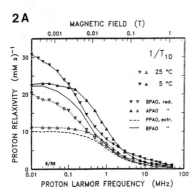

2 B

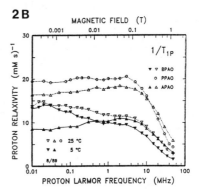

3

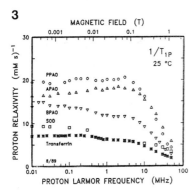

Figure 2. (A) Diamagnetic contributions T_{1d}^{-1} to the total relaxivities of amine oxidases at 5 and 25 °C per mM protein. Key: reduced BPAO, 5°C (▼), 25°C (▽); reduced APAO, 5°C (▲), 25°C (△); extrapolated PPAO, 25°C (--); extrapolated BPAO, 25°C (—). (B) Paramagnetic contributions T_{1P}^{-1} to the total relaxivities of amine oxidases at 5 and 25°C per mM protein. Key: BPAO, 5°C (▼), 25°C (▽); PPAO, 25°C (O); APAO, 5°C(▲), 25°C(△).

Figure 3. Comparison of the paramagnetic contributions T_{1P}^{-1} of PPAO, APAO, and BPAO to that of Cu_2Zn_2SOD and Cu transferrin at 25°C per mM protein.

NMRD measurements also provided new insights into the structure and dynamics of the Cu(II) sites in amine oxidases.[5] The key results are displayed in Figures 2 and 3. Note that the relaxivities for all the amine oxidases are high, but that the enzymes may be divided into two classes based on the temperature dependence. Specifically, the magnitude and dispersion of T_{IP}^{-1} are consistent with the presence of at least one liganded water molecule in rapid exchange with bulk solvent in amine oxidases from porcine plasma, porcine kidney (PKAO), bovine plasma, and *Arthrobacter* P1 (APAO). As is evident in Figure 3, the T_{IP}^{-1} values for amine oxidases display similar dispersions as that for Cu,Zn-superoxide dismutase (SOD) but have higher magnitudes. Since T_{IP}^{-1} of SOD arises from rapid exchange of an axial water, the NMRD profiles of amine oxidases suggest that the labile water molecule is also an axial ligand. The data in Figures 2 and 3 indicated that an additional exchangeable water ligand is present in PPAO and APAO. This water exchanges rapidly at 25 °C but relatively slowly at 5 °C, and is probably displaced upon binding anionic inhibitors. Exogenous anionic inhibitors, such as azide, are known to bind equatorially to tetragonal Cu(II) ions in amine oxidases; nonetheless, azide binding only slightly decreases T_{IP}^{-1} of BPAO and PKAO but significantly reduces T_{IP}^{-1} of PPAO. Collectively these results suggest that the equatorial water molecule is sufficiently labile to contribute to the NMRD profiles of the porcine plasma and *Arthrobacter* amine oxidases at 25 °C. Therefore the data accord fully with the proposed structural model for the Cu(II) sites in amine oxidases (Fig. 1A). Both equatorial and axial water molecules are ligands, but the dynamics of equatorial ligand substitution or exchange may vary.

Low-temperature MCD spectroscopy has been used to observe the ligand-field bands of amine oxidase Cu(II) and the data (Figures 4 and 5) are consistent with a tetragonal structure with N,O ligands. Low-temperature MCD spectra are dominated by transitions between paramagnetic states with appreciable spin-orbit coupling.[14] Consequently, the two intense, oppositely signed, bands in the visible are undoubtedly Cu(II) *d-d* transitions. The large shifts to higher energy induced by cyanide binding support this assignment (Figs. 4 and 5). Careful examination of the near-infrared region (to ~ 1800 nm) revealed no additional electronic transitions. Note that no intense features

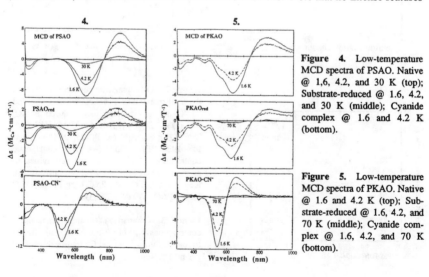

Figure 4. Low-temperature MCD spectra of PSAO. Native @ 1,6, 4.2, and 30 K (top); Substrate-reduced @ 1.6, 4.2, and 30 K (middle); Cyanide complex @ 1.6 and 4.2 K (bottom).

Figure 5. Low-temperature MCD spectra of PKAO. Native @ 1.6 and 4.2 K (top); Substrate-reduced @ 1.6, 4.2, and 70 K (middle); Cyanide complex @ 1.6, 4.2, and 70 K (bottom).

are found in the 350 - 600 nm region, where oxidized topa absorbs. Since topa quinone is diamagnetic no MCD intensity is expected for a topa-localized transition, *but the data also require that there are no charge-transfer transitions involving Cu(II) and topa, because such transitions would have significant MCD intensity.* No intense bands other than the Cu(II) *d-d* transitions are detected in the MCD spectra of the substrate-reduced enzymes. Therefore, equatorial coordination of the quinone or quinol state of topa to Cu(II) can probably be ruled out because such coordination should give rise to a charge-transfer transition in the near-uv to violet region. The visible MCD may be provisionally interpreted as follows. The ligand-field states of a tetragonal Cu(II) complex are generally ordered as $^2B_1 < {}^2A_1 < {}^2B_2 < {}^2E$. The derivative shaped MCD may arise from a "pseudo-A term" associated with transitions to the spin-orbit components of the $^2B_2 \rightarrow$ ^2E, with the transitions to the $^2A_1$ and $^2B_2$ states being forbidden (as in C_{4v} symmetry) or very weak.[15] These transitions can gain MCD intensity via spin-orbit coupling to the components of the ^2E state. Under certain circumstances the ^2E MCD intensity is mostly canceled and that from the $^2A_1$ and $^2B_2$ then dominates. Further work is needed to pin down the assignments, which should then permit a very detailed picture of Cu(II) electronic structure in amine oxidases.

Amine oxidase genes from *Hansenula polymorpha*,[16] lentil seedlings,[17] and *Arthrobacter* P1[18] were cloned and the cDNAs sequenced. A comparison of the translated protein sequences is shown in Figure 6. There appear to be four conserved histidines, one more than required by the spectroscopic data. It should be noted that an axial histidine ligand would not be inconsistent with the published ESEEM, EXAFS, or NMRD results.

```
        1                                              50
APAO    .........  .......LVG  VSHPLDPLSR  VEIARAVAIL  KEG.PAAAES
 YAO    MERLRQIASQ  ATAASAAPAR  PAHPLDPLST  AEIKAATNTV  K.S.YFAGKK
LSAO    .........  .....FTPLH  TQHPLDPITK  EEFLAVQTIV  QNKYPISNNK
                       •      H★★★★••    ★•  • ••• •    ••••
        (alignment residues 51-320 not shown)
        321                                            370
APAO    KNAFDSGEYN  IGNMANSLTL  GCDCLGEIKY  FDGHSVDSHG  NPWTIENAIC
 YAO    KHALDIGEYG  AGYMTNPLSL  GCDCKGVIHY  LDAHFSDRAG  DPITVKNAVC
LSAO    KTFFDSGEFG  FGLSTVSLIP  NRDCPPHAQF  IDTYIHSADG  TPIFLENAIC
        ★  •★ ★★••   ★  •  •★   •  ★C       ••  •★••  •   ★   •★   ••★★•C

        371                                            420
APAO    MHEEDDSILW  KHFDFRE..G  TAETRRSR.K  LVISFIATVA  NYEYAFYWHL
 YAO    IHEEDDGLLF  KHSDFRDNFA  TSLVTRAT.K  LVVSQIFTAA  NYEYVLYWVF
LSAO    VFEQYGNIMW  RHTETGIPNE  SIEESRTEVD  LAIRTVVTVG  NYDNVLDWEF
        •  ★•  ••••• •  •H •      •  •        ★•  •  ★•••  • ★••  ★Y★ •• ★ •

        421                                            470
APAO    FLDGSIEFLV  KATGIL...S  TAGQLPGEK.  NPYGQSLNND  GLYAPIHQHM
 YAO    MQDGAIRLDI  RLTGIL...N  TYILGDDEEA  GPWGTRVYPN  .VNAHNHQHL
LSAO    KTSGWMKPSI  ALSGILEIKG  TNIKHKDEIK  EEIHGKLVSA  NSIGIYHDHF
        •★ •   •   ★★★    ★   •★   •   ••  •    • H•H•
        (alignment residues 461-705 not shown)
```

Figure 6. Sequence Homology among Amine Oxidases. (★) perfect match, (•) well-conserved, (..) gaps, (H) conserved His, (C) conserved Cys, (Y) TOPA (modified Tyr). APAO, 639 residues; YAO, 692 residues; LSAO, 569 residues.

During the course of the NMRD experiments we found that extraordinary care is necessary to obtain valid and reproducible measurements on copper-depleted amine oxidases. The apo protein has a very high affinity for Cu(II) and efficiently scavenges copper from glass surfaces. The time required to obtain reproducible measurements of the specific activity, relaxivity, and absorbance depends on the identity of the enzyme and the method used to prepare the apo protein. Aggregation of the copper-depleted proteins is another problem; again reproducible results could only be obtained by adding Cu(II) to dilute solutions. Consequently, metal substitution in amine oxidases has been examined in order to test whether this technique is a useful probe of the role of copper. Such experiments acquired additional impetus by the claim that a Co(II)-substituted derivative had significant catalytic activity.[19] If correct, this result may be inconsistent with a redox role for copper in amine oxidases because Co(II) is expected to be redox-inert in a protein site. Our results to date are summarized in Table I.

Table I. Cobalt-Substituted BPAO

Sample[a]	% Activity	% Cu	% Co
CoCudep	20 ± 1	17 ± 2	57 ± 5
CoNidep[b]	13.4 ± 1	15 ± 1	11 ± 1
CoNidep	31 ± 3	23 ± 2	39 ± 5
CoCudep[c]	13 ± 1	8 ± 2	137 ± 10
CoCudep[c]	7.5 ± 2.5	4 ± 2	139 ± 10
CoCudep[c]	6 ± 1	19 ± 2	270 ± 30
CoNidep[c]	9 ± 1	22 ± 2	100 ± 10
CoCudep	16 ± 1.5	17 ± 2	104 ± 10
CoCudep	12.3 ± 1.2	9.3 ± 2	89 ± 9
CoCudep	13.6 ± 1.4	6 ± 3	111 ± 11
CoCudep	12.4 ± 1.2	4.2 ± 3	90 ± 9
CoCudep	6.4 ± 0.6	8.8 ± 0.4	61 ± 3
CoCudep	7.2 ± 0.7	6.5 ± 0.4	137 ± 6
CoCudep	12.8 ± 1.3	5.4 ± 0.3	48 ± 3
CoNidep	21.6 ± 2	21.3 ± 1	180 ± 9
CoNidep	7.7 ± 0.7	11.5 ± 0.6	71 ± 4
CoNidep	12 ± 1.2	9.2 ± 0.5	201 ± 10
CoNidep	13 ± 1.3	5.5 ± 0.3	127 ± 6

[a] CoCudep, prepared from Cu depleted BPAO; CoNidep, prepared from Ni depleted BPAO, by anaerobic dialysis vs. 2 mM Co²⁺, unless otherwise indicated. [b] Stoichiometric Co²⁺ added anaerobically and incubated 72 hours. [c] 100 - 400 fold excess Co²⁺ added anaerobically and incubated 72 hours. Some protein denaturation occurred whenever this incubation method was used.

Three points should be emphasized. First, despite considerable care at each stage of the metal substitution process some copper contamination is inevitable owing to the high affinity and selectivity of the apo protein for Cu(II). We found that copper-depleted samples of BPAO could bind trace Cu(II) even in the presence of relatively high concentrations of other metal ions. Second, for most of the samples the observed activity correlates with the copper content, but not the cobalt content. Third, Co(II) binds tightly (not removed by dialysis against 0.3 mM EDTA) to native BPAO, i.e. with 2 Cu/mole protein. Taken together, the data are inconsistent with true metal substitution but rather suggest that Co(II) binds to sites other than, or in addition to, the Cu(II) sites. Little, if any, activity can be unambiguously attributed to Co(II). The Ni(II)-containing protein is also inactive. It is possible that the Co(II)-substituted BPAO is slightly active but operates by a different mechanism than the native enzyme. Our findings call into question the validity of mechanistic, or structural conclusions, based on the properties of metal-substituted bovine plasma amine oxidase. Checking the generality of these results with other amine oxidases is necessary.

IDENTIFICATION OF TOPA QUINONE

One of the most difficult obstacles to developing a sound understanding of the structure and reactivity of amine oxidases was the uncertainty surrounding the identity of the active-site carbonyl group. Amine oxidases were originally identified as pyridoxal enzymes, despite considerable difficulties in reconciling some of the chemical, spectroscopic, and mechanistic data with the properties of pyridoxal. The suggestion in 1984 that amine oxidases contained "covalently-bound" pyrroloquinoline quinone (PQQ) rationalized all the available data and greatly stimulated additional research.[20] Ultimately Klinman and co-workers succeeded in isolating a labeled active-site peptide and structurally characterizing it via NMR and mass spectroscopy.[6] The carbonyl reagents phenylhydrazine and p-nitrophenylhydrazine were used to label the quinone. We have developed resonance Raman spectroscopy using these carbonyl reagents has a probe of the quinone cofactor in amine oxidases. Following the demonstration by Klinman and co-workers that topa quinone was the organic cofactor in BPAO, we subsequently established two additional, and critical, points.[21] First, comparisons of the resonance Raman spectra of the intact, derivatized proteins and the isolated, labeled peptide established that the topa structure was not formed during the isolation and purification of the active-site peptide. Second, the resonance Raman spectra of the phenylhydrazine and p-nitrophenylhydrazine derivatives of BPAO peptide were *identical* to the spectra of the corresponding derivatives of topa quinone, and very different from the spectra of PQQ

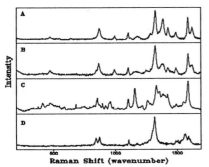

Figure 7. Comparison of the resonance Raman spectra of the p-nitrophenylhydrazones of topa-hydantoin (A), the active-site peptide of BPAO (B), PQQ (C), pyridoxal phosphate (D).

or pyridoxal (Figure 7). Hence, resonance Raman spectroscopy provided very strong, independent support for the presence of topa quinone in amine oxidases. Examination of the amine oxidases from the yeast *Hansenula polymorpha*[7] and from *Arthrobacter* (unpublished results with W. McIntire) established that both these enzymes contain topa and that a tyrosine codon corresponds to topa quinone at the active site. Therefore topa is produced in the active site via post-translational modification. We have now extended the resonance Raman method to peptides isolated from several amine oxidases; the results shown in Figure 8 establish that topa quinone is the cofactor in all the enzymes examined.[22] Sequences of the active site peptides are shown in Table II; the high degree of homology evident in Table II and Figure 6 provides further support for the conclusion that all of these amine oxidases contain topa. The consensus sequence deduced from these comparisons for topa is ASN-TYR-ASP/GLU,[22] which falls in a region of high homology among sequenced amine oxidases (Figure 6). It should be noted that mammalian lysyl oxidase is not homologous to this group of amine oxidases.[23] Although resonance Raman data are consistent with a quinone cofactor similar to topa, the lysyl oxidase spectrum is not identical to the spectra of the other amine oxidases.[24]

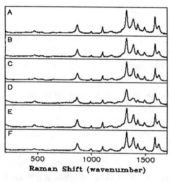

Figure 8. Resonance Raman spectra of the p-nitrophenylhydrazone labeled peptides and labeled topa-hydantoin. (A), PSAO; (B), PKAO; (C), PPAO; (D), BPAO; (E), SPAO; (F), topa-hydantoin.

Table II. Sequences for Active Site Thermolytic Peptides from Copper Amine Oxidases

BPAO	PPAO	PKAO	PSAO	YAO
		VAL	VAL	VAL
LEU	LEU	TYR	GLY	ALA
ASN	ASN	ASN	ASN	ASN
X	X	X	X	X
ASP	ASP	ASP	ASP	GLU
TYR	TYR	TYR	ASN	TYR
			VAL	VAL

COPPER-QUINONE INTERACTIONS AND THE SEMIQUINONE STATE

It has been clear for some time that substrate amines form covalent adducts with topa quinone. Subsequently an active-site base abstracts a proton from the imine leading to the two-electron reduction of the quinone to an aminophenol. The role of copper in amine oxidases has been uncertain since early EPR measurements failed to detect any change in the copper oxidation state following substrate addition.[25] Rapid freeze-quench EPR experiments appeared to rule out the possibility of a copper valence change at any stage of the catalytic cycle.[26] In 1991 we showed that substrate amines *do* reduce copper under anaerobic conditions, generating a Cu(I)-semiquinone state.[8] Figure 9 illustrates the key results. If the EPR spectra are obtained *at room temperature* a sharp signal at *g* ~ 2 is observed for all five amine oxidases, although the yield of this species is variable. The Cu(II) signals of PSAO and APAO are significantly decreased, and a minor decrease is apparent in the PKAO spectrum. Quantitatively, the Cu(II) signals of the pea seedling and *Arthrobacter* amine oxidases decrease by 40%, with the semiquinone signal intensity corresponding to about half of the copper change. The difference was attributed to disproportionation of the semiquinone; we have now in independent experiments confirmed that semiquinone does disproportionate under the conditions of Figure 9. CD spectra of APAO (Figure 10) and PSAO (not shown) are quantitatively consistent with copper reduction by substrates as inferred from the EPR. Collectively the data point to a temperature-dependent equilibrium:

$$\text{Cu(II)-reduced quinone} \rightleftharpoons \text{Cu(I)-semiquinone} \tag{1}$$

The Cu(II) state is favored at low temperatures, and thus the Cu(I)-semiquinone state was not observed in previous EPR experiments, which were invariably conducted at low temperature. Based on these and other results, and a wealth of chemical precedent, we proposed the catalytic cycle in Scheme 1.[8] These results suggest that copper has a redox function in amine oxidases; this re-opened possibility will require a great deal of new work to evaluate and thus pin down the role of copper in amine oxidases.

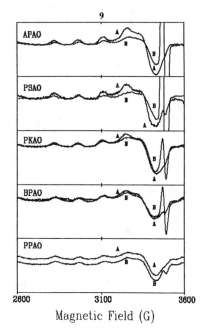

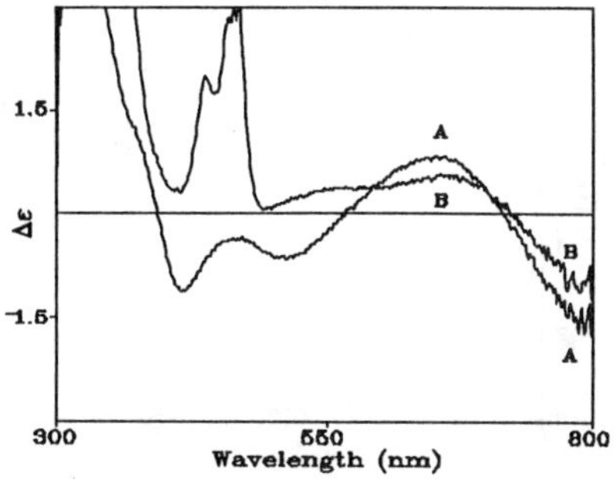

Figure 10. CD spectra ($\Delta\varepsilon = \varepsilon_L - \varepsilon_R$, $M^{-1}cm^{-1}$) of native APAO (A) and of APAO after anaerobic addition of substrate, 295 K.

Figure 9. Room-temperature EPR spectra of amine oxidases before (A) and after (B) anaerobic addition of substrate. The PPAO are offset for clarity.

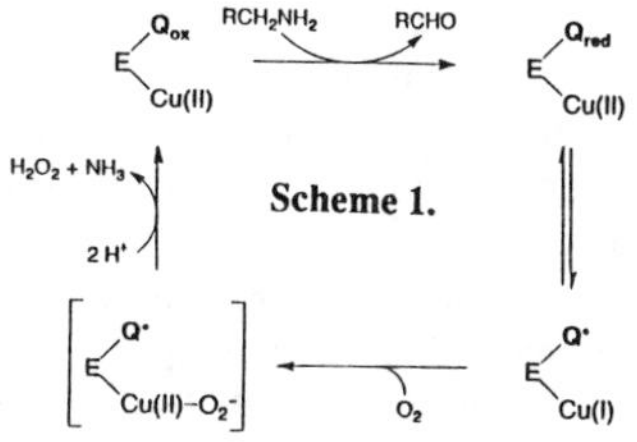

Scheme 1. Possible Catalytic cycle of copper-containing amine oxidases. The species in brackets is a hypothetical intermediate, shown here to emphasize the possibility of sequential one-electron steps in the reduction of O_2. Q_{ox} = topa quinone, $Q^{\bullet}$ = topa semiquinone, and Q_{red} = topa (6-hydroxydopa).

An essential first step is to check whether the semiquinone is formed at a rate sufficiently rapid so that its involvement in turnover is permitted. The observation of the temperature dependent equilibrium (Eq.1) suggested that temperature-jump relaxation experiments might provide the rate for intramolecular electron transfer between the reduced quinone and Cu(II). Since the semiquinone state is favored at higher temperatures a temperature jump should produce an increase in the absorbance at 460 nm. Our initial results, displayed as a decrease in transmittance at 460 nm for a temperature jump of about 1 K with substrate-reduced PSAO, are shown in Figure 11. Although the kinetics are not monophasic, most of the relaxation occurs within a very short time; the two fastest phases (representing 89% of the total absorbance change) correspond to k_{obs} values of 43,000 s^{-1} and 4,300 s^{-1}, respectively. Table III summarizes our initial fits assuming three parallel, reversible first-order reactions. For a reversible, first-order reaction, k_{obs} is the sum of the forward and reverse rate constants of the equilibrium (Eq. 1). Since the concentrations of the two species are approximately equal at 296 K, the equilibrium is ~1, and forward rate constants are about $0.5k_{obs}$. The rate constants for all three phases are larger (or much larger) than the measured first-order rate constant for enzyme reoxidation (about 100 s^{-1}, for the closely related lentil seedling enzyme).[27] Therefore formation of the semiquinone is kinetically competent and it may be involved in turnover.

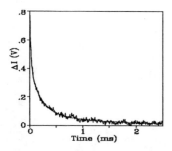

Figure 11. Relaxation behavior of substrate-reduced PSAO (296 K) after 1.1° temperature jump. Since ΔI refers to the transmittance intensity, a decrease in intensity corresponds to an increase in absorbance.

Table III. T-jump Kinetic Results

k_{obs}* (s^{-1})	$t_{\frac{1}{2}}$ (μs)	ΔI (%)
43,000±3,000	23±2	53±3
4,300±700	230±40	36±3
600±300	1,700±1,000	11±1

* Relaxation rate constant

The semiquinone state is at least an excellent candidate to be the species in the catalytic cycle that reacts with oxygen. Hence we have begun a series of experiments to characterize this state in detail. Pulsed EPR spectroscopy, specifically the electron spin echo envelope modulation (ESEEM) technique, has already proven informative.[28] In these experiments cyanide was used to stabilize the semiquinone state so that it could be examined at 4.2 K, as required by the ESEEM method. We have shown that the EPR signals of the radicals ± cyanide are identical. The pulsed EPR results confirm that the radicals in *Arthrobacter* and porcine kidney amine oxidases are identical. The ESEEM spectra provide strong evidence that at least one nitrogen nucleus is magnetically coupled to the paramagnet. Isotopic substitution experiments using [^{14}N]- and [^{15}N]-methylamine with *Arthrobacter* methylamine oxidase establish that the coupled nitrogen nucleus is derived from substrate. Magnetic coupling and hyperfine parameters obtained by computer simulation are consistent with previous formulations of this radical as an iminosemiquinone. The hyperfine coupling tensor elements are in line with what one would expect for a π-radical where covalent attachment of the substrate amine nitrogen to an aromatic group has occurred. These results effectively rule out a tyrosine radical (or a radical derived from any other amino acid) as the source of the g ~ 2 signal in substrate-reduced amine oxidases. Both the ESEEM and previous results support the aminotransferase mechanism. In conjunction with kinetic data the ESEEM data also suggest that the initial step in the oxidation of the reduced topa (and perhaps the first step in oxygen reduction) occurs prior to cleavage of the nitrogen-topa bond(s).

CONCLUDING REMARKS

Our knowledge of amine oxidases has advanced considerably in recent years, but important aspects of the active-site structure and mechanism of these enzymes remain uncertain. However, several productive approaches to the remaining questions have been identified that will generate considerable new information in the near future. Cloning and sequencing additional amine oxidase genes will be helpful, as will the construction and analysis of site-directed mutants. We are pursuing the crystallographic analysis of the pea seedling amine oxidase in collaboration with Hans Freeman at the University of Sydney. High-quality crystals have been obtained that diffract to a resolution of at least 2.5 Å with synchrotron radiation.[29] Successfully solving the crystal structure of an amine oxidase would begin a new and exciting phase of study of these important enzymes.

ACKNOWLEDGEMENTS

The research summarized here has been supported in part by the NIH (grant 27659 to DMD). DMD also gratefully acknowledges Rod Brown, Seymour Koenig, Marga Spiller, Susan Janes, Judith Klinman, Jack Peisach, John McCracken, Herb Kagan, Bob Scott, Hans Freeman, and Peter Knowles for their great insight and numerous contributions to the study of amine oxidases at Amherst.

REFERENCES

1.   B. Mondovi, Ed. *Structure and Function of Amine Oxidases* (CRC Press, Boca Raton, Florida, 1985).
2.   R.A. Cooper, P.F. Knowles, D.E. Brown, M.A. McGuirl, D.M. Dooley, *Biochem. J.*, in press.
3.   W.S. McIntire and C. Hartmann, in *Principles and Applications of Quinoproteins* (V. Davidson Ed.), in press.
4.   H.M. Kagan and P.C. Trackman, *Am. J. Resp. Cell Mol. Biol.* **5**, 206-210 (1991).
5.   D.M. Dooley, M.A. McGuirl, C.E. Cote, P.F. Knowles, I. Singh, M. Spiller, R.D. Brown III, S.H. Koenig, *J. Am. Chem. Soc.*, **113**, 754-761 (1991).
6.   S.M. Janes, D. Mu, D. Wemmer, A.J. Smith, S. Kaur, D. Maltby, A.L. Burlingame, J.P. Klinman, *Science*, **248**, 981-987 (1990).
7.   D. Mu, S.M. Janes, A.J. Smith, D.E. Brown, D.M. Dooley, J.P. Klinman, *J. Biol. Chem.*, **267**, 7979-7982 (1992).
8.   D.M. Dooley, M.A. McGuirl, D.E. Brown, P.N. Turowski, W.S. McIntire, P.F. Knowles, *Nature*, **349**, 262-264 (1991).
9.   R.A. Scott, and D.M. Dooley, *J. Am. Chem. Soc.*, **107**, 4348-4350 (1985).
10.   P.F. Knowles, R.W. Strange, N.J. Blackburn, S.S. Hasnain, *J. Am. Chem. Soc.*, **111**, 102-107 (1989).
11.   P.F. Knowles, personal communication.
12.   G.J. Barker, P.F. Knowles, K.B. Pandeya, J.B. Rayner, *Biochem. J.*, **237**, 609-612 (1986).
13.   J. McCracken, J. Peisach, D.M. Dooley, *J. Am. Chem. Soc.*, **109**, 4064-4072 (1987).
14.   D.M. Dooley and J.H. Dawson, *Coord. Chem. Rev.*, **60**, 1-66 (1984).

15. T.D. Westmoreland, D.E. Wilcox, M.J. Baldwin, W.B. Mims, E.I. Solomon, *J. Am. Chem. Soc.*, **111**, 6106-6123 (1989).

16. P.G. Bruienberg, M. Evers, H.R. Waterham, J. Kuipers, A.C. Arnberg, G. AB, *Biochim. Biophys. Acta*, **1008**, 157-167 (1989).

17. A. Rossi, R. Petruzzelli, A. Finazzi-Agrò, *FEBS Lett.*, **301**, 253-257 (1992).

18. W.S. McIntire, unpublished results.

19. L. Morpurgo, E. Agostinelli, B. Mondovi, L. Avigliano, *Biol. Metals.*, **3**, 114-117 (1990).

20. For a summary see: J.P. Klinman, D.M. Dooley, J.A.Duine, P.F. Knowles, B. Mondovi, J.J. Villafranca, *FEBS Lett.*, **282**, 1-4 (1991).

21. D.E. Brown, M.A. McGuirl, D.M. Dooley, S.M. Janes, D. Mu, J.P. Klinman, *J. Biol. Chem.*, **266**, 4049-4051 (1991).

22. S.M. Janes, M.M. Palcic, C.H. Scaman, A.J. Smith, D.E. Brown, D.M. Dooley, M. Mure, J.P. Klinman, *Biochemistry*, in press.

23. P.C. Trackman, A.M. Pratt, A. Wolanski, S.-S. Tang, G.D. Offner, R.F. Troxler, H.M. Kagan, *Biochemistry*, **29**, 4863-4870 (1990).

24. P.R. Williamson, R.S. Moog, D.M. Dooley, H.M. Kagan, *J. Biol. Chem.*, **261**, 16302-16305 (1986).

25. H. Yamada, K. Yasunobu, T. Yamono, H. Mason, *Nature*, **198**, 1092-1093 (1963).

26. J.Grant, I. Kelly, P. Knowles, J. Olsson, G. Pettersson, *Biochem. Biophys. Res. Commun.*, **83**, 1216-1224 (1978).

27. A. Bellelli, A. Finazzi-Agrò, G. Floris, M. Brunori, *J. Biol. Chem.*, **266**, 20654-20657 (1991).

28. J. McCracken, J. Peisach, C.E. Coté, M.A. McGuirl, D.M. Dooley, *J. Am. Chem. Soc.*, **114**, 3715-3720 (1992).

29. V. Vignevich, D.M. Dooley, J.M. Guss, I. Harvey, M.A. McGuirl, H.C. Freeman, *J. Mol. Biol.*, in press.

ASCORBATE OXIDASE STRUCTURE AND CHEMISTRY

Albrecht Messerschmidt

Max-Planck-Institut für Biochemie, D-8033 Martinsried, FRG

INTRODUCTION

The activation of dioxygen in biological systems has been the focus of interest of biochemists, bioinorganic chemists, and physiologists for many years. A whole body of literature exists on this topic. Furthermore, special conferences have been devoted to this subject and one of the last of these meetings should be mentioned here whose proceedings have been published by King et al..[1] Enzymes involved in direct oxygen activation are oxidases and oxygenases. Oxygenases introduce either one atom of dioxygen into substrate and reduce the other atom to water (monooxygenases) or transfer two oxygen atoms into substrate (dioxygenases). Oxidases can be divided in two-electron and four-electron transferring enzymes. The first group reduces dioxygen to hydrogen peroxide and the second one dioxygen to water. Most of the oxygenases as well as oxidases contain as prosthetic groups either flavin, iron (heme or non-heme) or copper.

Oxygen is used for respiration upon reduction to water by cytochrome-c oxidase, the terminal oxidase of the respiratory chain. Cytochrome-c oxidase is a complex metalloprotein containing two or three copper ions and two heme groups (a and a_3) that provides a critical function in cellular respiration in both prokaryotes and eukaryotes. The enzyme catalyzes the four-electron reduction of dioxygen with concomitant one-electron oxidation of cytochrome-c. Energy released in this exogonic reaction is conserved as a pH gradient and membrane potential across the membrane barrier, generated in part by H^+ consumption and also by proton translocation through the protein complex (see e.g. the review on cytochrome-c oxidase by Capaldi[2]).

Ascorbate oxidase (EC 1.10.3.3) is a blue multi-copper oxidase that catalyzes the four-electron reduction of dioxygen to water with concomitant one-electron oxidation of the reducing organic substrate.[3] Copper-dependent ascorbate oxidase is found only in higher plants.[4] The enzyme from *Cucurbita pepo medullosa* (green zucchini) is a dimer of 140,000 M_r[$] containing eight copper ions of three different spectroscopic forms classified as type-1, type-2 and type-3 according to Vänngard's proposal (see Fee[5]). The immunohistochemical localization of ascorbate oxidase in green zucchini reveals that

Footnotes:
[$]$M_r$, molecular mass; [⊗]FO, observed structures factor amplitude; [§]FC, calculated structure factor amplitude; [#]NATI, native; [£]PEOX, peroxide; [†]REDU, reduced.

From K.D. Karlin and Z. Tyeklár, Eds., *Bioinorganic Chemistry of Copper*
(Chapman & Hall, New York, 1993).

ascorbate oxidase is distributed in all specimens examined ubiquitously over vegetative and reproductive organs. At the cellular level the enzyme is linked with the cell wall and cytoplasm.[4] The in vivo role in plants of ascorbate and ascorbate oxidase is still under debate. As catechols and polyphenols are also substrates in vitro,[6] ascorbate oxidase might be involved in biological processes like fruit ripening. A role in a redox system, as an alternative to the mitochondrial chain, in growth promotion or in susceptibility to disease has also been postulated.[7] Primary structures of ascorbate oxidase from cucumber[8] and pumpkin[9] have recently been reported.

The preliminary three-dimensional X-ray structure of the fully oxidized form of ascorbate oxidase from zucchini has been published[10] and the polypeptide fold and the coordination of the mononuclear blue copper site and the unprecedented trinuclear copper site have been described. The structure has now been refined to a resolution of 1.9 Å (1 Å = 0.1 nm) and its detailed description and implications for the catalytic mechanism, electron transfer processes, and activation of dioxygen has been the subject for a further publication.[11] Recently, the X-ray structures of type-2 depleted ascorbate oxidase[12] as well as of the fully-reduced, peroxide and azide derivatives[13] were determined. These structure analyses revealed unexpected results which will be discussed in this chapter.

The structural relationship to the other blue copper oxidases, laccase (EC 1.10.3.2) and ceruloplasmin (EC 1.16.3.1), has been demonstrated by amino acid sequence alignment based on the spatial structure of ascorbate oxidase from zucchini.[14]

THREE-DIMENSIONAL STRUCTURE OF ASCORBATE OXIDASE

The crystal structure of the fully oxidized form of ascorbate oxidase from zucchini has been refined at 1.9 Å resolution, using an energy-restrained least-squares refinement procedure. The refined model, which includes 8764 protein atoms, 9 copper atoms and 970 solvent molecules, has a crystallographic R-factor of 20.3 % for 85,252 reflections

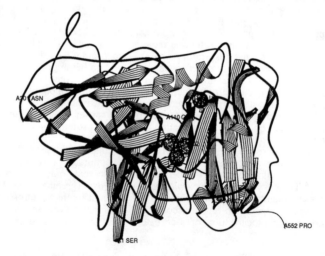

Figure 1. A schematic representation of the monomer structure of ascorbate oxidase. The copper ions are represented by spheres.

between 8 and 1.9 Å resolution. The root-mean-square deviation in bond lengths and bond angles from ideal values is 0.011 Å and 2.99°, respectively. The subunits of 552 residues (70,000 M_r) are arranged as tetramers with D2 symmetry. One of the diads is realized by the crystallographic axis parallel to the c-axis giving one dimer in the asymmetric unit. The dimer related about this crystallographic axis is suggested as the dimer present in solution. Two sugar moieties are N-linked per monomer[15] rather than three putative carbohydrate attachment sites apparent in the amino acid sequence of ascorbate oxidase from zucchini.[11] Asn 92 is the attachment site for one of the two N-linked sugar moieties which has defined electron density for the N-linked N-acetyl-glucosamine ring.[11] Each subunit is built up by three domains arranged sequentially on the polypeptide chain and tightly associated in space (see Fig. 1). The folding of all three domains is of a similar β–barrel type and related to plastocyanin and azurin.

Each subunit has four copper atoms bound as mononuclear and trinuclear species. The mononuclear copper has two histidine, a cysteine and a methionine ligand and represents the type-1 copper (see Fig. 2). It is located in domain 3. The bond lengths of the type-1 copper centre are comparable to the values for oxidized plastocyanin.

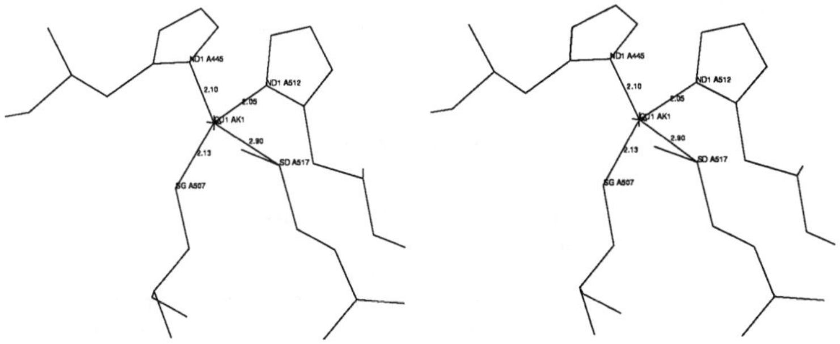

Figure 2. Stereo drawing of the type-1 copper site in domain 3. The displayed bond distances are for subunit A.

The trinuclear cluster has eight histidine ligands symmetrically supplied from domain 1 and 3 (see Fig. 3). It may be subdivided into a pair of copper atoms with histidine ligands whose ligating N-atoms (5 NE2 atoms and one ND1 atom) are arranged trigonal prismatic. The pair is the putative type-3 copper. The remaining copper has two histidine ligands and is the putative spectroscopic type-2 copper. Two oxygens are bound to the trinuclear species; as OH- or O²⁻ and bridging the putative type-3 copper pair,and as OH- or H_2O bound to the putative type-2 copper trans to the copper pair. The bond lengths within the the trinuclear copper site are similar to comparable binuclear model compounds.[16,17]

There has been experimental evidence in earlier studies that the type-2 copper is close to the type-3 copper and forms a trinuclear active copper site.[18-23] Solomon and associates based on spectroscopic studies of azide binding to tree laccase[22,23] addressed this metal binding site as trinuclear active copper site.

The putative binding site for the reducing substrate is type-1 copper.[11] Two channels providing access from the solvent to the trinuclear copper site, the putative binding site of the dioxygen, could be identified.[11]

X-RAY STRUCTURES OF FUNCTIONAL DERIVATIVES

Type-2 Depleted (T2D) Form[12]
Crystals of native oxidized ascorbate oxidase were anaerobically dialyzed in microcells against 50 mM sodium phosphate buffer, pH 5.2, containing 25% (v/v) methyl-pentane-diol (MPD), 1 mM EDTA, 2 mM dimethyl-glyoxime (DMG) and 5 mM ferrocyanide for 7 and 14 h. Thereafter, crystals were brought back to the aerobic 25% MPD solution

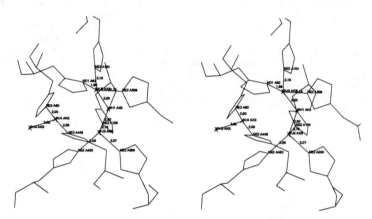

Figure 3. Stereo drawing of the trinuclear copper site. The displayed bond distances are for subunit A.

buffered with 50 mM sodium phosphate, pH 5.5. This procedure is based on the method of Avigliano et al.[24] to prepare T2D ascorbate oxidase in solution and was modified by Merli et al.[25] for use with ascorbate oxidase crystals.

The 2.5 Å resolution X-ray structure analysis by difference-Fourier techniques and crystallographic refinement shows that about 1.3 copper ions per ascorbate oxidase monomer are removed. The copper is lost from all three copper sites of the trinuclear copper species whereby the EPR-active type-2 copper is depleted somewhat preferentially (see Fig. 4). Type-1 copper is not affected. The EPR spectra from polycrystalline samples of the respective native and T2D ascorbate oxidase were recorded. The native spectrum exhibits the type-1 and type-2 EPR signals in a ratio of about 1:1 as expected from the crystal structure. The T2D spectrum reveals the characteristic resonancies of the type-1 copper centre as also observed for T2D ascorbate oxidase in frozen solution and the complete disappearance of the spectroscopic type-2 copper.

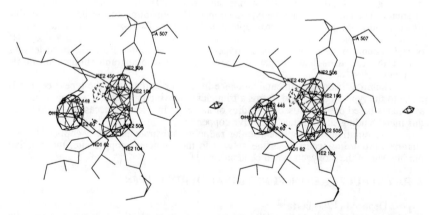

Figure 4. Averaged $FO^{\otimes}_{T2D}$-$FC^{\$}_{T2D}$-difference electron density map plus atomic model around the trinuclear copper site. Contour levels: -18.0 solid line, 18.0 dashed line, Magnitudes of holes less than -35.0.

Fully-Reduced Form[13]

Crystals of the reduced form of ascorbate oxidase had to be prepared and mounted in the glass capillary in a glove-box which was flushed with Argon gas and operated with a slight over-pressure of Argon. The degassed buffer solution was stored in the glove-box. Dithionite was added to the buffer solution to a concentration of 10 mM. The siliconized X-ray capillary was washed with the buffer. The crystals were soaked in the reducing buffer for half an hour. After 15 minutes the crystals had lost their blue colour. They were mounted in the X-ray capillary and carefully sealed with wax which had been degassed in the desiccator. Crystals mounted this way remained colourless and reduced over weaks.

The 2.2 Å resolution X-ray structure analysis by difference-Fourier techniques and crystallographic refinement delivers the following results. The geometry at the type-1 copper remains much the same compared to the oxidized form. The mean copper-ligand bond lengths of both subunits are increased by 0.04 Å in average which is insignificant but may indicate a trend. Similar results have been obtained for the reduced forms of poplar plastocyanin at pH 7.8,[26] azurin from *Alcaligenes denitrificans*[27] and azurin from *Pseudomonas aeruginosa*.[28] In reduced poplar plastocyanin at pH 7.8 a lengthening of the two CU-N(His) bonds by about 0.1 Å is observed. In reduced azurin, pH 6.0, from *Alcaligenes denitrificans*, the distances from copper to the axial methionine and the carbonyl oxygen each increase by about 0.1 Å. The same shifts are found in the refined structures of reduced azurin from *Pseudomonas aeruginosa* determined at pH values of 5.5 and 9.0.[28] The estimated accuracy of the copper-ligand bond lengths in the high resolution structures of above mentioned small blue copper proteins is about 0.05 Å. The type-1 copper sites in the small blue copper proteins as well as in ascorbate oxidase require little reorganization in the redox process.

A schematic drawing of the reduced form of ascorbate oxidase is shown in Fig. 5. The structural changes are considerable at the trinuclear copper site. Thus on reduction the bridging oxygen ligand OH1 is released and the two coppers CU2 and CU3 move towards their respective histidines and become three co-ordinate, a preferred stereochemistry for Cu(I). CU2 and CU3 are each trigonally planar co-ordinated by their respective histidine ligands with equal bond lengths and bond angles within the accuracy of this X-ray structure determination. The copper-copper distances increase from an average of 3.7 Å to 5.1 Å for CU2-CU3, 4.4 Å for CU2-CU4 and 4.1 Å for CU3-CU4. The mean values of the copper-ligand distances of the trinuclear copper site are comparable to native oxidized ascorbate oxidase and binuclear copper model compounds with nitrogen and copper ligands.[16,17] CU4 remains virtually unchanged between reduced and oxidized forms. Coordinatively unsaturated copper(I) complexes are known from the literature. Linear two-co-ordinated[29] and T-shaped three-co-ordinated[30] copper(I) compounds have been reported. The copper nitrogen distances for both linearly arranged nitrogen ligands are about 1.9 Å and about 0.1 Å shorter than copper nitrogen bond lengths in copper(II) complexes. In T-shaped copper(I) complexes the bond length of the third ligand is increased. Copper ion CU4 is in a T-shape threefold co-ordination not unusual for copper(I) compounds. The structure of the fully reduced trinuclear copper site is quite different therefore from that of the fully oxidized resting form of enzyme and its implications for the enzymatic mechanism will be discussed below.

Peroxide Form[13]

Native crystals of ascorbate oxidase were soaked in harvesting buffer solution (50 mM sodium phosphate, 25% MPD, pH 5.5) containing different H_2O_2 concentrations. H_2O_2 concentrations greater than 20mM caused the crystals to crack, to become brownish and finally to decompose within several hours depending on the H_2O_2 concentration. At 10 mM H_2O_2 concentration the crystals remained blue and did not get cracks for one and a half or two days. A native crystal was soaked for two hours in 10 mM H_2O_2 containing harvesting buffer solution, mounted in the X-ray capillary and immediately used for the X-ray intensity measurements.

The 2.6 Å resolution X-ray structure analysis by difference-Fourier techniques and crystallographic refinement reveals the following picture illustrated in Fig.6. The geometry at the type-1 copper site is not changed compared to the oxidized form. The

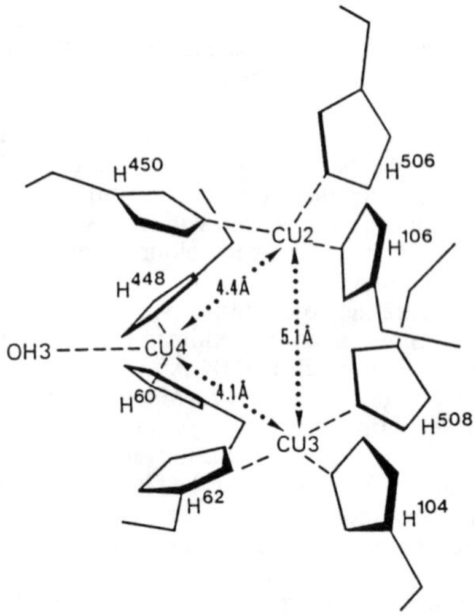

Figure 5. Schematic drawing of the reduced form of ascorbate oxidase around the trinuclear copper site. The included copper-copper distances are the mean values between both subunits.

copper-ligand bond distances averaged about both subunits show no significant deviations from those of the oxidized form. As in the reduced form, the structural changes are remarkable at the trinuclear copper site. The bridging oxygen ligand OH1 is absent, the peroxide binds terminally to the copper atom CU2 as hydroperoxide and the copper-copper distances increase from an average of 3.7 Å to 4.8 Å for CU2-CU3 and 4.5 Å for CU2-CU4. The distance CU3-CU4 remains at 3.7 Å. The mean values of the copper-ligand distances of the trinuclear copper site are again comparable to native oxidized ascorbate oxidase and corresponding copper model compounds.

Copper ion CU3 is three-fold co-ordinated as in the reduced form but the co-ordination by the ligating N-atoms of the corresponding histidines is not exactly trigonal-planar and the CU3 atom is at the apex of a flat trigonal pyramid. The co-ordination sphere around CU4 is not affected and similar in all three forms. Copper atom CU2 is fourfold co-ordinated to the NE2 atoms of the three histidines as in the oxidized form and by one oxygen atom of the terminally bound peroxide molecule in a distorted tetrahedral geometry. Its distance to CU3 increases from 4.8 Å in the oxidized peroxide derivative to 5.1 Å in the fully reduced enzyme. The bound peroxide molecule is directly accessible from solvent through a channel leading from the surface of the protein to the CU2-CU3 copper pair. This channel has already been described earlier[11] and its possible role as dioxygen transfer channel has been discussed. An interesting feature is the close proximity of the imidazole ring of histidine 506 to the peroxide molecule. Histidine 506 is part of one possible electron transfer pathway from the type-1 copper to the trinuclear copper site and could indicate a direct electron pathway from CU1 to dioxygen. It may also help to stabilize important intermediate states in the reduction of dioxygen.

The strong positive peaks at CU2 in both $FO_{NATI}{}^{\#}$-$FO_{PEOX}{}^{£}$ and $FO_{REDU}{}^{\dagger}$-FO_{PEOX} electron density maps could not be explained by a shift of CU2 alone. Occupancies of the copper atoms as well as of the oxygen atoms OH3 and of the peroxide molecule were refined. Type-1 copper CU1 is nearly not affected. Copper atoms CU3 and CU4 are only partly removed but copper atom CU2 is about 50% depleted. The oxygen ligands exhibit full occupancy. The treatment of crystals of ascorbate oxidase with hydrogen peroxide does not only generate a well defined peroxide binding but also a preferential depletion of copper atom position CU2. In the copper depleted molecules the ligating histidine 106 adopts an alternative side chain conformation as detected in the

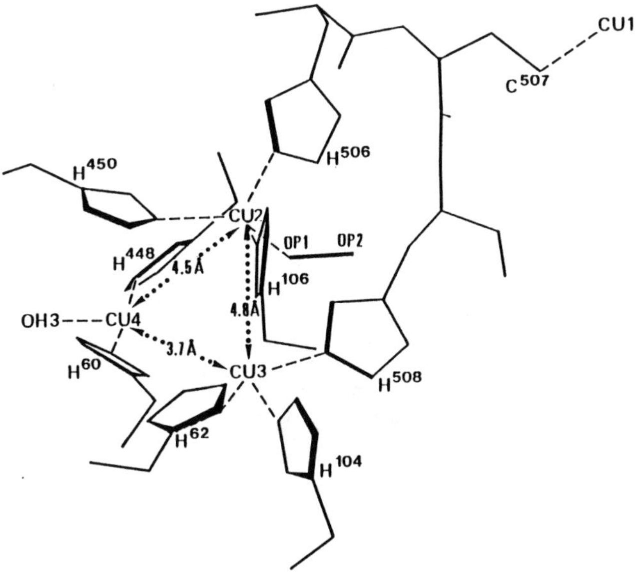

Figure 6. Schematic drawing of the peroxide form of ascorbate oxidase around the trinuclear copper site. The included copper-copper distances are the mean values between both subunits.

2FO-FC-map calculated with the final peroxide derivative model coordinates. This map shows that histidine 106 moves away when copper atom CU2 is removed and opens the trinuclear site even more. From the T2D crystal structure of ascorbate oxidase it is removed to different amounts. The movement of the histidine 106 side chain could explain how this process is accomplished.

Copper depletion may also cause instability of the protein against hydrogen peroxide. Reaction of hydrogen peroxide with ascorbate oxidase in solution in excess leads to a rapid degradation of the enzyme.[31] This can be monitored in the uv/vis PEOX-NATI difference spectrum by a negative band at 610 nm and a positive band at 305 nm. Adding four equivalents hydrogen peroxide per monomer ascorbate oxidase does not lead to enzyme degradation and gives a positive peak at 305 nm indicative for the peroxide binding. Unfortunately, it was not possible to to monitor a uv/vis spectrum of dissolved crystals after X-ray data collection because of the dissociation of the bound peroxide in solution.

The reaction of dioxygen with laccase or ascorbate oxidase has been investigated by several groups and is reviewed in a previous paper[11] where also the possible binding modes of dioxygen to binuclear and trinuclear copper centres are discussed. During its reaction with fully reduced laccase dioxygen binds to the trinuclear copper species and three electrons are very rapidly transferred to it resulting in the formation of an "oxygen intermediate" with a characteristic optical absorption near 360 nm[32,33] and a broad low temperature EPR signal near g = 1.7.[34,35] The type-1 copper is concomitantly reoxidized when the low temperature EPR signal is formed. The oxygen intermediate decays very slowly ($t_{1/2}$ ~ 1 to 15 s) correlated with the appearance of the type-2 EPR signal.[36]

Recently, Solomon and coworkers[37-39] have identified and spectroscopically characterized an oxygen intermediate during the reaction of either fully reduced native tree laccase or T1Hg-laccase with dioxygen. They concluded from their spectroscopic data that the intermediate binds as 1,1-μ hydroperoxide between either CU2 and CU4 or CU3 and CU4. As it is unlikely that the dioxygen migrates or rearranges coordination

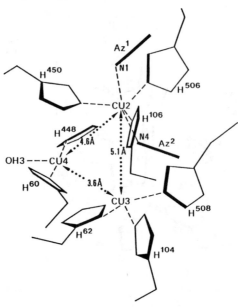

Figure 7. Schematic drawing of the azide form of ascorbate oxidase around the trinuclear copper site. The included copper-copper distances are the mean values between both subunits.

during reduction we propose that the binding site and mode determined in the peroxide derivative of ascorbate oxidase is representative for all reaction intermediates of dioxygen and by homology arguments valid in all blue oxidases. The relevance of this binding mode for the catalytic mechanism will be discussed later.

Azide Form[13]

The azide derivative was obtained by soaking the native crystals in the harvesting buffer solution containing 50 mM sodium azide for 24 hours. Binding of azide was indicated by a change of colour of the crystal from blue to brownish. After X-ray intensity data collection the crystals were dissolved in a solution containing 50 mM azide, 5 mM phosphate buffer, pH 5.5, and 8% MPD and the uv/vis-spectrum was recorded at room temperature. The results of the 2.3 Å resolution X-ray structure analysis by difference-Fourier techniques and crystallographic refinement are depicted in Fig. 7. The geometry at the type-1 copper site is not changed compared to the native form. The copper-ligand bond distances averaged about both subunits show no significant deviations from those of the native form. Again, the structural changes are large at the trinuclear copper site. The bridging oxygen ligand OH1 and water molecule 145 have been removed, CU2 moves towards the ligating histidines and two azide molecules bind terminally to it. The copper-copper distances increase from an average of 3.7 Å to 5.1 Å for CU2-CU3 and 4.6 Å for CU2-CU4. The distance CU3-CU4 is decreased to 3.6 Å. The mean values of the copper-ligand distances of the trinuclear copper site are again comparable to native ascorbate oxidase and corresponding copper model compounds.

The co-ordination of CU3 resembles that in the peroxide form. The threefold co-ordination by histidines is a very flat trigonal pyramid. The co-ordination sphere around CU4 is not affected. CU2 is fivefold co-ordinated to the NE2 atoms of the three histidines as in the reduced form and to the two azide molecules. The two azide molecules are terminally bound at the apexes of a trigonal bipyramid. Both azide molecules bind to the copper atom CU2 which is well accessible from the broad channel leading from the

478

surface of the protein to the CU2-CU3 copper pair. It is not unexpected that the second azide molecule (az^2 in Fig. 7) binds very similar to the peroxide molecule as azide is regarded as a dioxygen analogue. There is no azide molecule bound bridging either CU2 with CU4 or CU3 with CU4.

The binding of azide to laccase as well as to ascorbate oxidase has been studied extensively by Solomon and coworkers,[22,23,40] and by Marchesini and associates[41,42] by spectroscopic techniques. The derived spectroscopic models involve the binding of two azide molecules for laccase and three azide molecules for ascorbate oxidase with different affinities. As the binding of the high affinity azide molecules seemed to generate spectral features related to the type-2 and type-3 coppers, the spectroscopic data were interpreted as the binding of at least one azide molecule as 1,3-μ-bridge between the type-3 copper ions and the type-2 copper ion.

After the X-ray data collection of the azide derivative we dissolved the crystal in azide containing buffer and recorded a uv/vis spectrum to check the spectral properties of our sample. The spectrum is characterized by a broad increase of absorption in the 400-500 nm region and an intense absorption maximum at 425 nm very similar to the results of Casella et al.[41]

There are many structural studies of copper co-ordination compounds with azide ligands mainly of mononuclear and binuclear copper complexes but a few also of trinuclear copper complexes. A comprehensive review for copper coordination chemistry has been written by Hathaway.[43] Azide binds only terminally to mononuclear systems Fivefold co-ordination of nitrogen ligands including azide to Cu(II) is frequently found arranged as a trigonal bipyramid. In binuclear systems azide may bind terminally or bridging as 1,1-μ or 1,3-μ. Similarly two azides may bind di-1,1-μ or di-1,3-μ. The interaction with all three copper ions of a trinuclear complex may be either terminally or bridging as 1,1,1-μ or 1,1,3-μ. We see in the X-ray crystal structure that two azide molecules bind terminally to the type-3 CU2. Azide binding in ascorbate oxidase resembles therefore the binding of azide to an isolated copper ion. In fact there is little interaction of CU2 with CU3 and CU4 which are 5.1 Å and 4.6 Å away, respectively.

The co-ordination of copper ion CU4 in the native oxidized structure is of some interest. It has only three ligating atoms at close distances forming a T-shape co-ordination which is known for Cu(I) complexes (see discussion of reduced form). However, the ligand field is completed if we take into account the π-electron systems of the imidazole rings of histidines 62 and 450 (see Fig. 3). A ligand field with tetragonal-pyramidal symmetry around CU4 is then formed. The shortest distances of CU4 are to CD2 450 with 3.4 Å and to CG 62 with 3.6 Å. These distances are too long for strong copper-π-electron interactions but the histidines will contribute to the CU4 ligand field.

THE CATALYTIC MECHANISM

A tentative catalytic mechanism of ascobate oxidase has been proposed based on the refined X-ray structure and on spectroscopic and mechanistic studies of ascorbate oxidase and the related laccase. The results of these studies were discussed in detail.[11] The X-ray structure determinations of the fully reduced and peroxide derivatives define two important intermediate states during the catalytic cycle. A proposal for the catalytic mechanism incorporating this new information is presented in Fig. 8.

The catalytic cycle starts from the resting form (Fig. 8 (a)) in which all four copper ions are oxidized and CU2 and CU3 are bridged by an OH$^-$ ligand. CU2 and CU3 are most likely the spin-coupled type-3 pair of copper and CU4 is the type-2 copper. The first step is the reduction of the type-1 copper CU1 by the reducing substrate in a one-electron transfer step (Fig. 8 (b)). The electrons are transferred through the protein either to CU2 or CU3. Electron transfer may be through-bond, through-space or a combination of both. A branched through-bond pathway is available leading to CU2 with 9 bonds (including a hydrogen bond) and to CU3 with 11 bonds, respectively. The fully reduced enzyme requires four electrons to be transferred (Fig. 8 (c)). Its structure has been described in a previous section. The hydroxyl bridge between the copper pair has been released and the distance copper atom CU2 and CU3 has been increased to about 5.2 A.

Considerable

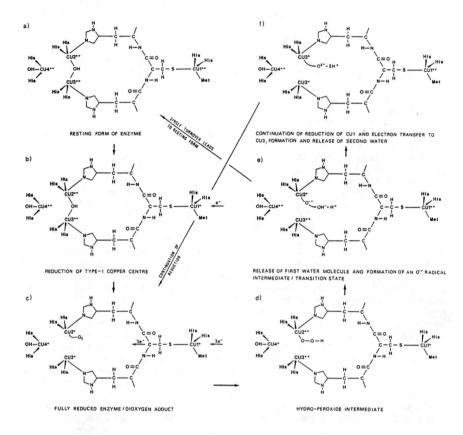

Figure 8. Proposal for the catalytic mechanism of ascorbate oxidase.

reorganisation energy will be necessary to reach this state from the resting form of enzyme. At this stage, dioxygen may bind to the enzyme at CU2 probably in the manner shown in the peroxide derivate described in the respective section. A transfer of two electrons from the copper pair to dioxygen leads to the formation of a hydro-peroxide intermediate (Fig. 8 (d)). A third electron may be transferred from CU4 to the hydro-peroxide intermediate, and a fourth electron from the type-1 copper to copper ion CU2. The O-O bond is broken at this stage and the first water molecule released (Fig. 8 (e)). An oxygen radical has been detected in laccase by EPR as mentioned before. The EPR spectrum indicated that the type-1 copper has been reoxidized and the EPR signals of the oxygen radical intermediate and type-1 copper are present. CU2 is in the reduced state whereas the oxidized copper atoms CU3 and CU4 may be spin-coupled and EPR silent. The reduced CU2 may facilitate O-O bond breakage and release of water. The catalytic cycle is continued by a further reduction of the type-1 copper centre by the reducing substrate. This electron may be transferred to CU3 of the copper pair via the 11 bond pathway. Now, the fourth electron may be transferred to the oxygen radical intermediate from copper atom CU2 and the second water molecule released (Fig. 8 (f)). In case of four electron equivalents supplied only, the reaction may lead to the resting form and the second water may remain bound as bridging ligand between CU2 and CU3 concomitant with a substantial rearrangement within the trinuclear copper site and its coordinating ligands. If turnover is continued this will not occur and the trinuclear copper site may maintain a structure very close to that found in the fully reduced form. Only minor rearrangements will take place at the trinuclear copper site during the catalytic cycle a prerequisite for facile electron transfer reactions. The four protons required for the formation of the two water molecules from dioxygen may be supplied from bulk water through the dioxygen channel via the water molecules bound in the vicinity of CU2 and CU3.

ELECTRON TRANSFER PROCESSES

Three kinds of electron transfer (ET) processes are involved in the catalytic cycle.

(i) Intermolecular Electron Transfer From The Reducing Substrate To The Type-1 Copper Redox Centre.
This takes place in a bimolecular second-order reaction with rates for electron-reduced nitroaromates (ArNO$_2$·⁻) compatible with the turnover number with ascorbate as reducing substrate.[44] The ET from ascorbate to the type-1 copper centre can be even faster and is completed within the deadtime of the stopped flow instrument.[45] It is therefore not the rate limiting step. One electron is transferred from ascorbate and the generated semihydro-ascorbate spontaneously dismutates in solution.[46]

(ii) Intramolecular Electron Transfer From The Type-1 Copper Centre To The Trinuclear Copper Centre.
Long-distance intramolecular ET can be described in the frame of the theory of Marcus.[47] The shortest distances between the type-1 copper (see Fig. 9) and the coppers of the trinuclear copper site are 12.2 Å and 12.7 Å. His506-Cys507-His508 serve as bridging ligand between the two redox centres providing a bifurcated pathway for ET from the type-1 copper centre to the trinuclear species. The difference in redox-potential of the type-1 copper centre and the type-3 coppers, the driving force, measured at 10°C, is $-\Delta G^0$ = 41 mV.[45] However, the binding of dioxygen to the partly reduced protein and the presence of reduction intermediates may affect this redox-potential (Avigliano et al.[48] have found a very slow equilibration between type-1 and type-3 coppers in ascorbate oxidase in the absence of dioxygen). For the reorganization energy, λ, and the electronic coupling factor, β, necessary magnitudes for the calculation of ET rates in the frame of Marcus theory, no estimates can be derived for ascorbate oxidase but reasonable values for proteins are $\lambda = 1$ eV and $\beta = 1.2$ Å⁻¹, according to Gray & Malmsström.[49] These values inserted into the Marcus ET rate equation yield $k_{ET} \sim 10^5$ s⁻¹. Changing β to 1.6 Å⁻¹ gives $k_{ET} \sim 4 \times 10^3$ s⁻¹ a value closer to the observed turnover number of 8×10^3 s⁻¹. McLendon[50] suggested that the ET in proteins may not be designed for very fast intramolecular ET with the exception of light induced ET in photosynthetic reaction centres. They could even be designed to slow down these rapid rates which might otherwise lead to biological "short circuits". Related to this point is the observation that maximal rates for intramolecular ET in organic donor-acceptor molecules with rigid spacers are significantly faster than those for Ru-labelled protein systems at similar distances.[51]

In the case of ascorbate oxidase, the observed ET rates for the reduction of the type-3 coppers with reductate as substrate are in the range of the turnover number.[45] Similar results have been obtained by Meyer et al.[52] who studied the anaerobic reduction and subsequent reoxidation of the type-1 copper by lumiflavin semiquinone using laser flash photolysis. The experimental rates are one order of magnitude slower than the turnover number. This could be due to a different redox potential of the trinuclear copper site under non-physiological anaerobic conditions. ET from the type-1 copper to the type-3 copper pair of the trinuclear copper site may be through-bond, through-space or a combination of both. A through-bond pathway is available for both branches, each with 11 bonds (see Fig. 9). The alternative combined through-bond and through-space pathway from the type-1 copper to copper ion CU2 involves a transfer from the SG atom of Cys507 to the main-chain carbonyl of Cys507 and through the hydrogen bond of this carbonyl to the ND1 atom of the His506.

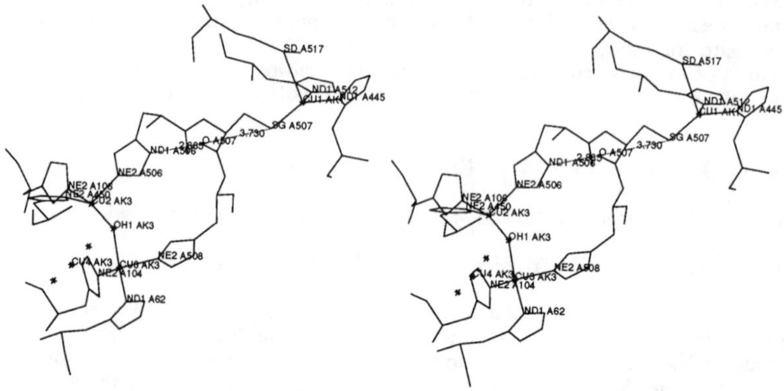

Figure 9. Stereo drawing of the region of the atomic model containing the type-1 copper centre in domain 3 and the trinuclear copper site between domain 1 and domain 3.

(iii) Electron Transfer Within The Trinuclear Copper Site
Electron exchange within the trinuclear copper site is expected to be very fast owing to the short distances between the copper atoms (from 3.7 Å to 5.2 Å in the reduced form) as will be ET to the bound dioxygen.

STRUCTURAL RELATIONSHIP TO LACCASE AND CERULOPLASMIN

Laccase is widely distributed in plants and fungi. The laccase of the Chinese or Japanese lacquer trees (*Rhus* species) is found in white latex, which contains phenols. In the presence of enzyme, these are oxidized by dioxygen to radicals, which spontaneously polymerize. Complete amino acid sequences are only available for fungal laccases (*Neurospora crassa*[53], *Aspergillus niger*[54], and *Phlebia radiata*[55]). However ,metal content and molecular, spectroscopic, catalytic and other properties are so similar that conclusions concerning the overall spatial structure and metal ligation can be applied to plant laccases, as well.

On the basis of the spatial structure of ascorbate oxidase an alignment of the amino acid sequence of the related blue oxidases, ascorbate oxidase, laccase and ceruloplasmin has been carried out.[14] This stronly suggests a three-domain structure for laccase closely related to ascorbate oxidase and a six-domain structure for cerulopasmin. The domains demonstrate homology with the small blue copper proteins. The relationships suggest that laccase, like ascorbate oxidase, has a mononuclear blue copper in domain 3 and a trinuclear copper site between domain 1 and 3 having the canonical copper ligands as in ascorbate oxidase except of the methionine ligand of the mononuclear copper which my be different.

REFERENCES

1. T.E. King, H.S. Mason and M. Morrison, Eds., *Progress in Clinical and Biological Research*, **274**, 1 (1988).
2. R.A. Capaldi, *Annu. Rev. Biochemistry,* **59**, 569 (1990).
3. R. Malkin and B.G. Malmström, *Adv. Enzymol.*, **33**, 177 (1970).
4. G. Chichiricco, M.P. Ceru, A. D'Alessandro, A. Oratore and L. Avigliano, *Plant Sci.*, **64**, 61 (1989).

5. J.A. Fee, *Struct. Bonding,* **23**, 1 (1975).
6. A. Marchesini, P. Cappalletti, L. Canonica, B. Danieli and S. Tollari, *Biochim. Biophys. Acta,* **484**, 290 (1977).
7. V.S. Butt, in *The Biochemistry of Plants. A Comprehensive Treatise. Metabolism and Respiration,* R. Davies, Ed., (Academic Press, London and New York, 1980), pp. 85-95.
8. J. Ohkawa, N Okada, A. Shinmyo and M. Takano, *Proc. Natl. Acad. Sci. U.S.A.,* **86**, 1239 (1989).
9. M. Esaka, T. Hattori, K. Fujisawa, S. Sakojo and T. Asahi, *Eur. J. Biochem.,* **191**, 537 (1990).
10. A. Messerschmidt, A. Rossi, R. Ladenstein, R. Huber, M. Bolognesi, G. Gatti, A. Marchesini, R. Petruzzelli and A. Finazzi-Agro, *J. Mol. Biol.,* **206**, 513 (1989).
11. A. Messerschmidt, R. Ladenstein, R. Huber, M. Bolognesi, L. Avigliano, R. Petruzzelli, A. Rossi and A. Finazzi-Agro, *J. Mol. Biol.,* **224**, 179 (1992).
12. A. Messerschmidt, W. Steigemann, R. Huber, G. Lang and P.M.H. Kroneck, *Eur. J. Biochem.,* in press (1992).
13. A. Messerschmidt, H. Luecke and R. Huber, *J. Mol. Biol.,* submitted (1992).
14. A. Messerschmidt, and R. Huber, *Eur. J. Biochem.,* **187**, 341 (1990).
15. G. D'Andrea, J.P. Bowstra, J.P. Kamerling and F.G. Vliegenthart, *Glycoconjugate,* **5**, 151 (1988).
16. K.D. Karlin, J.C. Hayes, Y. Gultneh, R.W. Cruse, J.W. McKnown, J.P. Hutchinson and J. Zubieta, *J. Amer. Chem. Soc.,* **106**, 2121 (1984).
17. P. Chaudhuri, D. Ventor, K. Wieghardt, E. Peters, K. Peters and A. Simon, *Angew. Chemie,* **97**, 55 (1985).
18. R. Bränden and J. Deinum, *FEBS Lett.,* **73**, 144 (1977).
19. C.T. Martin, R.H. Morse, R.M. Kanne, H.B. Gray, B.G. Malmström and S.I. Chan, *Biochemistry,* **20**, 5147 (1981).
20. L. Morpurgo, A. Desideri and G. Rotilio, *Biochem. J.,* **207**, 625 (1982).
21. M.E. Winkler, D.J. Spira, C.D. LuBien, T.J. Thamann and E.I. Solomon, *Biochem. Biophys. Res. Com.,* **107**, 727 (1982).
22. M.D. Allendorf, D.J.Spira and E.I. Solomon, *Proc. Natl. Acad. Sci. U.S.A.,* **82**, 3063 (1985).
23. D.J. Spira-Solomon, M.D. Allendorf and E.I. Solomon, *J. Amer. Chem. Soc.,* **108**, 5318 (1986).
24. L. Avigliano, A. Desideri, S. Urbanelli, B. Mondovi and A. Marchesini, *FEBS Lett.,* **100**, 318 (1979).
25. A. Merli, G.L. Rossi, M. Bolognesi, G. Gatti, L. Morpurgo and A. Finazzi-Agro, *FEBS Lett.,* **231**, 89 (1988).
26. J.M. Guss, P.R. Harrowell, M. Murata, V.A. Norris and H.C. Freeman, *J. Mol. Biol.,* **192**, 361 (1986).
27. W.E.B. Shepard, B.F. Anderson, D.H. Lewandowski, G.E. Norris, and E.N. Baker, *J. Amer. Chem. Soc.,* **112**, 7817 (1990).
28. Nar, H. *Ph.D. thesis*, Technische Universität München (1992).
29. M.J. Schilstra, P.J.M.W. Birker, G.C. Verschoor and J. Reedijk, *Inorg. Chem.,* **21**, 2637 (1982).
30. T.N. Sorrell and M.R. Malachowski, *Inorg. Chem.,* **22**, 1883 (1983).
31. A. Marchesini and P.M.H. Kroneck, *Eur. J. Biochem.,* **101**, 65 (1979).
32. L.-E. Andreasson, R. Bränden, B.G. Malmström and T. Vänngard, *FEBS Lett.,* **32**, 187 (1973).
33. L.-E. Andreasson, R. Bränden and B. Reinhammar, *Biochim. Biophys. Acta,* **438**, 370 (1976).
34. R. Aasa, R. Bränden, J. Deinum, B.G. Malmström, B. Reinhammar and T. Vänngard, *FEBS Lett.,* **61**, 115 (1976).
35. R. Aasa, R. Bränden, J. Deinum, B.G. Malmström, B. Reinhammar and T. Vänngard, *Biochem. Biophys. Res. Com.,* **70**, 1204 (1976).
36. R. Bränden and J. Deinum, *Biochim. Biophys. Acta,* **524,** 297 (1978).
37. J.L. Cole, P.A. Clark and E.I. Solomon, *J. Amer. Chem. Soc.,* **112**, 9534 (1990).
38. J.L. Cole, D.P. Ballou and E.I. Solomon, *J. Amer. Chem. Soc.,* **113**, 8544 (1991).

39. P.A. Clark and E.I. Solomon, *J. Amer. Chem. Soc.*, **114,** 1108 (1992).
40. J.L. Cole, L. Avigliano, L. Morpurgo and E.I. Solomon, *J. Amer. Chem. Soc.*, **113,** 9080 (1991).
41. L. Casella, M. Gullotti, G. Pallanza, A. Pintar and A. Marchesini, *Biochem. J.*, **251,** 441 (1988).
42. L. Casella, M. Gullotti, A. Pintar, G. Pallanza and A. Marchesini, *J. Inorg. Biochem.*, **37,** 105 (1989).
43. B.J. Hathaway, *Comprehensive Coordination Chemistry,* Vol. 5 (1987).
44. P. O'Neill, E. M. Fielden, A. Finazzi-Agro and L. Avigliano, *Biochem. J.*, **209,** 167 (1983).
45. P.M.H. Kroneck, F.A. Armstrong, H. Merkle and A. Marchesini in *Advances in Chemistry Series, No. 200. Ascorbic Acid: Chemistry, Metabolism, and Uses.* A. Seib and B.M. Tolbert, Eds., (Amer. Chem. Soc., New York, 1982), pp. 223-248.
46. I. Yamazaki and L.H. Piette, *Biochim. Biophys. Acta,* **50,** 62 (1961).
47. R.A. Marcus and N. Sutin, *Biochim. Biophys. Acta*, **811,** 265 (1985).
48. L. Avigliano, G. Rotilio, S. Urbanelli, B. Mondovi and A. Finazzi-Agro, *Arch. Biochem. Biophys.*, **185,** 419 (1978).
49. H.B. Gray and B.G. Malmström, *Biochemistry,* **28,** 7499 (1989).
50. G. McLendon, *Acc. Chem. Res.*, **21,** 160 (1988).
51. S.L. Mayo, W.R. Ellis, Jr, R.J. Crutchley and H.B. Gray, *Science,* **233,** 948 (1986).
52. T.E. Meyer, A. Marchesini, M.A. Cusanovich and G. Tollin, *Biochemistry,* **30,** 4619.(1991).
53. U.A. Germann, G. Müller, P.E. Hunzicker and K. Lerch, *J. Biol. Chem.*, **263,** 885 (1988).
54. R. Aramayo and W.E. Timberlake, *Nucleic Acid Res.*, **18,** 3415 (1990).
55. M. Saloheimo, M.-L. Niku-Paalova and K.C. Knowles, *J. Gen. Microbiol.*, **137,** 1537 (1991).

CYTOCHROME c OXIDASE: A BRIEF INTRODUCTION AND SOME NEW RESULTS FROM HIGH FIELD ENDOR STUDIES OF THE Cu_A AND Cu_B SITES

J. A. Fee[a], W. E. Antholine[b], C. Fan[c], R. J. Gurbiel[c], K. Surerus[a], M. Werst[c], and B. M. Hoffman[c]

[a]Division of Isotope and Nuclear Chemistry, Los Alamos National Laboratory, Los Alamos NM 87545; [b]National Biomedical ESR Center, Medical College of Wisconsin, Milwaukee WI 53226; [c]Department of Chemistry, Northwestern University, Evanston IL 60208

INTRODUCTION

A Mitchellian view of mitochondrial respiration is presented in Scheme I.[1,2] This illustrates electron transfer from an electron rich material, NH, through quinone, Q, through Complex III, into cytochrome c, and through Complex IV(cytochrome c oxidase) to O_2, with accompanying proton translocations across the lipid bilayer.

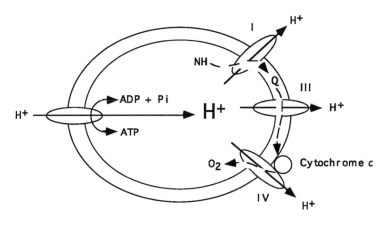

Scheme I

The generally accepted reaction catalyzed by the cytochrome c oxidase is

$$4 \text{ cytochrome } c^{2+}{}_{out} + O_2 + 8 \text{ H}^+{}_{in} \rightarrow 4 \text{ cytochrome } c^{3+}{}_{out} + 2 \text{ H}_2O_{in} + 4 \text{ H}^+{}_{out}$$

From K.D. Karlin and Z. Tyeklár, Eds., *Bioinorganic Chemistry of Copper* (Chapman & Hall, New York, 1993).

where in and out refer, respectively, to the plasma membrane of the bacterial cell or the inner membrane of the mitochondrion. The process is important from both qualitative and quantitative perspectives. Thus, coupling of vectorial reactions in this manner is a fundamental motif of energy transduction in cell respiration, and it accounts for the majority of biological O_2 reduction. To quote Bo Malmström,[3] "cytochrome c oxidase is not just an enzyme but a molecular machine..." that uses the free energy available in favorable electron transfers to physically move protons across the membrane.[2] Considerable progress has been made in elucidating the routes of electron transfer and some details of dioxygen reduction, but the mechanism of proton translocation remains a challenge.

The archetypal cytochrome c oxidase is that from bovine heart tissue, which has been extensively studied.[4-9] This is a complicated enzyme consisting of ~13 different protein subunits and four different metals bound at seven distinct sites; cooperative interactions occur between the various redox active metals, and the protein is capable of presenting several different structural states each characterized by a unique set of properties. The three largest of the subunits are encoded on mitochondrial genes while the remainder are encoded on nuclear genes.[10] For many years it was thought that the enzyme contained only two copper ions and two irons as heme A. However, in 1984, Einarsdóttir and Caughey[11] reported that zinc and additional copper are bound to the enzyme; magnesium was subsequently shown to be present.[12] These observations have been confirmed in other laboratories,[13,14] and the now generally accepted range of metal stoichiometries for the mammalian enzyme is $(2 \leq n \leq 3)$Cu/2Fe/1Zn/1Mg. As will be discussed below, the disposition of the 'extra' copper remains controversial.

Work begun in the early 1980's on the purification and characterization of bacterial cytochromes aa_3 has now led to the concept of a super family which includes all presently known heme and copper containing terminal oxidases.[15-23] In general, bacterial cytochrome and quinol oxidases are isolated as complexes consisting of two, three, or four protein subunits (I, II, III, and possibly IVb). It is now generally accepted that all subunits

Subunit I: the Conserved Histidines

```
Hum   ...VIVTAHAFVMI. .FWFFGHPEVYI. .FIVWAHHMFTVG. .LDIVLHDTYYVVAHFHYV...
Sc    ...VLVVGHAVLMI. .FWFFGHPEVYI. .FLVWSHHMYIVG. .LDVAFHDTYYVVGHFHYV...
Bj    ...VFVTSHGLIMI. .FWFFGHPEVYI. .FVVWAHHMYTVG. .VDRVLQETYYVVAHFHYV...
Tt    ...QILTLHGATML. .FWFYSHPTVYV. .TMVWAHHMFTVG. .LDYQFHDSYFVVAHFHNV...
         68      *       245     *      294    **      372    #     *  *
```

Subunit II: Putative Cu$_A$ Motif

```
Hum   ...PIEAPIRMMITSQDVLHSWAVPTLCLKTDAIPGRLNQTTFTATRPGVYYGQCSEICGANHSFMP.
Sc    ...PVDTHIRFVVTAADVIHDFAIPSLGIKVDATPGRLNQVSALIQREGVFYGACSELCGTGHANMP.
Tt    ...PAGVPVELEITSKDVIHSFWVPGLAGKRDAIPGQTTRISFEPKEPGLYYGFCAELCGASHARML.
Ec    ...PANTPVYFKVTSNSVMNSFFIPRLGSQIYAMAGMQTRLHLIANEPGTYDGISASYSGPGFSGMK.
Ps    .....(NosZ).....DVSHGFVVVNHGVSMEISPQQTSSITFVADKPGLHWYYCSWFCHALHMEMV.
         146               *                          *   *   *   *
```

Fig. 1. (Upper) Partial amino acid sequences of subunits I from selected species showing the conserved histidine residues (*). (Lower) Sequence of the 'CuA motif' in subunits II of selected species and nitrous oxide reductase (NosZ) showing conserved residues that are potential ligands to copper. Hum = *Homo sapiens*, Sc = *Saccharomyces cerevisciae*, Bj = *Bradyrhizobiom japonicum*, Tt = *Thermus thermophilus*, Ec = *Escherichia coli*, Ps = *Pseudomonas stutzeri*. See Refs. 25 and 26 for additional data (*Thermus* numbering).

I are homologous, all subunits II are homologous, and all subunits III are homologous, meaning that each of these proteins evolved from a common ancestor; there are potentially three or more common ancestors for the heme copper oxidase family [cf. Ref. 24]. Fragmentary sequence information in support of this concept is presented in Fig. 1, which shows portions of sequence from subunits I and II from selected organisms.

The answer to the question of how many Cu atoms are associated with the bacterial enzymes is not as firmly established as with the bovine enzyme. In one of our laboratories (JAF), both cytochromes caa_3 and ba_3 from *Thermus thermophilus* appear to contain only two Cu per cytochrome a_3 unit;[17,27] however, broad spectrum metal analyses have not been carried out on these proteins, and Buse [G. Buse, personal communication] finds extra Cu in his preparations of *Thermus* cytochrome caa_3. Bombelk et al.[28] have reported that the two subunit enzyme from *Paracoccus denitrificans* contains an extra equivalent of Cu and Steffens et al.[13] report a metal stoichiometry of $3Cu2Fe0.3Zn(0.5 - 1)Mg$ for the two subunit form of this enzyme. Sone, however, reports that the two-subunit cytochrome caa_3 from *Bacillus* PS3 lacks Zn (see Ref. 29). It thus appears possible that all the aa_3-type cytochromes contain 3Cu/2Fe and possibly other metals. Of concern here is the microstructural features of the bound metals.

Disposition of the Additional Copper
There are several extant views on the nature (and amount) of the additional copper bound to the oxidases. First, it is possible that it is simply accumulated at various sites and is truly extraneous. This must be true to some extent because inclusion of EDTA in purification buffers greatly reduces the amount of copper in the final product (for example, see the work of Malmström and co-workers[30]). In a variation on this, Chan and co-workers suggest that an additional ~0.5 Cu lies at the interface between two aa_3 monomers, where it may be responsible for dimerization, but it is largely reduced and heterogeneous in nature.[31] The most controversial idea suggests that the excess copper is uniquely bound to the protein and acts to form a binuclear arrangement at the Cu_A site (discussed in some detail below).

Zinc and magnesium
Naqui et al.[29] utilized extended X-ray absorbance fine structure to demonstrate that the Zn in bovine oxidase is bound to 3 S and 1 N(O) atoms and pointed out that Zn binding may be to mammalian subunit Vb which contains three conserved cysteine residues (see also Refs. 32 and 33). Yewey and Caughey[34] have suggested Mg may participate in binding of ATP to mammalian subunit IV. However, the roles of these intrinsic, non-redox metals in enzyme function remain to be established.

Redox Active Metals.
A large number of spectroscopic and redox studies indicate the presence of four distinct metal binding sites able to undergo one electron oxidation reduction reactions (See Refs. 3, 4, 7, 8, 35, 46 for review). These are designated cytochrome $a(b)$, Cu_A, and the cytochrome $a_3(o)$ / Cu_B pair. The first two are magnetically isolated from all the others while the last two form a special pair of metals, close together in three dimensional space, which is the site of dioxygen reduction. Cytochrome $a(b)$ appears to always be low-spin and probably resides in subunit I (see below). The Cu_A site is most likely on subunit II (see Ref. 35 for a summary of the rationale on this point). The dominant features of oxidized cytochrome a_3 are that the Fe is high-spin and undergoes spin coupling with the nearby Cu_B; in various states of the enzyme, Cu_B^{2+} can be observed by spin resonance techniques, but it is usually obscured by the spin of the nearby iron and is often refered to as the EPR silent Cu (see below).

Ligand Binding Metals
Two of the cannonical sites appear not to participate in ligand binding reactions; these are Cu_A and cytochrome $a(b)$. The metals of the special pair, $a_3(o)$/Cu_B do, however, coordinate a number of small, exogenous ligands, and this interaction has been widely

used to explore the structure and reactivity of this site. Indeed, Keilin and Hartree[36] demonstrated already in 1939 that CO interacted with cytochrome a_3 but not with cytochrome a, leading to a plethora of studies on and uses of this interaction.[37,38] Alben and co-workers[39] demonstrated that CO could be 'photolytically transfered' from reduced cytochrome a_3 to Cu_B^+ at low temperatures, and Woodruff and Dyer and their co-workers[40] showed that this occured transiently at room temperature as well. van Gelder and co-workers[41,42] demonstrated that NO binds at the special pair, making possible a variety of elegant studies of this site (see Ref. 43 and 44). Caughey and co-workers[45] used infrared spectroscopy to demonstrate that CN^- could bind directly to Cu_B^{2+} in the oxidized form of the protein. Many studies have demonstrated that this special pair of metals is the site of dioxygen reduction (see Refs. 4 and 46), and the oxygen analogs CO, CN^- and NO remain effective probes of this site's structure and reactivity.

Enzyme Mechanism
Much work has been done on mechanism of the enzyme (see Ref. 46 and references therein). Suffice it to say here that electrons flow from cytochrome c into the Cu_A and cytochrome $a(b)$ sites from whence they are transfered to the special pair and into dioxygen. Recent work from Wikström's laboratory suggests that coupling to proton translocation occurs at the site of dioxygen reduction, although the detailed mechanism of this remains unknown (see Refs. 46, 47 and references therein).

Histidine Coordination
Histidine residues have long been suspected to play important roles in redox metal binding. Indeed, physico-chemical measurements indicate 8 to 9 histidines are involved with redox metal coordination: 2 to cytochrome $a(b)$, 1 to cytochrome a_3 as the proximal ligand, 2 to the Cu_A site (see below), and 3 or 4 to Cu_B. The currently large number of known amino acid sequences combined with recent site directed mutagenesis studies, have made it possible to assign the individual, conserved histidine residues to the four spectroscopically distinct centers (see below). As shown in Fig. 1, there are six conserved histidine residues in subunit I and an additional histidine residue that is present in all sequences excepting that of the bacterium *Bradyrhizobium japonicum*.[48,49] There are two histidines conserved in subunit II sequences, amongst a number of other conserved residues, in a region we refer to as the 'Cu_A motif'. Notably, subunit II from the *E. coli* cytochrome *bo* lacks these conserved histidines as well as the two conserved cysteines, and spectroscopic studies show that it lacks a Cu_A site as well. Subunit III contains two conserved histidine residues, but because subunit III can be selectively removed from mammalian cytochrome oxidase and some bacterial enzymes lack this subunit when purified, these have not been implicated in redox metal binding.

Distribution of the Cannonical Metals and a Rough Three Dimensional Model
The redox active metals are almost certainly bound to subunits I and II. Recently, much work has been done on sequence analysis and site-directed mutagenesis that has led to tentative ligand assignments and to a preliminary three dimensional model. The progression of this is idicated in Figs. 2 and 3. Fig. 2(upper) shows the sequence of the*Thermus* subunit I molecule arranged to present the 12 conserved hydrophobic stretches as transmembrane helices. Similarly, Fig. 2 shows the sequence of the *Thermus* subunit II sequence revealing its two putative transmembrane helices and its Cu_A motif (for simplicity the covalently attached cytochrome c domain has not been shown.[25] The Cu_A site is thus well outside the membrane, and the copper may be coordinated by the conserved His, Cys, and Met residues in the Cu_A motif (see below). The six conserved histidine residues in subunit I are indicated as squares while the nearly conserved 'seventh' histidine residue is indicated as a hexagon between helices IX and X. If the hydrophobic helices are constrained to reside within the ~ 80 Å of the lipid bilayer, the conserved histidine residues will be near the outer surface of the membrane. Elegant site-directed mutagenesis studies from the laboratories of Gennis,[50] Ferguson-Miller,[51] and Anraku[52] as well as the extensive sequence analyses of Mather et al.[26] permit one to arrange helices

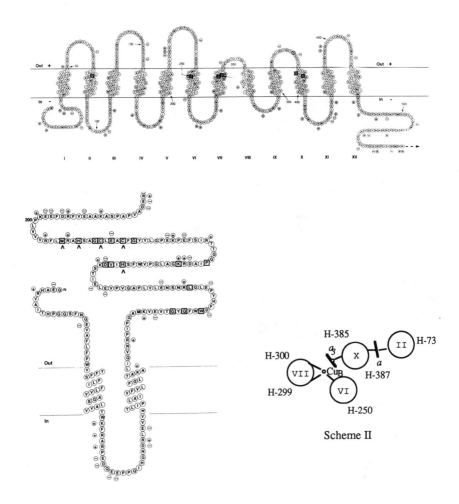

Fig. 2. Topological models for subunits I and II of *Thermus* cytochrome *caa₃*. Both are truncated because in this organism subunit I is apparently fused to subunit III[25] and subunit II is fused to a cytochrome *c*.[26] Conserved residues are denoted in squares. The 'seventh' histidine is between helices IX and X and is denoted in a hexagon. Putative ligands to the Cu$_A$ site are underscored by a carat.

II, VI, VII, and X of subunit I in two dimensions as shown in Scheme II, in which one is looking from the outside of the membrane along the helical axes. Here we indicate the coordination of His-250, His-299, His-300 to Cu$_B$, His-385 to cytochrome *a₃*, and His-73 and His-387 to cytochrome *a* (*Thermus* numbering[26]). One of us (JAF) has constructed a rough three-dimensional model of this arrangement, and a view of the hemes is shown in the stereo photo of Fig. 3. This shows very approximately how the three redox metals of subunit I may be distributed. Holm et al.[53] described a rough three dimensional model for subunit II that involved two hydrophobic transmembrane helices (see Fig. 2).

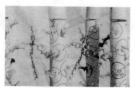

Fig. 3. Stereo photo model of the cytochrome *a* and *a3* / Cu_B portion of *Thermus* cytochrome *caa3* (see Ref. 35 for details).

Electrochemical studies of intact mitochondria are consistent with the idea that Cu_A is close to the membrane surface,[54] and a variety of spin interaction studies suggest that the Cu_A is some $10 \pm \sim 3$ Å away from the cytochrome *a*. [55,56]

THE Cu_A CENTER

The defining features of the Cu_A center are its unusual EPR spectrum and its intense (for copper) absorption band in the near infrared (both illustrated in Fig. 4). A unique MCD spectrum centered at ~500 nm has been identified by Thomson and co-workers.[57] The EPR spectrum is unique in not exhibiting any copper hyperfine structure at X-band frequency $(A(Cu) \leq 100$ MHz$)$[59] and having one of its g-values below that of the free electron $(g_z = 2.18, g_y = 2.03,$ and $g_x = 1.99)$.[30] These g-values are remarkably similar to those observed with thiyl radicals (see references in Ref. 60). The intensity of the Cu_A signal corresponds to some 30 to 40 % of the total copper present (*cf.* Ref. 30). The ~800 nm absorption band was assigned to a ligand-to-metal charge transfer transition between Cu^{2+} and a sulfur ligand[61]; such bands are common in Type 1 'blue' copper proteins.[62] At present a consensual explanation of the EPR properties does not exist.

The first attempt to rationalize the properties of the EPR observable Cu involved spin interaction between two Cu^{2+} sites (see comments of R. H. Sands in Foot-

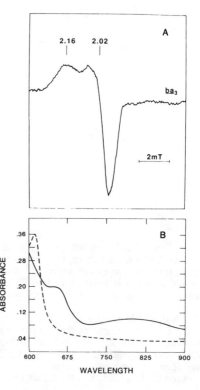

Fig. 4. Demonstration of the unique spectral properties of the Cu_A site in cytochrome *ba3*. Taken from the Ph.D. thesis of B. Zimmermann.[58]   →

note 4 of Ref. 61). However, all other ideas have involved a combination of copper and sulfur in the Cu$_A$ site. Based on early compositional analyses,[30,61] it is still widely thought that there are two Cu ions in the protein, one of which can be observed by EPR, Cu$_A$, and the EPR silent Cu$_B$; the two being magnetically isolated from each other. The late Peter Hemmerich[63] suggested that Cu^{2+} might interact with a disulfide in such a way that the "valence would be indeterminable either if two copper nuclei should interact with one disulfide of if copper should stabilize disulfide as a one-electron acceptor, i.e., if it should stabilize 'sulfur radical', by what is called 'valence isomerism'." This was later modified by Peisach and Blumberg[64] to a Cu$^+$-•SR type of structure. These ideas are still viable and have set the tone for research in this area to the present day. Key to the development of Cu/S based hypotheses was the observation that there existed a short region of sequence in subunit II which contains two conserved cysteine residues, one conserved methionine, and two conserved histidine residues and having some overall similarity to the binding sites of 'blue' copper proteins.[65] This so-called 'Cu$_A$ motif' is shown in Fig. 1. In recent years, several authors hypothesized that a nominally tetrahedral (His)$_2$(Cys)$_2$ coordination could account for the observed results. Indeed, Chan and co-workers have gone so far as to develop a molecular orbital scheme based on this idea.[66] Covello and Gray[67] have suggested that methionine may also be a ligand to copper in the Cu$_A$ site. Overall, however, the supporting data for these models have not been as strong as one would like, and they remain structural hypotheses.

Kroneck, Antholine, Zumft and their co-workers[68-70] have caused considerable excitement in this area by suggesting a bi-nuclear Cu site in which one is formally cuprous and one is formally cupric, with the electronic spin being distributed equally over both metals (Cu$^{1.5+}$-Cu$^{1.5+}$). Their arguments are based to a large extent on analysis of and comparison to a similar center in nitrous oxide reductase[38,71-75] which also contains a 'Cu$_A$ motif' in its amino acid sequence (see Fig. 1); however, note the two additional histidine residues in this sequence. These are intriguing analyses, but the interpretation would be more readily accepted if it were unequivocally established that the Cu$_A$-type site in nitrous oxide reductase was a binuclear center. This a rather difficult determination, because the enzyme contains multiple copper ions. Recently, the availability of a catalytically inactive form of N$_2$O reductase containing, on average, ~2 Cu / 120 kDa and

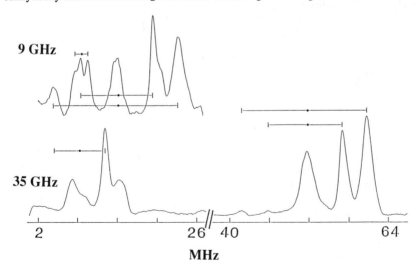

Fig. 5. Comparison of ENDOR spectra of the Cu$_A$ site in bovine cytochrome *aa$_3$* recorded at 9 GHz and 35 GHz.

having only the Cu_A type signal should facilitate testing this hypothesis.[76,77] The 'binuclear' hypothesis accounts for the 'extra' copper in cytochrome c oxidase, permits rather accurate simulations of the EPR spectrum, and it should be pursued (see Kroneck et al. in this volume).

ENDOR (electron nuclear double resonance) spectroscopy is a powerful tool for detecting ^1H and ^{14}N nuclei in contact with the electronic spin of protein bound paramagnetic centers.[78] Scholes and his co-workers pioneered ENDOR studies of Cu_A, making the initial observations, suggesting assignments of observables, and working with the Chan laboratory in examining isotopically labeled yeast cytochrome c oxidase.[66,79-81] Hoffman and his co-workers[60] have identified ENDOR signals arising from $^{63,65}Cu$. Unfortunately, the interpretation of 9 GHz ENDOR spectra is severely complicated by the overlap of ^1H and ^{14}N signals, as shown in Fig. 5 (9 GHz). Nevertheless, these efforts have contributed greatly to our understanding of the Cu_A site, and the salient features of this work may be summarized as follows: (1) The multiple ^1H signals arise from protons that are very strongly coupled to the electronic spin of the Cu^{2+} having coupling constants, $A(^1H)$ between 12 and 20 MHz, similar to those observed in Type 1 'blue' copper proteins.[82] As in the latter, these were assigned to β-CH_2 protons of cysteine thiolate ligands. Isotopic substitution of β-C^2H_2 for β-C^1H_2 and β-$^{13}CH_2$ for β-$^{12}CH_2$ yielded evidence for at least one cysteine in the Cu_A site[66,80]. (2) There are also multiple ^{14}N resonances in the envelope between 2 and 16 MHz and one was assigned having $A(^{14}N) = 17$ MHz. Substitution of L-[δ,ε-$^{15}N_2$]histidine for L-[δ,ε-$^{14}N_2$]histidine provided evidence for at least one histidine coordinated to Cu in the Cu_A site.[80] The coupling constants are some 20 - 40 % less than the ^{14}N resonances observed in Type 1 'blue' copper proteins sites, even though both evidently arise from histidine residues. If

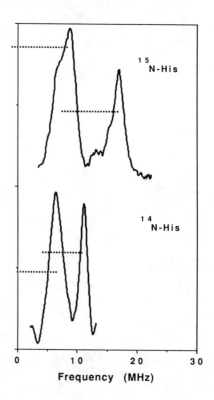

the electrons are shared equally in a bi-nuclear site, then the value for $A(^{14}N)$ is half that expected in a monomeric site. Thus, a value of ~17 x 2 = 34 MHz suggests that the proposed bi-nuclear site should be comprised of 'blue' type sites for which the range of $A(^{14}N)$ is 22 - 40 MHz (see Kroneck et al., this volume). (3) At much higher frequencies than shown in Fig. 5 one finds additional, much broader signals, that have been assigned to $^{63,65}Cu$. Field dependent studies have yielded the principle values of the |A|-tensor: $A_z = 90$, $A_y = 98$, and $A_x = 68$ MHz.[60] ENDOR measurement of the major hyperfine couplings within the Cu_A EPR envelope is precedent to complete simulation of the EPR spectrum.

We have been using the *Thermus* cytochromes *caa3* and *ba3* to explore the nature of this site through a combination of stable isotope labeling and high frequency ENDOR spectroscopy. At 35 GHz, the ^1H and ^{14}N signals are widely separated in

Fig. 6. 35 GHz ^{14}N and ^{15}N ENDOR spectra of Cu_A in *Thermus* cytochrome *caa3*. The upper spectrum was obtained from cells grown in the presence of [U-^{14}N]histidine while the lower spectrum was obtained from a His⁻ strain of *T. thermophilus* grown in the presence of [δ,ε-$^{15}N_2$]histidine. Details are given in Gurbiel et al., in preparation.   →

frequency (*cf.* Fig. 5) with no loss of spectral resolution. While this advantage is obvious, an additional advantage is the extreme sensitivity of high-frequency ENDOR spectroscopy. For example, it is possible to obtain high signal-to-noise spectra on as little as 50 µl of 50 µM Cu_A. This is ideal for working with the expensive and often hard-to-obtain, isotopically labeled bacterial oxidases.

While our work with these systems is far from complete, we provide one notable result and show ^1H spectra that are provocative but not completely interpreted. Fig. 6 compares ^{14}N ENDOR obtained from *Thermus* cytochrome caa_3 containing L-[U-^{14}N]histidine with the same enzyme enriched with L-[δ,ε-$^{15}N_2$]histidine. These spectra provide unequivocal evidence for the presence of two, unequal histidine residues in the Cu_A site. Moreover, careful field-dependence studies can yield detailed structural information about the environments of the two ligands, and these are underway (R. J. Gurbiel et al., in preparation).

Fig. 7. 35 GHz ^1H spectra obtained from bovine cytochrome aa_3 (top), *Thermus* cytochrome caa_3, and *Thermus* cytochrome ba_3 (bottom).   →

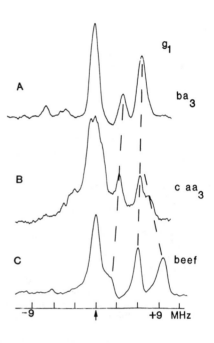

Fig. 7 shows the ^1H spectra for bovine cytochrome aa_3 and the two *Thermus* enzymes. At first glance, these spectra suggest that more than 4 strongly coupled protons may be involved in the aa_3 type enzymes and fewer than 3 in the cytochrome ba_3. Because they all arise from a fundamentally similar site, sorting out these ^1H signals will require ^2H substitutions, possibly requiring direct observation of ^2H.

THE Cu_B CENTER

EPR spectra of oxidized cytochrome c oxidases fail to reveal the presence of one copper, accordingly there is relatively little information available about this EPR silent Cu, which is also called Cu_B or Cu_{a3}. In the fully oxidized form of the protein, the $S = 1/2$ spin of Cu_B^{2+} is coupled to the nearby $S = 5/2$ ferric ion of cytochrome a_3. The exact nature of this coupling is not yet defined.[83,84] Upon partial reduction of the mammalian enzyme, Cu_B^{2+} is reduced in preference to cytochrome a_3, and the heme is revealed by a g6 EPR signal arising from the now isolated high-spin ferric ion. Three methods were discovered to break the spin coupling between the heme and Cu_B^{2+}, thus permitting its EPR signal to be recorded.

The first of these involves reoxidation of the enzyme by dioxygen at 173 K, at which temperature binding of oxygen and subsequent electron transfer reactions take place.[85] The signal can only be observed at low temperature and high microwave power, and, while the spectrum contains four distinct copper hyperfine lines, it does not arise from a typical isolated Cu^{2+} center. When reoxidation is carried out with [$^{17}O_2$]dioxygen, each of these lines undergoes significant broadening, indicating that ^{17}O is bound to the Cu or is in very close proximity. Hansson et al.[85] have interpreted these observations in terms of an ionically coordinated Cu^{2+} in weak spin coupling with an $S = 2$, Fe(IV)=O form of the nearby cytochrome a_3. The second method also reveals Cu_B that is spin coupled to the nearby heme. Stevens et al.[86] discovered that the addition of N_3^- to the NO-complex of oxidized cytochrome c oxidase resulted in the reduction of the ferric a_3-NO complex to the ferrous state followed by binding of NO; the ferrous-a_3-NO complex should have $S = 1/2$ and be readily observed by EPR. However, a triplet EPR spectrum is observed having both $\Delta M = 1$ and $\Delta M = 2$ transitions. The so-called half-field resonances show clear evidence of Cu hyperfine, which were attributed to Cu_B.

The third method involves reoxidation of the fully reduced enzyme at room temperature.[87,88] This reveals a fairly typical Cu^{2+} EPR signal that has been quantified, simulated and examined by ENDOR spectroscopy. The experiment can be carried out in such a way as to account for 0.45 Cu per aa_3 unit. It's relevant parameters are $g_z = 2.28$, $g_y = 2.11$, $g_x = 2.05$, and $A_z = \sim100$ gauss, which are similar to those obtained from isolated Type 3 Cu^{2+} sites in half-met hemocyanin and in the laccases (see Ref. 87); they are remarkably similar to those of Cu^{2+} in the Zn^{2+} site of bovine superoxide dismutase,[89] which has $g_z = 2.31$ and $A_z = \sim100$ gauss, and which is clearly rhombic in nature. The latter site consists of three coordinating imidazole rings and one carboxylate ion.[90] Cline et al.[91] examined the ^{14}N ENDOR signals obtainable from this signal with a 9 GHz spectrometer. They report the presence of at least 3 distinct ^{14}N residues with coupling constants close to those reported earlier for $[Cu(imidazole)_4]^{2+}$.[47] These observations are consistent with Cu_B being coordinated to at least three histidine residues.

Working with cytochrome ba_3 from *Thermus thermophilus*, we have found yet another way to reveal Cu_B^{2+} (Ref. 92). The protein contains a functional Cu_A site, cytochrome b replaces the cytochrome a of aa_3 and Cu_B appears to be functionally the same as in cytochrome aa_3.[17] Thus, the designation cytochrome ba_3 originally described as a single subunit analog of cytochrome aa_3.[17] However, recent results from molecular genetic studies in one of our laboratories (JAF) reveal that ba_3 is more likely a homolog of the heme-copper oxidase family. We have cloned and obtained partial sequences from two genes encoding subunits I and II of this enzyme (J. A. Keightley et al., unpublished). Our present analyses of these sequences suggest that cytochrome ba_3 contains at least two subunits that are homologous to subunits I and II of the heme-copper oxidase family; subunit I contains the 7 nearly conserved histidines and subunit II has a typical Cu_A motif. In contrast to mammalian cytochrome aa_3, *Thermus* cytochrome ba_3 reacts with cyanide yielding low-spin ferrous cytochrome a_3 and revealing an $S = 1/2$ Cu_B^{2+} signal (see below).

Oxidized bovine cytochrome aa_3 (and *Thermus* cytochrome caa_3) reacts with cyanide to form a low-spin ferric heme that is spin-coupled to Cu_B^{2+}. Infrared spectral studies show that CN^- is bound to the Cu but suggest that CN^- is not bound to the heme.[45] However, Mössbauer studies[93,94] and recent resonance Raman studies (W. A. Oertling et al., in preparation) clearly indicate that CN^- is bound to the heme. Finally, both MCD[95] and Mössbauer[93] results are consistent with ferromagnetic coupling between the Cu_B^{2+} and the Fe_{LS}^{3+}. As recently reviewed,[35] the best interpretation of the extant data suggests a schematic structure

$$Fe_{LS}^{3+}-CN^--[(His)_{n \geq 3}]Cu_B^{2+}-CN^-$$

in which the Fe_{a3}^{3+}, a bridging CN^-, and Cu_B^{2+} are co-linear and an additional CN^- is bound to Cu_B^{2+}, but it need not be co-linear; at least three histidine residues appear to be contributing ligands to the Cu. Not surprisingly, this cyano complex does not exhibit an EPR signal.

The reaction of cyanide with cytochrome ba_3 is different primarily because some form of auto-reduction occurs (probably with oxidation of CN^- to cyanogen). Thus,[92] optical and resonance Raman studies indicate that the A heme of ba_3 is reduced and converted to its low-spin form, and resonance Raman spectra are consistent with the presence of a linear Fe-CN arrangement. EPR spectra show that the Cu_A site is unperturbed by this treatment and that an additional Cu^{2+} signal appears. This signal has

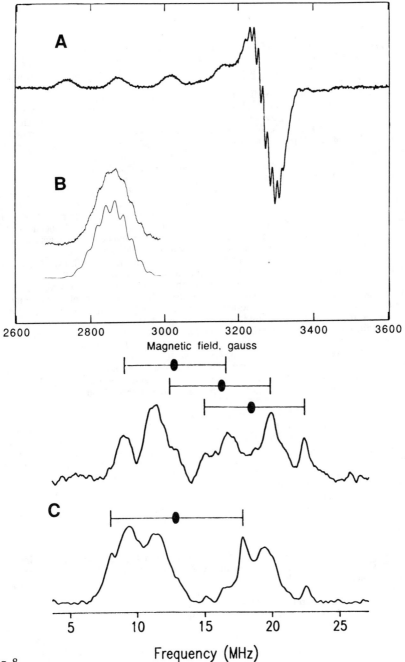

Fig. 8

495

parameters $g_\parallel = 2.28$, $g_{x,y} = 2.07$, and $A_\parallel = 140$ gauss, typical of a Type 2 copper, represents ~1 Cu per functional ba_3 unit (see Fig. 8A), and it was assigned to Cu_B^{2+}. Upon addition of dithionite the EPR signals from low-spin ferric cytochrome b and the Cu_A site disappear, but that from the new Cu_B^{2+} signal is unchanged (W. A. Oertling et al., in preparation).

The numerous superhyperfine lines near $g_{x,y}$ indicate the coupling of the Cu_B^{2+} spin to several ^{14}N nuclear spins. This region is notoriously difficult to analyze because $A(^{14}N)$ may ~ $A(^{63,65}Cu)$. However, more concentrated samples of ba_3, from which very high signal-to-noise EPR can be obtained, reveal well-resolved ^{14}N superhyperfine on the $M_I = -3/2$ line of the Cu_B^{2+} signal (Fig. 8B). This structure arises uniquely from interactions of the cupric electronic spin with ^{14}N nuclei, and we have analyzed this spectrum to 'count' the number of ^{14}N nuclei bound to Cu_B^{2+} in the ba_3-cyano complex. As a first step, we examined the 35 GHz ENDOR spectrum and identified at least three ^{14}N signals in the region of 15 to 25 MHz having $A(^{14}N_i)$ ~37, 32, and 28 MHz (Fig. 8C). There is significant overlap in this region, and additional, underlying signals cannot be excluded. Nevertheless, we attempted to simulate the low-field Cu-hyperfine line with these parameters. While not shown here, we obtained an inadequate fit because the computed line was much too narrow and the number of superhyperfine lines was smaller than what we observe. Next, we included an additional ^{14}N with a coupling constant equal to $A(^{14}N)$ ~ 37 MHz and obtained the calculated spectrum shown in Fig. 8B, which is quite a good fit to the experimental spectrum - any A-value between 28 and 37 MHz gave an acceptably similar result. The results of these calculations strongly suggest that Cu_B^{2+} in the cyano complex of ba_3 is coordinated to 4 ^{14}N atoms. Further, because these coupling constants are very similar to those observed with $[Cu(imidazole)_4]^{2+}$ (Ref. 47) we suggested the ligands may be imidazole rings of histidine residues.

Additional ENDOR experiments involved the detection of a ^{13}C signal from $^{13}C^{14}N^-$ but not a ^{15}N signal from $^{12}C^{15}N^-$ (see Fig. 8C). The coupling constant (~ 12 MHz) to ^{13}C is very much less than observed with superoxide dismutase and copper carbonic anhydrase cyano complexes[96,97] which have coupling constants of ~143 and 345 MHz, respectively. We concluded from this large difference that the cyanide bound to Cu_B^{2+} was an axial ligand, whereas CN^- is probably an equatorial ligand in the other two systems (see Refs. 98, 99).

One further point needs to be made regarding the structure of the ba_3-cyano complex. Cyanide is considered to be a very strong ligand to Cu, and generally defines the equatorial plane of a Cu-CN tetragonal complex.[41] However, it is possible to fix the equatorial plane in a cyclic chelate in such a way that CN^- is constrained to remain an axial ligand. This is evident in the work of Anderson and co-workers (see Ref. 100) who have described various of these in some detail. While we are lacking a structure, it is clear that CN^- is not an equatorial ligand in the ba_3-cyano complex. This implies that a very strong tetragonal ligand field is supplied by endogenous ligands, with CN^- in an apical position. Such an arrangement is fully consistent with the observation (W. A. Oertling et al., in preparation) that cyano-ba_3 is not reduced by dithionite; this complex must have a very low redox potential.

← Fig. 8. Portion A shows the X-band EPR spectrum which appears on reaction of oxidized *Thermus* cytochrome ba_3 with HCN. The spectrum of the Cu_A center is unperturbed and has been subtracted. The inset (B) shows an expanded view of the $M_I = -3/2$ line, (upper) experimental trace, (lower) computed (see text). Portion C shows 35 GHz ENDOR spectra from Cu_B^{2+}; signals below 15 MHz arise primarily from Cu_A while signals above 15 MHz arise from Cu_B^{2+}. The upper trace was recorded from enzyme treated with $HC^{15}N$ (there are no signals attributable to ^{15}N); the three assigned ^{14}N resonances are marked. The lower trace was recorded from enzyme treated with $H^{13}CN$ and shows a new signal arising from ^{13}C superimposed on the ^{14}N signals. Adapted from Ref. 92.

Present data are rationalized by the following schematic structure

$$Fe_{LS}^{2+}\text{-}CN^- ::: [(His)_4Cu_B]^{2+}\text{-}CN^-$$

where the Fe-CN bond is likely to be linear, the ::: symbols indicate that the two metals are not strongly bridged by the cyanide but are probably very close to each other, the Cu_B^{2+} is equatorially coordinated to 4 imidazole rings of histidine, and another CN^- is likely to be an axial ligand to the copper; the formal charge on such a complex would be zero.

These observations raise some interesting questions about the differences between cytochromes aa_3 and ba_3. We are currently analyzing resonance Raman data which indicate that subtle differences between the environment of a_3 in aa_3 and ba_3 may contribute to a higher redox potential for the a_3 of ba_3. These data suggest that when exposed to a (weak) reducing agent as well as a strongly coordinating ligand, ba_3 is more likely to become reduced and annated. In addition, the Fe-CN bond in aa_3 is probably not linear whereas the Fe-CN bond in ba_3 appears to be linear. These differences are presented as tentative reasons for the obvious difference in chemical behavior of the two enzymes (W. A. Oertling et al., in preparation). Note that the proposed structures for the two cyano complexes are not greatly different from each other.

An additional question concerns the number of histidines coordinated to Cu_B in the native enzyme. The site-directed mutagenesis results of Shapleigh et al.[51] have been interpreted to mean that only His-250, His-299, His-300 are coordinated to Cu_B while His-377 is not involved in metal coordination. However, there are pitfalls in these limited observations and one should not dismiss the fact that only one of some 36 different subunit I sequences is missing this residue (substitution with glutamine). Importantly the enzyme from *B. japonicum* has neither been purified nor carefully characterized *in situ*; perhaps it has significantly different properties? Further, it is possible that His-377 binds to Cu_B only in certain states of the enzyme. Thus, in the cyano complexes, His-377, which lies only a short distance from Cu_B in our model may have swung in to coordinate the Cu. If this is the case, the three sets of data are reporting on different aspects of the problem. Future studies will need to address these differences.

ACKNOWLEDGEMENTS

This work was supported National Institutes of Health Grants GM35342 (JAF), HL13531 (BMH), and RR 01008 (WEA) and by National Science Foundation grant DMB-9105519 (WEA); by a LANL Director's Postdoctoral Fellowship to KKS; the ^{15}N labeled histidine used in this work was supplied by the National Stable Isotope Resource at Los Alamos (NIH grant RR02231); work done at Los Alamos was carried under the auspices of the U. S. Department of Energy. We thank our collaborators W. A. Oertling, R. B. Dyer, and W. H. Woodruff for permission to mention unpublished results and Tatsuro Yoshida for assistance in photographing the model.

REFERENCES

1.   P. Mitchell *Science* **206**, 1148 (1979).
2.   P. Mitchell *J. Biochem. (Tokyo)* **97**, 1 (1985).
3.   B.G. Malmström *Chem. Revs.* **90**, 1247 (1990).
4.   M. Wikström, K. Krab, M. Saraste *Cytochrome Oxidase: A Synthesis*; (Academic Press, London, 1981), pp 198.
5.   B.G. Malmström *Arch. Biochem. Biophys.* **280**, 233 (1990).
6.   B.G. Malmström *FEBS Lett.* **250**, 9 (1989).
7.   S.I. Chan and P.M. Li *Biochemistry* **29**, 1 (1990).
8.   G. Palmer *Pure & Appl. Chem.* **59**, 749 (1987).
9.   R.A. Capaldi *Arch. Biochem. Biophys.* **280**, 252 (1990).

10. B. Kadenbach, L. Kuhn-Nentwig, U. Büge *Current Topics in Bioenergetics* **15**, 113 (1987).
11. O. Einarsdóttir and W.S. Caughey *Biochem. Biophys. Res. Commun.* **124**, 836 (1984).
12. O. Einarsdóttir and W.S. Caughey *Biochem. Biophys. Res. Commun.* **129**, 840 (1985).
13. G.C.M. Steffens, R. Biewald, G. Buse *Eur. J. Biochem.* **164**, 295 (1987).
14. L.-P. Pan, Q. He, S.I. Chan *JBC* **266**, 19109 (1991).
15. J.A. Fee, M.G. Choc, K.L. Findling, R. Lorence, T. Yoshida *Proc. Nat'l. Acad. Sci. U.S.A.* **77**, 147 (1980).
16. B. Ludwig and G. Schatz *Proc. Nat'l. Acad. Sci. U. S. A.* **77**, 196 (1980).
17. B.H. Zimmermann, C.I. Nitsche, J.A. Fee, F. Rusnak, E. Münck *Proc. Nat'l. Acad. Sci.* **85**, 5779 (1988).
18. K. Hon-nami and T. Oshima *Biochemistry* **23**, 454 (1984).
19. N. Sone and Y. Yanagita *Biochim. Biophys. Acta* **682**, 216 (1982).
20. V. Chepuri, L. Lenieux, D.C.-T. Au, R.B. Gennis *J. Biol. Chem.* **265**, 11185 (1990).
21. K. Kita, K. Konishi, Y. Anraku *J. Biol. Chem.* **259**, 3368 (1984).
22. K. Shoji, T. Yamazaki, T. Nagano, Y. Fukumori, T. Yamanaka *J. Biochem. (Tokyo)* **111**, 46 (1992).
23. M. Saraste, H. L., L. Lemieux, J. van der Oost *Biochemical Society Transactions* **19**, 608 (1991).
24. R.F. Doolittle *Of URFS and ORFS: A primer on how to analyze derived amino acid sequences*; (University Science Books, Mill Valley, California, 1986), pp 3.
25. M.W. Mather, P. Springer, J.A. Fee *J. Biol. Chem.* **266**, 5025 (1991).
26. M.W. Mather, P. Springer, S. Hensel, G. Buse, J.A. Fee *J. Biol. Chem.* **In press.**, (1993).
27. T. Yoshida, R.M. Lorence, M.G. Choc, G.E. Tarr, K.L. Findling, J.A. Fee *J. Biol. Chem.* **259**, 112 (1984).
28. E. Bombelka, F.W. Richter, A. Stroh, B. Kadenbach *Biochem. Biophys. Res. Commun.* **140**, 1007 (1986).
29. A. Naqui, L. Powers, M. Lundeen, A. Constantinescu, B. Chance *J. Biol. Chem.* **263**, 12342 (1988).
30. R. Aasa, S.P.J. Albracht, K.E. Falk, B. Lanne, T. Vänngård *Biochem. Biophys. Acta* **422**, 260 (1976).
31. L.-P. Pan, Z. Li, R. Larsen, S.I. Chan *J. Biol. Chem.* **266**, 1367 (1991).
32. R.A. Scott *Annu. Rev. Biophys. Biophys. Chem.* **18**, 137 (1989).
33. R. Rizzuto, D. Sondonà, M. Brini, R. Capaldi, R. Bisson *Biochem. Biophys. Acta* **1129**, 100 (1991).
34. G.L. Yewey and W.S. Caughey *Biochem. Biophys. Res. Commun.* **148**, 1520 (1987).
35. J.A. Fee, T. Yoshida, K. K. Surerus, M.W. Mather *J. Bioenerg. Biomemb.* **In press,** (1993).
36. D. Keilen and E.F. Hartree *Proc. R. Soc. London Ser. B* **127**, 167 (1939).
37. Q.H. Gibson and C. Greenwood *Biochem. J.* **86**, 541 (1963).
38. B. Chance, C. Saronio, J. Leigh J. S. *Proc. Nat'l. Acad. Sci. U. S. A.* **72**, 1635 (1975).
39. J.O. Alben, P.P. Moh, F.G. Fiamingo, R.A. Altschuld *Proc. Nat'l. Acad. Sci. U. S. A.* **78**, 234 (1981).
40. R.B. Dyer, K.A. Peterson, P.O. Stoutland, W.H. Woodruff *J. Am. Chem. Soc.* **113**, 6276 (1991).
41. M.F.J. Blokzijl-Homan and B.F. van Gelder *Biochim. Biophys. Acta* **234**, 493 (1971).
42. R. Wever, G. van Ark, B.F. van Gelder *FEBS Lett.* **84**, 388 (1977).
43. T. Stevens and S.I. Chan *J. Biol. Chem.* **256**, 1069 (1981).

44. R. Boelens, H. Rademaker, R. Wever, B.F. van Gelder *Biochim. Biophys. Acta* **765**, 196 (1984).
45. S. Yoshikawa and W.S. Caughey *J. Biol. Chem.* **265**, 7945 (1990).
46. G.T. Babcock and M. Wikström *Nature* **356**, 301 (1992).
47. M.F.K. Wikström *Nature* **338**, 776 (1989).
48. M. Bott, M. Bolliger, H. Hennecke *Mol. Microbiol.* **4**, 2147 (1990).
49. C. Gabel and R.J. Maier *Nucleic Acids Research* **18**, 6143 (1990).
50. L.J. Lemieux, M.W. Calhoun, J.W. Thomas, W.J. Ingledew, R.B. Gennis *J. Biol. Chem.* **267**, 2105 (1992).
51. J.P. Shapleigh, J.P. Hosler, M.M.J. Tecklenburg, Y. Kim, G.T. Babcock, R.B. Gennis, S. Ferguson-Miller *Proc. Nat'l. Acad. Sci. U. S. A.* **89**, 4786 (1992).
52. J. Minagawa, T. Mogi, R.B. Gennis, Y. Anraku *J. Biol. Chem.* **267**, 2096 (1992).
53. L. Holm, M. Saraste, M. Wikström *EMBO J.* **6**, 2819 (1987).
54. P.R. Rich, I.C. West, P. Mitchel *FEBS Lett.* **233**, 25 (1988).
55. C.P. Scholes, R. Janakiraman, H. Taylor, T.E. King *Biophys. J.* **45**, 1027 (1984).
56. G. Goodman and J. Leigh *J. S. Biochemistry* **24**, 2310 (1985).
57. A.J. Thomson, C. Greenwood, J. Peterson, C.P. Barret *J. Inorg. Biochem.* **28**, 195 (1986).
58. B.H. Zimmermann Thesis, University of Michigan, 1988.
59. W. Froncisz, C.P. Scholes, J.S. Hyde, Y.-H. Wei, T.E. King, R.W. Shaw, H. Beinert *J. Biol. Chem.* **254**, 7482 (1979).
60. B.M. Hoffman, J.E. Roberts, M. Swanson, S.H. Speck, E. Margoliash *Proc. Nat'l. Acad. Sci.* **77**, 1452 (1980).
61. H. Beinert, D.E. Griffiths, D.C. Wharton, R.H. Sands *J. Biol. Chem.* **237**, 2337 (1962).
62. E.I. Solomon, M.J. Baldwin, M.D. Lowery *Chem. Revs.* **92**, 521 (1992).
63. P. Hemmerich, in *The Biochemistry of Copper*; J. Peisach, P. Aisen,W.E. Blumberg, Ed., (Academic Press, New York, 1966), pp. 15.
64. J. Peisach and W.E. Blumberg *Arch. Biochem. Biophys.* **165**, 691 (1974).
65. G.J. Steffens and G. Buse *Hoppe-Seyler's Zeit. physiol. Chem.* **360**, 613 (1979).
66. C.T. Martin, C.P. Scholes, S.I. Chan *J. Biol. Chem.* **263**, 8420 (1988).
67. P.S. Covello and M.W. Gray *FEBS Lett.* **268**, 5 (1990).
68. P.M.H. Kroneck, W.A. Antholine, J. Riester, W.G. Zumft *FEBS Lett.* **248**, 212 (1989).
69. P.M.H. Kroneck, W.E. Antholine, D.H.W. Kastrau, G. Buse, G.C.M. Steffens, W.G. Zumft *FEBS Lett.* **268**, 274 (1990).
70. P.M. Li, B.G. Malmström, S.I. Chan *FEBS Lett.* **248**, 210 (1989).
71. D.M. Dooley, R.S. Moog, W.G. Zumft *J. Am. Chem. Soc.* **109**, 6730 (1987).
72. P.M.H. Kroneck, W.A. Antholine, J. Riester, W.G. Zumft *FEBS Lett.* **242**, 70 (1988).
73. R.A. Scott, W.G. Zumft, C.L. Coyle, D.M. Dooley *Proc. Nat'l. Acad. Sci. U. S. A.* **86**, 4082 (1989).
74. H. Jin, H. Thomann, C.L. Coyle, W.G. Zumft *J. Am. Chem. Soc.* **111**, 4262 (1989).
75. J.A. Farrar, A.J. Thomson, M.R. Cheesman, D.M. Dooley, W.G. Zumft *FEBS Lett.* **294**, 11 (1991).
76. W.G. Zumft, A. Viebrock-Sambale, C. Braun *Eur. J. Biochem.* **192**, 590 (1990).
77. W.E. Antholine, D.H.W. Kastrau, G.C.M. Steffens, G. Buse, W.G. Zumft *Eur. J. Biochem.* **209**, 875 (1992).
78. B.M. Hoffman *Acc. Chem. Res.* **24**, 164 (1991).
79. H.L. van Camp, Y.H. Wei, C.P. Scholes, T.E. King *Biochim. Biophys. Acta* **537**, 238 (1978).
80. T.H. Stevens, C.T. Martin, H. Wang, G.W. Brudwig, C.P. Scholes, S.I. Chan *J. Biol. Chem.* **257**, 12106 (1982).
81. C. Fan, J.F. Bank, R.G. Dorr, C.P. Scholes *J. Biol. chem.* **263**, 3588 (1988).
82. M.M. Werst, C.E. Davoust, B.M. Hoffman *J. Am. Chem. Soc.* **113**, 1533 (1991).

83. J.S. Griffith *Mol. Physics* **21**, 141 (1971).
84. F.M. Rusnak, E. Münck, C.I. Nitsche, B.H. Zimmermann, J.A. Fee *J. Biol. Chem.* **262**, 16328 (1987).
85. Ö. Hansson, B. Karlsson, R. Aasa, T. Vänngård, B.G. Malmström *EMBO J.* **1**, 1295 (1982).
86. T.H. Stevens, G.W. Brudvig, D.F. Bocian, S.I. Chan *Proc. Nat'l. Acad. Sci. U. S. A.* **76**, 3320 (1979).
87. B. Reinhammar, R. Malkin, P. Jensen, B. Karlsson, L.-E. Andréasson, R. Aasa, T. Vänngård, B.G. Malmström *J. Biol. Chem.* **255**, 5000 (1980).
88. B. Karlsson and L.-E. Andréasson *Biochim. Biophys. Acta* **635**, 73 (1981).
89. B. Chance, C. Saronio, J. Leigh J. S. *J. Biol. Chem.* **250**, 9226 (1975).
90. E.D. Getzoff, J.A. Tainer, M.M. Stempien, G.I. Bell, R.A. Hallewell *Proteins* **5**, 322 (1989).
91. J. Cline, B. Reinhammar, P. Jensen, R. Venters, B.M. Hoffman *J. Biol. Chem.* (1983).
92. K.K. Surerus, W.A. Oertling, C. Fan, R.J. Gurbiel, E. Ò., W.E. Antholine, R.B. Dyer, B.M. Hoffman, W.H. Woodruff, J.A. Fee *Proc. Nat'l. Acad. Sci.* **89**, 3195 (1992).
93. T.A. Kent, E. Münck, W.R. Dunham, W.F. Filter, K.L. Findling, T. Yoshida, J.A. Fee *J. Biol. Chem.* **257**, 12489 (1982).
94. T.A. Kent, L.J. Young, G. Palmer, J.A. Fee, E. Münck *J. Biol. Chem.* **258**, 8543 (1983).
95. A.J. Thomson, M.J. Johnson, C. Greenwood, P.E. Gooding *Biochem. J.* **193**, 687 (1981).
96. G. Rotilio, L. Morpurgo, C. Giovagnoli, L. Calabrese, B. Mondovi *Biochemistry* **11**, 2187 (1972).
97. P.H. Hafner and J.E. Coleman *J. Biol. Chem.* **248**, 6626 (1973).
98. B.J. Hathaway *Essays Chem.* **2**, 61 (1971).
99. R.A. Lieberman, R.H. Sands, J.A. Fee *J. Biol. Chem.* **257**, 336 (1982).
100. O.P. Anderson and A.B. Packard *Inorg. Chem.* **19**, 2941 (1980).

INDEX